Bioluminescence & Chemiluminescence

Proceedings of the 11th International Symposium on

BIOLUMINESCENCE & CHEMILUMINESCENCE

Asilomar Conference Grounds, Pacific Grove, Monterey, California, USA
6 – 10 September 2000

Editors

JF Case
Marine Science Institute, UCSB, USA

PJ Herring
Southampton Oceanography Centre, UK

BH Robison
Monterey Bay Aquarium Research Institute, USA

SHD Haddock
Monterey Bay Aquarium Research Institute, USA

LJ Kricka
University of Pennsylvania, USA

PE Stanley
CRTT Ltd., Cambridge, UK

World Scientific
Singapore • New Jersey • London • Hong Kong

Published by

World Scientific Publishing Co. Pte. Ltd.

P O Box 128, Farrer Road, Singapore 912805

USA office: Suite 1B, 1060 Main Street, River Edge, NJ 07661

UK office: 57 Shelton Street, Covent Garden, London WC2H 9HE

British Library Cataloguing-in-Publication Data
A catalogue record for this book is available from the British Library.

BIOLUMINESCENCE AND CHEMILUMINESCENCE 2000

ISBN 981-02-4679-X

Printed in Singapore by World Scientific Printers (S) Pte Ltd

PREFACE

These Proceedings of the 11th International Symposium on Bioluminescence and Chemiluminescence result from a meeting at the Asilomar Conference Grounds, Monterey, California USA, 6–10 September 2000. Abstracts of the Symposium were published in *Luminescence* Vol. 15(4). Not all abstracts of posters or oral presentations are necessarily represented by papers in this volume. These biannual Symposia are now sponsored by the International Society for Bioluminescence and Chemiluminescence. The 2002 Symposium will be held at Robinson College, Cambridge, England on 5-9 April, 2002 (see www.lumiweb.com).

In developing the Symposium program, the Symposium Co-Directors, James Case and Bruce Robison, and the Editors and Officers of the ISBC are particularly indebted to:

The Program Committee:

Robert Allen – Emory Univ. Medical School (USA)
Fernand Baguet – Catholic Univ. of Louvain (Belgium)
Gustav Bernroider – Univ. of Salzburg (Austria)
Anselm Berthold – Berthold Detection Systems (Germany)
Christopher H. Contag – Stanford Univ. School of Medicine (USA)
Steven H. D. Haddock – Monterey Bay Aquarium Research Inst. (USA)
J. Woodland Hastings – Harvard Univ. (USA)
Peter Herring – Southampton Oceanography Centre (UK)
Larry J. Kricka – Univ. of Pennsylvania Medical School (USA)
Michael I. Latz – Scripps Institution of Oceanography (USA)
John Lee – Univ. of Georgia (USA)
Arne Lundin – BioThema (Sweden)
Margaret McFall-Ngai – Univ. of Hawaii (USA)
James Morin – Cornell Univ. (USA)
Kenneth Nealson – NASA-JPL (USA)
Dennis J. O'Kane – Mayo Clinic (USA)
Aldo Roda – Univ. of Bologna (Italy)
A. Paul Schaap – Lumigen Inc. (USA)
Dinesh O. Shah – Abbott Laboratories (USA)
Philip Stanley – Cambridge Research and Technology Transfer (UK)
A. A. Szalay – Loma Linda Univ. Medical School (USA)
Akio Tsuji – Showa Univ. (Japan)
Keith B. Ward – Office of Naval Research (USA)
William W. Ward – Rutgers Univ. (USA)
Edith A. Widder – Harbor Branch Oceanographic Inst. (USA)
Keith Wood – Promega Corp. (USA)

Local Committee:

Christopher Contag – Stanford Univ. School of Medicine
Mark Moline – California Polytechnic State Univ., San Luis Obispo
Steven Haddock – Monterey Bay Aquarium Research Inst.
Carrie McDougall – Univ. of California Santa Barbara
Christen Herren –Univ. of California Santa Barbara
Michael Latz – Scripps Institution of Oceanography
Cyndi Stubbs – Monterey Bay Aquarium Research Inst.

Sponsors:

Abbott Laboratories
Berthold GmbH & Co. KG (PerkinElmer)
Electron Tubes Ltd.
Hamamatsu Photonic Systems Corp.
Knight Scientific Ltd.
Lumigen, Inc.
Monterey Bay Aquarium Research Inst.
Promega Corp.
Roper Scientific
Univ. of California Santa Barbara
U.S. Office of Naval Research
Xenogen Corp.

Exhibitors:

Berthold Detection Systems
Berthold GmbH & Co. KG (PerkinElmer)
Chemicon International
Hamamatsu Photonic Systems Corp.
U.S. Office of Naval Research
Optronics
Prolume, Ltd.
Promega Corp.
Roche Molecular Biochemicals
Spherotech, Inc.
Turner Designs, Inc.
Xenogen Corp.

Symposium Administrative Group:

Marine Science Institute – University of California Santa Barbara
Carrie McDougall, Marie Ciluaga, Marisol Hernandez, Shanna Bowers

Editorial Note: This volume was compiled without peer review from camera-ready manuscripts of talks and posters presented at the Symposium. The editors have, in the interest of rapid publication made only minor stylistic changes. They take no responsibility for scientific or priority matters. We thank Andrea Henderson for secretarial assistance.

The Editors: JF Case, PJ Herring, BH Robison, SHD Haddock, LJ Kricka, PE Stanley

CONTENTS

PART 4. APPLICATIONS OF BIOLUMINESCENCE

PART 6. IMMUNOASSAY AND DNA PROBE ASSAYS

PART 7. CELLULAR LUMINESCENCE

PART 8. REPORTER GENES IN CELL BIOLOGY AND ANALYSIS

PART 9. LUMINESCENCE IN SCIENCE AND EDUCATION

INTRODUCTION TO THE 11TH SYMPOSIUM

James F. Case and Bruce H. Robison
Symposium Co-Directors

This Symposium expresses the intention of the membership of the 8th Symposium, at its meeting in Cambridge, England, to broaden the scope of these meetings to more adequately cover the full range of bioluminescence research, and particularly to include physiology, behavior, ecology and oceanography. With ISBC President Peter J. Herring, we have worked to bring this concept to life. The effort was rewarded by an ample registration of 325 delegates from 26 countries presenting material from at least 40 major topical areas.

One reason to hold the meeting at Asilomar was because in 1954 *The Conference on Bioluminescence* was held there. Those proceedings, *The Luminescence of Biological Systems* (1955), edited by Frank H. Johnson of Princeton University, represented the first time that the entire panoply of bioluminescence and chemiluminescence research was represented in a single conference (of only 34 persons). The continuum was reinforced by the fact that two participants at that landmark conference, J. W. Hastings and F. I. Tsuji were with us at this 11th Symposium. In the intervening years the Monterey region has grown even more appropriate for such a program. Building on the early presence of the Hopkins Marine Station, the establishment of the Monterey Bay Aquarium and the Monterey Bay Aquarium Research Institute, the Moss Landing Marine Laboratories and many commercial enterprises focused on chemi- and bioluminescence makes the Monterey region a vital center for bioluminescence research and development.

In addition to the program of exhibits, posters and talks, there were three special occasions:

At the Opening Convocation greetings to the delegates were conveyed by the three primary sponsors, Vice Chancellor France Cordova, of the University of California Santa Barbara, CEO Marcia McNutt, of the Monterey Bay Aquarium Research Institute, and Dr. James Eckman, representing the Office of Naval Research. UCSB and MBARI helped with early financial support and with administrative services. The Office of Naval Research has a long history of interest in bioluminescence. As early as 1970 it was supporting the aequorin research of Frank Johnson and it has recently supported these Symposia. The Plenary Speaker was Kenneth Nealson whose work on bioluminescence ranks high among his many research accomplishments. As a leader in exobiology research at JPL/NASA, he brought these topics to bear on the search and recognition of life in his talk, "Following the Light–From Inner to Outer Space."

On the next evening there was a video session dealing with marine life and marine bioluminescence. Bruce Robison (MBARI) presented video records of his long-term studies of life in the Monterey Bay Submarine Canyon using remotely operated underwater vehicles. Peter Herring dealt with bioluminescence in the abyss, showing new data from a benthic lander equipped with a low-light level camera. Edith Widder presented a synthesis of her own work on bioluminescence from the Johnson-Sea-Link submersibles along with research by other investigators, in a video titled "Secret lights in the sea: how marine organisms use bioluminescence to survive". All three presentations strongly reinforced the concept that the seas are the primary reservoir of bioluminescence on earth and emphasized the exotic beauty and remarkable behavior of many marine bioluminescent organisms.

At the Closing Convocation incoming President Aldo Roda discussed his hopes for the future of the ISBC and its symposia. The delegates then paid tribute to William D. McElroy who on February 17, 1999, died at age 82 in La Jolla, California. His was a career of remarkable achievements in science relating to bioluminescence, and of distinguished administration as a Director of the National Science Foundation and Chancellor of the University of California, San Diego. With his wife, Marlene DeLuca, he did pioneering research establishing the mechanism of firefly luminescence and applications of bioluminescence to molecular biology and medicine. J. Woodland Hastings presented a warm and thoughtful appreciation of his life and career.

The DeLuca Prizes for investigators not more than 35 years of age were awarded. Fritz Berthold at the 5th Symposium in Florence, Italy, established the Prizes in 1988, in memory of Marlene DeLuca, for her many contributions to bioluminescence research. We were fortunate to have him sketch the history of the awards and to introduce James Grand of Berthold GmbH & Co. KG (PerkinElmer) who presented the awards. A committee appointed by ISBC President Peter Herring selected the winners from 50 eligible candidates. The prize winners were:

Raphaël De Cock, University of Antwerp, Belgium
Steven H.D. Haddock, Monterey Bay Aquarium Research Institute, USA
Marko Virta, University of Turku, Finland

IN MEMORIAM

pH 14 SCIENCE: McELROY AND THE PURSUIT OF FUNDAMENTAL RESEARCH

J. WOODLAND HASTINGS
Department of Molecular and Cellular Biology, Harvard University
16 Divinity Avenue, Cambridge, MA 02138 USA
E-mail: Hastings@FAS.Harvard.edu

With the passing of Bill McElroy in February 1999, the field of luminescence has lost one of its most ardent and colorful members. Much more than that. He was an innovative internationally prominent scientist, a science promoter and administrator. Imaginative and energetic, he was a mover and shaker, with a continuing agenda for experimental projects and research support for all areas of science, both basic and applied.

To mention a few of his major activities and honors, Mac served as president of three of the largest scientific organizations: the American Society of Biological Chemists; the American Institute of Biological Sciences and the 116,000-member American Association for the Advancement of Science. He served on the President's Science Advisory Committee under both Kennedy and Johnson, and was Director of the National Science Foundation under Nixon. He was the recipient of honorary degrees from some twelve institutions, including Johns Hopkins and the University of California, San Diego. The Rumford Prize, recognizing important discoveries concerning heat or light, was conferred upon him by the American Academy of Arts and Sciences, and he was also a member of the US National Academy of Sciences and the American Philosophical Society.

While extraordinarily active in administrative and advisory activities, his first love and his continuing devotion was to basic research in the field of bioluminescence. He attended the first meeting on the subject held here at Asilomar in 1954, the published volume from which is titled "The luminescence of biological systems", edited by Frank Johnson.[1] With his wife Marlene DeLuca, he attended the first of the present series of international symposia on the subject in Brussels in 1978, and together they hosted the second in 1980 at La Jolla,[2] where he was Chancellor of the University of California at San Diego. His impact on the field was profound and enduring. He considered basic research to be the well-spring of applications. His vision in this regard is vindicated by the great number and diversity of analytical and reporter bioluminescent systems now developed, from luciferases and aequorin to green fluorescent protein.

As an undergraduate at Stanford, Mac was a football player. Maybe football was an outlet for his aggressiveness; or maybe this is where it was spawned. He earned a Master's degree from Reed College in 1941 before entering Princeton, where he obtained his doctorate with the father of bioluminescence, E. Newton Harvey.[3] It was there that he conceived the idea that ATP might be involved in bioluminescence, evidently stimulated by the provocative 1941 review by Fritz Lipman on ATP as a high energy intermediate.[4] Together with Bob Ballentine he published a paper reporting a release of inorganic phosphate concomitant with the luminescence of *Cypridina*.[5]

Although it turned out much later that this release of phosphate in *Cypridina* was unrelated to light emission, the further pursuit of the idea was high on his agenda when he joined the faculty at Johns Hopkins in 1946 and found himself in firefly heaven during the next summer. Grinding up the light organ in cold buffer and allowing the light to decay, he showed that luminescence could be restored simply by adding ATP; the hot extract was not needed.[6] Thus, the factor that had been used up in the cold fraction of the luciferin-luciferase test was not luciferin itself, but ATP. With that experiment, he showed that the original DuBois experiment had been interpreted incorrectly!!

It also turned out that ATP alone could not be the energy source for light emission, as suggested by the title of the publication.[6] The energy of a photon in the yellow-green is more than five times greater than that obtained from ATP hydrolysis. It later became clear that the role of ATP in the light emitting reaction is analogous to the activation of amino acids in protein synthesis, emphasizing the relatedness of different biochemical systems and the importance of understanding basic biochemistry.

One should not recount this experiment without recalling that ATP was not found in the freezer or in a catalog in those days, when few biochemical reagents could be purchased. So Mac had been obliged to purify the ATP from rabbit muscle. As it happened, the extraction and sale of ATP provided much of the business that led to the founding of Sigma Chemical Co. in St. Louis, originally a side activity of Dan Broida's shoe polish company, now a premier biochemical supplier.

I joined Mac's lab in the early summer of 1951 after completing my Ph.D. at Princeton on the role of oxygen in bioluminescence, also with E. Newton Harvey. Mac put me in charge of firefly collection, the now-legendary operation in which children were paid a penny each for fireflies collected. Mac instructed me in the fundamentals of negotiation and bluffing, which he later made use of at the higher levels of government and university administration, and always at the poker table. "When they tell you they have 800 fireflies and it's clear that there are no more than 300, offer to give them double pay if there are 400 on counting, and nothing if there are fewer." After a few losses the word got around and the children claimed more reasonable numbers.

The firefly collection operation had an enormous positive impact on the perception of and interest in science nationwide. There were innumerable articles in newspapers and magazines explaining the importance of such research in basic science, even though significant applications were not very evident at the time. Unfortunately, there were concurrent political efforts to ridicule basic science, for example with "golden fleece" awards, directed at studies based on perhaps less appealing organisms, such as snakes. Years later, as director of the National Science Foundation, Mac strongly opposed such political intervention and was a strong and effective spokesman for the support of basic research in all areas.

A question of great interest to us was how the flash is controlled. Oxygen had been implicated in earlier work, notably the pseudoflash discovered by John Buck.[7] In my work with the partially purified luciferase, I found that a flash with a peak intensity about 100 times the baseline intensity could be obtained by the removal and rapid readmission of oxygen in a reaction mixture after its decay to the inhibited level.[8] This was attributed to the anaerobic accumulation of the luciferyl-adenylate intermediate and its rapid utilization upon the admission of oxygen, with kinetics that approached those of the in vivo flash. This revealed a basic biochemical mechanism compatible with oxygen control of flashing.

An alternate mechanism emerged from other studies. We found that the inhibited luciferase in a reaction mixture could be reactivated by the addition of inorganic pyrophosphate.[9] Mac perceived that pyrophosphate could be produced subsequent to neurotransmitter release and thereby trigger the flash. In the scheme envisioned, the acetate formed in the hydrolysis of acetylcholine would react with ATP and Coenzyme A with the formation of pyrophosphate.[10] This led to our serendipitous discovery of the ability of CoA to cause a striking increase of light emission of product-inhibited luciferase and foreshadowed the discovery of the evolutionary affinities of firefly luciferase, which came decades later. The flash mechanism in the firefly remains unsolved,[11] and it is a challenging one, now being re-addressed after all these years by several groups. It would gratify Mac to know that advances and answers are finally forthcoming.

Photometers were also not available in 1951, so with Ted MacNicol I designed and built one with high voltage supply and DC amplifier integrated, which later became the first such instrument to be marketed. Baltimore's humid climate favored fireflies but not photomultipliers, so Mac agreed that I could buy a window air conditioner for my lab. These had just become generally available, in fact this was the first and only air conditioned room in all of the Biology department, probably in all of Johns Hopkins. So it came about that on certain evenings my lab was reserved for a poker game hosted by Mac, with the Dean of Faculty, the Provost and a few other key players as guests. I thus titled my salutation on the occasion his honorary degree from Bologna "Firefly flashes and royal flushes: life in a full house.[12] His was indeed a full life.

Later Mac transferred the poker game custom to Thursday nights in Woods Hole, where in 1956 he was appointed director of the renowned summer course in Physiology at the Marine Biological Laboratory. His impact at the MBL was profound; he was elected a trustee and participated actively in laboratory governance. The Thursday poker game still continues there, and derivative games are located in many places around the country. Veterans attest to Mac's uncanny ability to regularly finish the evening with sizable winnings, and he typically wore a cap with a green celluloid visor with magic marker writing, "Make checks payable to W. D. McElroy." It is also said that a bottle of champagne was transferred each week to the home of the host, so as to be prepared for a celebration should he ever lose, which he eventually did.

As a young assistant professor in the late 1940s Mac learned of a gift proffered to Johns Hopkins by a well-to-do farmer named Pratt in honor of Elmer V. McCollum, a Hopkins scientist who had done early research on the role of vitamins and trace metals on plant nutrition, and credited with the discovery vitamin A. There were no Hopkins faculty working in the area of plant nutrition, and no one seemed to have a good idea as to what might be done with the money. Mac came forward and proposed to establish and direct a research institute for the study of plant nutrition as a part of the Biology department. Never mind that he had no credentials or knowledge in the area; good basic research, he argued, would inevitably lead to advances in an understanding of plant nutrition. Who were his first appointments? Not plant nutritionists, but three of the best young biochemists in the country, Sidney Colowick and Nate Kaplan, who shortly thereafter initiated the series Methods in Enzymology, which is still going strong today, and Al Nason, a plant biochemist.

Mac was keen on meetings; he wanted everybody to go and exchange results and ideas. In those years $100 was equal to $1000 today, and I recall him passing out $100 bills to any graduate students who was planning to go to the Federation meeting in Chicago that year. No accounting needed. He was also one of the first to organize topic-oriented symposia, initially on subjects closely related to vitamins and nutrition, in keeping with the mission of the McCollum-Pratt Institute. The first symposia in the early 50s were on copper, nitrogen and phosphorus metabolism, but all were focused on basic science. By 1960 he had sufficient funding and the freedom to venture into other areas of basic science, as exemplified by the "Light and Life" symposium,[13] which included chemiluminescence and bioluminescence, as well as the many other ways in which light impacts on living organisms.

Mac was a true pro at obtaining funding for research. The National Science Foundation was started while I was in his lab, and I recall well sitting with him for about 30 minutes while he wrote his first proposal, for $10,000 I think. And it was successful. Wouldn't it be wonderful if we could write successful proposals in 30 minutes??!! In the years thereafter, especially after he became chairman of Biology at Hopkins in 1956, he obtained large grants from NSF, NIH, ONR and other agencies in support of research by Hopkins faculty. It was he, by all accounts, who first put a part of the term-time faculty salaries on grants, thereby liberating University funds and allowing him to build a strong department.

He was so imaginative and effective, one might infer, that NSF decided that it would be wise to hire him as Director, a post that he assumed in 1969. As the NY Times reported in its obituary,[14] he was the second choice, after the offer to Franklin Long of Cornell was withdrawn when it was learned that Long opposed Nixon's proposed antiballistic missile system. McElroy was similarly opposed but quiet about it, and viewed government service (even under Nixon) as a way to promote science and to get increased funding for research. As Director, he made bold moves; in a meeting to discuss research funding at Cal Tech an undergraduate asked why students could not apply for grants. His reply? He asked what the research project was, took a piece of paper and on the spot signed a grant in support of the work.

In 1972 Mac accepted a call to become chancellor at the fledgling campus of the University of California San Diego in La Jolla, which had been established only 12 years before. He oversaw the tripling of the budget, the expansion of arts, humanities and social sciences, and, significantly, he involved community leaders in the governance of the University by establishing a Board of Overseers. A distinguished endowed yearly lecture in biology is named in his honor, as is a garden and terrace adjacent to the Mandeville Auditorium.

In 1980 he returned to full time research and teaching. With his wife Marlene he directed work leading to the cloning of the firefly luciferase gene and its expression in several organisms, including tobacco,[15] with the obvious implications. Queried about that on National Public Radio, Marlene joked that her next project would be to clone it in yeast and produce Lite beer. Keith Wood and other students from that era can certainly recount many details of activities in the lab in that decade, which included the discovery of different genes coding for luciferases eliciting different colors of bioluminescence.[16] They themselves continue to be leaders in the biochemistry and molecular biology of firefly bioluminescence.

In dedicating the Proceedings from this meeting to Bill McElroy, the now very large community of researchers in bioluminescence and chemiluminescence recalls with great pleasure his achievements and leadership in science, and the special place he will always have in our own field.

Acknowledgements

Support for research on basic mechanisms and properties of bioluminescent systems has been generously and wisely provided to many scientists for many decades by the Office of Naval Research and the National Science Foundation.

References

1. Johnson FH. ed *The Luminescence of Biological Systems*. Washington, DC: AAAS Press, 1955
2. DeLuca M, McElroy WD. editors. *Bioluminescence and Chemiluminescence: Basic Chemistry and Analytical Applications.* New York: Academic Press, 1981
3. Harvey EN *Bioluminescence.* New York: Academic Press, 1953
4. Lipmann F. *Adv Enzymol* Metabolic generation and utilization of phosphate bond energy. 1941; **1**: 99-162.
5. McElroy WD, Ballentine R. The mechanism of bioluminescence. *Proc Natl Acad Sci USA* 1942; **30**: 377-382.
6. McElroy WD. The energy source for bioluminescence in an isolated system. *Proc Natl Acad Sci USA* 1947; **33**: 342-345.
7. Hastings JW, Buck JB. The firefly pseudoflash in relation to photogenic control. ***Biol Bull*** 1956; **111**: 101-113.
8. Hastings JW, McElroy WD Coulombre J. The effect of oxygen upon the immobilization reaction in firefly luminescence. *J Cell Comp Physiol* 1953; **42**: 137-150.
9. McElroy WD, Hastings JW, Coulombre J, Sonnenfeld V. The mechanism of action of pyrophosphate in firefly luminescence. *Arch Biochem Biophys* 1953; **46**: 399-416.

10. McElroy WD, Hastings JW. Biochemistry of firefly luminescence. In: Johnson FH, ed. *The Luminescence of Biological Systems*. Washington, DC: AAAS Press, 1955: 161-198
11. McElroy WD, Hastings JW. Initiation and control of firefly luminescence. In: Prosser CL. ed *Physiological Triggers* Englewood Cliffs, NJ: Ronald Press: 1956: 80-84
12. Hastings JW. Firefly flashes and royal flushes: life in a full house. *J Biolumin. Chemilumin* 1989; **4**: 29-30
13. McElroy WD, Glass B. editors. *Light and Life*. Baltimore, MD: The Johns Hopkins Press, 1961.
14. New York Times, Sunday February 21, 1999. page 34
15. Ow DW, Wood KV, DeLuca M, deWet JR, Helinski DR, Howell SH. Transient and stable expression of the firefly luciferase gene in plant cells and transgenic plants. *Science* 1986; **234**: 856-859.
16. Wood KV, Lam YA, Seliger HH, McElroy WD. Complementary DNAs encoding click beetle luciferases can elicit bioluminescence of different colors. *Science* 1989; **244**: 700-702.

PART 1

MARINE BIOLUMINESCENCE

HETERODIMERIZATION BETWEEN BLUE AND GREEN FORMS OF *AEQUOREA VICTORIA* GFP

C. M. BARBIERI, C. M. THOMSON, AND W. W. WARD
Dept. of Biochemistry and Microbiology, Rutgers University, Cook College, New Brunswick, NJ 08901, USA
E-mail: crebb@rci.rutgers.edu

1. INTRODUCTION

Native and recombinant forms of *Aequorea victoria* GFP each exists as an equilibrium mixture of monomeric and dimeric molecules. The degree of dimerization is influenced by protein concentration and by salt concentration and composition. The wild-type recombinant GFP (wtGFP) is predominantly dimeric at concentrations greater than 5 mg/mL.[1] In crystals of GFP dimer, monomers are reported to associate by interactions of three hydrophobic residues and fourteen hydrophilic residues along the surface of the β-can.[2,3] Such associations within GFP crystals are strongly influenced by the chemical composition of the crystallization medium. Under high salt conditions, hydrophobic associations dominate. In organic solvents, electrostatic associations dominate..

The aim of this project is to determine whether green and blue GFP variants will form a heterodimer under conditions of moderate protein concentration. Gel-filtration chromatography on G-75 Sephadex was chosen as a method to distinguish monomeric protein from heterodimeric protein. The apparent molecular weights of three green variants of GFP at low concentrations (wtGFP,[4] EGFP,[5] and GFPuv[6]) were determined with and without the continuous presence of a relatively high concentration (5 mg/mL) of EBFP,[7] a blue variant of GFP.

2. MATERIALS AND METHODS

Protein Purification: Initial EBFP (Clontech) isolation was performed as previously reported.[8] Three-phase protein extraction was subsequently performed with t-butanol and 1.6 M ammonium sulfate.[9] This was followed by hydrophobic interaction chromatography using a Phenyl Sepharose column (AP Biotech). EBFP carrying a histidine tag was then purified to homogeneity using a 10 mL TALON® metal-ion affinity column (Clontech). GFP variants were purified by methods described previously.[8]

Gel Filtration: A 10 mL Sephadex G75 superfine column equilibrated with a buffer of 10 mM Tris, 10 mM EDTA pH 8.0 was used for molecular weight analysis of each GFP variant. The elution profile of each green variant was measured at low concentration (0.35 mg/mL wtGFP, 0.043 mg/mL EGFP, and 0.35 mg/mL GFPuv), concentrations favoring the monomer. Chromatography was repeated with each green variant (same concentration) in the continuous presence of EBFP at 5 mg/mL.

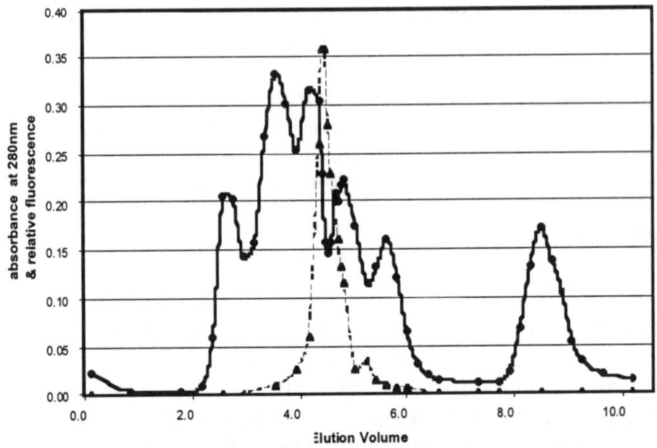

Fig. 1. Gel filtration of molecular weight standards (1 through 6) and a representative GFP (7) in its monomeric form. Absorbance peaks (1 through 6) correspond to blue dextran 2,000,000 Da (1), bovine serum albumin 66,200 Da (2), ovalbumin 42,000 Da (3), carbonic anhydrase 31,000 Da (4), myoglobin 16,700 Da (5), and vitamin B12 1,200 Da (6). Peak (7) GFPuv was measured fluorometrically. Peaks 1, 5, and 6 were confirmed using absorbance readings at 610 nm, 405 nm, and 360 nm, respectively.

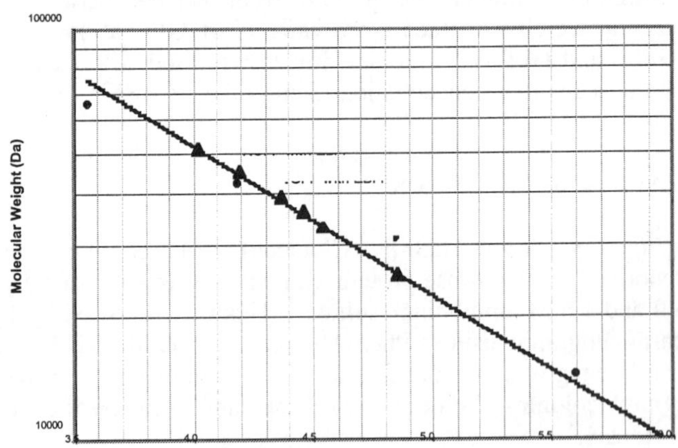

Fig. 2. Elution volumes of GFP variant combinations and corresponding molecular weights. GFP variant combinations are: GFPuv with EBFP, wtGFP with EBFP, EGFP with EBFP, GFPuv, wtGFP, EGFP. The volumes of elution for wtGFP, EGFP, and GFPuv were referenced to the internal standards.

Spectroscopy and Fluorometry: Spectrophotometric assays of gravimetrically-measured fractions in tared tubes were made using a Genysis 2 UV-Vis spectrophotometer, as shown in figure legends. Fluorometric measurements were via a Turner 111 Fluorometer with broad excitation band (from 410 to 480nm) and narrow emission band at 514nm.

3. RESULTS AND DISCUSSION

Each green variant that was chromatographed, individually, at low protein concentration (Fig. 1 & 2) eluted at a position consistent with its monomeric molecular weight (≈ 30,000 Da). However, the approximately two-fold increase in apparent molecular weight of each of the green variants in the presence of EBFP (Fig. 2) provides substantial evidence for heterodimerization between each of the green variants and EBFP. The heterodimerization of multiple color GFP molecules is cause for concern, particularly when using variants of GFP for fluorescence resonance energy transfer (FRET) analysis.[10] Heterodimerization would bring the two GFP chromophores close enough together for resonance energy transfer. Electronically excited EBFP would transfer energy efficiently to any one of the green variants while the two proteins are associated in the heterodimer. In FRET applications, in drug discovery for example, a blue-emitting form of GFP could be genetically linked to a protein of interest (protein 1) while a green-emitting variant of GFP is genetically linked to a second protein (protein 2). FRET signal could be interpreted to mean that protein 1 interacts specifically with protein 2 in response to the added drug. The results we present here raise the question of whether false positive FRET signals, created by self-association of the two variants of GFP, may occur in some FRET assays. In transformed cells, GFP is often biologically microconcentrated by targeting to membranes or organelles. GFP, microconcentrated in such a way, is difficult to quantitate. It is likely that membrane or organelle-targeted GFP reaches concentrations in excess of the threshold for heterodimer formation, inadvertently generating a false FRET signal. Careful consideration of the local concentration of the GFP used in such experiments is key to avoid false positive signals. As heterodimer formation is assumed to be species specific, the use of fluorescent proteins from two different species might eliminate the false positive FRET signals that could occur in a FRET pair when both GFPs are color variants of the same species.

References

1. Ward WW. Biochemical and physical properties of green fluorescent protein. In: Chalfie M, Kain S, eds. *Green Fluorescent Protein: Properties, Applications, and Protocols,* New York: Wiley-Liss, Inc., 1998: 45-75.
2. Ormo M, Cubitt AB, Kallio K, Gross LA, Tsien RY, Remington J. Crystal structure of the *Aequorea victoria* green fluorescent protein. *Science* 1996; **273**: 1392—1395.

3. Yang F, Moss LG, Phillips GN. The molecular structure of green fluorescent protein. *Nat Biotechnol* 1996; **14**: 1246—1251
4. Prasher, DC, Eckenrode,VK, Ward,WW, Prendergast,FG, Cormier, MJ. Primary Structure of the *Aequorea victoria* green-fluorescent protein. *Gene* 1992; **111**: 229-233.
5. Yang,TT, Sinai P, Green G, Kitts, PA, Chen YT, Lybarger L, Chervenak R, Patterson GH, Piston D, Kain SR. Improved fluorescence and dual color detection with enhanced blue and green variants of the green fluorescent protein. *J Biol Chem* 1998; **14**: 8212-8216.
6. Crameri A, Whitehorn EA, Tate E, Stemmer WP. Improved green fluorescent protein by molecular evolution using DNA shuffling. *Nat Biotechnol* 1996; **14**: 315-319.
7. Heim R, Prasher DC, Tsien RY. Wavelength mutations and posttranslational autooxidation of green fluorescent protein. *Proc Natl Acad Sci USA* 1994; **91**: 12501-12504.
8. Gonzalez DG, Ward WW. Large-scale purification of recombinant green fluorescent protein from *Escherichia coli*. *Methods Enzymol* 2000; **305**: 212-223.
9. Dennison C, Loviren R. Three phase partitioning: concentration and purification of proteins. *Protein Expression Purificat* 1997; **11:** 149-161.
10. Pollok BA, Heim R. Using GFP in FRET-based applications. *Trends Cell Biol* 1999; **2**: 57-60.

THE BIOLUMINESCENT BAYS OF VIEQUES - A MESSAGE TO THE SCIENTIFIC COMMUNITY

BARBARA BERNACHE-BAKER
The Vieques Conservation and Historical Trust, 138 Calle Flamboyan, Vieques, PR 00765, USA

I wish to draw your attention to an issue of vital interest to any and all interested in preserving unique and irreplaceable natural environments as well as one of concern to the health and well-being of a community of fellow human beings.

The Island of Vieques lies approximately ten miles off the eastern coast of Puerto Rico. Vieques possesses one of the last surviving, truly magnificent bioluminescent bays in the world. I wish to bring Vieques to your attention as a potential site for you and/or your graduate students to visit and study bioluminescence as it occurs in nature and stand ready to assist you in any way that I possibly can concerning travel to Vieques, housing, etc., as well as to make available my admittedly very limited laboratory facilities in at the Vieques Conservation and Historical Trust (VCHT), of which I am a Director. Both short and long term residences are often available on Vieques and, since I have lived in Vieques for more than 25 years, I know the Island well.

Vieques currently faces a problem of immense sociological and political consequences for its citizens because of the occupation of two thirds of the Island by the United States Navy, beginning in 1940 just prior to World War II. The residential space allotted to the displaced native Viequenses in the middle one-third of the Island is not large enough to be a viable economic entity even today, of Puerto Rico's 78 municipalities, Vieques ranks last economically and many of the displaced Viequenses were forced to emigrate. Vieques recently came into the news because of the accidental killing of a local Viequense, David Sanes Rodriguez, by an errant bomb (the Navy uses the eastern end of Vieques as a target range for offshore artillery from ships and dive bombers, as well as for International military maneuvers involving landings on the beaches). The VCHT Board of Directors recently voted to request that the Navy cease the bombing, clean up the environmental damage it has caused in the target and ammunition storage areas, and return the Island to its people.

I wish make it perfectly clear that my primary purpose here is merely to bring Vieques to your attention as a place anyone interested in bioluminescence, chemiluminescence, marine biology or island ecology should surely at least consider visiting, if not to carry out some field studies on the Island. The political situation vis a vis the United States Navy must be left for you to decide after you have visited the Island and talked with whichever of its citizens you might choose, both military and non-military . . . indeed the membership of the VCHT remains divided on this issue.

INTRACELLULAR IMAGING OF BIOLUMINESCENCE IN *PYROCYSTIS NOCTILUCA*

GUDRUN BEHRMANN AND R. HARDELAND

Institute of Zoology and Anthropology, University of G ttingen, Berliner Str. 28, D-37073 G ttingen, Germany
E-mail: rhardel@gwdg.de

W. STICKAN AND INA SIEBERT

Institute for Scientific Film GmbH G ttingen, Nonnenstieg 72, D-37075 G ttingen, Germany
E-mail: ina.siebert@iwf.de

1. INTRODUCTION

Among dinoflagellates, members of the genus *Pyrocystis* belong to the species capable of emitting highest intensities of light. Moreover, many of them are relatively large; the more or less spherical coccoid life stage of *P. noctiluca*, e.g., attains diameters up to 300 micrometers. For these reasons, this organism is particularly suitable for studying intracellular aspects of bioluminescence *in vivo*.

The bioluminescent system of dinoflagellates consists of a multitude of discrete microsources. Their function and regulation has been studied most in detail in a phylogenetically related, but much smaller and usually mastigote species, *Gonyaulax polyedra.*[1-3] The light-emitting organelles, termed scintillons in that organism, seem to be ultrastructurally different from those in *Pyrocystis*, and an essential component of the *Gonyaulax* luminescent system, the proton-sensitive luciferin-binding protein (LBP), has not yet been identified in *Pyrocystis.*[4] Nevertheless, the role of an acidic vacuole in controlling light emission may be similar, since protons can stimulate light emission in *Pyrocystis*, too.[5]

Another aspect of the bioluminescent system relates to circadian rhythmicity. *P. noctiluca*, showing a prominent rhythm of stimulated bioluminescence,[5] is regulated differently in this regard, compared to *Gonyaulax.*[4] Apart from a lack of circadian oscillations in luciferase concentration and the missing evidence for the existence of LBP,[4] *P. noctiluca* exhibits additionally a conspicuous rhythm of plastid movements,[5] which is much more expressed than in *Gonyaulax*. Branches of chloroplasts are retracted at onset of night from peripheral cytoplasmic strands, to become concentrated at a small region of the cell located in the vicinity of nucleus and other organelles such as the PAS body.[5-7] This enables the cell to largely avoid absorption of luminescence by chloroplasts.

2. MATERIALS AND METHODS

Pyrocystis noctiluca was grown at 20°C, under a light/dark cycle of 12:12 h, in a modified f/2 seawater medium as described earlier.[5] Cells were investigated in the night stage, when plastids had been retracted and bioluminescent microsources were clearly visible by fluorescence microscopy. Since spontaneous flashing is, contrary to other species such as *P. lunula*, *P. elegans* or *P. acuta*,[8] relatively scarce in undisturbed cells of *P. noctiluca*, bioluminescence was stimulated by adding a small volume of concentrated citric acid. Intracellular microscopy of bioluminescence was recorded using an Axioplan microscope (Zeiss, Oberkochen, Germany) equipped with a videocamera (CF8/1FMC)/image intensifier (BIV) sys-tem from Kappa opto-electronics (Gleichen, Germany).

3. RESULTS AND DISCUSSION

After stimulation by citric acid, luminescence can be visualized inside cells of *P. noctiluca*. In Fig. 1, a series of pictures is presented showing the rapid changes in intracellular flashing. Individual microsources are discernable from which flashes originate in a spatio-temporal pattern reflecting waves of excitation spreading along the vacuolar membrane (more details shown in videofilm ref. 7).

These waves presumably represent conducted intracellular potentials as known from other dinoflagellates.[2,3] Since the microsources of *P. noctiluca* are predom-inantly located, like strings of pearls, at the above-mentioned cytoplasmic strands bordering on the acidic vacuole,[7] the intracellular pattern of flashing seems to largely reflect the propagation of the conducted potentials at these strands. Such cytoplasmic strands also represent tracks for movements of plastid branches, which, therefore, have to be withdrawn at the onset of darkness, for avoiding proximity between light-absorbing thylacoids and light-emitting microsources during the phase of nocturnal luminescence. Consequently, patterns of biolumi-nescence, as demonstrated here, can be only obtained if cells are investigated after having chloroplasts sufficiently retracted. In *P. noctiluca*, the rhythm of plastid movements is controlled by two mechanisms, an endogenous circadian component,[5,6] and an exogenous direct light effect (unpubl. observations); plastid expan-sion can be

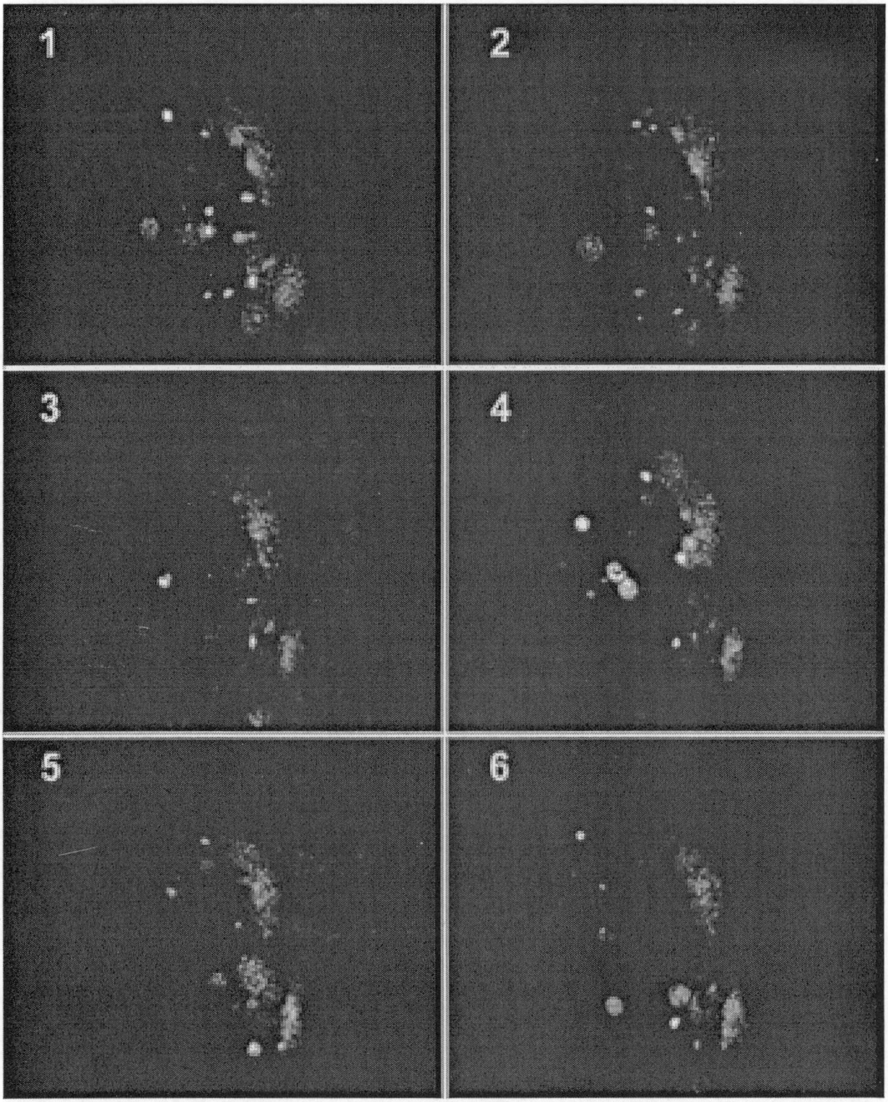

Fig. 1. A series of pictures taken from an individual cell showing rapid changes in the spatio-temporal pattern of flashing produced by distinct intracellular microsources. The cell is shown from a lateral aspect (on the right: area containing retracted plastids and nucleus, partially shading luminescence). Time difference between pictures: 0.5 seconds.

induced more rapidly via the exogenous than the endogenous route; therefore, even transient light exposure has to be avoided in cells from a nocturnal phase for obtaining a high-quality images of luminescence.

References

1. Dunlap JC, Taylor W, Hastings JW. The control and expression of bioluminescence in dinoflagellates. In: Nealson KH, ed. *Bioluminescence: Current Perspectives*, Minneapolis, Burgess 1981: 108-121.
2. Morse DS, Fritz L, Hastings JW. What is the clock? Translational regulation of circadian bioluminescence. *Trends Biochem Sci* 1990; **15**: 262-265.
3. Fritz L, Morse D, Hastings JW. The circadian bioluminescence rhythm of *Gonyaulax* is related to daily variations in the number of light-emitting organelles. *J Cell Sci* 1990; **95**: 321-328.
4. Knaust R, Urbig T, Li L, Taylor W, Hastings JW. The circadian rhythm of bioluminescence in *Pyrocystis* is not due to differences in the amount of luciferase: A comparative study of three bioluminescent marine dinoflagellates. *J Phycol* 1998; **34**: 167-172.
5. Hardeland R, Nord P. Visualization of free-running circadian rhythms in the dinoflagellate *Pyrocystis noctiluca. Mar Behav Physiol* 1984; **11**: 199-207.
6. Hardeland R, Nord P, Fischer J. Free-running circadian rhythms of plastid movements in four species of the dinoflagellate genus *Pyrocystis*. *Biometeorology* 1997; **14** (pt. 2, II): 286-291.
7. Behrmann G, Hardeland R. Bioluminescence and diurnal rhythmicity in dinoflagellates. *Videofilm C2013 (Institute for Scientific Film Gottingen)* 1999.
8. Hardeland R. Circadian rhythms of bioluminescence in two species of *Pyrocystis* (Dinophyta). Measurements in cell populations and in single cells. *J Interdiscipl Cycle Res* 1981; **13**: 49-54.

WEAK ELF MAGNETIC FIELD EFFECTS ON *Gonyaulax sp.* BIOLUMINESCENCE

MAJA BERDEN, ALEXIS ZRIMEC and IGOR JERMAN
Bion Institute, Stegne 21, 1000 Ljubljana, Slovenia, Europe
E-mail: maja.berden@bion.si

1. INTRODUCTION

Our knowledge about the mechanisms by which weak, extremely low frequency (ELF) magnetic fields (MFs) influence living systems is very poor. Although some physical theories have been proposed during the years, they are very difficult to examine empirically due to the lack of suitable biological systems which would be sensitive and at the same time reliable enough (in the sense of reproducibility of effects) for testing the theories. The bioluminescence of *Gonyaulax* sp. proved to be very useful for such experimentation, especially because it is very sensitive to external influences and at the same time reflects the state of the cell.[1,2] It is associated with many cell functions and linked to Ca^{2+} ion activity,[3] which is an important factor in biological response to weak, ELF MFs.[4] In addition, bioluminescence measurement is a non-invasive, easily implemented technique and MF effects can also be monitored in real-time. We tested the parametric resonance **model** (PRM) presented by Lednev in 1991.[5] It states that the target of weak ELF MFs influence in living matter is an ion bound to oxygen ligands in calcium-binding proteins (e.g. calmodulin). The possibility of this effect depends on the ratio between the field-densities of parallel dynamic (a.c.) and static MF (d.c., in the power range of the geomagnetic field) as well as on the dynamic field frequency, which is specific according to target ions. This model has already been successfully tested for Ca^{2+} and K^{+} ions.[6,7] We also tested it for Ca^{2+} ion.

2. MATERIALS AND METHODS

2.1. Cultures

The unicellular autotrophic dinoflagellate *Gonyaulax sp.* used was from the Gulf of Trieste and was isolated by Dr. Alfred Beran, Laboratorio di Biologia Marina, Trieste, Italy. Cultures were grown in f/2 medium in 200 mL Erlenmayer flasks at 24 ± 2° C, in alternating light : dark (L : D) conditions (L : D = 12 h : 12 h). The intensity in the light phase was 4500 lux (measured with digital light meter Iskra MI 7060) generated by a cool white fluorescent light Osram L 18W/20. For experiments, 10 ml aliquots of culture were pipetted into 30 ml borosilicate cups the day before the measurement in the late light phase.

2.2. Bioluminescence measurements

Bioluminescence was measured with a VALVO XP 2202 B photomultiplier tube (PMT) operating at -900V, coupled to an integrator (NTV, Ljubljana, Slovenia), which was connected to a computer through an AD/DA converter. To prevent influence of MF on PMT's performance, the light from cells was directed to the photomultiplier's photocathode through a 50 cm long and 4 cm thick light guide made of Plexiglas (Fig. 1). At that distance from the coil, the MF's intensity falls to 0.5%. Nevertheless, no measurable influence on the PMT's performance was observed.

2.3. Experiments

Generally, two types of tests were performed. In the first type, the real-time effects were observed during the late light phase when there is normally no measurable bioluminescence. The cells were exposed to the field during their ninth (L8-8.5) and tenth (L9-9,5) hour of the light phase. The signal was integrated for 10 s and amplified ten times.

In the second type of tests, the influence of ELF MF on circadian rhythmicity was observed. Different temporal patterns of exposure to the field were tested. According to the preliminary results, the most effective exposure was during their fourth and sixth hour of the dark phase (D3-4 and D5-6). This time pattern was used for further experiments. The signal was integrated for 10 s and amplified twice.

Cells were exposed to a d.c. MF of 45 μT, and a 35Hz a.c. MF with field-densities 83 μT and 244 μT where MF effects are predicted by PRM and also to 175 μT, where reverse effects are expected.

Before any exposure to ELF MF, a series of control experiments was performed where the circadian rhythmicity and other characteristics of light emission of our *Gonyaulax* sp. was studied.

3. RESULTS AND DISCUSSION

With the first type of experiments, the cells started to emit light in at the most 15 minutes after the fields were switched on where normally there was no observable bioluminescence for at least two and a half hours after the beginning of the measurements (Fig.2). There was no observable difference between the effects of 83μT and 244μT. At exposure to 175μT, where reverse effects were predicted, there was no triggering of bioluminescence.

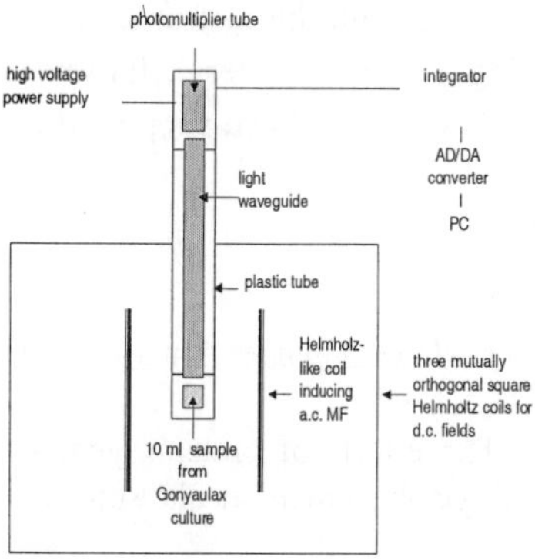

Fig. 1: Experimental setup.

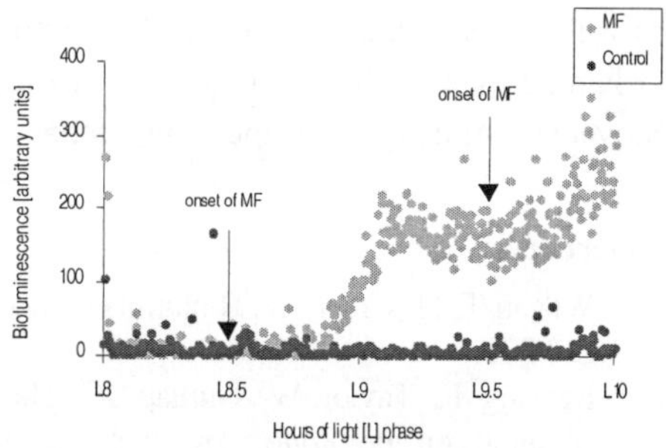

Fig. 2. Real-time effects of MF during the light phase

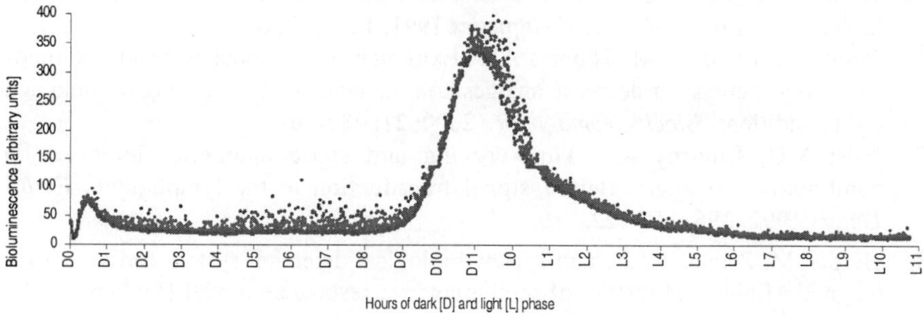

Fig.3. Bioluminescence circadian rhythm of *Gonyaulax sp.*

The observations of deviations from the normal circadian rhythmicity (Fig.3) of bioluminescence intensity proved to be very useful in determining the sensitivity of our new detection system to ELF MF,[8] but did not give consistent results with the PRM testing. The possible reason is that the cells were fit and could compensate the weak MF influence, but under mild stress the influence of the first maximum and minimum became more pronounced (data not shown).

Acknowledgements

Our thanks go to Dr. J. Woodland Hastings and Dr. Frank S. Prato for their valuable help as well as Dr. V.V. Lednev and I.Y.Belyaev for their comments. We would also like to thank our colleague Luka Drinovec for all his help with the instrumentation as well as Dr. Albert Beran and Laboratorio di Biologia Marina (Trieste, Italy) for providing us with the dinoflagellate *Gonyaulax* sp. This work has been supported by Slovenian Ministry of Science and Technology.

References

1. Wilson T, Hastings JW. Bioluminescence. *Ann Rev Cell Development Biol* 1998; **14**: 197-230.
2. Rensing L, Taylor W, Dunlap JC, Hastings JW. The effects of protein synthesis inhibitors on the *Gonyaulax* clock. II. The effect of cycloheximide on ultrastructural parameters. *J Compar Physiol* 1980; **138**: 9-18.
3. Hamman JP, Seliger HH. The mechanical triggering of bioluminescence in marine dinoflagellates: Chemical basis. *J Cell Physiol* 1972; **80**: 397-408.
4. Gapeyev AB, Chemeris NK: Nonlinear processes of intracellular calcium signaling as a target for the influence of extremely low-frequency fields. *Electro- Magnetobiol* 2000; **19**: 21-42.
5. Lednev VV. Possible mechanism for the influence of weak magnetic fields on biological systems. *Bioelectromagnetics* 1991; **12**: 71-75.
6. Prato FS, Kavaliers M, Thomas AW. Extremely low frequency magnetic fields can either increase or decrease analgesia in the land snail depending on field and light conditions. *Bioelectromagnetics* 2000; **21**: 287-301.
7. Yost MG, Liburdy RP: Time-varying and static magnetic fields act in combination to alter calcium signal transduction in the lymphocyte. *FEBS Letters* 1992; **296**: 117-122.
8. Berden M, Zrimec A, Jerman I. New biological detection system for weak ELF magnetic fields and testing of the parametric resonance model (Lednev 1991). Submitted to *Electro- Magnetobiol*

DIEL VARIATIONS OF *IN SITU* PLANKTONIC BIOLUMINESCENCE

A.S. CUSSATLEGRAS AND P. GEISTDOERFER
IRENav (Institut de recherche de l'École navale)
BP 600 - Lanvéoc Poulmic - 29240 Brest Naval
France
E-mail: cussat@ecole-navale.fr; geist@ecole-navale.fr

1. INTRODUCTION

Increase in planktonic bioluminescence at night has been observed since the first *in situ* measurements.[1,2] Two main mechanisms have been proposed as a source of diel variations. The first one based on laboratory experiment corresponds to the endogenous circadian rhythm of light emissions along with the photoinhibition of bioluminescence by daylight. Both mechanisms have been demonstrated on several species of autotrophic [3-5] and heterotrophic [6,7] dinoflagellates. The second source of day night changes is the vertical migration of bioluminescent zooplankton towards the surface at night, which can represent the main source of nycthemeral variations in some places.[8]

If many studies have been made by comparing day and night data, only a few studies have observed the precise cycle with regular *in situ* measurements during 24 hours. Therefore, the aim of this research was to study the day night cycle of planktonic bioluminescence and determine which mechanisms were responsible for the observed diel pattern, this knowledge being necessary for better sampling strategies and future modeling.

2. MATERIALS AND METHODS

Bioluminescence is measured *in situ* by a bathyphotometer designed and built at the Naval Academy.[9] It is composed of a dark chamber separated by a glass window from a photomultiplier which can detect light intensities from 10^{-4} to 10^{-9} $\mu W.cm^{-2}$. A pump with a constant flow rate (0.4 L/s) ensures the incoming flow of seawater in the dark chamber. A grid at the entrance of the chamber restricts entry of organisms smaller than 2 mm and provokes a turbulent flow, inducing a mechanical and synchronic stimulation of the organisms entering into the chamber. The bathyphotometer is coupled to a CTD probe, providing vertical profiles of temperature, salinity, fluorescence and bioluminescence. Bioluminescence signal is obtained in volt and converted in $\mu W.m^{-2}$ with a calibration formula issued from a reference photomultiplier.

Two different types of data have been collected. A three day sampling along a 120 miles long transect in Alboran Sea (south-west Mediterranean Sea) allowed a continuous sampling of bioluminescence every 2 hours. In addition, we measured bioluminescence at a fixed station every 1 or 2 hours in a shallow zone of the Iroise Sea, in July 1999 and June 2000. Data acquisition frequency was set up to 4 Hz; all time hours are expressed in UT +1.

3. RESULTS AND DISCUSSION

Time series measurements of bioluminescence in the first 100 m along the 3 days-long transect indicate higher bioluminescence between 23 h and 6 h. Bioluminescence data always display cyclic day-night variations even though strongly different water masses were crossed during the transect. Total number of flashes increased in the first part of the day from 10-12 h to 0-2 h and decreased during the other part of the 24 h cycle. Amounts of both intensity and number of flashes are higher during nighttime hours than during the daylight hours (around 5 times higher). Below 150 m, there is no longer day-night differences, all profiles display only a few bright flashes probably due to zooplankton.

In June in the Iroise Sea, repetitive sampling at the same station during 24 h every hour allowed a precise visualization of the bioluminescence *in situ* day night cycle (Fig. 1). Bioluminescence is minimum between 8 h and 18 h (0.1-0.2 $\mu W.m^{-2}$). After 18 h, light emissions increase slowly during 2 hours and dramatically from 20 h to 2 h where mean bioluminescence reaches 1.5 $\mu W.m^{-2}$, representing an increase by 3.6 times.

A decrease occurs from 2 h to 8 h with a nearly symmetrical curve. The highest variations occur between 20 h and 21 h for the increasing phase and between 4 h and 5 h for the decreasing phase, which is less than an hour after sunrise and fall (4:20 h and 20:18 h). The observed cycle of *in situ* bioluminescence is really close to the circadian endogenous rhythm displayed *in vitro* by dinoflagellate cultures with similar transition hours.[10,11] In July 1999, the same cycle was observed but both the overall bioluminescence and the amplitude of the diurnal cycle were lower.

Bioluminescence profiles (Fig. 2) show that day night changes concern the whole water column. The highest level of bioluminescence is located between 5-15 m. There is no movement of a bioluminescent layer towards the surface at night which could occur in case of vertical migration of bioluminescent organisms.

Considering that only organisms smaller than 2 mm penetrate inside the instrument and that the whole water column is sampled, these results suggest that enhanced bioluminescence at night is, in this particular case, rather due to the endogenous rhythm combined with the release of dinoflagellates photoinhibition, than to vertical migration of zooplankton towards the surface. The importance of photoinhibition

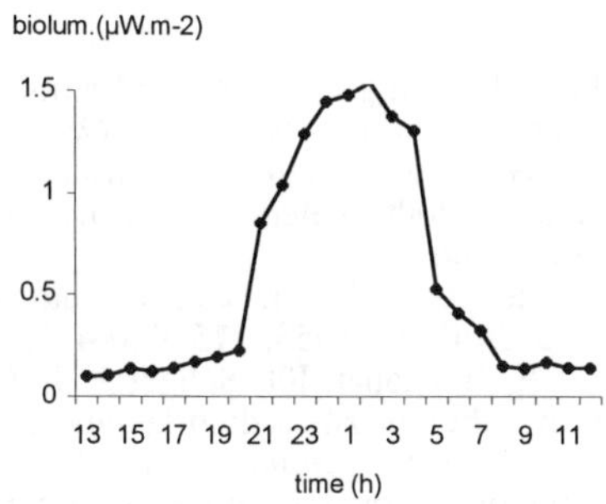

cannot be estimated but is suggested by the profiles displaying no flashes during the brightest hours of the day (12-14 h) (Fig. 2).

Fig. 1. Nycthemeral cycle of *in situ* bioluminescence observed in June 2000 in the Iroise Sea. Bioluminescence values are means for the whole water column (0-40 m).

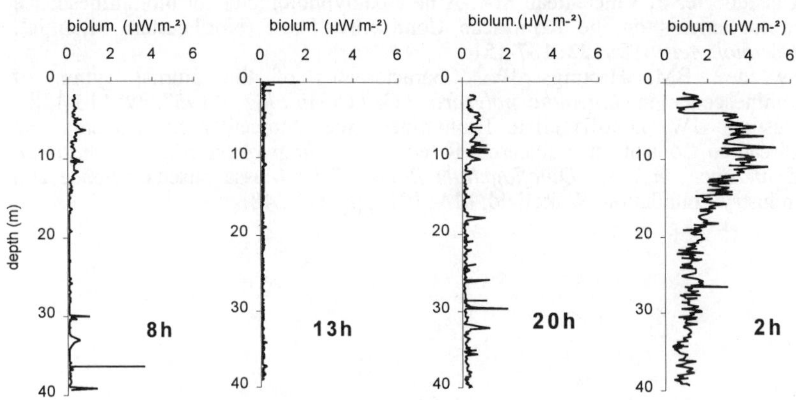

Fig. 2. Raw bioluminescence profiles measured at a fixed station in Iroise Sea, in June 2000, at different hours (8 h, 13 h, 20 h and 2 h).

Measurements made with similar bathyphotometer filtering big organisms could reveal the presence of autotrophic or heterotrophic dinoflagellates by the importance of nycthemeral variations of *in situ* bioluminescence.

References

1. Yentsh CS, Backus RH, Wing A. Factors affecting the vertical distribution of bioluminescence in the euphotic zone. *Limnol Oceanogr* 1964; **9**: 519-524.
2. Clarke L, Kelly MG. Measurements of diurnal changes in bioluminescence from the sea surface to 2000 meters using a new photometric device. *Limnol Oceanogr* 1965; **10**: 54-66.
3. Hastings JW, Sweeney BM. A persistent diurnal rhythm of luminescence in *Gonyaulax polyedra. Biol Bull* 1958; **115**: 440-458.
4. BiggleyWH, Swift E, Buchanan RJ, Seliger HH. Stimulable and spontaneous bioluminescence in the marine dinoflagellates, *Pyrodinium bahamense, Gonyaulax polyedra, Pyrocystis lunula. J Gen Physiol* 1969; **54**: 69-122.
5. Sweeney BM. Variations in the bioluminescence per cell in dinoflagellates. In: Nealson KH (ed) *Bioluminescence : Current Perspectives*, Burgess Publ., Minneapolis 1981 pp 90-94.
6. Buskey EJ, Coulter CJ, Brown SL. Feeding, growth and bioluminescence of the heterotrophic dinoflagellate *Protoperidinium huberi. Mar Biol* 1994; **121**: 373-380.
7. Li Y, Swift E, Buskey EJ. Photoinhibition of mechanically stimulable bioluminescence in the heterotrophic dinoflagellate *Protoperidinium depressum. J Phycol* 1996; **32**: 974-982.
8. Buskey EJ. Epipelagic planktonic bioluminescence in the marginal ice zone of the Greenland Sea. *Mar Biol* 1992; **113**: 689-698.
9. Geistdoerfer P, Vincendeau MA. A new bathyphotometer for bioluminescence measurements on the Armorican Continental Shelf (Northeastern Atlantic). *Oceanol. Acta* 1999; **22**: 137-151.
10. Sweeney BM, Hastings JW. Characteristics of the diurnal rythm of luminescence in *Gonyaulax polyedra. J Cell Comp Physiol* 1957; **49**: 115-128.
11. Hastings JW. Dinoflagellate Bioluminescence: Molecular Mechanisms and Circadian Control. In: Locicero VR (ed) *Proceedings of the First International Conference on Toxic Dinoflagellate Blooms*. The Massachusetts Science and Industry Foundation, Wakefield, MA. 1975, pp 235-248.

ON THE FUNCTION AND EVOLUTION OF BACTERIAL LUMINESCENCE

AGATA CZYZ

Laboratory of Molecular Biology (affiliated with the University of Gdansk), Institute of Biochemistry and Biophysics, Polish Academy of Sciences, Kladki 24, 80-822 Gdansk, Poland

E-mail: czyz@biotech.univ.gda.pl

GRZEGORZ WEGRZYN

Department of Molecular Biology, University of Gdansk, Kladki 24, 80-822 Gdansk, Poland, and Marine Biology Center, Polish Academy of Sciences, Sw. Wojciecha 5, 81-347 Gdynia, Poland

E-mail: wegrzyn@biotech.univ.gda.pl

1. INTRODUCTION

There are many bacterial species that are able to emit light. Luminescent bacteria seem to be the most abundant and widespread of the light-emitting organisms.[1] However, although the genetics and biochemistry of bacterial luminescence have been investigated extensively, the biological role of this phenomenon remains unclear. In fact, the process of luminescence is found in symbiotic, saprophytic, parasitic, as well as in free-living bacteria.[1] Although bacteria consume several percent of cell energy for luminescence, it is hard to understand what specific ecological benefit these simple organisms derive from producing light. The higher organisms can use light for attraction of prey, escape from predators or intraspecies communication. However, bacteria can be able to produce light but cannot detect such a signal. Bacteria living in symbiotic association with higher organisms may find better life conditions relative to free living microbes, but it seems unlikely that establishment of such a symbiosis could be the main evolutionary drive to develop very complicated and energy-consuming light-emitting systems.

Our recent studies revealed that one of luminescent bacteria, *Vibrio harveyi*, is very sensitive to UV irradiation when its light-emitting system is defective.[2] It appears that luminescence of this bacterium is used as an internal source of light, which stimulates DNA repair by photolyase.[2] Repair of DNA is such an important process that a need for development of a system ensuring its high efficiency might lead to appearance of bacterial luminescence.

On the other hand, it was recently proposed that primary roles of luminescent systems could be detoxification of the deleterious oxygen derivatives.[3] If this hypothesis were true, one might assume that bacterial luciferase precursors and perhaps also the present-day enzymes could be involved in the detoxification of the toxic metabolites (e.g. H_2O_2, aldehydes) generated when cells are subjected to oxidative stress.

In this work we investigated whether *V. harveyi* luciferase is involved in the protection of bacterial cells against oxidative stress. Our results indicated that the luciferase can play such a role indeed. Therefore, it seems likely that development of luminescnet bacterial systems was driven in evolution due to needs of cells to both efficient DNA repair and detoxification of oxygen derivatives.

2. MATERIALS AND METHODS

2.1 Bacterial strains, culture medium and cultivation conditions

V. harveyi wild-type strains (BB7 and BB392)[4,5] and *luxA*::Tn*5kan*, *luxB*::Tn*5kan* (derivatives of BB7)[6] and *luxD*(Gly75Glu) (derivative of BB392)[5] mutants were used. Bacteria were cultivated in the BOSS medium (described previously)[7] at 30°C in a water bath shaker.

2.2 Measurement of sensitivity of bacteria to H_2O_2

Bacteria were grown in the BOSS medium at 30°C to mid-log phase. Then the culture was divided, one half was not specifically treated, and to the second half H_2O_2 was added to indicated concentration. At different times after this treatment, samples of the cultures were withdrawn and number of living cells was estimated by plating on the BOSS plates at 30°C. The bacterial survival (expressed as percentage) was calculated by dividing the number of living cells in the H_2O_2-treated culture by the number of living cells in the untreated culture, and multiplying the resultant value by 100%.

3. RESULTS

We investigated sensitivity of wild-type bacteria and otherwise isogenic *luxA* and *luxB* mutants to H_2O_2. The *luxA* and *luxB* genes code for two subunits of luciferase, thus dysfunction of one of them results in a lack of activity of this enzyme.

We found that *V. harveyi luxA* and *luxB* mutants are more sensitive to H_2O_2 than the wild-type, luminescent, bacteria (Table 1). It is noteworthy that at H_2O_2 concentrations different than those presented in Table 1, and at different times after treatment of bacterial cultures, the differences in survival of *V. harveyi* cells were less pronounced (data not shown).

Table 1 Sensitivity of *V. harveyi* wild-type and *luxA* and *luxB* mutants to H_2O_2. Samples were withdrawn one hour after addition of H_2O_2.

H_2O_2 concentration (%)	Bacterial survival (%)		
	lux$^+$	*luxA*	*luxB*
0	100	100	100
0.001	35	22	18
0.025	5.7	0.0055	< 0.0001

To answer the question whether increased sensitivity of the *luxA* and *luxB* mutants to H_2O_2 results from the loss of luminescence or the loss of the luciferase enzyme itself, we repeated the experiments using the *luxD* mutant and an otherwise isogenic wild-type strain. The *luxD* gene codes for the acetyltransferase enzyme producing fatty acids for the luminescence reaction, thus *luxD* mutants are dark without affecting luciferase. We found no significant differences in the sensitivity of *lux*$^+$ and *luxD* cells to H_2O_2 (Table 2). When H_2O_2 concentrations and times of cultivation different from those shown in Table 2 were employed, we also did not observe significant differences between wild-type and *luxD* bacteria (data not shown).

Table 2 Sensitivity of *V. harveyi* wild-type and the *luxD* mutant to H_2O_2. Samples were withdrawn one hour after addition of H_2O_2.

H_2O_2 concentration (%)	Bacterial survival (%)	
	lux$^+$	*luxD*
0	100	100
0.001	104	106
0.010	< 0.0001	0.0007

4. CONCLUSIONS

The results presented in this report strongly suggest that *V. harveyi* luciferase is involved in the detoxification of H_2O_2, thus playing a role in the protection of cells against oxidative stress. This conclusion supports the previously proposed hypothesis that this could be one of the functions of bacterial luciferase precursors.[3] Another function of the luciferase is production of light that can be used for stimulation of DNA repair by photolyase.[2] Therefore, we propose that the need of bacteria to posses efficient systems for protection against oxidative stress and for DNA repair led to appearance and development of bacterial luminescence.

Acknowledgments

We thank Katarzyna Potrykus and Joanna Potrykus for critical reading of the manuscript. This work was supported by the University of Gdansk (grant BW/1190-5-0120-0) and by Marine Biology Center of the Polish Academy of Sciences (topic 4.2). G.W. acknowledges also a financial support from the Foundation for Polish Science (subsydium 14/2000).

References

1. Meighen EA. Genetics of bacterial luminescence. *Annu Rev Genet* 1994; **28**: 117-139.
2. Czyz A, Wrobel B, Wegrzyn G. *Vibrio harveyi* bioluminescence plays a role in stimulation of DNA repair. *Microbiology* 2000; **146**: 283-288.
3. Rees J-F, De Wergifosse B, Noiset O, Dubuisson M, Janssens B, Thompson EM. The origins of marine bioluminescence: turning oxygen defense mechanisms into deep-sea communication tools. *J Exp Biol* 1998; **201**: 1211-1221.
4. Belas R, Mileham A, Cohn D, Hilmen M, Simon M, Silverman M. Bacterial luminescence: isolation and expression of the luciferase genes from *Vibrio harveyi*. *Science* 1982; **218**: 791-793.
5. Miyamoto CM, Boylan M, Graham AF, Meighen EA. Organization of the *lux* structural genes of *Vibrio harveyi*. Expression under the T7 bacteriophage promoter, mRNA analysis, and nucleotide sequence of the *luxD* gene. *J Biol Chem* 1988; **263**: 13393-13399.
6. Bassler B, Wright M, Silverman MR. Multiple signaling systems controlling expression of luminescence in *Vibrio harveyi*: sequence and function of genes encoding a second sensory pathway. *Mol Microbiol* 1994; **13**: 273-286.
7. Klein G, Zmijewski M, Krzewska J, Czeczatka M, Lipinska B. Cloning and characterization of the *dnaK* heat shock operon of the marine bacterium *Vibrio harveyi*. *Mol Gen Genet* 1998; **259**: 179-189.

BIOLUMINESCENCE IN THE BRITTLE STAR *AMPHIPHOLIS SQUAMATA* (ECHINODERMATA): AN OVERVIEW OF 10 YEARS OF RESEARCH

DIMITRI D. DEHEYN

Scripps Institution of Oceanography, University of California, San Diego, La Jolla, CA 92093-0202, USA

E-mail: ddeheyn@ucsd.edu

1. INTRODUCTION

Bioluminescence is a common feature among echinoderms with only sea urchins lacking luminous representatives. In most cases data are solely qualitative and largely restricted to the anatomical distribution of light production, except for a few limited species of brittlestars,[1] where bioluminescence has been investigated for its function,[2] structural localization,[3] neuroconduction[4] and chemistry.[5]

The luminous brittlestar *Amphipholis squamata* (Delle Chiaje, 1828) has been of interest to ecologists because of its remarkable widespread geographic and bathymetric distribution.[6,7] Only in the past decade has this ophiuroid been investigated for the morphological, neurophysiological, behavioral, and ecological aspects of its bioluminescence.[8,9]

Specimens of *A. squamata* are small (disc diameter of ca. 3 mm and arm length of ca. 12 mm, adult specimen), with the arms being the only body parts to produce light.[3,10] In the arm, light production is particularly intense at the base of each spine where there are neural ganglia composed of five different cell types.[11] Based on an ultrastructural investigation at various stages of the luminous reaction, only cells of one type exhibit obvious ultrastructural changes associated with light production. They were isolated and shown to produce light, confirming their identity as photocytes.[12] Isolated photocytes are a useful complement to studies using entire arms that investigate the signaling pathways leading to light production.[13-15]

The photocytes are branched in shape and make contact with the nerve cord running along the arm, and with the basiepidermic nervous plexus.[11] There appears to be two mechanisms of neural triggering of light production and clearly intensity of the luminous reaction depends on intensity of the stimulus, with a weak contact triggering a local luminous reflex and a strong one resulting in propagation of the luminescence along the arm.[11,17] One mechanism involves the basiepithelial nervous plexus that develops into a local luminous reflex in response to the mechanical stimuli, as described based on the pattern of light production in reaction to contact with other organisms.[17] The other mechanism involves the activation of the nerve cord in the arm, as supported by the fact that the neuro-mediator modulating the light production is contained in the nerve cord.[18-20]

2. BRITTLE STAR BIOLUMINESCENCE

Beside a defense function, bioluminescence might also be associated with an intra-specific function used for courtship and mating. Indeed the ophiuroid light production shows a seasonal variation that is correlated with a combination of factors, both of environmental and biological origin.[20-22] The light production increases in intensity in summer and winter as a result of the alternate influence of two factors, an exogenous factor (temperature) causing the luminescence to increase in summer, and an endogenous factor linked to the reproductive cycle causing the luminescence to increase in winter. Under laboratory conditions, the intensity of light production increases with temperature, and changes of photoperiod (regulating the reproductive cycle in the ophiuroid) induce changes of bioluminescence.[22] *A.squamata* reproduce via a copulation[6] and one can then propose the bioluminescence to be used for defensive purposes, except during the reproduction period (i.e., winter) when it may be used for aggregation and reproduction.[22]

Bioluminescence is reduced by anthropogenic factors such as the amount of heavy metals in the field.[23] Therefore, the bioluminescence in *A. squamata* appears modulated by environmental factors as well as the physiological condition of an individual. For example, bioluminescence increases during the time when the ophiuroid broods its offspring in bursal pockets.[24] This may be linked to a neuro-hormonal factor involving the bursa. Indeed, a similar (and even greater) stimulatory effect of the bioluminescence is also obtained when parasitic copepod inhabits the bursa rather than offspring.[25]

A. squamata is a polychromatic species meaning that individuals, even within the same population, belong to different color varieties. The intensity of bioluminescence is always different for each color variety, which is not true for the kinetics of light production.[20,24] It indicates that the varieties have the same chemistry of the luminous reaction but different amount of the luminous reagent. Populations from different habitats display different varieties and the same level of bioluminescence is always found for a given variety, even if individuals are from different populations.[26] Therefore the difference in bioluminescence between color varieties appears genetically driven rather than having a somatic or environmental basis.[26,27] The linkage between color variety and bioluminescence leads to questions about the taxonomic status of the *A. squamata* varieties.[27] Continued intensive research on *A. squamata* will provide new insights on biochemical, physiological and evolutionary processes governing its bioluminescence.

Acknowledgements

I am grateful to Dr MI Latz for his constructive comments on the manuscript. I thank for research support the Free University of Brussels, the FRIA, Belgian American Educational Foundation, NATO and TSR&TP (University of California).

References

1. Herring PJ. Bioluminescent echinoderms: Unity of function in diversity of expression? In: Emson RH, Smith AB, Campbell AC, eds. *Echinoderm Research*. Balkema AA: Rotterdam, 1995: 9-17.
2. Grober MS. Brittle-star bioluminescence functions as an aposematic signal to deter crustacean predators. *Anim Behav* 1988; **36**: 493-501.
3. Brehm°P,°Morin°JG. Localization and characterization of luminescent cells in *Ophiopsila californica* and *Amphipholis squamata* (Echinodermata: Ophiuroidea). *Biol Bull* 1977; **152**: 12-25.
4. Brehm P. Electrophysiology and luminescence of an ophiuroid radial nerve. *J Exp Biol* 1977; **71**: 213-227.
5. Shimomura O. Bioluminescence of the brittle star *Ophiopsila californica. Photochem Photobiol* 1986; **44**: 671-674.
6. Johnson J. *The Biology of Amphipholis squamata Delle Chiajei (Echinodermata: Ophiuroidea)*. PhD, Newcastle upon Tyne, UK, 1972: 186 p.
7. Emson RH, Jones MB, Whitfield PJ. Habitat and latitude differences in reproductive pattern and life-history in the cosmopolitan brittle-star *Amphipholis squamata* (Echinodermata). In: Ryland JS, Tyler PA, eds. *Reproduction, genetics and distributions of marine organisms*. Fredensborg: Olsen & Olsen, 1989: 75-81.
8. Deheyn D. Study of bioluminescence in the ophiuroid *Amphipholis squamata* (Delle Chiaje, 1828) (Echinodermata): structure of photogenous sites; characteristics of expression and function of the bioluminescence. PhD, Free University of Brussels, Belgium, 1999: 167 p.
9. De Bremaeker N. *Nervous control of the luminescence in the ophiuroid Amphipholis squamata (Delle Chiaje, 1828) (Echinodermata)*. PhD, Catholic University of Louvain-la-Neuve, Belgium, 1999: 147 p.
10. Mallefet J, Vanhoutte P, Baguet F. Study of *Amphipholis squamata* luminescence. In: Scalera-Liaci L, Canicatti C, eds. *Echinoderm Research*. Balkema AA: Rotterdam, 1992: 125-130.
11. Deheyn D, Alva V, Jangoux M. Fine structure of the photogenous areas in the bioluminescent ophiuroid *Amphipholis squamata* (Delle Chiaje) (Echinodermata). *Zoomorphology* 1996; **116**: 195-204.
12. Deheyn D, Mallefet J, Jangoux M. Cytological changes during light production process in dissociated photocytes from the ophiuroid *Amphipholis squamata* (Echinodermata). *Cell Tissue Res* 2000; **299**: 115-128.
13. Mallefet J, Chabot B, De Bremaeker N, Baguet F. Evidence for calcium requirement in *Amphipholis squamata* (Ophiuroidea) luminescence. In: Mooi R, Telford M, eds. *Echinoderms: San Francisco*. Balkema AA: Rotterdam, 1998: 387-392.
14. Mallefet J. Physiology of bioluminescence in echinoderms. In: Candia Carnevali MD, Bonasoro F, eds. *Echinoderm Research 1998*. Balkema AA: Rotterdam, 1999: 93-102.

15. De Bremaeker N, Baguet F, Mallefet J. Effects of catecholamines and purines on luminescence in the brittlestar *Amphipholis squamata* (Echinodermata). *J Exp Biol* 2000; **203**: 2015-2023.
16. De Bremaeker N, Baguet F, Thorndyke MC, Mallefet J. Modulatory effects of some amino acids and neuropeptides on luminescence in the brittlestar *Amphipholis squamata. J Exp Biol* 1999; **202**: 1785-1791.
17. Deheyn D, Mallefet J, Jangoux M. Expression of bioluminescence in *Amphipholis squamata* (Ophiuroidea: Echinodermata) in presence of various organisms: a laboratory study. *J Mar Biol Ass UK* 2000; **80**: 179-180.
18. De Bremaeker N, Mallefet J, Baguet F. Luminescence control in the brittle-star *Amphipholis squamata*: effects of cholinergic drugs. *Comp Biochem Physiol* 1996; **115C**: 75-82.
19. De Bremaeker N, Deheyn D, Thorndyke MC, Baguet F, Mallefet J. Localization of S1- and S2-like immunoreactivity in the nervous system of the brittle star *Amphipholis squamata* (Delle Chiaje 1828). *Proc R Soc Lond* 1997; **264B**: 667-674.
20. De Bremaeker N, Baguet F, Mallefet J. Characterization of acethylcholine-induced luminescence in *Amphipholis squamata* (Echinodermata: Ophiuroidea). *Belg J Zool* 1999; **129**: 353-362.
21. Deheyn D, Mallefet J, Jangoux M. Variation in bioluminescence with ambient illumination and diel cycle in a cosmopolitan ophiuroid (Echinodermata). *Cah Biol Mar* 1999; **40**: 57-63.
22. Deheyn D, Mallefet J, Jangoux M. Evidence of seasonal variation in luminescence performance of *Amphipholis squamata* (Ophiuroidea, Echinodermata): effect of environmental factors. *J Exp Mar Biol Ecol* 2000; **245**: 245-264.
23. Deheyn D, Jangoux M, Warnau M. Alteration of bioluminescence in *Amphipholis squamata* (Ophiuroidea: Echinodermata) by heavy metal contamination: a field study. *Sci Tot Env* 2000; **247**: 41-49.
24. Deheyn D, Mallefet J, Jangoux M. Intraspecific variations of bioluminescence in a polychromatic population of *Amphipholis squamata* (Delle Chiaje, 1828) (Echinodermata; Ophiuroidea). *J mar biol Ass UK* 1997; **77**: 1213-1222.
25. Deheyn D, Watson N, Jangoux M. Symbioses in *Amphipholis squamata* (Echinodermata, Ophiuroidea, Amphiuridae): geographical variation of infestation and effect of symbionts on the host s light production. *Int J Parasitol* 1998; **28**: 1413-1424.
26. Deheyn DD, Mallefet J, Jangoux M. Evidence from polychromatism and bioluminescence that the cosmopolitan ophiuroid *Amphipholis squamata* (Echinodermata) might not represent a unique taxon. *C R Acad Sci Paris* 2000; **323**: 499-509.
27. Deheyn D, Jangoux M. Colour varieties as sibling species in the polychromatic ophiuroid *Amphipholis squamata* (Echinodermata): Evidence from inheritance of body colour and luminescence characters. *J Exp Mar Biol Ecol* 1999; **234**: 219-234.

INVOLVEMENT OF CALCIUM IN THE LUMINESCENCE CONTROL OF THREE OPHIUROID SPECIES (ECHINODERMATA): A COMPARATIVE STUDY

YANNICK DEWAEL AND JEROME MALLEFET

Laboratory of Animal Physiology, Catholic University of Louvain, Batiment Carnoy, 5, Place Croix du Sud, B-1348 Louvain-la-Neuve, Belgium

1. INTRODUCTION

Among echinoderms, luminous phenomena have been mostly studied in ophiuroids. Although the eco-ethological aspect is well documented,[1,2] little is known about the physiological mechanisms of the light emission. Recent works on the brittlestar *Amphipholis squamata* have showed that photogenesis is under nervous control. Acetylcholine is the main transmitter triggering light production.[3,4] Moreover, it has been shown that calcium is necessary for potassium chloride and acetylcholine-induced light emission on isolated arms and on dissociated photocytes of *A.* squamata.[5,6]

The aim of this study was to investigate the involvement of calcium in the KCl-induced luminescence in three ophiuroid species: *Ophiopsila aranea* (Forbes 1843), *O. californica* (Clarck 1921), and *Amphiura filiformis* (O.F. M ller).

2. MATERIAL AND METHODS

Ophiopsila aranea (Banyuls-sur-Mer, France), *Ophiopsila californica* (Santa Barbara, USA) and *Amphiura filiformis* (Fiskebackskil, Sweden) were transported to our laboratory in Belgium where they are kept in aquaria filled with recirculating natural and artificial sea water at 12¡C.

After anaesthetizing the animals in 3,5% w/w $MgCl_2$ in artificial sea water (ASW), arms were isolated from the disc and divided into sections of 8 articles (*Ophiopsila*) or 20 articles (*Amphiura*), which were then rinsed in ASW (NaCl 400.4mM, $CaCl_2$ 9.9mM, KCl 9.6mM, $MgCl_2$ 52.3mM, Na_2SO_4 27.7mM, Tris 20mM, pH 8.3).

Stimulations were performed by injection of potassium chloride (KCl) 200mM (osmolarity equivalent to ASW). For each experimental protocol, one arm segment was stimulated in normal ASW, as control, while the other segments of the same row were treated for 10 minutes with the tested drug or calcium concentration before stimulation with KCl. Light emission was measured with a FB12 Berthold luminometer linked to a personal computer. Statistical analyses (ANOVA) were performed using SAS program (Statistic Analysis System).

3. RESULTS

Maximal intensities of light emission on isolated arms of the three species tested decreased gradually with lowered concentrations of Ca^{++} in ASW (Fig. 1).

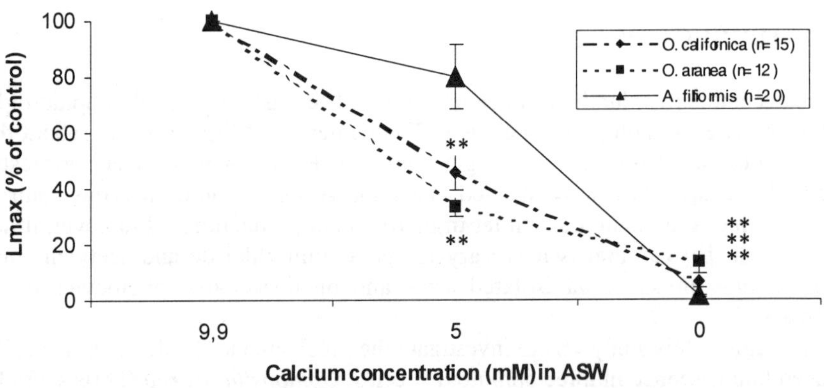

Fig. 1 Effect of calcium concentration on KCl-induced light emission of *Ophiopsila californica, O. aranea* and *Amphiura filiformis* arm segments. MEAN ± SEM. Lmax = the maximum of light emission; ** = p<0.01; n = number of stimulated segments.

Photogenesis was also kinetically altered by the decreasing of calcium concentration: light production was either slowed or progressively abolished (number of luminous responses decreased) by low calcium concentration.

Several inorganic calcium substitutes (Ba^{++} and Sr^{++}), calcium blockers (Ni^{++} and Co^{++}) and organic L-type voltage-gated calcium channels blockers (diltiazem, nifedipine, trifluoroperazine and verapamil) were used before KCl stimulation. Effects of these drugs are shown in Table 1.

Table 1 Effect of inorganic and organic drugs on KCl-induced luminescence. - : inhibition; 0: no effect.

Drugs	*O. californica*	*O. aranea*	*A. filiformis*
Ba^{++} 9.9mM	-	-	0
Sr^{++} 9.9mM	-	-	0
Ni^{++} 5mM	-	-	-
Co^{++} 5mM	-	-	-
Diltiazem 1mM	-	-	-
Nifedipine 0.1mM	-	0	0
Trifluoroperazine 0.1mM	-	0	-
Verapamil 0.1mM	-	0	-

4. DISCUSSION

Our experiments revealed that removal of calcium from the external medium induces an inhibition of the KCl-induced luminescence from isolated arms of *Ophiopsila aranea, Ophiopsila californica* and *Amphiura filiformis*. Calcium appears to be a necessary factor in the luminescence control of the three species of ophiuroids. A similar observation has already been made on *Ophiopsila californica*[7] and *Amphipholis* squamata.[5]

Our results showed that barium and strontium, two calcium substitutes,[8] have exerted inhibitory effects on the light emission of both *O. aranea* and *O. californica* but not of *A. filiformis*, suggesting that they are not reliable substitutes for calcium in these species. Two inorganic Ca^{++}-channels blockers, cobalt and nickel were found to partially inhibit photogenesis of all the three species. These results are in agreements with literature.[9]

We have also shown that KCl-induced luminescence of isolated arms of *O. californica* is reduced by L-type voltage-gated calcium channel blockers diltiazem (benzodiazepine), nifedipine (1,4-dihydropyridine) and verapamil (phenylalkylamine) and by calmodulin inhibitor trifluoroperazine. On the contrary, only diltiazem has an inhibitory effect on *O. aranea* while diltiazem and verapamil nearly completely abolished luminescence in *A. filiformis*. Our results support then the hypothesis that Ca^{++} channels involved in the luminescence of *O. californica* and *A. filiformis* but not in *O. aranea* are likely to be of the L-type.

In conclusion, it has been shown from this study that calcium movements are necessary to activate the luminous reaction. The precise role of calcium in ophiuroid light emission remains unknown. We suggest that control mechanisms of light emission appear to show differences from one species to another. Further studies, using *patch-clamp* techniques should be developed to characterize precisely the calcium channels types involved in the control of the luminescence in these three ophiuroid species.

Acknowledgements

We thank I. Cogneau for technical assistance. We acknowledge financial supports to Y.D. and J.M. from a EEC Large Scale Facility at Kristineberg Marine Station, Fiskeb ckskil, Sweden; Fonds L opold III at Arago Laboratory, Banyuls-sur-Mer, France and F.N.R.S. short stay to J.M. at Marine Sciences Institute, U.C.S.B., California, USA. Research supported by a FRIA grant to Y.D.; J.M. is a Research Associate of the National Fund for Scientific Research (FNRS, Belgium).

References

1. Sk ld M. Escape response in four epibenthic brittle stars (Ophiuroidea: Echinodermata). *Ophelia* 1998; **49**: 163-179.
2. Basch LV. Bioluminescent anti-predator defense in a subtidal ophiuroid. In: Burke *et al.*, eds. *Echinoderm Biology*. Rotterdam: Balkema, 1988: 503-515.
3. De Bremaeker N, Mallefet J, Baguet F. Luminescence control in the brittlestar *Amphipholis squamata*: effect of cholinergic drugs. *Comp Biochem Physiol* 1996; **115C**: 75-82.
4. De Bremaeker N, Baguet F, Mallefet J. Characterization of acetylcholine-induced luminescence in *Amphipholis squamata* (Echinodermata: Ophiuroidea). *Belg J Zool* 1999; **129**: 353-362.
5. Mallefet J, Ajuzie CC, Baguet F. Aspects of calcium dependence of light emission in the ophiuroid Amphipholis squamata (Echinodermata). In David B, Guille A, F ral J-P and Roux M, eds. *Echinoderm through Time*. Rotterdam: Balkema, 1994: 455-460.
6. Mallefet J, Chabot B, De Bremaeker N, Baguet F. Evidence for a calcium requirement in *Amphipholis squamata* (Ophiuroidea) luminescence. In Mooi, R and Telford M eds. *Echinoderms: San Francisco*. Rotterdam: Balkema, 1998: 387-392. Balkema, Rotterdam.
7. Brehm P. Electrophysiology and luminescence of an ophiuroid radial nerve. *J Exp Biol* 1977; **71:** 213-227.
8. Berrios A, Brink D, del Castillo D. Some properties of the action potentials conducted in the spines of the sea urchin *Diadema antillarum*. *Comp Biochem Physiol* 1985; **81A**: 15-23.
9. Cobb JLS, Moore A. Studies on the ionic basis of the action potential in the brittle-star *Ophiura ophiura*. *Comp Biochem Physiol* 1988; **91A**: 821-825.

BIOLUMINESCENCE AND GENETIC VARIABILITY IN *AMPHIPHOLIS SQUAMATA*

S. DUPONT AND J. MALLEFET

Laboratoire de Physiologie Animale, Universite catholique de Louvain,5 Place Croix du Sud, 1348 Louvain-la-Neuve, Belgium
E-mail: dupont@bani.ucl.ac.be

P. BARET

Unit de Genetique, Universite catholique de Louvain, 2 Place Croix du Sud, 1348 Louvain-la-Neuve, Belgium
E-mail: baret@gena.ucl.ac.be

1. INTRODUCTION

The small cosmopolitan ophiuroid *Amphipholis squamata* presents unique biological properties. Thirteen years ago, Binaux and Bocquet showed that the species is polychromatic.[1] Since adult body colour remains and that juveniles always present the same body colour as its parent, they suggest that colour morphs are genetically determined and that most of the reproduction occurs by selfing.

The species is also bioluminescent and recently, Deheyn *et al.* have demonstrated that each colour morph possesses their own luminous capabilities.[2] In laboratory conditions, they have confirmed the results of Binaux and Bocquet on inheritance of body colour and they have showed that luminous capabilities are also transmitted to their offspring.[3] On the basis of these observations, they suggest that *Amphipholis squamata* is a complex of sibling species, each colour morph constituting a species genetically different and isolated for reproduction.

First genetic analyses were performed by Poulin *et al.*[4] Using RAPDs markers, they demonstrated that selfing and outcrossing are both possible for reproduction in *A. squamata*. This method was also used by Dupont *et al.* in order to prove that body colour and luminous capabilities are genetically determined.[5] Therefore, polychromatism and bioluminescence are good indicators of genetic variability in this species. Nevertheless, luminescence is a more powerful indicator since luminescence is an objective quantitative parameter when body colour is more subjective and qualitative.

The aim of this work was to use bioluminescence as a tool in order to determine the proportions of selfing and outcrossing within a polychromatic population. On the other hand, genetic basis of luminous traits has been assessed by an estimation of the heritability.

2. MATERIAL AND METHODS

In March 2000, ophiuroids were randomly collected in four populations from the lagoon system of Oliveri-Tindari, Sicily. After anaesthesia, several parameters were identified under binocular microscope for each individuals: (i) The colour morph (dark-brown or spotted as described by Dupont & Mallefet[6]). (ii) Adult individuals (disc diameter>1.45 mm[6]) were chosen. *Amphipholis* brood juveniles in genital bursae located into the disc. To determine the brooding condition, bursae were dissected. If present, juveniles were isolated from the adults. The number of articles of the arms were measured for each one.

For the characterization of brooding adults light capabilities, two arms were removed from the disc while entire individuals were used for juveniles due to their small sizes (discal diameter<0.95 mm[6]). Light responses were induced by KCl injection and characterized by the maximal intensity of light in megaquanta per millimeter of arm (Lmax in $Mq.s^{-1}.mm^{-1}$)[6].

Parameters of a non linear model were estimated by a least square procedure. Variance components are estimated under a general linear mixed model by residual maximum likelihood. Intensity is log transformed to normalize the data.

3. RESULTS

A part of the tested juveniles don't emit light in response to KCl. Figure 1 presents the proportion of luminescent juveniles according to the size of their arms expressed in articles for the dark-brown and the spotted variety. Luminescence changes with the age of the juvenile. The frequency of luminous ophiuroids follows a sigmoide curve. Therefore, the following analyses were restricted to the oldest juvenile stages (number of article higher than 9 for the dark-brown variety and higher than 7 for the spotted variety corresponding to more than 99% of luminous juveniles). A total of 94 individuals collected on 32 adults were studied for dark-brown morph and a total of 89 individuals collected on 18 adults are studied for spotted morph.

Adults of the dark-brown morph produce light less intense than the one of the spotted morph (76.56 and 698.8 $Mq.s^{-1}.mm^{-1}$ respectively). These differences could also be observed for the juveniles. We can then separate the brooded juveniles in four classes according to their luminous capabilities (Fig. 2): A, juveniles from dark-brown parent with 'dark-brown' capabilities; B, juveniles from spotted parent with 'spotted' capabilities; C, juveniles from dark-brown parent with 'spotted' capabilities; and D, juveniles from spotted parent with 'dark-brown' capabilities. The part of the variance explained by brooding parent for maximal intensity of emitted light were 0.45±0.12 and 0.19±0.12 for dark-brown and spotted variety respectively.

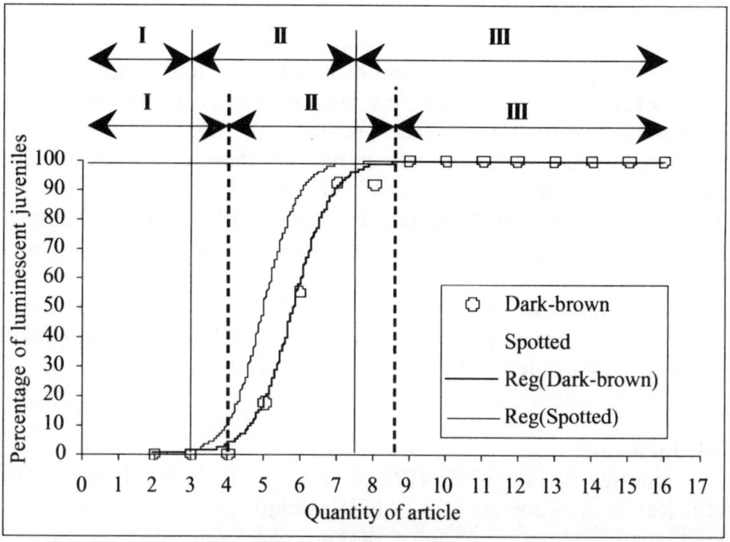

Fig. 1 Percentage of luminescent juveniles from brooding adults of dark-brown and spotted morphs according to the length of their arms expressed in article. (Reg = non linear sigmoid regression; Class I= 100% of non luminescent juveniles; Class II=between 0 and 99% of luminescent juveniles; and Class III=more than 99% of luminescent juveniles).

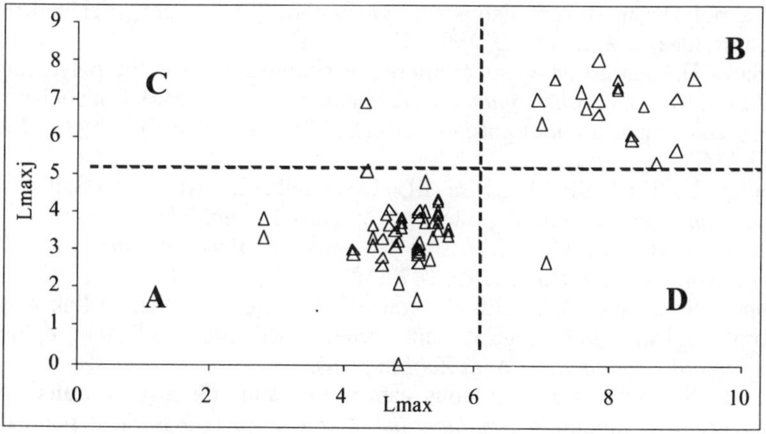

Fig. 2 Relation between the maximal intensity of emitted light of juveniles and adults (Lmaxj and Lmax in log of $Mq.s^{-1}.mm^{-1}$ respectively).

4. DISCUSSION

The intensity of light emitted by juveniles is partially determined by the brooding parent. This observation strengthens the idea of genetic determinism of luminous capabilities in *Amphipholis squamata*. Therefore, the huge observed differences for luminous capabilities between the colour morphs indicate genetic variations. Our results also confirm the possibility of outcrossing since juveniles from dark-brown/spotted parent present luminous capabilities of spotted/dark-brown morph. Nevertheless, outcrossing is a rare phenomenon (2 on 183 juveniles). This could be explained by differences in the population dynamics of these 2 morphs: fertilization mainly occurs in autumn for spotted individuals but only in spring for the dark-brown.[6] As a consequence of the possibility of outcrossing, *A. squamata* colour morphs can not be considered as different sibling species.

Acknowledgements
Research supported by FRFC 2.4573.00 and a FRIA grant to S. Dupont. J. Mallefet is Research Associate of the FNRS (Belgium).

REFERENCES

1. Binaux R, Bocquet C. Sur le polychromatisme de l'ophiure *Amphipholis squamata* (Delle Chiaje). *C R Hebd Acad Sc Paris* 1971; **273**: 1618-1619.
2. Deheyn D, Mallefet J, Jangoux M. Intraspecific variations of bioluminescence in a polychromatic population of *Amphipholis squamata* (Echinodermata: Ophiuroidea). *J Mar Ass UK* 1997; **77**: 1213-1222.
3. Deheyn D, Jangoux M. Colour varieties as sibling species in the polychromatic ophiuroid *Amphipholis squamata* (Echinodermata): evidence from inheritance body colour and luminescence characters. *J Exp Mar Biol Ecol* 1999; **234(2)**: 219-234.
4. Poulin E, F ral JP, Florensa M, Cornudella L, Alv V. Selfing and outcrossing in the brood protecting ophiuroid *Amphipholis squamata*. In: Candia Carnevali MD, Bonasoro F, eds. *Echinoderm Research 1998*, Rotterdam: A.A.Balkema, 1999: 147-150.
5. Dupont S, Chaufer S, Poulin E, F ral JP, Mallefet J. Is there a link between morphological, physiological and genetic variability of the ophiuroid *Amphipholis squamata*. *J Neth Zool* in press.
6. Dupont S, Mallefet J. Luminous capabilities and life history traits of the polychromatic ophiuroid *Amphipholis squamata* in two isolated populations from Oliveri-Tindari lagoon system, Sicily. *Hydrobiologia* in press.

ORIGIN AND CONSEQUENCES OF AGGREGATION ON BIOLUMINESCENCE DISTRIBUTION IN THE BRITTLE STAR *AMPHIPHOLIS SQUAMATA*

S. DUPONT, F. DUMONT AND J. MALLEFET

Laboratoire de Physiologie Animale, Université catholique de Louvain, 5 Place Croix du Sud, 1348 Louvain-la-Neuve, Belgium

E-mail: dupont@bani.ucl.ac.be

1. INTRODUCTION

The small ophiuroid species *Amphipholis squamata* is constituted of dense local populations in various biotopes: under stones of tide pools, in phanerogam meadows (*Zostera, Cymodocea,* etc.), in seaweeds (*Chaetomorpha, Corallina*, etc.), endosymbiotic of sponges, in coral reefs, etc.[1] In rocky shore populations, such as Langrune-sur-mer, it presents an aggregative distribution, each aggregate corresponding to all the ophiuroids living under the same stone.[2]

Recent genetic studies indicate that high genetic differences between individuals are observed between and within populations.[3-4] Moreover, the species is polychromatic, bioluminescent and huge differences were observed within the species. For example, Deheyn *et al.* demonstrate that each colour morphs exhibits its own capabilities to produce light[5] and Dupont *et al.* show that genetic structure is homogenous for each colour morphs within a population.[6] Therefore, polychromatism and/or bioluminescence are parameters that one has to take into account to study spatial and genetic structure of the species and are good indicators of genetic variability within population.

The aim of this work was to study the microdistribution of *A. squamata* within the rocky shore population of Langrune-sur-mer using bioluminescence as an indicator of genetic variability

2. MATERIALS AND METHODS

Ophiuroids were collected under stones of the tide pools at Langrune-sur-mer (Normandy, France). Sampling took place at very low tides between March 1999 and February 2000. Natural aggregates (more than 3 individuals under the same stone) were isolated in small aquaria for the descriptive part of this work. Artificia aggregates were obtained in aquarium. Ten individuals of each of the six colour morphs (Orange, Beige, Dark-brown, Grey, Black or Spotted as described by Deheyn *et al.*,[5] each possessing its own luminous capabilities) were placed in an aquarium containing six litres of aerated sea water and six stones of the sample site. After one month, each aggregate was isolated in a small aquarium. The same experiment was performed on both natural and artificial aggregates. For each animal, the colour morphs was identified under binocular microscope. Then, after anaesthesia in 3.5% w/w $MgCl_2$ in artificial sea water, two arms were excised for the measurement of light capabilities. Each isolated arm was stimulated by 200 mM KCl solution and light response was measured using a photomultiplier phototube (see Mallefet *et al.*[7]). The maximal intensity expressed in megaquata per second and per millimetre of arm length (Lmax in $Mq.s^{-1}.mm^{-1}$) was used to characterize the monophasic light response. A logarithmic transformation was performed to normalize the data. Analysis of variance (ANOVA) was used to estimate the part of variance explained by the aggregative distribution. Analyses were performed using Statistical Analysis System (SAS Institute).

3. RESULTS

For the descriptive part of this study, 360 individuals were collected under 63 stones. The experimental part of this work was performed two times and 58 individuals were collected under 12 stones. ANOVA analysis was performed for the maximal intensity of light with the aggregate number as class variable in order to estimate the part of the variance explained by the aggregation. For the natural aggregates, a significant effect ($p<0.0001$) is observed indicating that the variability within the aggregate was lower than in between. Opposite results were obtained for the experimentally induced aggregates where the analysis of variance is not significant (Table 1).

Table 1 Analysis of variance on the maximal intensity of emitted light ($\log(Mq.s^{-1}.mm^{-1})$) using the aggregate number as class variable: F value (Significant at $p<0.0001$***) and n, number of animals.

1.AGGREGATE	1.F	p	n
Natural	4.25	0.0001***	360
Experimental	1.68	0.1743	58

4. DISCUSSION

Field observations revealed that animals are not randomly distributed in aggregates. Under a same stone, ophiuroids present similar luminous capabilities since the variability is lower within than between an aggregate. Previous works indicate that the luminous capability is a good indicator of genetic variability.[6] Therefore, genetic strucure is homogenous within an aggregate. On the other hand, opposite results were obtained in laboratory. When animals with various luminous capabilities are placed in an aquarium, they rapidly form aggregates. Nevertheless, bioluminescence abilities are randomly distributed. This apparent paradox could be explained by the light of the reproductive biology of the species. The low genetic variability within an aggregate is a consequence of the low dispersal of the animal rather than a selective aggregation. *Amphipholis squamata* is a simultaneous hermaphrodite[2] and both selfing and outcrossing are possible.[4] The species possesses a weak potential of dispersion related to the lack of larval stage.[2] The species is also brood protecting since eggs, larvae and juveniles are kept in the genital bursae located into the disc for up to seven month before being released.[1-2] Therefore, released juveniles stay within the aggregate of the parents. The low genetic variability within an aggregate observed in Langrune-sur-mer is thus a consequence of a genetic drift rather than a selective aggregation.

Acknowledgements

Research supported by a FRIA grant to S. Dupont. J. Mallefet is Research associate of the FNRS (Belgium).

References

1. Dupont S, Mallefet J. Population dynamics of *Amphipholis squamata* (Echinodermata: Ophiuroidea) in the Oliveri-Tindari's lagoon system, Sicily. In: Candia Carnevali MD, Bonasoro F, eds. *Echinoderm Research 1998*, Rotterdam: A.A.Balkema, 1999: 65-70.
2. Alvà V. Reproduction, développement, incubation et dynamique de population de l'ophiure *Amphipholis squamata* (Echinodermata) en Baie de Seine. PhD thesis. Bruxelles: Université Libre de Bruxelles, 1996.
3. Sponer R, Roy MS, Mladenov PV. Molecular evolution of *Amphipholis squamata* (Echinodermata: Ophiuroidea): High genetic differences among populations. In: Candia Carnevali MD, Bonasoro F, eds. *Echinoderm Research 1998*, Rotterdam: A.A.Balkema, 1999: 405-408.
4. Poulin E, Féral JP, Florensa M, Cornudella L, Alvà V. Selfing and outcrossing in the brood protecting ophiuroid *Amphipholis squamata.* In: Candia Carnevali MD, Bonasoro F, eds. *Echinoderm Research 1998*, Rotterdam: A.A.Balkema, 1999: 147-150.
5. Deheyn D, Mallefet J, Jangoux M. Intraspecific variations of bioluminescence in a polychromatic population of *Amphipholis squamata* (Echinodermata: Ophiuroidea). *J Mar Biol Ass UK* 1997; **77**: 1213-1222.

6. Dupont S, Chaufer S, Poulin E, Féral JP, Mallefet J. Is there a link between morphological, physiological and genetic variability of the ophiuroid *Amphipholis squamata*. *J Neth Zool* in press.
7. Mallefet J, Vanhoutte P, Baguet F. Study of *Amphipholis squamata* luminescence. In: Scalera-Liaci L, Canicatti C, eds. *Echinoderm Research 1991*, Rotterdam: A.A.Balkema: 125-130.

EFFECT OF HABITAT ON INTRASPECIFIC DIVERSITY OF BIOLUMINESCENCE OF THE OPHIUROID *AMPHIPHOLIS SQUAMATA*

S. DUPONT, G. VANDEMEULEBROECKE AND J. MALLEFET

Laboratoire de Physiologie Animale, Universite catholique de Louvain, 5 Place Croix du Sud, 1348 Louvain-la-Neuve, Belgium
E-mail: dupont@bani.ucl.ac.be

M.-T. COSTANZO AND L. SALPIETRO

Universita di Messina, 31 Salita Sperone, 98166 Messina, Santa Agata, Italy
E-mail: mariaped@interfree.it

1. INTRODUCTION

Amphipholis squamata is an amazing ophiuroid species. It appeared more than 100 million years ago. At the present time, it is the only ophiuroid considered as cosmopolitan and it occupies a range of habitats from the intertidal zone to 1330 m depth.[1] Previous work has demonstrated that the species presents a huge inter- and intrapopulational variability at all the studied levels from gene to behaviour (genetic markers, physiological control, morphology, life-history traits, etc.).[2-4] The species is polychromatic and bioluminescent. Populations are constituted by several colour morphs, each possessing its own luminous capabilities (intensity and kinetic of emitted light). For example, the Spotted morph produces light more than 400 times more intense than the Orange one.[3]

Populational obervations indicate that great differences were observed in the frequency of the colour morphs according to the populations. In Langrune-sur-mer (France), 6 colour morphs were described when only 2 were observed in Porto Vecchio (Italy). In Langrune-sur-mer, 82% of the individuals produce low intensity light (<100 $Mq.s^{-1}.mm^{-1}$)[3] when 94% of the population is constituted by individuals producing high intensity light (<500 $Mq.s^{-1}.mm^{-1}$) in Porto Vecchio.[5]

Bioluminescence is a mode of communication between the luminous individual and its biotic environment. In *Amphipholis*, an anti-predatory effect has been proposed.[6] Production of light during interaction with crusteceans together with arm autotomy suggest that light could be used as a defensive strategy.[5]

The aim of this work was to compare the effect of bioluminescence on predator behaviour in two different biotopes: (i) Langrune-sur-mer (France) where animals are located in a confined habitat under stones; and in (ii) Porto Vecchio (Sicily) where animals live in an open habitat in green seaweed.

2. MATERIAL AND METHODS

Amphipholis squamata and two abundant co-occuring invertebrates were collected in two populations. (i) Ophiuroids and the Galatheidae *Galathea squamifera* under stones of the tide pools at Langrune-sur-mer (France); (ii) Ophiuroids and the shrimp *Palaemon serratus* in the phanerogam meadow of the Porto Vecchio lagoon (Sicily). In order to test the effects of the intensity of bioluminescence on predator behaviour, two morphs with contrasting luminous capabilities have been tested in each biotope: (1) The Beige and the Spotted morphs in Langrune-sur-mer; the first one producing 170 time less intense light than the second one; (2) The Dark-brown and the Spotted morphs in Porto Vecchio; the first one producing 20 time less intense light than the second one.

Experimental studies were performed in a small aquarium (11 x 17.5 x 11.5 cm) containing a stone in order to reproduce the confined habitat of Langrune-sur-mer. Three individuals of *A. squamata* of the same colour variety and one of the investigated invertebrates were placed in this aquarium. Control comprised ophiuroids alone in an aquarium. Each treatment was repeated nine times. During the first three days, the stone was turned over every twelve hours to check the integrity of the animals (number of discs and arms of *A. squamata*), this manipulation is then repeated at the end of the experimental period after seven days. Two parameters were defined in order to determine the intensity of interaction between ophiuroid and other invertebrates: The predation coefficient, which represents the cumulative intensity of predation, was calculated by:

$$Pc=1-(Ai+Di)/(Ao+Do)$$

Where Ai is the number of arms and Di, the number of discs at the time ti; Ao, the initial number of arms and Do, the initial number of discs. The predation activity, which represents the daily intensity of predation, was calculated by:

$$Pa=1-(Ai+Di)/(Ai\text{-}1+Di\text{-}1)$$

Where Ai-1 is the number of arms and Di-1 the number of discs at the time t-1. Analysis of variance (ANOVA) and t-test were used to determine the signifiance of the observed differences; each mean value is expressed with its standard error of mean (mean±SEM). All statistical analyses were performed on data using Statistical Analysis System (SAS Institute).

3. RESULTS

Aquaria experiments peformed on both predators showed similar results. (a) On average, the predation activity (Pa) is significantly higher during the night for both predators as shown in Table 1. (b) Figure 1 presents the predation coefficient through time for both predators. The coeffcient is higher ($p<0.01$ up to 2 days) for the morphs producing high intensity light.

4. DISCUSSION

Amphipholis squamata is a bioluminescent ophiuroid which emits light when mechanically stimulated.[5] An increase of the predation activity during the night suggests that both tested crustaceans are nocturnal predators. Therefore, they are able to see the light emission of the ophiuroids. Moreover, predation coefficient is higher when the prey emit high intensity light. This observation suggests that light emission of the ophiuroid is attractive for the predator. As a consequence of these observations, we propose that the structure of the biotope could determine the frequency of the colour morphs and then explain the observed differences between

Table 1 Average of the Predation activity of *Galathea squamifera* and *Palaemon serratus* during night and day (mean±SEM)

	Day	night
Galathea squamifera	0.255±0.074	0.531±0.100
Palaemon serratus	0.181±0.082	0.422±0.068

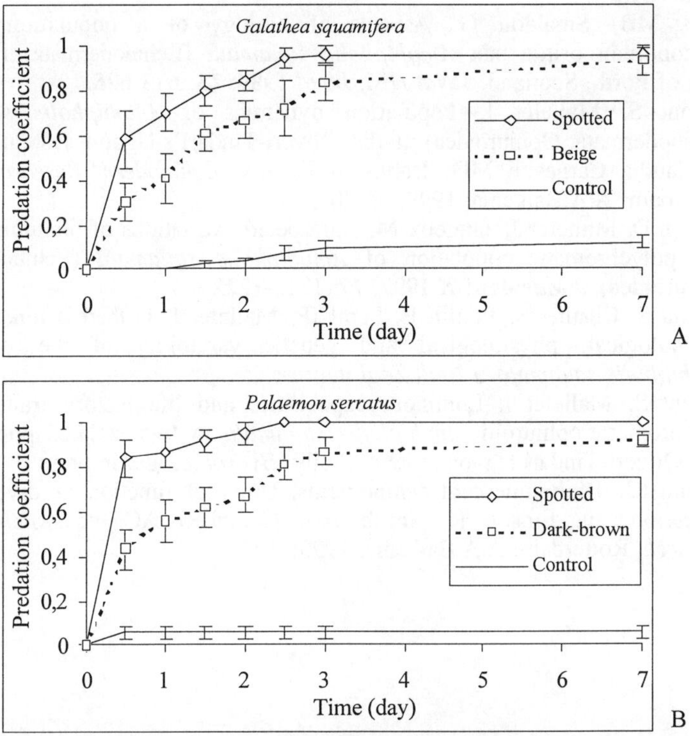

Fig. 1 Effect of colour morphs of *Amphipholis squamata* (Spotted vs Beige or Dark-brown) on predation by *Galathea squamifera* (A) or *Palaemon serratus* (B) in aquarium

the two studied populations. (i) In an open habitat, such as in Porto Vecchio in Sicily, light of high intensity could be used as an anti-predatory signal. Two different ways to deter predation are proposed: the decoy strategy (autotomy of one luminescent arm attracts the predator while the non-luminescent ophiuroid escapes) or the burglar alarm hypothesis (the light emitted by the ophiuroid attracts secondary predators).[6] In this kind of biotope, morphs producing high intensity light will be selected. (ii) In confined habitats, such as in Langrune-sur-mer, these functions can not be used since the luminous ophiuroid have no possibility to escape from its predators. Therefore, light emission will have negative impact on ophiuroid fitness and animals producing light of lowest intensity will be selected.

AKNOWLEDGEMENTS

Research supported by a FRIA grant to S. Dupont. J. Mallefet is Research associate of the FNRS (Belgium).

REFERENCES

1. Jones MB, Smaldon G. Aspects of biology of a population of the cosmopolitan brittle star *Amphipholis squamata* (Echinodermata) from the Firth of Forth, Scotland. *J Nat Hist, Lond* 1989; **23**: 613-625.
2. Dupont S, Mallefet J. Population dynamics of *Amphipholis squamata* (Echinodermata: Ophiuroidea) in the Oliveri-Tindari's lagoon system, Sicily. In: Candia Carnevali MD, Bonasoro F, eds. *Echinoderm Research 1998*, Rotterdam: A.A.Balkema, 1999: 65-70.
3. Deheyn D, Mallefet J, Jangoux M. Intraspecific variations of bioluminescence in a polychromatic population of *Amphipholis squamata* (Echinodermata: Ophiuroidea). *J Mar Ass UK* 1997; **77**: 1213-1222.
4. Dupont S, Chaufer S, Poulin E, Feral JP, Mallefet J. Is there a link between morphological, physiological and genetic variability of the ophiuroid *Amphipholis squamata*. *J Neth Zool* in press.
5. Dupont S, Mallefet J. Luminous capabilities and life history traits of the polychromatic ophiuroid *Amphipholis squamata* in two isolated populations from Oliveri-Tindari lagoon system, Sicily. *Hydrobiologia* in press.
6. Herring PJ. Bioluminescent echinoderms: Unity of function in diversity of expression? In: Emson R, Smith AB, Campbell AC, eds. *Echinoderm Research*, Rotterdam: A.A.Balkema, 1995: 9-17.

SENSITIVITY OF OBELIN AND APOOBELIN TO DIGESTION BY SOME PROTEASES

L.A. FRANK, V.S. BONDAR, AND E.S.VYSOTSKI
Photobiology Lab, Institute of Biophysics, Russian Academy of Sciences, Siberian Brunch, Krasnoyarsk 660036, Russia
Email: Eugene.Vysotski@usa.net

1. INTRODUCTION

The Ca^{2+}-regulated photoproteins have been used as reporter proteins both to monitor gene expression in cells[1] and to monitor the process of protein translation in a cell-free system.[2,3] The application of any gene as a reporter depends on the sensitivity of detection of the product being synthesized, and by the sensitivity of the reporter protein to environmental conditions. This is of particular importance when, to estimate the amount of a protein synthesized, its enzymatic activity is used. The latter may be affected by numerous factors: digestion by proteases, temperature, pH, ionic strength etc. The influence of such factors needs to be taken into account for any reporter enzyme, for example, a photoprotein, the production of which is evaluated by luminescence.

In a cooperation with Bayer AG, recombinant obelin was developed as a reporter protein for monitoring gene expression in mammalian cells. To facilitate the interpretation of light signals obtained from cells, the influences of some environmental conditions on the luminescent properties of photoproteins were investigated. To estimate possible consequences of digestion by proteases on luminescent activity, we tested proteases with a variety of different substrate specificities: trypsin that digests proteins in Arg-Y and Lys-Y sites; papain with its wide substrate specificity that digests proteins in Arg-Y, Lys-Y, Y-Phe, and Phe-Y sites; *Staphylococcus aureus* strain V8 protease digesting proteins in Glu-Y and Asp-Y sites; aminopeptidase M digesting proteins beginning at the N-terminus of a protein; and carboxypeptidase Y that digests proteins beginning from the C-terminus.

2. MATERIALS AND METHODS

<u>Chemicals</u>. Immobilized aminopeptidase M, immobilized papain, immobilized TPCK-trypsin, *Staphylococcus aureus* strain V8 protease were purchased from Pierce, carboxypeptidase Y was from Fluka.

<u>Apophotoprotein charging with coelenterazine and luminescent measurements.</u> Apoobelin was purified according to procedures described elsewhere[4] and was

charged by adding a molar excess (approx. 10 times) of coelenterazine to the apoprotein solution in a solution containing 10 mM DTT, 5 mM EDTA, 20 mM Tris-HCl pH 7.0. The apoobelin charging was done at 4°C overnight. To measure the photoprotein luminescence 5 μL of the mixture was added to 0.5 mL of an assay solution containing 10 mM EDTA, 100 mM Tris-HCl pH 8.8, and placed in the monitoring cuvette of a luminometer. To trigger the luminescence 0.2 mL of a solution containing 100 mM $CaCl_2$ in 100 mM Tris-HCl pH 8.8 was injected into the cuvette.

Apo- and photoprotein digestion by proteases. Apo- and photoprotein digestion was carried out in a solution containing 150 mM KCl, 5 mM PIPES pH 7.1 at 37°C and 4°C. When the charged photoprotein was digested the buffer contained additionally 2 mM EDTA. When the immobilized proteases were used the reaction mixture was incubated in a rapidly shaking water bath. The charging buffer was replaced for a digestion solution by dialysis. To estimate the influence of proteases digestion the 5 μL aliquots were taken out and the light emission was measured as described above. In the case of apophotoproteins the luminescence was tested after charging the sample with coelenterazine for 4 h at 4°C. Protein concentration in all experiments was approximately 0.1-0.2 mg per mL.

3. RESULTS AND DISCUSSION

Fig. 1 shows the luminescence activity changes in apoobelin and obelin after digestion by trypsin and papain at 4°C. The digestion by trypsin is rapid both the apoprotein and the charged photoprotein. After 2 h of incubation the residual luminescent activity of apoobelin constitutes about 40% and by 17 h — around 20% of the initial value. The rate of apoobelin digestion by papain is somewhat slower. After 3 h of incubation the residual activity of apoobelin is around 80% of the initial level. By 17 h of incubation the residual luminescent activity drops by 20% more and constitutes 60% of initial value. Coelenterazine binding to the protein slows photoprotein digestion by both trypsin and papain. In contrast to apoobelin, for which luminescent activity declined during the whole time studied, the luminescent activity of the charged photoprotein leveled off after decreasing rapidly during the first 3 h of incubation in trypsin or papain (Fig.1).

Fig. 2 shows the luminescence activity changes in apoobelin and obelin resulting from digestion by aminopeptidase M and *Staphylococcus aureus* strain V8 protease at 4°C. A surprising effect was observed in case of apoobelin digestion by aminopeptidase M. During the first hour of incubation, the luminescent activity of apoobelin remained unchanged, but in the next 2 h of incubation it increased by about of 30%. However, by the 17th hour of incubation the luminescent activity returned to the initial value. The binding of coelenterazine to the protein, in contrast to the case of trypsin and papain, does not slow digestion of obelin by aminopeptidase M, but rather accelerates this process. Apoobelin is practically stable to digestion by *S.aureus* strain V8 protease at 4°C. By the 17th hour of incubation with this protease the apoobelin luminescent activity had decreased only 10%. As with aminopeptidase M the binding of coelenterazine to apoobelin hastens the digestion of obelin by *S.aureus* strain V8 protease. Fig. 3 shows the luminescent activity changes in apoobelin and obelin after digestion by carboxypeptidase Y at 4°C. Even after 30 min of hydrolysis the apoobelin practically loses the ability to form an active complex with coelenterazine — the residual activity is only 3%. A distinctly different picture is observed for obelin treatment with carboxypeptidase Y. The activity of obelin remains essentially constant during 18 h of incubation. The data

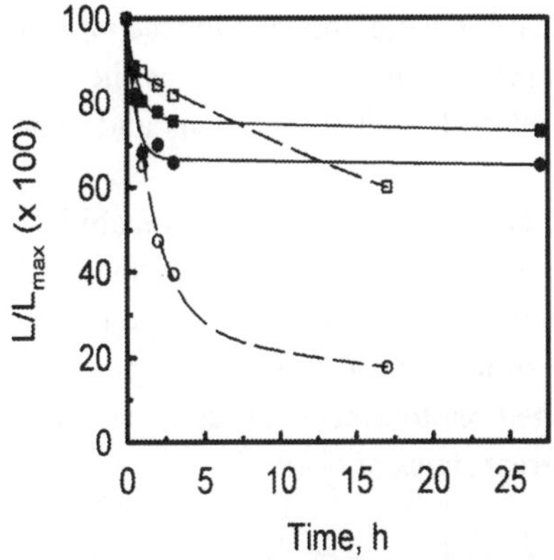

Fig.1. The digestion of apoobelin (dashed line) and obelin (solid line) by trypsin (circles, solid and open) and papain (squares, solid and open) at 4°C

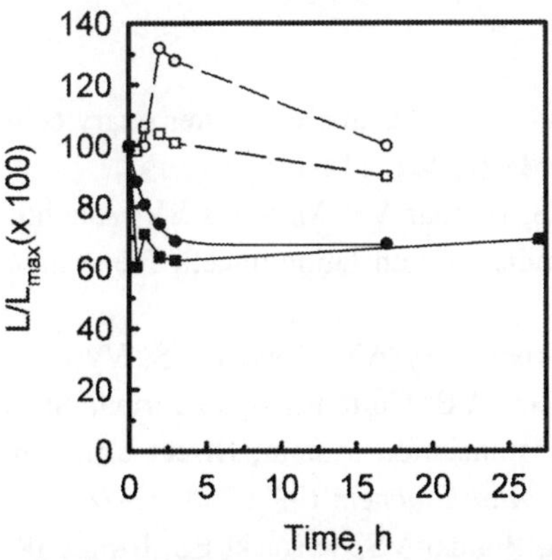

Fig.2. The digestion of apoobelin (dashed line) and obelin (solid line) by aminopeptidase M (circles, solid and open) and *Staphylococcus aureus* strain V8 protease (squares, solid and open) at 4°C.

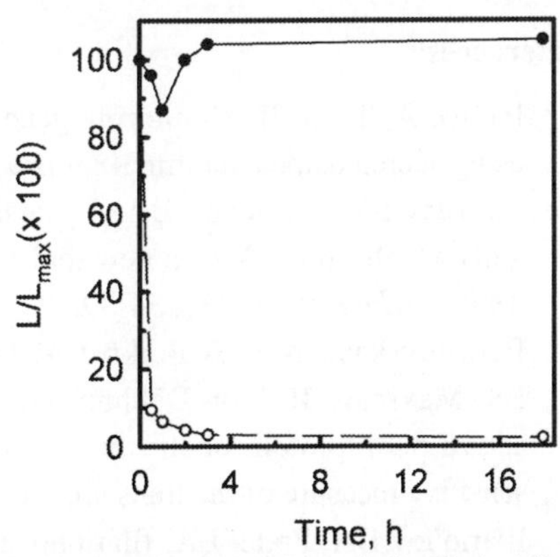

Fig.3. The digestion of apoobelin (dashed line) and obelin (solid line) by carboxypeptidase Y at 4°C.

obtained show that the C-terminus of photoproteins is of great importance in the formation of a photoprotein from apoprotein, coelenterazine, and oxygen.

Very similar results on digestion by trypsin, papain, aminopeptidase M, *S.aureus* strain V8 protease, and carboxypeptidase Y were obtained for recombinant apoaequorin and aequorin. An increase in temperature to 37°C enhances the effect of protease digestion on luminescent activity of both apophotoproteins and photoproteins.

The difference between the apo- and charged photoproteins in stability to digestion by proteases indicates that the binding of coelenterazine to the apoprotein causes changes in the protein moiety of the molecule. As a result of these changes some of the peptide bonds accessible for digestion by proteases in the apoprotein, become inaccessible for proteases in the charged photoprotein. These observations are correlated with NMR data obtained for different states of obelin.[5]

Acknowledgments

We thank John Blinks from University of Washington (Friday Harbor Laboratories) for critical comments and discussions at manuscript preparation. This work was supported by Bayer AG (Germany).

References

1. Inouye A, Tsuji FI. Monitoring gene expression in Chinese hamster ovary cells using secreted apoaequorin. *Anal Biochem* 1992; **201**: 114-118.
2. Matveev SV, Illarionov BA, Vysotski ES, Bondar VS, Markova SV, Alakhov YuB. Obelin mRNA — a new tool for studies of translation in cell-free systems. *Anal Biochem* 1995; **231**: 34-39.
3. Berestovskaya NG, Shaloiko LA, Gorokhovatsky AY, Bondar VS, Vysotski ES, Maximov JE, von Doehren H, Alakhov YB. Cotranslational formation of active photoprotein obelin in a cell-free translation system: Direct ultrahigh sensitive measure of the translation course. *Anal Biochem* 1999; **268**: 72-78.
4. Illarionov BA, Frank LA, Illarionova VA, Bondar VS, Vysotski ES, Blinks JR. Recombinant obelin: Cloning and expression of cDNA, purification, and characterization as a calcium indicator. *Methods Enzymol* 2000; **305**: 223-249.
5. Lee J, Glushka JN, Markova SV, Vysotski ES. Protein conformational changes in obelin shown by ^{15}N-HSQC nuclear magnetic resonance. (this volume).

BIOLUMINESCENT PATROLLING OF MARINE ECOSYSTEMS - BIOALARM

J.I. GITELSON, L.A. LEVIN, A.P. SHEVIRNOGOV AND R.N. YTYUSHEV
Institute of Biophysics of SB RAS, Akademgorodok, Krasnoyarsk, 660036 Russia
E-mail: root@gitelson.ibp.krasnoyarsk.su

1. INTRODUCTION

Studies done in the 1960s-1980s in expeditions in all oceans from the North Pole to Antarctic waters and from surface to a depth of many kilometers showed that excited bioluminescence is recorded in situ with a frequency sufficient for characterization of marine ecosystems. Especially rich in bioluminescence are ecosystems of the photic layer, the upwelling zones and coral lagoons. Using this approach, the vertical and horizontal structure of the plankton ecosystem can be monitored with profiling and routine bathyphotometers with high spatial and temporal resolution: 1-3 m, 5-6 min. ("Romashka" kit). We have adapted the bioluminescent profiling for ecological patrolling and early warning of deleterious changes in the marine ecosystem, particularly due to human intervention.

2. STATEMENT OF THE PROBLEM

The World Ocean on the whole, its marginal seas, shelves and water above it are subject to increasing impact of diverse anthropogenic activities. For the open ocean it is pollution by ships, impoverishment by fishing; for the shelf it is removal of marine organisms and extraction of mineral resources (gas and oil). Human activities have the greatest effect on coastal waters via effluent discharged immediately into the sea or through river estuaries of industrially polluted and eutrophicated agricultural and household waters, contamination of water area in ports. Adverse anthropogenic impact on marine ecosystems is permanent. The technical pressure on the sea life at the end of the 20th century and forecast for increase of this stress in the 21st century necessitate permanent monitoring of marine waters and the condition of organisms inhabiting them especially in the most dangerous areas. The techniques available for analysis of marine ecosystems' condition are highly accurate, specific, sensitive, but insufficiently quick to detect rapid changes, insufficiently integrated to cover vast water areas and insufficiently simple and cheap for rapid application in ecological monitoring. We propose a fast biophysical method based on marine bioluminescence that would quickly detect changes in the plankton communities over vast water areas (up to hundreds miles a day) using a moving ship with a continuously operating sensor with an immediate alarm signal. Bioluminescence (capacity to emit visible light as a result of specific enzymatic reactions) is inherent to many marine organisms - from bacteria and Protozoa to fish and cephalopods. Not all species inhabiting the sea emit light, but they are represented in each marine ecosystem amounting from 2% to 80% and more of the total number of its members. Clarke[1] and Backus, Boden and Kampa[2] were the first to investigate bioluminescence in situ using instruments called bathyphotometers, and these studies have been continued by Hastings and others[3].

3. METHODS

The profiling bathyphotometers we have developed[4] can be used for fast probing of the photic layer of the ocean. Submerging in 2-3 minutes to a depth of 150-200 m in their course they excite bioluminescence which can be continuously recorded by the instrument. The fast-action of the equipment recording the light signals makes it possible to obtain information with high spatial and time resolution (1-3 m, 5-10 min) impossible for conventional methods of collecting and processing plankton samples. Correlation between the number of recorded light flashes and total plankton concentration makes possible a rapid evaluation of its abundance and variability of the spatial location under the effect of external factors. For many areas of the Central Pacific, Sea of Japan and Okhotsk Sea the correlation factor between bioluminescence intensity and total concentration of plankton is 0.65-0.75. The highest correlation factors (0.85-0.90) are observed in the coastal areas, in lagoons of coral atolls. In this manner the vertical structure of the ecosystem is determined and target samples for biological and chemical analysis are obtained using bathometers comprising the "Romashka" bathyphotometer kit[5] (Fig. 1).

Another bathyphotometer version called "route" permits measurement of the bioluminescent signal in subsurface waters continuously at the speed of the ship carrying the instrument. With this instrument we have covered thousands of miles in the Indian Ocean, Red Sea, the Mediterranean and Black Seas[6]. The "route" bathyphotometer has been shown to record distinctly the alternation of water composition, heterogeneity of plankton distribution, its diurnal dynamics, and the effect of outflow of fresh waters at large river estuaries, etc[7].

4. PROPOSED METHODS OF OPERATION

The vessel with the installed "route" meter of bioluminescence patrols the coastal waters of assigned area continuously replenishing the database about the

Fig. 1. The outward appearance of the profiling bathyphotometer. (It comprises an inflow measuring chamber; container of electronics; sensor of pressure and temperature; 4 - assembly for connection with a cable; guard frame; bathometer - a rosette consisting of 6 bathometers with the electromagnetic valves for closure).

bioluminescent signal, its diurnal, tidal and seasonal dynamics. An alarm signal is produced when there is a statistically reliable deviation of the bioluminescent signal from that in the database for that particular location and time. The anomalous area is then outlined by route measurements and its coordinates are generalized for detailed chemical and biological analysis of the cause of the anomaly.

5. RESULTS

Implementation of the project should result in a meter for bioluminescence and/or a complex consisting of the routine meter and profiling bathyphotometer and development of methods for bioluminescent patrolling of waters from mobile carriers for rapid detection of ecological anomalies and providing an alarm signal.

Acknowledgements Funding provided by the Russian Ministry of Education and U.S. Civilian Research and Development Foundation for the Independent States of the Former Soviet Union (CRDF), Grant REC-002 "Yenisei".

References

1. Clarke GL, Backus RH. Measurements of light penetration in relation to vertical migration and records of luminescence in deep-sea animals. *Deep-Sea Res* 1956; **4**: 1-14.
2. Boden BP, Kampa EM. Records of bioluminescence in the ocean. *Pacif Sci* 1957; **11**: 229-235.
3. Hastings JW. Bioluminescence in marine dinoflagellates. *Proc Nati Biophys Conf* 1959; **1**: 427-434.
4. Gitelson II. ed. Marine Bioluminescence. Moscow: Nauka, 1969.
5. Gitelson II, Levin LA, Shevirnogov AP, Utyushev RN, Artemkin AS. Bathyphometric sounding of the pelagic zone of the ocean and its possible use for the studies of spatial structure of a biocenosis. In: Vinogradov ME, ed. *Functioning of Pelagic Communities in the Tropical Regions of the Ocean*. Moscow: Nauka, 1971: 50-64.
6. Gitelson II, Levin LA, Rudjakov JA and Utyushev RN. The estimation of sea bioluminescence by mesozooplankton biomass distribution in the upper layer of the Central Atlantic. In: Hastings JW, Kricka LJ, Stanley PE, eds. *Bioluminescence and Chemiluminescence. Molecular Reporting with Photons*. Chichester: John Wiley,1997, 165-168.
7. Gitelson II, Utyushev RN, Shelegov AV. Underway measurement of bioluminescent field during biological expedition. *Oceanology* 1991; **31**: 631-637.

LUMINOUS BACTERIA CULTURE COLLECTION IBSO: NATURAL BIODIVERSITY AND APPLIED ASPECTS

J.I. GITELSON, G.A. VYDRYAKOVA, A.M. KUZNETSOV, E.K. RODICHEVA, S.E. MEDVEDEVA AND YU.V CHUGAEVA

Institute of Biophysics, Siberian Branch of Russian Academy of Sciences, Krasnoyarsk, 660036, Russia, E-mail: ccibso@ksc.krasn.ru

1. INTRODUCTION

Luminous bacteria occur commonly in waters of all regions of the World's oceans that have been studied and from tropical to polar latitudes. The Culture Collection of the Institute of Biophysics (CC IBSO) contains about 700 strains of marine luminous bacteria (mesophylls and psychrophiles, free-living and associated with various marine inhabitants), 8 typical strains, *Escherichia coli* strains with *lux*-gene, dim and dark variants of luminescent forms. The vertical stratification and distribution of luminous bacteria in oceanic waters and their association with marine organisms is considered. The species composition of luminous bacteria found in photophores is the same as from gastroenteric tracts in organisms that have been studied. The microflora of some marine organisms may be composed of several species: *Photobacterium.phosphoreum* and *Vibrio harveyi*, or *P. leiognathi* and *V. harveyi*. The application of luminous bacteria in ecology, medicine and biotechnology is discussed.

Large groups of bacteria capable to emitting light are examples of microbial diversity. The majority of luminous bacteria are from marine organisms, but fresh water vibrios and those symbiotic on insect nematodes are known. Recent studies have demonstrated the important role of luminous bacteria in marine biocenoses. The Culture Collection of the Institute of Biophysics of the Siberian Branch of the Russian Academy of Sciences (CC IBSO) is a specific Russian collection which maintains about 700 strains of luminous bacteria isolated from different regions of the World Ocean, type strains of 8 luminous species and genetically modified Escherichia coli strains with marker lux-gene.

The present work makes an analysis of natural diversity, the main characteristics and possible use of the luminous bacteria especially from the Culture Collection IBSO.

2. NATURAL BIODIVERSITY

The luminous bacteria of the CC IBSO belong to different ecological groups: (1) free-living, and (2) associated with marine fauna. Free-living luminous bacteria grow on detritus, dead fish, molluscs, and other sea organisms.

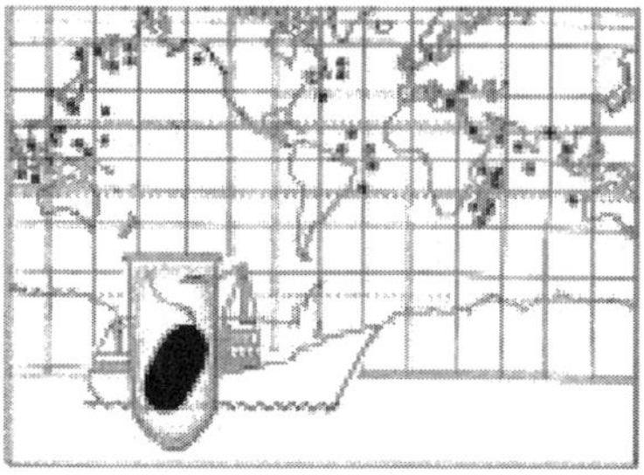

Figure 1. The luminous bacteria occur regularly in waters of all regions in the World's oceans which have been studied from tropical to polar latitudes and from the surface water layers to depths of several thousand meters

Luminous bacteria constitute 0.1-10% of the total number of bacterial-saprophytes in the waters of the open areas of the Atlantic and Pacific Oceans and 60-70% in the places with high concentrations of organic substances, e.g. lagoons of atolls, reefs, shoals. About half of the seawater samples from depths down to 1000 m contain free-living luminescent bacteria, and they constitute up to 50% of water microflora. These data are consistent with the hypothesis that the concentration of the dissolved organic constituents is an important factor in the distribution of luminous bacteria. Analysis of the composition of luminous bacteria species shows their discrete distribution with depth and change of species with variation in depth. The composition of the vertical stratification of luminous bacteria species apparently reflects the different interaction of certain species with the change of physical-chemical and biological factors in the environment. For example, in the waters of the open areas of the Pacific bacteria *V. harveyi* was isolated from depths 100-400 m, *P. phosphoreum*- 200-1200 m, *P. leiognathi* - 300-500 m. It should be noted that 400-800 m depths (where luminous bacterium *P. phosphoreum* is dominant) are their normal habitat because this depth does not have the optimal temperature or dissolved organics content. It may be that the source of sea water infection on these depths by *P. phosphoreum* are deep-water fish forming the specific symbiosis with this species and also fish of middle depths with luminous bacteria in their enteric tract. Moreover, it is known that *P. phosphoreum* species is the most psychrotrophic of all luminous bacteria and has apparently the best possibility for development at these depths. The strains of *V. harveyi* were isolated, as a rule, from warm coastal waters (over thermoclines) and didn't occur at depths below 400 m. It is not known whether luminous bacteria division between mesophylls and psychrophiles coincides with their latitude distribution but there is no reason to refute such connection.

Symbiosis between luminous bacteria and marine organisms is one of the most unique possibilities in evolution. The luminous bacteria associated with sea fauna are saprophytes (from the surface of various marine organisms), symbionts (from photophores) and commensals (from gastroenteric tracts). Bacterial symbiosis is most developed in fish. At present more than 200 luminescent fishes species have been described. Symbionts are represented by *P. leiognathi* and *P. phosphoreum*. All isolates from light organs were luminous. The strains of the species *P. phosphoreum* have been found in symbiotic association with deep-water fish and fish of medium depths and were dominant from a depth of 100 m to abyss. The feature of symbiosis of fish and molluscs with bacteria is the maintenance of pure culture of the one bacterial species in light organs. Luminous bacteria from light organs of deep-water fish and fish from middle depths were represented by only *P.phosphoreum.*[1] The light organ of mollusc from 70 m depth contained *P. leiognathi.*[2] It should be noted that species composition of luminous bacteria from photophores is the same as from gastroenteric tracts of organisms which have been studied. So, the composition of luminous bacteria species of marine animals is determined by such factors as host depth location, species composition of free-living luminous bacteria in the region of spawning and the presence of factors of recognition — lectins (they determine the species composition of associated microflora).

The other favorite ecological niche for luminous bacteria is the gastroenteric tract of sea animals and luminous bacteria-commensals have been isolated in 20-50% cases. Up to 60% of colonies grown on agar medium were from fish that were taken from depths of 600-1000 m. Analysis of the microflora of benthos sea organisms, the inhabitants of coastal water and coral reefs of the Indian Ocean, shows that the luminous bacteria isolate with high frequency from the intestines of marine inhabitants: fish and molluscs, not so frequently from shrimps holothurians, ophiuroids, sea stars, sea urchins, crinoids, etc. Luminous bacteria *P. leiognathi, P. phosphoreum, V. fischeri* and *V. harveyi* were isolated from gastroenteric tracts of marine organisms in the north-west part of the Indian Ocean. Most isolates were *P. phosphoreum*. Sometimes *V. harveyi* was isolated from the intestine of fish simultaneously with *P. phosphoreum*. *V. fischeri* was found only in a few cases. The strains of *V. harveyi* were found as a part of shrimps and molluscs microflora in tropical regions of the Indian Ocean.

The commensal *P. leiognathi* were found in the gastroenteric tract of non-luminous fish, and they were the only luminous bacteria there.

The Indian Ocean shrimps microflora was characterized by the highest diversity of luminous bacteria (*P. phosphoreum* and *V. harveyi*, or *P. leiognathi* and *V. harveyi* occurred simultaneously).

Vibrio fischeri isolates from intestinal microflora of fishes of the Black (mesophylls) and Okhotsk Seas (psychropiles). Bacteria of *Vibrio* genus produce non-luminous forms more quickly than *Photobacterium* during dissociation.

3. APPLIED ASPECTS

Luminous bacteria have been studied intensively in different ways: distribution in the World s oceans, symbiotic models, molecular-structural organization of luminescent system and genetic mechanism of it s regulation. Luminous bacteria are producers of many enzymes and biologically active substances. The main ones are luciferase, oxidoreductase and different dehydrogenases. It is possible to obtain the *E. coli* recombinant strains with a high yield of the luminescent system basic enzyme - luciferase, and to increase the enzyme synthesis 5-10 times as compared to the most productive strains of luminous bacteria. The producers of the following endonucleases of restriction: BanI, HaeIII, MboI, AflII, Asu I, Pst I, Pph 14I, are found among the strains of the CC IBSO. The majority of cultures produce such enzymes of the lyase class such as L-ornithine-, L-arginine-, L-lysine-decarboxylases. The luminous bacteria produce lipases, amylase, neuraminidase, chitinase, cellulase, alginate lyase. The most active cultures were isolated from water or algae surface.

Biosensors based on luminous bacteria *P. phosphoreum* (Microbiosensor B17 677F) and *Escherichia coli* with pPHL7-plasmid (Microbiosensor ECK) are designed to estimate the total toxicity of waters, to test for heavy metals and different chemical substances such as quinones, phenols, pesticides, etc.[1] They are simple and convenient to work with, can be readily standardized and made semi-quantitative, have rapid response to action of different substances and high sensitivity to environmental pollutants. It takes not more than 30 min to do the biotest (the other biotests take 48-96 h). Biotests were used to measure the integral biological toxicity of different substances, to control the water quality in such Siberian rivers as the Yenisei, the Angara, some rivers and the underground and surface waters of the Altai Territory, waste water of wood-pulp and paper production facilities and other industries. A kit of reagents for analytical bioluminescence (KRAB) based on the coupled enzyme system: luciferase - NAD(P)H:FMN-oxidoreductase can be used in medicine and ecological monitoring.

References

1. Rodicheva EK, Vydryakova GA, Medvedeva SE. *Catalogue of Luminous Bacteria Cultures*. 1997; Novosibirsk, Nauka. 125p.
2. Vydryakova GA, Kuznetsov AM, Primakova GA, Chugaeva YuV, Fish AM Luminous bacteria —symbionts and commensals of luminous and non-luminous representatives of marine fauna of the Indian Ocean. *Mikrobiol* 1995; 64:692-695.

OBSERVATIONS OF BIOLUMINESCENCE ON THE ABYSSAL SEAFLOOR

P.J. HERRING
Southampton Oceanography Centre, Waterfront Campus, European Way, Southampton SO14 3ZH, U.K.
E-mail: p.j.herring@soc.soton.ac.uk

I.G. PRIEDE, P.M. BAGLEY AND S. WAY
University of Aberdeen, Department of Zoology, Tillydrone Avenue, Aberdeen AB24 2TZ, Scotland, U.K.
E-mail: i.g.priede@abdn.ac.uk

J.C. PARTRIDGE
School of Biological Sciences, University of Bristol, Woodland Road, Bristol BS8 1UG, U.K.
E-mail: j.c.partridge@bristol.ac.uk

1. LIGHT AND VISION IN THE DEEP SEA

Many meso- and bathypelagic fishes have both luminous organs and eyes, and their visual systems are adapted to the detection of bioluminescent signals and/or very dim down-welling daylight.[1] At abyssal depths there is neither any light from the surface nor do many near-bottom fishes have luminous organs; nevertheless they do have eyes whose relative size is often little different from those of shallower species.[2] Many are scavengers, detecting deep-sea food falls by chemoreception, [3] so the function of their eyes is not immediately obvious. Can the bioluminescence of other organisms at these depths aid in finding food?

2. MEASUREMENTS OF BIOLUMINESCENCE

Measurements of bioluminescence in the ocean are usually made by inducing a turbulent flow which stimulates organisms to luminesce.[4] Such bioluminescence diminishes with increasing depth.[5,6] Unstimulated (natural) bioluminescence in midwater can be detected only by arrays of sensors viewing an undisturbed volume of water, or from neutrally buoyant manned submersibles.[7] We have imaged the bioluminescence at sea floor sites in the northeast Atlantic with an ISIT camera and pre-programmed videorecorder mounted on an autonomous lander which descends to the sea floor by free-fall. Other instruments record flow, pressure and temperature. In 1999, during cruise 243 of RRS *Discovery*, a mesh screen was placed at the focus of the camera. This provides scope for quantitative comparisons of bioluminescence, assuming all luminous organisms impacting on the screen are stimulated to emit light.[8] The camera was operated in three modes (Fig. 1).

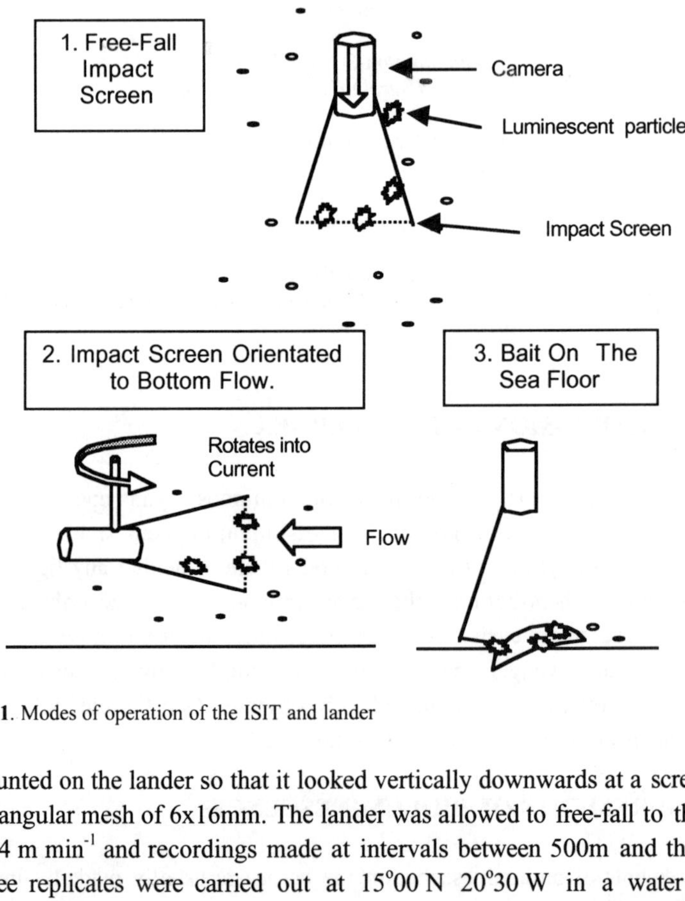

Fig. 1. Modes of operation of the ISIT and lander

1. Mounted on the lander so that it looked vertically downwards at a screen with a rectangular mesh of 6x16mm. The lander was allowed to free-fall to the bottom at 34 m min^{-1} and recordings made at intervals between 500m and the bottom. Three replicates were carried out at 15°00 N 20°30 W in a water depth of 4046m. There was a tenfold decrease from 1 bioluminescent target m^{-3} at 500m to $0.1m^{-3}$ at 2000m. Between 3000m and the sea floor the mean value was 0.015 m^{-3}, with no significant change over the depth range (Figure 2).
2. Mounted 0.5m above the bottom on a turntable looking horizontally at a 2mm mesh screen. The camera rotated automatically so that it faced into the bottom water current. This system was deployed twice at 4040m at the site of the free-fall experiments and twice at 3200m on the Cape Verde Terrace at 17° 45 N 20° 00 W. Water velocities ranged between 0.01 and 0.09m s^{-1}. Up to 0.45 luminescent events min^{-1} were observed on the screen. The mean density of

bioluminescent particles was estimated to be 2.04 m^{-3} at 3200m and 0.44 m^{-3} at 4040m, more than ten times that in the overlying water (Figure 2).

3. Mounted looking vertically downwards at bait (mackerel and prawns) on the sea floor. 165s recordings were made every 9 min followed by a 30s period in which the scene was illuminated with a very dim light to help identify any active fauna.

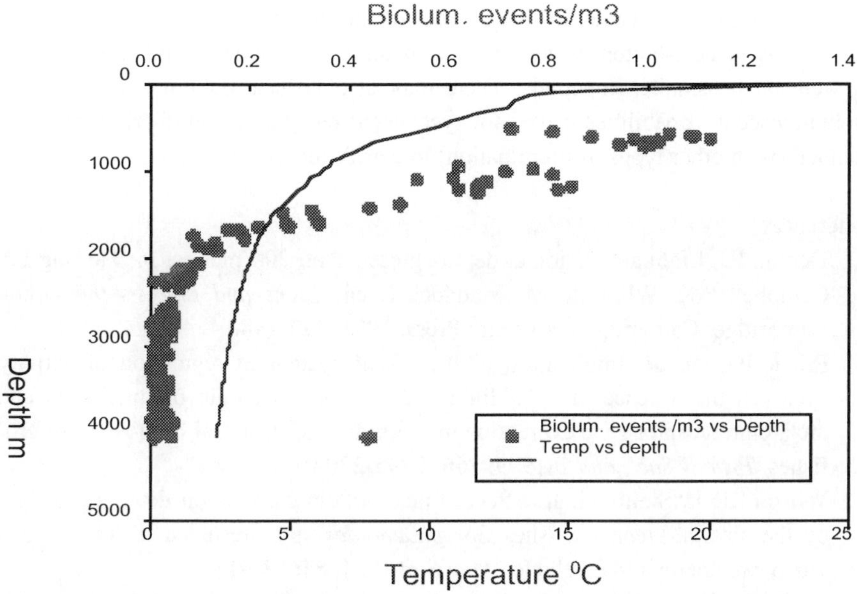

Fig. 2 Profiles of bioluminescence events and temperature from 3 three free-fall and two seafloor deployments to 4050m.

Luminescence was observed at the bait within 35min of its arrival on the sea floor and approximately 2 luminous events min^{-1} were observed thereafter. Single flashes were of 0.5-1s duration and occasionally repeated 2-3 times in ~2s. Rapid movements of the light source occurred on several occasions. Diffuse luminescent material released into the water was also observed sweeping across the field of view. Large (~30cm) shrimp (*Plesiopenaeus armatus*) and two large (~70cm) grenadiers (*Coryphaenoides armatus*) were seen at the bait. Neither animal is known to be luminescent. Small animals (<10mm), probably scavenging amphipods, were visible in many of the illuminated periods.

Most observations of marine bioluminescence obtain quantitative information from the effects of turbulence on the organisms. However, our recordings of bioluminescence around the bait appear to result from spontaneous interactions. For olfactory location of food falls the animals generally congregate in the downstream odour plume. In our study a number of the light emissions occurred upstream of the bait, but we do not know whether they were truly spontaneous or induced by the turbulence of another animal close by.

We cannot yet identify the sources of the luminous events. The luminous macrofauna in near-bottom waters include medusae, ctenophores and holothurians, as well as the smaller copepods. There may also be near-bottom species whose bioluminescent capabilities have not yet been recognized. Further studies are planned (with other types of illumination) to clarify this issue.

References

1. Denton EJ. Light and vision at depths greater than 200 metres. In: Herring PJ, Campbell AK, Whitfield M, Maddock L eds *Light and Life in the Ocean* Cambridge: Cambridge University Press, 1991; 127-148.
2. Priede IG, *et al.* Implication of the visual system in regulation of activity cycles in the absence of solar light: 2-[^{125}I]iodomelatonin binding sites and melatonin receptor gene expression in the brains of demersal deep-sea gadiform fishes. *Proc R Soc Lond B* 1999; **266:** 2295-2302.
3. Wilson RR Jr, Smith KL Jr. Effect of near bottom currents on detection of bait by the abyssal grenadier fishes *Coryphaenoides* spp., recorded in situ with a video camera on a free vehicle. *Mar Biol* 1984; **84:** 83-91
4. Case JF, Widder EA, Bernstein S, Ferer K, Young D, Latz MI, Geiger M, Lapota D. Assessment of marine bioluminescence. *Naval Res Rev* 1993; **45**:31-41.
5. Clarke GL, Hubbard CJ Quantitative records of bioluminescent flashing of oceanic animals at great depths. *Limnol Oceanogr,* 1959; **4**: 163-180.
6. Webster M.S. *et al* Mechanical stimulation of bioluminescence in the deep Pacific Ocean. *Deep-Sea Res* 1991; **38:** 201-217.
7. Widder EA, Bioluminescence in Monterey submarine canyon: image analysis of video recordings from a midwater submersible. *Mar Biol* 1989; **100:** 541-551.
8. Widder EA, Johnsen S. 3D spatial patterns of bioluminescent plankton: a map of the minefield . *J Plankton Res* 2000; **22**: 409-420.

LARGE SCALE BIOLUMINESCENT PHENOMENA IN THE OCEAN

P.J. HERRING
Southampton Oceanography Centre, Waterfront Campus, European Way, Southampton SO14 3ZH, U.K.
E-mail: p.j.herring@soc.soton.ac.uk

1. INTRODUCTION

Bioluminescence (BL) occurs from the surface to the seafloor, but is most often observed at the surface. Seafarers criss-crossing the oceans provide a huge potential corps of observers,[1] although their accounts are not always easy to interpret. Meteorological reports from ships are routinely collated by national agencies; ancillary BL observations may also be archived. These data are used to assess the geographical and seasonal occurrence of BL at the sea surface, and its correlation with hydrographic data. At the same time specialized bathyphotometers (deployed from research ships, submersibles, ROVs and AUVs) provide details of the BL of particular ocean regions and depths. Four large scale phenomena have been widely reported, namely flashing or glowing patches, erupting BL, Milky Seas and phosphorescent wheels.

2. GLOWING AND FLASHING PATCHES

The most frequently observed large scale BL is of glowing or sparkling patches of ocean, whose dimensions range from a few m to hundreds of m. Glowing patches usually represent the sites of very dense accumulations (or blooms) of luminous dinoflagellates, particularly species of *Noctiluca, Pyrocystis, Pyrodinium, Ceratium, Lingulodinium* (*Gonyaulax)* and *Protoperidinium*. In daylight such patches are often visible as reddish discolourations of the sea[2] and in coastal regions are known as red tides. BL is stimulated by the turbulence induced by wind or by the movement of other organisms. These glowing patches are often visible from the shore and may persist for long periods in specific locations where dinoflagellate blooms are maintained (e.g. Puerto Mosquito, Vieques, Puerto Rico). Sparkling patches are more indicative of aggregations of larger dinoflagellates (e.g. *Noctiluca*) whose flashes may be separately distinguished, or of patches which also contain large numbers of small luminous zooplankton, e.g. copepods or jellies. Glowing patches in the NW Indian Ocean can also be produced by mating swarms of ostracods. In the open ocean crescentic subsurface **glowing** patches are usually linked with dense concentrations of euphausiid shrimps, and produced either by their own luminescence or by their turbulent movements within a patch of dinoflagellates.

Flashing patches (the patch appears to flash as a unit) and pulsating patches (the patch brightness changes rapidly) are probably different accounts of the same phenomenon. Both are uncommon and usually refer to many small patches visible at the same time. They are usually encountered in relatively shallow water and some of them at least are likely to be surface mating aggregations of animals such as luminous syllid worms (fireworms) whose displays have a marked lunar rhythm.

3. ERUPTING LUMINESCENCE

Reports of this striking phenomenon describe luminous balls, bubbles or columns, 1m or less in diameter, moving very rapidly up to the surface, where they expand (or explode) into an ephemeral patch many m across. It is often interpreted as indicating the upwelling of deeper water, but the circumstances in which such upwelling might occur so locally and so rapidly are difficult to imagine. Even if water were to be upwelled it would still have to generate sufficient turbulent shear to stimulate the luminous organisms. It is more likely in most cases that what is being observed is the upward movement of a tightly aggregated shoal or school of fish, squid or shrimp, which then spreads out as it reaches the surface. Lanternfishes have been observed to produce this effect[2] but their individual luminous tracks are only distinguishable at very close range; most observers will see only an upwelling glow. Some reports describe a brightening as the luminescence reaches the surface; this is entirely consistent with the school moving into increased numbers of buoyant luminous dinoflagellates right at the surface.

Upwelling of bubbles of methane from deep water, expanding as they rise through luminous plankton near the surface, might produce some of the reported observations. There are many regions of the shallower seafloor through which methane is vented; under the deep water conditions of low temperature and pressure solid methane ice can accumulate. Seismic events could cause some of this material to float towards the surface where a phase change would occur, liberating potentially large amounts of gas as bubbles. However, expansion of the erupting BL when the bubbles burst at the surface is unlikely, except by propagation of the resulting pressure wave. It would be visible at best only as a very localised spreading ring of light, not a large-scale expansion of the luminous area.

4. MOVING BANDS AND PHOSPHORESCENT WHEELS

Phosphorescent wheels are the most spectacular - and least understood - large scale phenomena. Observers report huge luminous wheels with spokes, sometimes curved, sometimes straight, often extending as far as the eye can see and rotating at high speeds. They are often first seen as moving underwater strobes or searchlights (an underwater lighthouse is frequently used as an analogy). The kaleidoscopic effect then becomes more organised into moving parallel bands of light, or concentric rings, which finally change to rotating bands or spokes.[3] Some observers report a single huge wheel, others several separate wheels which appear to maintain station with the moving vessel and rotate in different directions. Most reports describe 1-2 spokes passing the observer per second. The cause is still unknown. The very large scale and the rapid movement indicate that some physical phenomenon is stimulating the luminous plankton. It is likely that some at least of the rotary motion is an optical illusion; an early (1879) account noted that rotating bands observed from on deck became parallel bands when seen from 50 ft aloft.[3] It has been argued that the parallel luminous bands could be generated by turbulence at the intersection of bow waves and parallel wave fronts, or the focusing effect of cylindrical waves,[4] but this does not account for their apparent (perhaps illusory?) speed. Sightings occur only in shallow waters, never beyond the continental shelf. They are most frequent in the Arabian Gulf area, the Bay of Bengal and south China Sea. This led to the ingenious hypothesis that they result from multiple reflections of seismic waves between the seafloor and the surface, whose interference patterns might account for the spoked wheels and their apparent rotation.[5] However their distribution does not match that of sea floor seismic activity, nor can it so easily explain rapidly moving parallel bands of light, or multiple wheels apparently associated with a vessel. The latter strongly suggests that the vessel plays a part in the stimulus, and the frequency of the bands is indeed similar to the frequency of most ships engines. There are no historical accounts from sailing vessels and such a spectacular phenomenon would surely have been noted by early seafarers.

5. MILKY SEAS

In the BL phenomenon known as a Milky Sea the ocean has a uniform glowing white surface, often extending to the horizon. The ship appears to be sailing through milk or over a snow-covered field. The scene has the reversed features of a film negative, with the sky dark and the ocean surface bright. The disturbed water of the wake may be darker than elsewhere, suggesting that the light emission occurs right at the surface. Other kinds of BL are usually absent. Occasionally the observing vessel passes across a sharp boundary between the milky sea conditions and the more usual dark ocean surface. A significant number of reports describe an apparent reduction in the sea state in the milky sea region.[6]

Milky Seas are commonest in the NW Indian Ocean during the SW Monsoon (June-Sept) but have also been seen regularly in the Banda Sea region. Bacteria are the only marine BL organisms which glow continuously and the phenomenon is widely interpreted as an unusually dense growth of luminous bacteria on an oily surface scum derived from dead plankton. The SW Monsoon causes intense surface growth in the region and upwelling cold water can produce high mortality of other organisms. Such scums are occasionally noted in daytime reports, and their calming effect could account for the apparent reduction in surface sea state. In the one case in which this phenomenon was investigated by a research ship, luminous bacteria (*Vibrio harveyi*) were present on algal debris taken in near-surface water samples.[7] The circumstantial evidence for bacteria is strong, but the described brightness of the observations requires a huge and stable density of bacteria, and luminous bacteria do not glow until a high enough concentration of a self-secreted chemical, or autoinducer, accumulates in the water. In most conditions turbulent diffusion would disperse this material too rapidly for it to initiate luminescence (the sea conditions in many Milky Sea reports exceed Beaufort Force 3). Bacteria provide the most likely cause but it is not yet clear how they could accumulate over such large areas in the densities necessary to induce BL.

References

1. Turner RJ. Marine bioluminescence. *Mar Obsvr* 1966; **36**: 20-29.
2. Herring PJ Red tides , water discolouration and bioluminescence in the Gulf of Oman and Strait of Hormuz. *Mar Obsvr* 1998; **68**: 19-25.
3. Herring PJ, Horsman P. Phosphorescent wheels: fact or fiction? *Mar Obsvr* 1985; **55**:194-201.
4. Verploegh G. The phosphorescent wheel. *Dt hydrogr. Z. Hamburg* 1968; 21: 152-162.
5. Kallc K. Die ratselhalfte und unheimliche Naturerscheinung des explodieren und des rotierenden Meeresleuchtens —eine Folge Seebeben? *Dt hydrogr Z Hamburg* 1960; **13**: 49-77.
6. Herring PJ, Watson M. Milky seas: a bioluminescent puzzle. *Mar Obsvr* 1993; **63**: 22-30.
7. Lapota, D *et al.* Observations and measurements of planktonic bioluminescence in and around a milky sea. *J Exp Mar Biol Ecol* 1988; **119**: 55-81.

ALL THREE CALCIUM-BINDING SITES PARTICIPATE IN THE REGULATION OF OBELIN LUMINESCENCE

B.A. ILLARIONOV, V.A. ILLARIONOVA, V.S. BONDAR, AND E.S. VYSOTSKI
Photobiology Laboratory, Institute of Biophysics, Krasnoyarsk 660036, Russia
Email root@biolum.ibp.krasnoyarsk.su

JOHN R. BLINKS
Friday Harbor Laboratories, University of Washington, Friday Harbor WA 98250, USA
E-mail blinks@u.washington.edu

1. INTRODUCTION

All known calcium-regulated photoproteins contain three Ca^{2+}-binding sites of the EF-hand type. However, it has been suggested on the basis of luminescence titration measurements that only two of them may be involved in the regulation of aequorin luminescence.[1,2] Implicit in this conclusion are the assumptions that Ca^{2+} functions as an indispensable reagent in the luminescent reaction, and that it does not recycle. An alternative view is that Ca^{2+} serves a purely regulatory role, increasing the rate of a reaction that would proceed more slowly in its absence. In that case, one would expect all Ca^{2+}-binding sites to be filled only when the reaction rate was maximal. This condition was probably never realized in the titration experiments, the design of which made sense only if Ca^{2+} was assumed to function as a reagent, not a regulator. The latter possibility seemed to us far more likely, so we approached the question in another way – by determining the influence on Ca^{2+}-dependent luminescence of disabling the Ca^{2+}-binding sites, individually and in all possible combinations. The results indicate that all three sites are critically involved in the control of photoprotein luminescence, and give some insights into their interaction.

2. METHODS

Wild-type recombinant obelin (cDNA from *Obelia longissima*) and seven different mutants were used in this study. Two amino acid substitutions were made in the individual EF-hand binding sites in order to interfere with Ca^{2+} binding. The most important change was to remove the highly conserved glutamic acid that normally occupies the last (twelfth) position of the E-F hand binding loop, replacing it with glycine. In numerous other systems regulated by Ca^{2+}-binding proteins, this substitution has been shown to lead to a drastic reduction of the Ca^{2+}-affinity of the mutant EF-hand site. However, we also removed the aspartic acid (D) in position 11, because it seemed possible that its free carboxyl group might be capable of substituting for that of the glutamic acid (E) in position 12. Hence the designation "DE mutants" for mutants in which the D in position 11 had been changed to A, and the E in position 12 had been changed to G. The EF-hand Ca^{2+}-binding sites are designated by the Roman numerals I, II, and III, numbering in the direction N- to C-terminus. Mutagenesis, cultivation of transformed *E. coli*, isolation and purification of the apoproteins, and charging with coelenterazine were carried out as in the past.[3]

Full Ca^{2+} concentration-effect curves (C-E curves) were determined at 20°C and plotted as explained elsewhere.[3] Peak luminescence was measured and expressed as fractional rate of discharge -- *i.e.*, normalized to the time integral (total light yield) of the signal from a comparable sample of the mutant protein exposed to high $[Ca^{2+}]$. Ca^{2+} concentrations were established with a combination of 2 mM Ca-EGTA buffers (low $[Ca^{2+}]$) and simple dilutions of $CaCl_2$ (high $[Ca^{2+}]$) in a Chelex-treated solution of 150 mM KCl, 5 mM PIPES, pH 7.0. Ionic strength was kept constant by adjusting [KCl]; hence the curves were not carried above 50 mM Ca^{2+}.

3. RESULTS AND DISCUSSION

Figure 1 shows C-E curves determined on wild-type obelin (WTO) and the seven DE mutants. Like that for aequorin, the curve for WTO has a maximum slope of about 2.5 and spans slightly more than 4 log units on the $[Ca^{2+}]$ scale. This span is somewhat more than would be expected for identical sites from the Langmuir adsorption isotherm, suggesting (along with the maximum slope of 2.5 instead of 3) that the Ca^{2+}-affinities of the individual binding sites may not all be the same.

It is immediately obvious from Fig. 1 that the DE mutation cannot be made in any one of the three E-F hand sites of obelin without drastically modifying the properties of the photoprotein. The curves for the 3 single mutants are clustered below the curve for WTO, with maximum luminescence levels between 1 and 2 log units lower. That in itself is strong evidence against the position that only 2 Ca^{2+} binding sites are involved in the regulation of luminescence. But there is more than that to be learned from these curves. Consider next the curves in groups: compare (1) WTO, (2) the cluster of three single mutants, (3) the cluster of three double mutants, and (4) the triple mutant. To a rough approximation (more later about some apparent exceptions), it appears that the DE mutation in any one site reduces the maximum luminescence by 1 to 2 orders of magnitude. The mutation in any two sites reduces the maximum luminescence by 3 to 4 orders of magnitude, and the mutation in all three sites reduces the maximum luminescence by almost 5 orders of magnitude. Thus the three binding sites seem to contribute similarly to the up-regulation of luminescence, they are more or less interchangeable, and the effects of their binding Ca^{2+} are multiplicative, suggesting that the three sites might be linked to separate structures, all of which must be in a favorable conformation for luminescence to occur. The probability of all three being simultaneously in the favorable state would be the product of the individual probabilities, each of which would be greatly increased by the binding of Ca^{2+}. It must be assumed that these

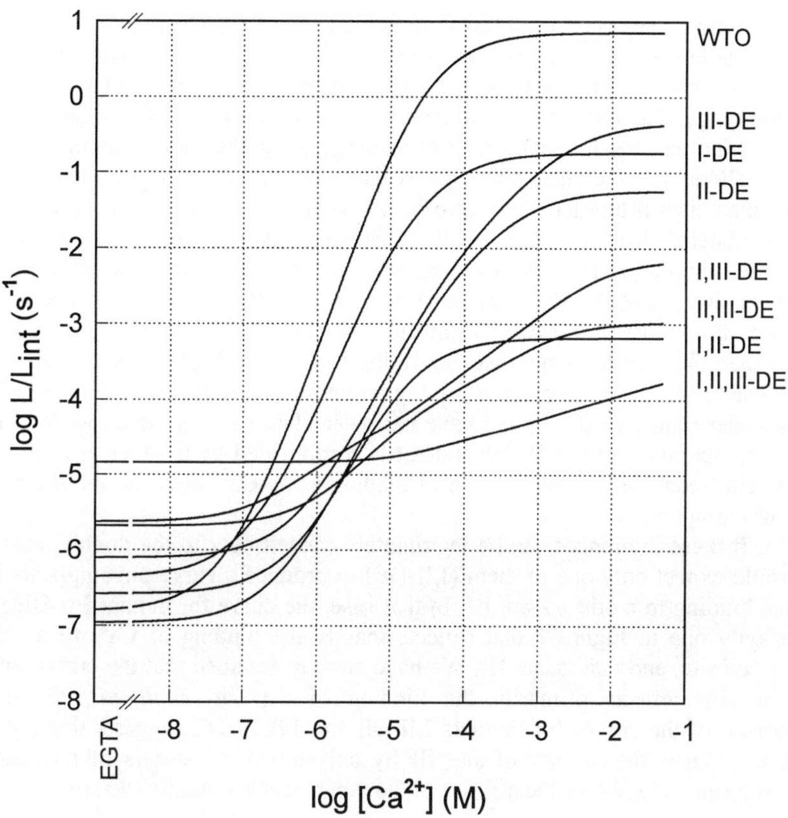

Fig.1. Ca^{2+} concentration-effect curves for wild-type obelin (WTO) and mutants in which the three putative Ca^{2+}-binding sites have been disabled, singly and in all possible combinations. Lines were computer-fitted to the experimental points. Temperature 20°C.

structures are capable of assuming the favorable state in the absence of Ca^{2+} [hence the Ca^{2+}-independent luminescence (CIL)], but that the probability of their doing so is very low. Note that the CIL for most of the DE mutants (especially the triple mutant) is higher than that of WTO, suggesting that the DE modification tends to reduce the stabilizing effect of the Ca^{2+}-free binding sites.

A closer inspection of the C-E curves for the various single mutants reveals some possibly significant differences. The curves for the I-DE and II-DE mutants have well-defined plateaus (from 10^{-4} M for I-DE, and about 10^{-3} M for II-DE). On the other hand, the curve for III-DE is still rising at 50 mM Ca^{2+}, and at that point the luminescence is considerably above the level of the plateaus reached by the other two single mutants. This suggests that in the III-DE mutant the affinity of the modified binding site is reduced less than in the I-DE and II-DE mutants, and that it is still capable of binding some Ca^{2+} when $[Ca^{2+}]$ is in the millimolar range. This would also account for the fact that the triple mutant does respond slightly to Ca^{2+} in the same range. The flat plateaus of the curves for I-DE and II-DE suggest that the modified binding sites are not capable of binding Ca^{2+} in the range examined.

There is also evidence to suggest that unmodified site I has a lower Ca^{2+}-affinity than sites II and III. The curve for the I-DE mutant has a steeper slope and broader plateau than that for II-DE. This suggests that the affinities of the unmodified binding sites in the I-DE mutant (*i.e.*, sites II and III) are more nearly alike than sites I and III (II-DE mutant). If sites II & III are alike, and I & III are not, then site I must be different from the other two. Site I probably has a lower affinity than the other two, because modifying it does not displace the C-E curve as much. The curves for both mutants having an unmodified site I (II-DE and III-DE) span broader ranges of $[Ca^{2+}]$ and have narrower plateaus than the curve for I-DE. However, the curve for the III-DE mutant is complicated by the fact that the third binding site seems still to be capable of binding some Ca^{2+} after the mutation (see previous paragraph).

If these inferences can be legitimately carried over to the double mutants, one would expect only one of them (I,II-DE) to produce a curve uncomplicated by residual binding to modified site III. In that case, the curve for mutant I,II-DE may be the only one in Figure 1 that reflects mainly the binding of Ca^{2+} to a single *unmodified* site, and that is site III. We have already assumed that the curve for the triple mutant reflects primarily the binding of Ca^{2+} to *modified* site III. A comparison of the curves for mutants I,II-DE and I,II,III-DE suggests that the DE mutation reduces the affinity of site III by only about 2-3 orders of magnitude, whereas it seems to reduce the affinity of the other sites essentially to zero.

Acknowledgments

Supported by NIH grants HL12186 and TW00412 (FIRCA), Washington Sea Grant, and Grant # 99-04-48452 from the Fundamental Research Foundation of the Russian Academy of Sciences.

The authors are indebted to Michelle Herko for expert technical assistance.

References

1. Shimomura O. Luminescence of aequorin is triggered by the binding of two calcium ions. *Biochem Biophys Res Comm* 1995; **211**: 359-363.
2. Shimomura O, Inouye S. Titration of recombinant aequorin with calcium chloride. *Biochem Biophys Res Comm* 1996; **221**: 77-81.
3. Illarionov BA, Frank LA, Illarionova VA, Bondar VS, Vysotski ES, Blinks JR. Recombinant obelin: Cloning and expression of cDNA, purification and characterization as a calcium indicator. *Meth Enzymol* 2000; **305**:223-249.

WHERE DOES MANGANESE BIND WHEN IT TRIGGERS THE LUMINESCENCE OF CALCIUM-REGULATED PHOTOPROTEINS?

B.A. ILLARIONOV, V.A. ILLARIONOVA, V.S. BONDAR, AND E.S. VYSOTSKI

Photobiology Laboratory, Institute of Biophysics, Krasnoyarsk 660036, Russia
Email root@biolum.ibp.krasnoyarsk.su

JOHN R. BLINKS

Friday Harbor Laboratories, University of Washington, Friday Harbor WA 98250, USA
E-mail blinks@u.washington.edu

1. INTRODUCTION

Mn^{2+} is capable of stimulating photoprotein luminescence, though to lower maximum levels than Ca^{2+}.[1,2] All known calcium-regulated photoproteins contain three ion binding sites of the EF-hand type, and Ca^{2+} ions interact with all of them.[3] However, the results of electron paramagnetic resonance (EPR) studies on both aequorin[4,5] and obelin[6] have been interpreted as indicating that only one Mn^{2+} binds per molecule of photoprotein. If Mn^{2+} stimulates luminescence through an action on only one of the EF-hand binding sites, and if Mn^{2+} binding is interfered with by the same mutations that disable Ca^{2+} binding, it should be straightforward to establish which of the three sites binds Mn^{2+}. To this end we examined the effects of Mn^{2+} on wild-type obelin and on two sets of obelin mutants in which the individual Ca^{2+}-binding sites had been disabled in different ways.

2. METHODS

Two different amino acid substitutions were made in the individual EF-hand binding sites in order to interfere with Ca^{2+} binding. In the first series (referred to as the DE mutants) the main substitution was to remove the highly conserved glutamic acid that normally occupies the last (twelfth) position of the E-F hand binding loop, replacing it with glycine. In numerous other systems regulated by Ca^{2+}-binding proteins, this substitution has been shown to lead to a drastic reduction of the Ca^{2+} affinity of the mutant EF-hand site. However, in the DE mutants we also removed the aspartic acid (D) in position 11, because it seemed possible that its free carboxyl group might be capable of substituting for that of the glutamic acid (E) in position 12. Hence the designation "DE mutants" for mutants in which the D in position 11 had been changed to A, and the E in position 12 had been changed to G. The EF-hand Ca^{2+}-binding sites are designated by the Roman numerals I, II, and III, numbering in the direction N- to C-terminus.

In the second set of mutants (designated the D1 series), a different critical carboxyl group was removed from the coordination sphere – the highly conserved

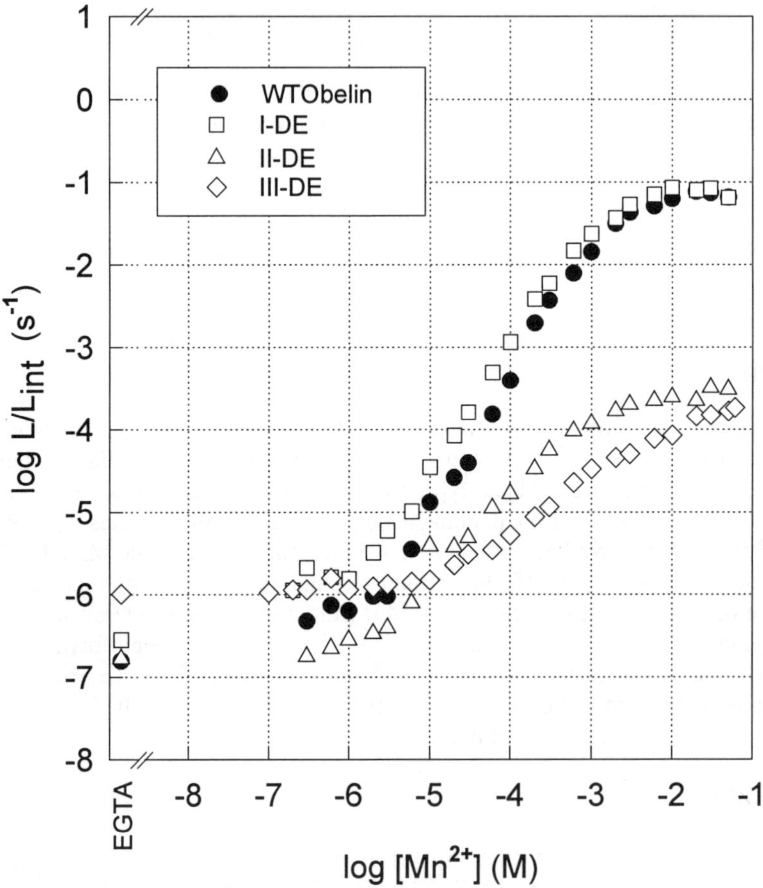

Fig.1. Mn^{2+} concentration-effect curves for wild-type obelin (WTObelin) and mutants of the DE series, in each of which a single Ca^{2+}-binding site had been disabled. Temperature 20°C.

aspartic acid in the first position of the twelve amino-acid EF-hand binding loop was replaced with arginine. Mutagenesis, cultivation of transformed *E. coli*, isolation and purification of the apoproteins, and charging with coelenterazine were carried out as has been described previously.[7]

Full concentration-effect curves (C-E curves) for Mn^{2+} were determined at 20°C and plotted as explained elsewhere.[7] Peak luminescence was measured and expressed as fractional rate of discharge -- i.e., normalized to the time integral (total light yield) of a comparable sample of the mutant protein exposed to high [Ca^{2+}].

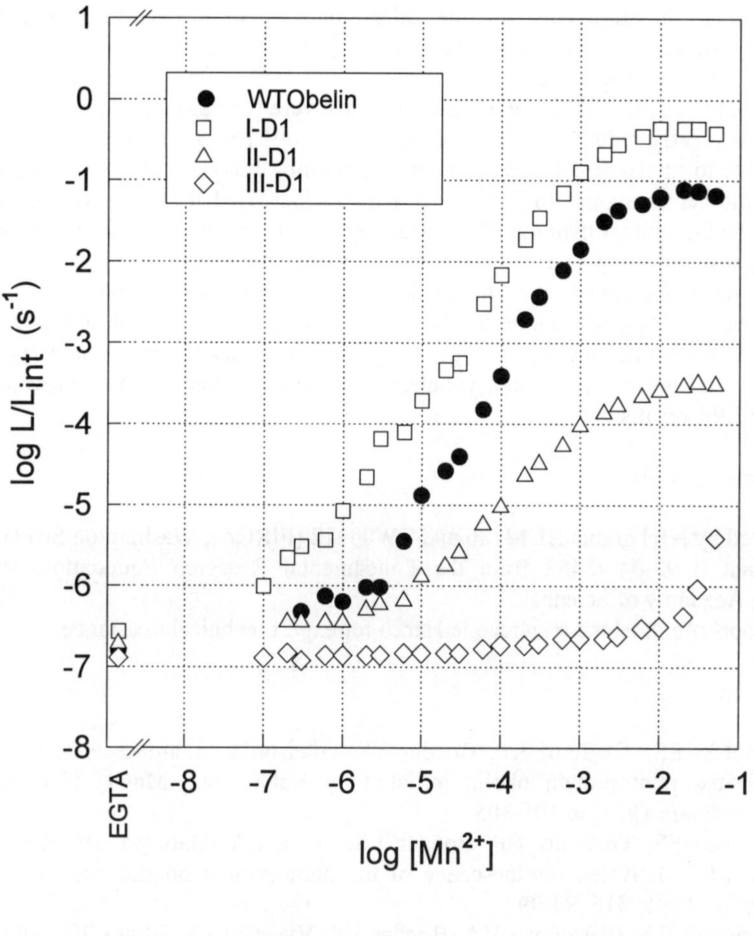

Fig. 2. Mn^{2+} concentration-effect curves for wild-type obelin (WTObelin) and mutants of the D1 series, in each of which a single Ca^{2+}-binding site had been disabled. Temperature 20°C.

Mn^{2+} concentrations were established by simple dilution of $MnCl_2$ in a Chelex-treated solution of 150 mM KCl, 5 mM PIPES, pH 7.0. Ionic strength was kept constant by adjusting [KCl], hence the curves were not carried above 50 mM Mn^{2+}.

3. RESULTS AND DISCUSSION

While the maximum rate of the light-yielding reaction is approximately two orders of magnitude less under the influence of Mn^{2+} than under that of Ca^{2+}, the quantum yield is approximately the same for the two cations. This was found to be true for the mutants in which site I had been disabled, as well as for wild-type obelin.

(Comparable measurements for the other mutants were attempted, but were probably not accurate in the presence of Mn^{2+} because of the exceedingly long integration times required.)

The results with the two sets of mutants were similar in that the response to Mn^{2+} was greatly reduced when either site II or site III was modified in ways that are known to interfere with Ca^{2+} binding. Disabling binding site I did not reduce the luminescent response to Mn^{2+}, and actually increased it slightly (more in the case of the D1 mutant than the DE). Disabling site III greatly reduced the ability of Mn^{2+} to stimulate luminescence (also more in the case of the D1 mutant than the DE) but did not eliminate it. This made us think at first that site III might be the single Mn^{2+}-binding site suggested by the EPR data. However, disabling site II produced an intermediate result, suggesting that Mn^{2+} may be able to interact with site II as well as site III. It is not yet clear to us how this finding can be reconciled with the EPR results.

Acknowledgments

Supported by NIH grants HL12186 and TW00412 (FIRCA), Washington Sea Grant, and Grant # 99-04-48452 from the Fundamental Research Foundation of the Russian Academy of Sciences.

The authors are indebted to Michelle Herko for expert technical assistance.

References

1. Vysotski ES, Trofimof KP, Bondar VS, Gitelson JI. Luminescence of Ca^{2+}-activated photoprotein obelin initiated by NaOCl and $MnCl_2$. *J Biolumin Chemilumin* 1993, **6**: 301-305.
2. Vysotski ES, Trofimov CP, Bondar VS, Frank LA, Markova SV, Illarionov BA. Mn^{2+}-activated luminescence of the photoprotein obelin. *Arch Biochem Biophys* 1995; **316**: 92-99.
3. Illarionov BA, Illarionova VA, Bondar VS, Vysotski ES, Blinks JR. All three calcium-binding sites participate in the regulation of obelin luminescence. This volume.
4. Kemple MD, Ray BD, Jarori GK, Rao BDN, Prendergast FG. Electron paramagnetic resonance of spin-labelled aequorin. *Biochemistry* 1984; **23**: 4383-4389.
5. Kemple MD, Lovejoy ML, Ray BD, Prendergast FG, Rao BDN. Mn(II)-EPR measurements of cation binding by aequorin. *Eur J Biochem* 1990; **187**:131-135.
6. Bondar VS., Hagen WR. Unpublished observations.
7. Illarionov BA, Frank LA, Illarionova VA, Bondar VS, Vysotski ES, Blinks JR. Recombinant obelin: Cloning and expression of cDNA, purification and characterization as a calcium indicator. *Meth Enzymol* 2000, **305**:223-249.

DIVERGENCE OF THE PROTEINS FOR BACTERIAL BIOLUMINESCENCE WAS DRIVEN BY GC RATIO REQUIREMENTS FOR BACTERIAL GENOMES

TIMOTHY C. JOHNSTON[1] AND JAMES R. COX[2]

[1]Department of Biological Sciences and [2]Department of Chemistry, Murray State University, Murray, KY 42071, USA E-mail: tim.johnston@murraystate.edu

1. CODON USAGE AND CODON BIAS IN THE LUX GENES OF *VIBRIO HARVEYI* AND *V. FISCHERI*

Organisms show a bias for the codons used in genes. The bias is seen primarily at the redundant third position of codons and has been explained by several hypotheses. Ikemura[1] proposed that codons are chosen to match tRNA abundance which then affects translational efficiency.[2] Parker, et al.,[3] showed that rare codons are more easily misread by near-cognate tRNA and therefore codons are chosen to reduce translational errors. Grantham, et al.,[4] suggested that codon choice is related to the kind of genome that the organism falls into, e.g. bacterial, phage, plant, animal, etc. Bacteria have very different biases in codon usage, and the idea that codon usage is related to genome is the focus of this investigation. Further, we have discovered that the nucleotide content of the genome has influenced codon choice at all three positions, resulting in the divergence of protein sequences.

The *luxA* and *luxB* genes for the α and β subunits of the luciferase of *V. harveyi* were the first sequenced and the codon usage was striking if not disturbing. The very codons that had been shown to be error prone in *Escherichia coli* (AAU, asn) and therefore avoided, were used frequently in these genes. Codon usage in the *luxA* and *luxB* genes of *V. fischeri* actually prefer AAU codons to AAC codons. The codon usage of these genes are given below (Table 1). In most cases, *V. harveyi* shows at least a slight bias for U or A in the third position. *V. fischeri* shows a marked preference for U or A in the third position. The six leucine codons are particularly interesting. The codons for leucine are roughly evenly distributed among the six codons in *V. harveyi* with a slight preference for UUG. *V. fischeri* uses UUA to encode leucine 29 out of 45 times, with the other 16 distributed among the other codons. These observations suggest that codon choice in these organisms is driven by a requirement to avoid GC.

2. GC CONTENT AND THE STRUCTURE OF THE ENCODED PROTEINS

Closer observation of the codon table shows the α and β subunits of *V. fischeri* have fewer amino acids encoded by G or C in the first and second positions of codons.

Table 1 Codon Usage in luxA + luxB of *V. harveyi* and *V. fischeri*

		V. h.	V. f.			V. h.	V. f.			V. h.	V. f.			V. h.	V. f.
UUU	Phe	24	34	UCU	Ser	12	9	UAU	Tyr	9	24	UGU	Cys	12	8
UUC	Phe	16	4	UCC	Ser	2	4	UAC	Tyr	17	5	UGC	Cys	2	4
UUA	Leu	7	29	UCA	Ser	9	10	UAA	End	2	2	UGA	End	0	0
UUG	Leu	17	2	UCG	Ser	5	4	UAG	End	0	0	UGG	Trp	8	7
CUU	Leu	6	2	CCU	Pro	6	12	CAU	His	11	17	CGU	Arg	8	8
CUC	Leu	5	3	CCC	Pro	0	2	CAC	His	12	3	CGC	Arg	5	3
CUA	Leu	10	6	CCA	Pro	11	8	CAA	Gln	14	27	CGA	Arg	8	7
CUG	Leu	7	3	CCG	Pro	3	0	CAG	Gln	13	5	CGG	Arg	0	1
AUU	Ile	14	31	ACU	Thr	12	18	AAU	Asn	20	33	AGU	Ser	7	9
AUC	Ile	17	6	ACC	Thr	8	6	AAC	Asn	19	7	AGC	Ser	4	3
AUA	Ile	0	10	ACA	Thr	7	19	AAA	Lys	22	29	AGA	Arg	4	3
AUG	Met	18	21	ACG	Thr	9	7	AAG	Lys	13	11	AGG	Arg	1	2
GUU	Val	15	21	GCU	Ala	11	9	GAU	Asp	26	38	GGU	Gly	22	15
GUC	Val	11	4	GCC	Ala	11	3	GAC	Asp	24	7	GGC	Gly	12	3
GUA	Val	7	15	GCA	Ala	16	16	GAA	Glu	38	36	GGA	Gly	4	11
GUG	Val	14	3	GCG	Ala	20	10	GAG	Glu	12	13	GGG	Gly	5	10

For example, there are 43 glycines, encoded by GGN, in *V. harveyi* luciferase compared to 39 in *V. fischeri*. The largest difference between the luciferases of these bacteria is in the use of alanine, encoded by GCN. The *V. harveyi* proteins have 58 alanines while the *V. fischeri* proteins have only 38. Changes to codons in the first position are more conservative than changes to the second position. Most first position changes result in hydrophilic for hydrophilic or hydrophobic for hydrophobic substitutions. Second position changes may result in a significant change in the character of the encoded amino acid causing alterations to the secondary and tertiary structure of the protein. We aligned the subunits of these proteins to determine what differences exist between the codons that encode aligned amino acids (data not shown). As expected, there are more first position changes than second position changes to codons of non-conserved positions in these proteins and *V. harveyi* is more likely to have a G or C in the first or second positions of *luxA* and *luxB* codons than *V. fischeri*. There are 254 codons with G or C in the first and/or second positions of non-conserved amino acids in *V. harveyi* proteins and 200 codons with G or C in those of *V. fischeri*.

Divergence of the primary structure must conserve the secondary and tertiary structure to retain activity. Table 2 shows the amino acids in the *V. fischeri* luciferase that replace alanine residues in the *V. harveyi* enzyme. We have shown the *V. fischeri* residues as "replacements" since we have the crystal structure of the *V. harveyi* enzyme.[5] All of the alanine residues listed in Table 2 are in alpha-helical regions in the *V. harveyi* structure (Fig. 1) and all of the *V. fischeri* replacements conserve the tendency to exist in a helix. Further, the positions that are exposed in the *V. harveyi* protein are replaced by polar residues in *V. fischeri* and those that are buried are replaced by nonpolar residues. The genes for these proteins diverged so that structure of the encoded protein was conserved.

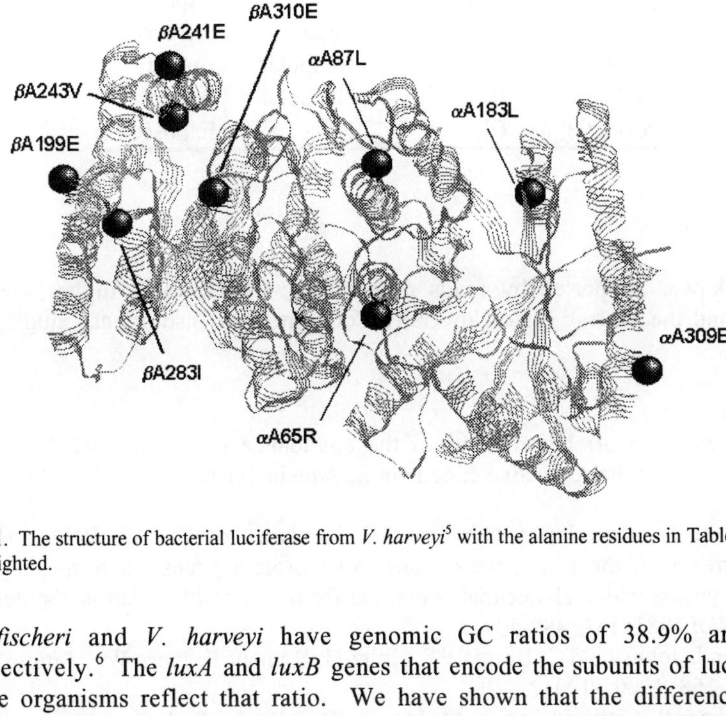

Fig. 1. The structure of bacterial luciferase from *V. harveyi*[5] with the alanine residues in Table 2 highlighted.

V. fischeri and *V. harveyi* have genomic GC ratios of 38.9% and 46.4% respectively.[6] The *luxA* and *luxB* genes that encode the subunits of luciferase in these organisms reflect that ratio. We have shown that the differences in GC content in these genes have not been restricted to the redundant third position of codons, but is also extended to the first and second positions. While some first and most second position alterations to codons alter the character of the encoded amino acid, the differences in the primary structures of these luciferases have maintained the conformational preferences of side chains.[7] Additionally, the *V. fischeri* enzyme has polar residues in positions where the *V. harveyi* enzyme has small, but nonpolar residues on the surface of the molecule. Similarly, the *V. fischeri* enzyme has nonpolar residues in positions where the *V. harveyi* enzyme has nonpolar residues buried in the interior. These data also suggest that the structural elements in the *V. harveyi* enzyme will be present in the *V. fischeri* structure.

Table 2 Selected Alanine Replacements in *V. fischeri.*

V. harveyi	*V. fischeri*	Polarity	Location
αAla-65	Arg	Polar	Exposed
αAla-87	Leu	Nonpolar	Buried
αAla-183	Leu	Nonpolar	Buried
αAla-309	Glu	Polar	Exposed
βAla-199	Glu	Polar	Exposed
βAla-241	Glu	Polar	Exposed
βAla-243	Val	Nonpolar	Buried
βAla-283	Ile	Nonpolar	Buried
βAla-310	Glu	Polar	Exposed

Acknowlegments

This work was supported by funds provided by the Howard Hughes Medical Institute and the Murray State University Committee for Institutional Studies and Research.

References

1. Ikemura T. Correlation between the abundance of E. coli tRNA and the occurrence of the respective codons in its protein genes. *J Mol Biol* 1981; **146**: 1-21.
2. Ikemura T. Correlation between the abundance of E. coli tRNA and the occurrence of the respective codons in its protein genes: a proposal for a synonymous codon choice that is optimal for the E. coli translational system. *J Mol Biol* 1981; **151**: 389-409.
3. Parker J, Johnston TC, Borgia PT, Holtz G, Remaut E, Fiers W. *J Biol Chem* 1983; **258**: 10007-10012.
4. Grantham R, Gautier C, Gouy M, Mercier R, Pave A. Codon catalog usage and the genome hyposthesis. *Nucleic Acids Res* 1980; **8**: 49-61.
5. Fisher AJ, Raushel FM, Baldwin TO, Rayment I. Three-dimensional structure of bacterial luciferase from *Vibrio harveyi* at 2.4 Angstrom resolution. *Biochemistry* 1995; **34**: 6581-6586.
6. Haygood M. Relationship of the luminous bacterial symbiont of the caribbean flashflight fish, *Kryptonphanaron alfredi* (family Anomolopidae) to other luminous bacteria based on bacterial luciferase (lux A) genes. *Arch Microbiol* 1990; **154**: 496-503.
7. Williams RW, Chang A, Juretic D, Loughran S. Secondary structure predictions and medium range interactions. *Biochim Biophys Acta* 1987; **916**: 200-204.

PURIFICATION AND PROPERTIES OF THE LUCIFERASE FROM THE MARINE OSTRACOD *Vargula hilgendorfii.*

KOJI KOBAYASHI AND YOSHIHIRO OHMIYA
Graduate School of Electronic Science and Technology, Shizuoka University, Shizuoka 422-8529, Japan

DAISAKU SHINOHARA AND YOSHIHIRO OHMIYA
Faculty of Education, Shizuoka University, Shizuoka 422-8529, Japan

TAKUJI NABETANI AND HARUKI NIWA
Department of Applied Physics and Chemistry, The University of Electro-Communications, Chofu, Tokyo 182-8585, Japan

1. INTRODUCTION

The small marine ostracod crustacean *Vargula hilgendorfii* (Vh), the sea firefly, is found around the coast of Japan.[1] It lives in sand at the bottom of shallow waters in the daytime whereas it becomes an active feeder at night. When disturbed or attacked, it produces a brilliant luminous cloud by ejecting luciferin (substrate) and luciferase (enzyme) from separate glands into the seawater. The characteristic blue luminescence is due to a highly specific, enzyme-substrate reaction in which a small organic molecule, luciferin (Mr, 478), an imidazopyrazine compound, is oxidized by molecular oxygen in a reaction catalyzed by luciferase.[1-3] The products of the reaction are light, oxyluciferin, and carbon dioxide.[4,5] Excited-state oxyluciferin bound to luciferase is the emitter in the reaction.[6] The primary structure of Vh

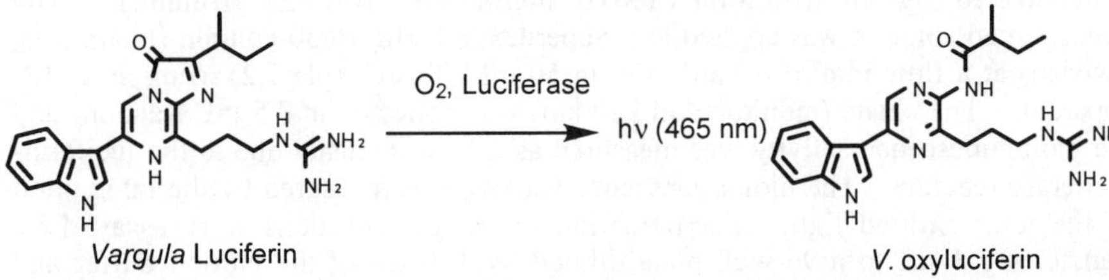

cloned luciferase consists of a single 555 amino acid polypeptide chain (MW 62,171) deduced from the nucleotide sequence.[7] This luciferase has a secretory signal, and mammalian cells transfected with cDNA for Vh luciferase secrete the enzyme. Mixing an aliquot of the medium with luciferin and measuring the light intensity can readily assay the activity of luciferase in the cell culture medium. When the transformed cells, which were transfected with the mammalian expression vector have a cDNA for Vh luciferase, the luciferase activities in the culture medium showed their promoter activities.[8] Stable transformants of Chinese hamster ovary cells carrying the Vh luciferase gene secreted luciferase from discrete sites on the cell surface, and this secretion could be monitored in real time by the luciferase activities as light using an image-intensifying technique.[9] Thus, Vh luciferase may be used as a convenient and sensitive reporter enzyme for studying the gene expression in mammalian cells. However, the properties of Vh luciferase are still not well established, especially with respect to structure and function. The luciferase is not very easy to purify from the natural materials as the native enzyme or from the cultured cells as a recombinant protein. Also, it is difficult to collect a large amount of the living organism as its population has started to decrease. We describe a new procedure to purify native luciferase in which about 80 μg of the purified luciferase could be obtained from 600 living organisms within 24 h.

2. MATERIALS AND METHODS

Animals *Vargula hilgendorfii* were collected in Shimoda (Izu Peninsula, Pacific coast of Central Japan). The living material was kept in an aquarium after collection.

Materials 2-Amino-2-hydroxymethyl-1, 3-propanediol was from Wako Pure Chemical Co. (Osaka, Japan), sodium chloride was from Nacalai tesque, Inc. (Kyoto, Japan), and PNGase F was from BioLabs Inc. (New England, USA). *Vargula* luciferin was kindly provided by Prof. Shimomura. All other reagents were of analytical grade.

Purification of luciferase The collected Vh was kept in filtered seawater. About 500 living specimens were transferred to 200 mL of fresh filtered seawater in a small tank connected to an electrode. A brilliant luminous cloud by ejecting luciferin and luciferase from the separate glands in the sea water was obtained when the living organisms were charged by electricity at 20 – 50 volts between 0.1 – 0.2s for approx. 20 times. Ejected luciferin and luciferase solution (200 mL) was concentrated by ultrafiltration (YM10 membrane, Amicon, Ireland). The concentrated solution was applied to a Superdex 200 HR 10/30 column (Pharmacia, Sweden) at a flow rate of 0.5 mL/min in 50 mM Tris/Cl (pH 7.2) using an FPLC apparatus. The eluate (monitored at 280 nm) was collected in 0.5 mL fractions and the bioluminescence activity was measured as a light intensity due to the luciferin-luciferase reaction. The bioluminescence activity was measured by the integration of the total emitted light. Each fraction or sample solutions were assayed by transferring 1 μL to a 96-well plate diluted by 150 μL of the elution buffer and injected with a luciferin solution (50 μL of 0.7 ng/mL luciferin, prepared in H_2O). The integrated light intensity for 20 s was measured with a Luminescencer AB-2000 luminometer (ATTO). Protein concentrations were measured by a Bio-Rad protein assay (Bio-Rad, USA) according to the instructions of the manufacturer. Bovine serum albumin (BSA) was used as a standard.

3. RESULTS AND DISCUSSION

Previously luciferase was purified from an aqueous extract of the dried Vh (approx. 10 mg of purified protein from 500 mg dried organism (ca. 500,000 organisms), by ammonium sulfate precipitation and chromatography on a DEAE column.[2] Now it is impossible to collect large quantitites of Vh, and instead we have now devised a method based on the purirication of enzyme ejected by Vh into seawater. About 600 specimens of the living organisms of Vh were transferred to 200 mL of newly filtered seawater in small tank attached to an electrode. The living organisms were charged by electricity at 20 – 50 volts between 0.1 – 0.2 s for approx. 20 times. This electric stress did not injure the living organisms, and almost all of them were alive. After 1 week, the luciferase could be obtained using the same living organisms. To purify the luciferase, the concentrated solution was applied to size-exclusion FPLC. The first peak at 280 nm coincided with the bioluminescence activity, and the SDS/PAGE pattern of this first fraction showed a single band corresponding to that of a 62-kDa protein (data not shown), indicating that the first fraction must be the purified luciferase. About 80 μg of enzyme was purified from 600 specimens in only three steps, i.e., electric stress, concentration using ultrafiltration and size-exclusion chromatography within 24h. This procedure is a convenient and efficient method for the purification of the secreted enzyme. The molecular weight of Vh luciferase is 62,171 deduced from the nucleotide sequence of 555 amino acid residues[7]. However, the molecular weight and N-terminal residue of the native luciferase having a secreted signal sequence was not determined. The N-terminal residue of the purified native luciferase could not be determined because of the modified of it (data not shown). The MALDI/TOF profile of the purified protein indicated its molecular mass 61.9 kDa with an accuracy of 1 % (Fig. 1).

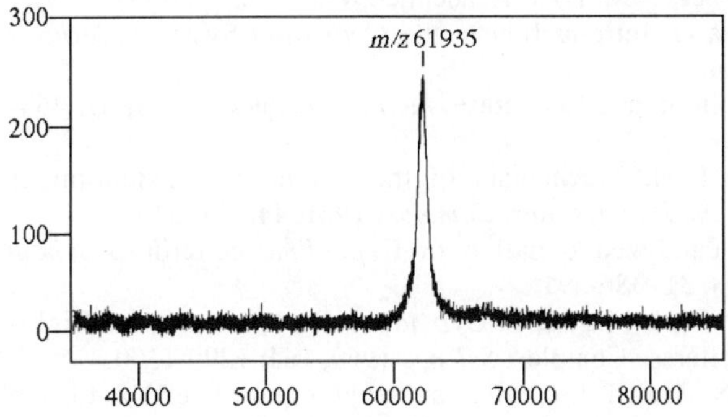

Fig.1 MALDI/TOF profile of the purified enzyme. A time of flight (TOF) mass spectrometer was used to obtain the matrix-assisted laser desorption/ionization (MALDI) mass spectra for determining the molecular mass in a linear TOF mode.

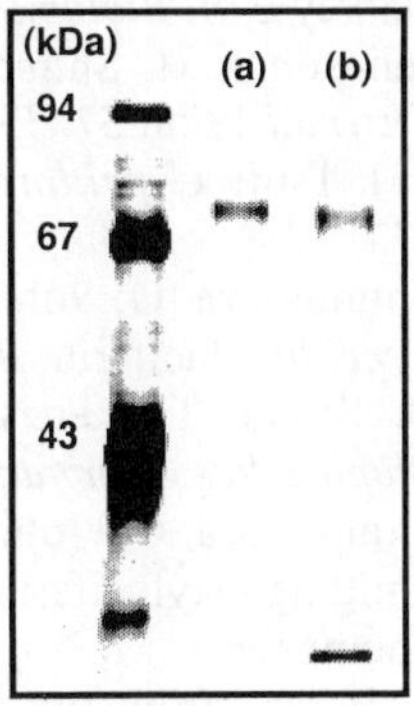

Fig.2 SDS/PAGE patterns of the purified luciferase (A) and the PNGaseF treated luciferase (B).

The difference in size between the wild type and theoretical value deduced from the nucleotide is about 300-500 corresponding to 3 – 5 amino acid residues. In general, a secreted protein having a signal sequence (15 - 25 amino acid residues) is a glycoprotein. The native luciferase could be modified at a specific site by the glycosylation after the translation and then deleting a signal sequence. The nucleotide sequence of the enzyme suggests two potential N-glycosylation sites (amino acid residues 186-188 and 408-410). PNGaseF can cleave between the innermost GlcNAc and asparagine residues of high mannose, hybrid, and complex oligosaccharides from N-linked glycoproteins. After PNGaseF, the electrophoresis mobility of the treated enzyme was slightly changed to a lower molecular weight corresponding to about 2000 – 3000 (Fig. 2). These results indicate the Vh luciferase could be an N-linked glycoprotein, which is cleaved at the signal sequence.

Acknowledgements
Financial support from the Ministry of Education, Science, Sports and Culture (Grant-in-Aid for Scientific Research on Priority Areas (A) and No. 11680586) and the Sumitomo Foundation is gratefully acknowledged. We thank Prof. O. Shimomura for providing the Vh luciferin.

References

1. Harvey E N. *Bioluminescence.* New York: Academic Press, 1952: 297-331.
2. Johnson F H, Shimomura O. Introduction to the *Cypridina* System. *Methods Enzymol* 1978; **57**: 331-364.
3. F. I. Tsuji. *Cypridina* Luciferin and Luciferase. *Methods Enzymol* 1978; **57**: 364-372 .
4. Shimomura O, Johnson F H. Mechanism of the luminescent oxidation of *Cypridina* luciferin. *Biochem. Biophys. Res. Commun.* 1971; **44**: 340-346 .
5. H. Stone. The enzyme catalyzed oxdation of *Cypridina* luciferin. *Biochem Biophys Res Commun* 1968; **31**: 386-391.
6. Shimomura O, Johnson F H, Masugi T. *Cypridina* Bioluminescence:Light-Emitting Oxyluciferin-Luciferase Complex. *Science* 1969; **164**: 1299-1300.
7. Thompson E M, Nagata S, Tsuji F I. Cloning and expression of cDNA for the luciferase from the marine ostracod *Vargula hilgendorfii*. *Proc Natl Acad Sci USA* 1989; **86**: 6567-6571.
8. Thompson E M, Nagata S, Tsuji F I. *Vargula hilgendorfii* luciferase: a secreted repoter enzyme for monitoring gene expression in mammalian cells. *Gene* 1990; **96**: 257-262.
9. Inouye S, Ohmiya Y, Toya Y, Tsuji F I. Imaging of the luciferase secretion from transformed Chinese hamster ovary cells. *Proc Natl Acad Sci USA* 1992; **89**: 9584-9587.

ON THE ORIGIN OF BIOLUMINESCENT SYSTEMS

Y. A. LABAS
Institute of Ecology and Evolution, Russian Academy of Sciences, Leninsky Prospect 33, 117071 Moscow, Russia
E-mail: yulii_labas@mail.ru

M.V. MATZ
Institute of Bioorganic Chemistry, Russian Academy of Sciences, Miklukho-Maklaya 16/10, 117871 Moscow, Russia
E-mail: matz@mail.ibch.ru

V.A. ZAKHARTCHENKO
Center for Craniofacial Molecular Biology, University of Southern California, 2250 Alcazar Street, CSA 103, Los Angeles, CA 90033, USA
E-mail: zakhartc@hsc.usc.edu

1. THE PROBLEM OF GLOWING MUTATIONS

Bioluminescent systems (BS) are present in aerobic organisms on different levels of phylogenesis from bacteria to fishes. There are no bioluminescent species among autotrophs (except dinoflagellates), several groups of sea and terrestrial animals, and for some unknown reason -among freshwater organisms (except -the gastropod *Latia neritoides* and some species of parasitic photobacteria).[1]

Charles Darwin mentioned that the origin of bioluminescence is the problematic questions of his theory.[2] The function of bioluminescence is directly associated with the visual behavior of organisms. Its role may be be chasing the predators with flashes, camouflage, communication, and attraction with steady glowing (fungi, non-symbiotic photobacteria).[3-5] It means that the initial material for selection for development of new BS were only such neutral mutations of non-luminescent organisms which resulted in well visible glowing, associated (in case of impulse mode) with certain behavioral reactions, e.g. escape. It suggests preexistence in non-bioluminescent precursors from luminescent species all the major components of the newly-developed BS: luciferine (L), luciferase (E) or their analogs, triggering mechanisms and secondary emitters (e.g.GFP-s from coelenterates, blue fluorescent proteins from photobacteria, red fluorescent protein from deep-sea fish *Malacosteus niger*). Different phyla have different L and E. Taking in account dissimilarity between L,E and other components of luminescence mechanism in different phyla, it is reasonable to assume that about 30 types of BS originated independently from preexisting non-bioluminescent reactions.[6] There are some reports on conditionally bioluminescent species. In some non-bioluminescent species of Copepoda, e.g. *Nannocalanus minor and Oithona plumifera* some individuals can occasionally be found that generate flashes under stimulation.[7] There are both luminous and non-luminous species in the same genus: e.g. *Obelia,* among Hydrozoa,[8] and *Oncaea*, *Oithona*, among *Cyclopoida*.[7] The close taxonomic relations of several luminous organisms with non-luminous provide evidence in favor of the recent genesis of some BS.

2. THE PROBLEM OF ANCESTRAL FUNCTION

There is increasing evidence available to prove that an ancestral function of L and some E was detoxification of reactive oxygen species (ROS), based on the antioxidant abilities of L.[5,9-11] Coelenterazine occurs in many luminous marine eucaryotes as well as in non-luminescent species.[12] Some L are able to emit light with ROS in the absence of E.[13-15] The luminescence of E-L photoprotein complexes is also be activated by ROS rather than molecular oxygen in *Pholas dactylus*,[16] and also in Polynoidae which have a specific reaction to superoxide.[17] The calcium-activated photoprotein obelin is able to emit light in the absence of calcium but in the presence of singlet oxygen.[18] An activation of synthesis s well as antioxidant enzymes such as SOD and catalase during hyperoxygenation may indicate the antioxidant origin of E in fireflies.[10] Bacterial E can produce a luminescent reaction *in vitro* in the presence of H_2O_2 or other ROS and in the absence of aliphatic aldehyde.[11]

In some eukaryotes an impulse luminescent flash, triggered by external stimulation, initiates an endogenous release of ROS. In fact, in earthworms *Diplocardia longa* and ascidians *Clavelina miniata* BS is localized in phagocytes - cells of the respiratory burst , which glow on the animal surface under lysis.[19,20] For earthworms, H_2O_2-dependent nature of biolumiscent reaction has been proved. In non-bioluminescent earthworm *Lampito mauritii* phagocytes also glow in the presence of H_2O_2.[21] In Polynoidae, intracellular luminescence is the result of enzymatic generation of superoxide.[17] . There are some other mechanisms of bioluminescence triggering in Dinoflagellata through H^+ or calcium in Anthozoa with an induced release of L from the complex with luciferin-binding protein (LBP).[22,23] Based on this data, we propose that the ancestral function of the L and some E might have been protection of prophotogenic cells from the ROS secreted by themselves (in the majority of eukaryotes, in the impulse mode). The original function of LBP might have been protection from antioxidants: proluciferins from self-oxidation prior the respiratory burst. The generation of ROS in low concentration is one of the prerequisite conditions for normal vitality (in connection with theirs functions of antiparasitic or antipredator defense, secondary messengers, etc). An absence of BS in freshwater organisms could be explained by fact that many pro-BS were scavengers of $HOCl^-$ ion that could be the terminal product in ROS secretion only at a high concentration of Cl^- ion in the environment. Anthozoa and Hydrozoa have different types of E and relatively close GFPs and L. It implies that their BS originated independently whereas GFPs and L were inherited from a common non-bioluminescent ancestor. Prediction of that fact resulted in a discovery of a new family of GFP-like proteins in non-bioluminescent Anthozoa.[25] We also suggest that photocytes in Anthozoa originated from an entodermal pigment cells which are able to secrete ROS for phagocytosis. The proposed functions of GFP-like proteins could be photoactivation of several enzyme reactions and tissue photoreception.

Acknowledgement Supported by Russian Foundation for Fundamental research (grant 99-04-48873).

References

1. Harvey EN. *Bioluminescence.* New York: Academic Press, 1952: 542.
2. Darwin CR. *The origin of species by means of natural selection.* London: John Murray, 1859: 502.
3. Hastings JW. Bioluminescence. In: Speralakis N, ed. *Cell Physiology.* New York: Academic Press, 1995: 651-681.
4. Sivinski JM. Phototropism, bioluminescence, and the Diptera. *Florida Entomologist* 1998; **81**: 282-292.
5. Wilson T, Hastings JW. Bioluminescence. *Annu Rev Cell Dev Biol* 1998; **14**: 197-230.
6. Buck JB. Function and evolution of bioluminescence. In: Herring PJ, ed. *Bioluminescence in Action.* London: Academic Press, 1978: 419-460.
7. Evstigneev PV, Bitjukov EP. *Bioluminescencija morskich copepod.* Kiev: Naukova dumka, 1990: 145.
8. Morin JG. Coelenterate bioluminescence. In: Muscatine L, Lenhoff H, eds. *Coelenterates Biology. Review of New Perspectives.* New York: Academic Press, 1974: 397-438.
9. Rees JF, de Wergifoss B, Noiset O, Dubuisson M, Janssens B, Thompson EM. The origins of marine bioluminescence: turning oxygen defense mechanisms into deep-sea communication tools. *J Exp Biol* 1998; **201**: 1211-1221.
10. Barros MP, Bechara EJ. Bioluminescence as a possible auxiliary oxygen detoxifying mechanism in elaterid larvae. *Free Radic Biol Med* 1998; **24**: 767-777.
11. Watanabe H, Nagoshi T, Inaba, H. Luminescence of bacterial luciferase intermediate by reaction with H2O2: The evolutionary origin of luciferase and source of endogenous light-emission. *Biochem Biophys Acta* 1993; **1141**: 297-302.
12. Thompson CM, Herring PJ, Campbell AK. The widespread occurrence and tissue distribution of the imidazolopyrazine luciferins. *J Biolumin Chemilumin* 1997; **12**: 87-91.
13. Skatchkov MP, Sperling D, Hink U, Anggard E, Munzel T. Quantification of superoxide radical formation in intact vascular tissue using a Cypridina luciferin analog as an alternative to lucigenin. *Biochem Biophys Res Commun* 1998; **248**: 382-386.

14. Akutsu K, Nakajima H, Katoh T, Kino S, Fujimori K. Chemiluminescence of Cipridina luciferin analogs. 2. Kinetic studies on the reaction of 2-methyl-6-phenylimidazo(1,2-a)pyrazin-3(7H)-one(CLA) with superoxide-hydroperoxyl radical is an actual active species used in initiate the reaction. *J Chem Soc - Perkin Trans* 1995; **2**: 1699-1706.
15. Shimomura O. Superoxide triggered chemiluminescence of the extract of luminous mushroom Pannelus stripticus after treatment with methylamine. *J Exp Bot* 1991; **41**: 555-560.
16. Roberts PA, Knight J, Campbell AK. Pholasin--a bioluminescent indicator for detecting activation of single neutrophils. *Anal Biochem* 1987; **160**: 139-148.
17. Bassot JM, Nicolas MT. Bioluminescence in scale-worm photosomes: the photoprotein polynoidin is specific for the detection of superoxide radicals. *Histochem Cell Biol* 1995; **104**: 199-210.
18. Vysotsky ES, Trofimov KP, Bondar VS, Gitelson JJ. Luminescence of Ca2+-activated photoprotein obelin iniciated by NaOCl and MnCl2. *J Biolumin Chemilumin* 1993; **6**: 301-305.
19. Belissario R, Spencer TE, Cormier MJ. Isolation and properties of luciferase, a non-heme peroxidase from the bioluminescent earthworm, Diplocardia longa. *Biochemistry* 1972; **11**: 2256-2266.
20. Chiba K.,Hoshi M.,Isobe M.,Hirose E.Bioluminescence in the tunic of the colonial ascidian, Clavelina miniata: identification of luminous cells in vitro. J.of Exper.Zool.1998. **281**:546-553
21. Sanhanam KSV, Limaye NM. Electrobioluminescence of cells extracted from Lampito mauritii. *Biochem Bioenerg* 1989; **22**: 219-229.
22. Lee DH, Mittag M, Sczekan S, Morse D, Hastings JW. Molecular cloning and genomic organization of a gene for luciferin-binding protein from the dinoflagellate Gonyaulax polyedra. *J Biol Chem* 1993; **268**: 8842-8850.
23. Anderson JM, Carbonneau H, Cormier MJ. Mechanism of calcium induction of Renilla bioluminescence. Involvement of calcium-trigged luciferin binding protein. *Biochemistry* 1974; **13**: 1195-1200.
24. Kumar S, Harrylock M, Walsh KA, Cormier MJ, Charbonneau H. Amino acid sequence of the Ca2(+)-triggered luciferin binding protein of Renilla reniformis. *FEBS Letters* 1990; **268**: 287-290.
25. Matz MV, Fradkov AF, Labas YA, Savitsky AP, Zaraisky AZ, Markelov ML, Lukyanov SA. Fluorescent proteins from nonbioluminescent Anthozoa species. *Nat Biotech* 1999; **17**: 969-973.

FLOW-STIMULATED DINOFLAGELLATE BIOLUMINESCENCE

MICHAEL I. LATZ
Scripps Institution of Oceanography, University of California San Diego,
La Jolla, CA 92093-0202 USA
E-mail: mlatz@ucsd.edu

JIM ROHR
SPAWAR Systems Center, San Diego, CA 92152-5000 USA
E-mail: rohr@spawar.navy.mil

1. INTRODUCTION

Whereas the nocturnal movements of copepods, fish, seals, dolphins, divers, and ships[8] are all capable of stimulating dinoflagellate bioluminescence, typical oceanic motion is not stimulatory with a few exceptions, most notably breaking waves. Low background bioluminescence levels allow luminescent signatures of moving objects to be readily discernable from planes, e.g., fish schools and surface ships, such as the last German U-boat sunk in World War I. In order to know what flow fields stimulate bioluminescence, it is necessary to identify the hydrodynamic features that best serve as an index for bioluminescence response. It has been suggested that steady laminar flow is not stimulatory, and that flow-stimulated bioluminescence is indicative of turbulence. Although an appealing demarcation for flow stimulation, this cannot be the entire story because most oceanic fluid motion is turbulent. There are several reasons to consider shear as the primary flow feature associated with stimulated dinoflagellate bioluminescence: (1) flow-stimulated bioluminescence is associated with boundary layers and flow separation;[8] (2) at the small spatial scales of individual plankton, oceanic turbulence is experienced as laminar shear; (3) shear affects other physiological processes in dinoflagellates, such as the inhibition of population growth;[11] and (4) the physiology of mammalian flow-sensitive cells is affected by fluid shear. Because of fluid viscosity, fluid shear is manifested as a shear force (stress) across the cell. Although some 250 years ago bioluminescence was thought to be due to friction between particles of water and salt, it has only been in the last thirty years that experiments have begun to address the nature of the hydrodynamic friction or shear stress associated with flow stimulated bioluminescence. Because it is difficult to re-create oceanic turbulence in the laboratory, laboratory approaches tend to focus on aspects of the flow field that can be studied quantitatively. Initially, these studies yielded contradictory interpretations of the flow conditions associated with bioluminescence.

Christianson and Sweeney,[2] using a coiled capillary tube, were the first to quantify bioluminescence as a function of flow rate. Although they did not directly determine a response threshold, they observed stimulation in laminar flow.

However, when the same flow field was directly imaged, flashes were observed only near the inlet where there was an abrupt transition.[1] This result, together with corroborative measurements obtained in other flow fields, led Anderson et al.[1] to conclude that bioluminescence is not stimulated by steady laminar flow. Subsequent pipe flow studies[10] suggested there was an abrupt increase in plankton bioluminescence upon transition from laminar to turbulent flow, further supporting the view that laminar flow was not stimulatory. The dramatic increase in bioluminescence through transition could be due to an order of magnitude increase in shear stress or the fundamentally different nature of turbulent flow. Latz et al.[3] were the first to measure response thresholds for both cultured dinoflagellates and freshly collected plankton samples as a function of quantifiable flow conditions. Using steady Couette shear generated in the gap between concentric cylinders, they measured a consistent response threshold in laminar flow occurring at a shear stress value of about 0.1 N m^{-2}. Above threshold, the level of bioluminescence was proportional to the shear stress with no evidence for any change in response due to the laminar or turbulent nature of the shear stress.

Because of the discrepancies between these initial studies, an extended effort over the past five years has been undertaken to answer fundamental questions such as: Is laminar flow stimulatory? What are threshold levels for flow-stimulated bioluminescence? Are there fundamental differences in the response of luminescent dinoflagellates to laminar and turbulent flows? These questions have been methodically addressed for both cultured dinoflagellates and freshly collected mixed plankton samples dominated by dinoflagellates.

2. EXPERIMENTAL APPROACH AND RESULTS

Subsequent studies with fully developed pipe flow support the original finding of a response threshold in laminar flow. The response threshold, which occurs at wall shear stress values of about 0.1-0.3 N m^{-2}, is similar for cultured dinoflagellate species and natural plankton samples,[5,8] although there is some variability in sensitivity across species.[7] The threshold is much greater than shear levels in most oceanic flows but is consistent with an antipredation function of dinoflagellate bioluminescence.[3]

The pipe flow experiments have examined the response of individual cells (rate of flashing, flash intensity) as well as the population response (mean bioluminescence). Results support the hypothesis that the luminescent response is correlated with the shear stress in the flow.[5] Experiments using cells injected at either the pipe wall or at centerline in laminar flow demonstrate that the flash response occurs only where adequate shear stress levels are present. By using an overlapping range of laminar and forced turbulent pipe flows it has been possible to show that bioluminescence is related to the shear stress levels, not the laminar/turbulent nature of the flow.

Further experiments using pipe flow have examined changes in bioluminescence with flow conditions. Above threshold, mean bioluminescence increases as a function of shear stress level in the flow, due primarily to an increase in the number of cells responding rather than an increase in flash intensity.[5] Upon transition from laminar to turbulent flow, there is a further increase in the number of cells responding but no increase in flash intensity. The maximum flash intensity of individual cells increases with wall shear stress values from threshold up to 1 N m^{-2} in laminar flow, but remains constant thereafter regardless of the laminar or turbulent nature of the flow or the length scales of the turbulence.

Many of the Couette and pipe flow results have been verified in a series of independent experiments using a converging laminar flow field.[4] In this completely laminar flow field, regions of high shear are spatially separated from those of high acceleration. Flashes are associated only with regions of high shear near the wall, at shear stress levels known to stimulate luminescence in Couette and pipe flows. Thus there is a consistent picture of bioluminescence flow stimulation as a function of fluid shear stress.

Dinoflagellates can be considered tracer particles that flash only when suitable levels of shear stress are present, whether the flow field is laminar or turbulent. Furthermore, their response is essentially instantaneous, of short duration, and sufficiently bright to be used as a flow visualization tool.[6,8] Natural flow conditions that stimulate dinoflagellate bioluminescence (e.g., swimming animal, whitecaps) are relatively energetic compared to typical levels of oceanic turbulence. However, much lower levels of shear stress, such as those present in oceanic turbulence in near-surface waters on a windy day, are sufficient to affect other aspects of dinoflagellate physiology, including cell division[11] and morphology.[12]

For a moving object, the intensity of any luminescent signature will be a function of boundary layer thickness, degree of flow separation, and bioluminescence potential (related to the species abundance of luminescent plankton). For example, through computer simulations the bioluminescent wake of an aircraft carrier moving at 8 knots is estimated to be several km long,[9] consistent with anecdotal accounts of World War II pilots relocating their ships at night by their long bioluminescent wakes.

Unfortunately, the stimulation of oceanic bioluminescence is not necessarily sufficient for its observance. Detecting a relatively dim bioluminescence signal among brighter background illumination is a particularly attractive candidate for emerging multispectral technology because of the relatively narrow spectral distribution of bioluminescence compared to that of star- or moonlight. A correctly tuned multispectral sensor will have one target band in which the bioluminescence signal is detected and one guard band that is correlated to the background noise.

Some 500 years after Leonardo da Vinci used pollen seeds to reveal many hitherto-invisible subtleties of water in motion, there is renewed interest in using naturally occurring dinoflagellate "particles" for flow visualization.

Acknowledgments

The authors are grateful for research support through SPAWARSYSCEN and the Office of Naval Research.

References

1. Anderson DM, Nosenchuck DM, Reynolds GT, Walton AJ. Mechanical stimulation of bioluminescence in the dinoflagellate *Gonyaulax polyedra* Stein. *J Exp Mar Biol Ecol* 1988; **122**: 277-288.
2. Christianson R, Sweeney BM. Sensitivity to stimulation, a component of the circadian rhythm in luminescence in *Gonyaulax*. *Plant Physiol* 1972; **49**: 994-997.
3. Latz MI, Case JF, Gran RL. Excitation of bioluminescence by laminar fluid shear associated with simple Couette flow. *Limnol Oceanogr* 1994; **39**: 1424-1439.
4. Latz MI, Juhl A, Elghobashi SE, Ahmed AM, Rohr J. Response of dinoflagellates in well-characterized laminar and turbulent flows: differentiating the effect of shear and acceleration. 1997 Meeting of the American Association of Limnology and Oceanography, Santa Fe, New Mexico. 1997.
5. Latz MI, Rohr J. Luminescent response of the red tide dinoflagellate *Lingulodinium polyedrum* to laminar and turbulent flow. *Limnol Oceanogr* 1999; **44**: 1423-1435.
6. Latz MI, Rohr J, Hoyt J. A novel flow visualization technique using bioluminescent marine plankton —Part I: Laboratory studies. *IEEE J Oceanic Eng* 1995; **20**: 144-147.
7. Nauen JC. Biomechanics of two aquatic defense systems. Ph.D. Dissertation, University of California San Diego. 1998.
8. Rohr J, Latz MI, Fallon S, Nauen JC, Hendricks E. Experimental approaches towards interpreting dolphin-stimulated bioluminescence. *J Exp Biol* 1998; **201**: 1447-1460.
9. Rohr J, Schoonmaker J, Losee J, Latz MI, and Hyman M. Flow visualization in the ocean — implications oflaboratory bioluminescence experiments. Oceans 99 IEEE/ITCMS. 1999.
10. Rohr JJ, Losee JR, Hoyt JW. Stimulation of bioluminescence by turbulent pipe flow. *Deep-Sea Res* 1990; **37**: 1639-1646.
11. Thomas WH, Gibson CH. Quantified small-scale turbulence inhibits a red tide dinoflagellate, *Gonyaulax polyedra* Stein. *Deep-Sea Res* 1990; **37**: 1583-1593.
12. Zirbel MJ, Veron F, Latz MI. The reversible effect of flow on the morphology of *Ceratocorys horrida* (Peridiniales, Dinophyta). *J Phycol* 2000; **36**: 46-58.

PROTEIN CONFORMATIONAL CHANGES IN OBELIN SHOWN BY ^{15}N-HSQC NUCLEAR MAGNETIC RESONANCE

JOHN LEE, JOHN N. GLUSHKA, SVETLANA V. MARKOVA
AND EUGENE S. VYSOTSKI
University of Georgia, Athens, GA 30602, USA, and Institute of Biophysics, Russian Academy of Sciences (SB), 660036 Krasnoyarsk, Russia.
E-mail: jlee@arches.uga.edu

1. INTRODUCTION

The calcium-regulated photoproteins are named because the addition of Ca^{++} alone triggers the emission of bioluminescence. These photoproteins have been isolated from several types of marine bioluminescent creatures and a number of photoproteins have been sequenced and inserted into efficient expression systems.[1] The extracted apoprotein is converted to photoprotein by a reaction with the substrate coelenterazine in the presence of molecular oxygen and dithiothreitol. The photoprotein is then purified by several steps of chromatography.

This present report concerns one of these photoproteins, obelin, named for its original source from the hydroid *Obelia longissima.* We demonstrate that the protein undergoes clear conformational changes through five states, the first one when Ca^{++} is included with the apoprotein, the second on charging the apoprotein (in the absence of Ca^{++}) with coelenterazine to produce the holoprotein obelin, then on the addition of Ca^{++} to cause the bioluminescence reaction, and finally on removal of Ca^{++} from the product. The technique employed is Heteronuclear Single Quantum Correlation (HSQC) NMR spectroscopy using the protein fully enriched with the magnetically active isotope of nitrogen, ^{15}N. It is well established that the NMR chemical shifts of the peptidic nitrogens and protons reflect the structural state of a protein in solution and can be used semi-quantitatively to determine the relative content of helix, sheet and loops.[2]

2. EXPERIMENTAL

Recombinant apo-obelin was obtained with a previously constructed *E. coli* expression system using the pET expression vector and as the S2A mutant. Selected clones were grown in a defined medium (per liter): 2 g glucose, 0.5 g NaCl, 1g $^{15}NH_4Cl$, 0.25 g $MgSO_4.7H_2O$, 0.14 g $CaCl_2.2H_2O$, 0.005 g $FeCl_2.6H_2O$, 3 g Na-acetate, 0.1 g ampicillin, 0.05 g thiamine, and 50 mM Na/K phosphate, with pH = 7.0. Growth was at 37°C with vigorous aeration and at a cell density corresponding to OD (600 nm) = 0.5-0.6, induction was made with 1 mM IPTG. After 3 h continued growth the cells were harvested by centrifugation. The cell cake was sonicated at 0°C and recentrifuged. The apo-obelin was extracted from the inclusion bodies with 6 M urea. The protein was purified by two steps of chromatography and shown to be homogeneous by HPLC-electrospray ionization mass spectrometry. Obelin was charged by procedures described elsewhere in this volume.

The addition of 10 mM Ca^{++} results in the bioluminescence reaction and the resulting product has a high fluorescence, hence named in the case of aequorin, the blue fluorescence protein (BFP). For the study here the removal of the Ca^{++} from the (obelin) BFP was by gel chromatography on BioGel P2. The samples for NMR were concentrated to 0.5 mM, limited to this concentration since it had been determined by a sedimentation study that for the case of obelin, aggregation occurred at about 0.9 mM.

The NMR measurements were made with 0.5 ml of the protein solution in a 5 mm NMR tube at 20°C in 10 mM Tris-HCl, pH 7.0, 10% D_2O, 1 mM Na azide, and either 1 mM Ca^{++} or 1 mM EDTA. The NMR experiments used a Varian Unity 800 MHz Spectrometer and a 5 mm H(CN) probe. An ^{15}N Watergate-HSQC experiment, FHSQC",[3] was used. A dataset of 2048 x 400 complex points was collected, with acquisition times of 171 ms and 62 ms for t2 and t1, respectively. Data were processed using Felix95 software (Biosym), by applying gaussian functions and zero-filling in both dimensions. The proton dimension was calibrated by referencing to HDO at 4.89 ppm, and in the nitrogen dimension by calculating the carrier frequency to be 119.3 ppm.[2]

3. RESULTS AND DISCUSSION

The result of an HSQC experiment is a two dimensional plot of the chemical shifts of each covalently bonded hydrogen and nitrogen. HSQC experiments usually focus on the region of the peptide N and H chemical shifts, i.e., the ranges displayed in Fig. 1 here. For small peptides in solution which are expected to have random structure, the H-shifts cluster around 8.1 ppm, and the N-shifts around 120 ppm, except Gly around 110 ppm. Side chain N-shifts are found near 113 ppm and can be distinguished here in Fig. 1 as pairs of H-shifts at precisely the same N-shift (top right quadrant).

The HSQC spectrum for apo-obelin in Fig. 1 implies that the protein is not folded, the shifts are typical of small polypeptides. However the circular dichroism (CD) (not shown) of the apo-protein indicates that it has a definitely folded structure in the 10 microM concentration range used for CD measurement, so it is more likely that the apo-protein is aggregated at the higher concentrations required for the less

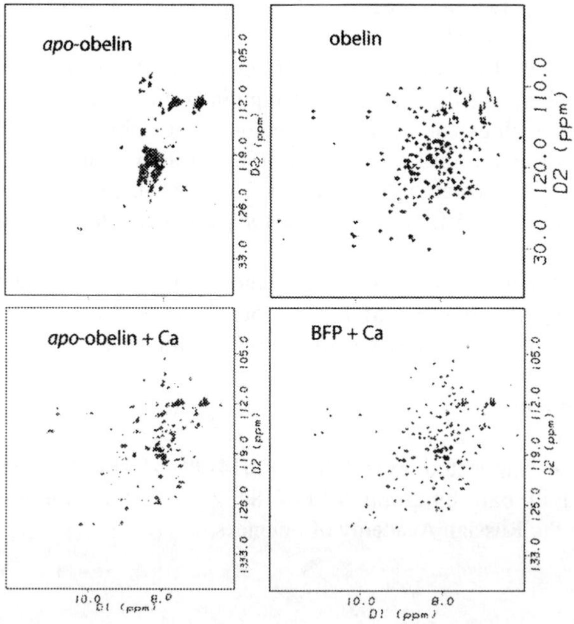

Fig. 1 ^{15}N-HSQC spectra: for apo-obelin (0.5 mM) and after adding Ca^{++}; and for obelin and after adding Ca^{++} to form the "blue fluorescence protein" product.

sensitive NMR measurement.

On Ca^{++} addition to the apo-protein the chemical shifts become more dispersed, two resonances appear downfield as far as 11.0 ppm. This dispersion is typical of a folded protein, the disaggregation induced evidently by the specific binding in the Ca^{++} sites. The sample is monomeric because the number of resonances is approximately that expected from the number of peptide bonds in the sequence. It is also noted that the resonances have variable intensity and this can be attributed to dynamical fluctuations of the spacial structure leading to variable exchange rates of the amide proton with the solvent water.

The panels on the right of Fig. 1 are (upper) for the holoprotein obelin and (lower) after adding Ca^{++} to the sample to trigger the bioluminescence reaction whereupon it goes to the product BFP. Both spectra show good dispersion with little overlap of the resonances, unusual for a protein of this size (22 kDa). About 212 resonances can be counted as expected from the 192 peptide bonds, less the 5 prolines plus the 12 side-chain pairs. Obelin shows very uniform intensities for the resonances whereas the BFP has a more wide distribution of levels. It is clear that obelin forms a very rigid structure, perhaps the reason for its facile crystallizability for X-ray study.[5] On the other hand, at this time of writing, BFP shows no inclination to form crystals.

About 40% of the resonances shift only slightly between obelin and BFP implying that there is little structural change in some sections of the protein in the two forms. An analogy can be drawn to calmodulin where the binding of Ca^{++} folds a dumb-bell structure into a more globular form without much change in the secondary structure overall.[6] The BFP in Fig. 1 is in the presence of Ca^{++}. On removing the Ca^{++} the HSQC picture (not shown) is further altered evidence of an additional conformational shift.

The NMR experiments described here show that the conformation of apo-obelin is controlled by the binding of the various ligands, Ca^{++}, coelenterazine, and coelenteramide (in BFP), and five distinct conformations are noted.

Acknowledgements

This work was supported by ONR N00014-99-1-0414, the Georgia Research Alliance, and in part by grant 99-04-48452 from the Fundamental Research Foundation of the Russian Academy of Sciences.

References

1. Illarionov BA, Frank LA, Illarionova VA, Bondar VS, Vysotski ES, Blinks JR. Recombinant obelin: cloning and expression of cDNA, purification, and characterization as a calcium indicator. *Methods Enzymol* 2000; **305**: 223-249.
2. Wishart DS, Sykes BD. Chemical shifts as a tool for structural determination. *Methods Enzymol* 1994; **239**: 363-392.
3. Mori S , Abeygunawardana C, Johnson MO, van Zijl PCM. Improved sensitivity of HSQC spectra of exchanging protons at short interscan delays using a new fast HSQC (FHSQC) detection scheme that avoids water saturation. *J Magn Res* 1995; **108B**: 94-98.
4. Liu ZJ, Vysotski ES, Chen CJ, Rose J, Lee J, Wang BC. Structure of the Ca^{2+}-regulated photoprotein obelin at 1.7 Angstrom resolution determined directly from its sulfur substructure. *Protein Sci* 2000; **9**: 000
5. Vysotski ES, Liu ZJ, Rose J, Wang BC, Lee J. Preparation and preliminary study of crystals of the calcium-regulated photoprotein obelin from the bioluminescent hydroid *Obelia longissima*. *Acta Crystallogr* 1999; **D55**: 1965-1966.
6. Tjandra N, Kuboniwa H, Ren H, Bax A. Rotational dynamics of calcium-free calmodulin studied by N-15-NMR relaxation measurements. *Eur J Biochem* 1995; **230**: 1014-1024.

KINETICS AND FLUORESCENCE OF A BACTERIAL LUCIFERASE MUTANT WITH ONLY TWO TRYPTOPHANS

LEO LIN, ROSE SZITTNER AND EDWARD MEIGHEN

Dept of Biochemistry, McGill University, 3655 Promenade Sir William Osler, Montreal, PQ, Canada, H3G1Y6

E-mail: meighen@med.mcgill.ca

1. INTRODUCTION

The intrinsic fluorescence of bacterial luciferase acts as a natural sensor of the microenvironment near the tryptophan residues. Previous studies have shown that FMN will quench the fluorescence of *Xenorhabdus luminescens* luciferase in the presence of myristic acid.[1] Analysis of Trp mutants missing one of the eight Trp residues in the heterodimeric luciferase($\alpha\beta$) has shown that only mutation of Trp250 to Tyr caused a major reduction in flavin and aldehyde binding; mutation of any of the other Trp residues to Tyr had relatively little effect on interactions with substrates or products although the W194Y mutant had a major change in its rate of luminescence decay in the standard luminescence assay.[2]

As the major decrease in interactions with FMN for the W250Y mutant implicates this residue in binding flavin, it would be predicted that mutants with most of the Trp residues changed to Tyr but retaining W250 would still effectively interact with the substrates and the relative degree of quenching by FMN would be enhanced. To test this hypothesis, the properties of a mutant with all the tryptophan residues converted to Tyr except for the two Trp residues (Trp250 and Trp194) directly implicated in function were investigated.

2. MATERIALS AND METHODS

Intrinsic fluorescence spectra of the parental and chimeric luciferases in 0.05 M phosphate, 50 μM myristic acid, in the presence and absence of 10 μM FMN were measured by excitation at 296 nm with a wavelength dispersion of 5 nm for excitation and emission. Luminescence intensities were measured by injection of 50 μM $FMNH_2$ in the standard assay.

3. RESULTS AND DISCUSSION

The dependence of the light intensity on aldehyde (decanal) concentration provides a sensitive measurement of the interaction of luciferase with its substrates. Fig. 1

shows that a mutant containing only two tryptophans (W250/W194) still strongly interacts with decanal in a manner similar to the wild type luciferase while a mutant

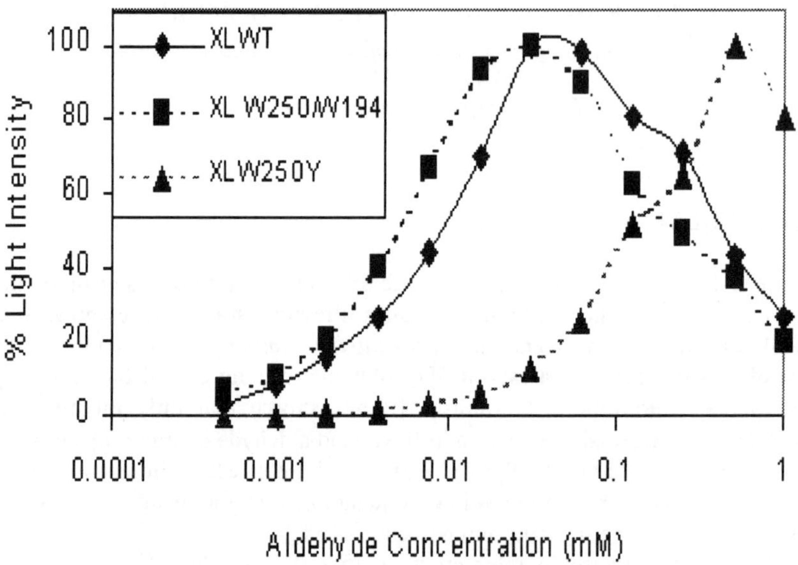

Fig. 1 Dependence of luminescence intensity on decanal concentration

missing only Trp250(W250Y) requires a much higher aldehyde concentration to reach maximum activity. These results support the previous conclusion that only Trp250 is needed for interaction with the substrates of the luminescent reaction and that mutation of the other seven Trp residues to Tyr will not have major effects on substrate binding.

To measure directly the interaction with flavin, the fluorescence spectra of the W250/W194 mutant was determined with myristic acid in the presence and absence of 10 μM FMN. A relatively low concentration of flavin was used so that the correction for self absorption by flavin would be insignificant. Fig. 2 demonstrates that the degree of quenching is even smaller than that measured for the wild type luciferase under the same conditions. These results indicate that the primary quenching of the intrinsic fluorescence of the wild type bacterial luciferase on binding flavin is not due to the quenching of the fluorescence of Trp250 or Trp194 but arises due to quenching of the fluorescence of other Trp residues in the molecule.

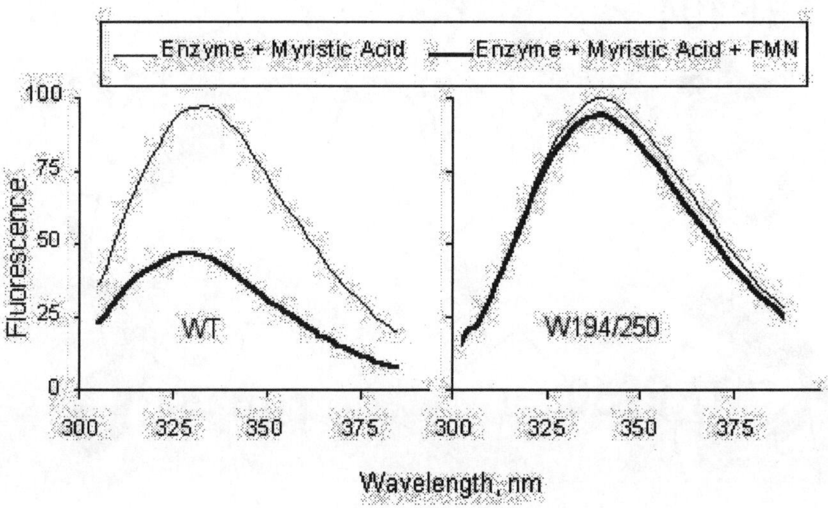

Fig. 2 Intrinsic fluorescence spectra of wt luciferase and the W194/250 mutant in the presence and absence of FMN.

Fig. 3 shows the relative locations of five of the six Trp residues in the α subunit and the distances from the anion binding site proposed to be the location of the phosphate moiety of FMN[3]. A fully extended FMN molecule is about 15 in length and is proposed to interact with W250 and W194.[2,3] Consequently, FMN could not be in direct contact with the other Trp residues at the same time that it interacts with Trp250. As mutation of the other residues does not affect the binding of the flavin or aldehyde substrates, it would suggest that binding of FMN causes a major conformational change resulting in the quenching of the intrinsic fluorescence of some or all of the other Trp residues in luciferase. Analysis of the ability to quench the fluorescence of other mutants containing Trp250 and one of the other tryptophan residues should define those Trp residues whose fluorescence are primarily quenched and indicate the regions undergoing a major conformational change.

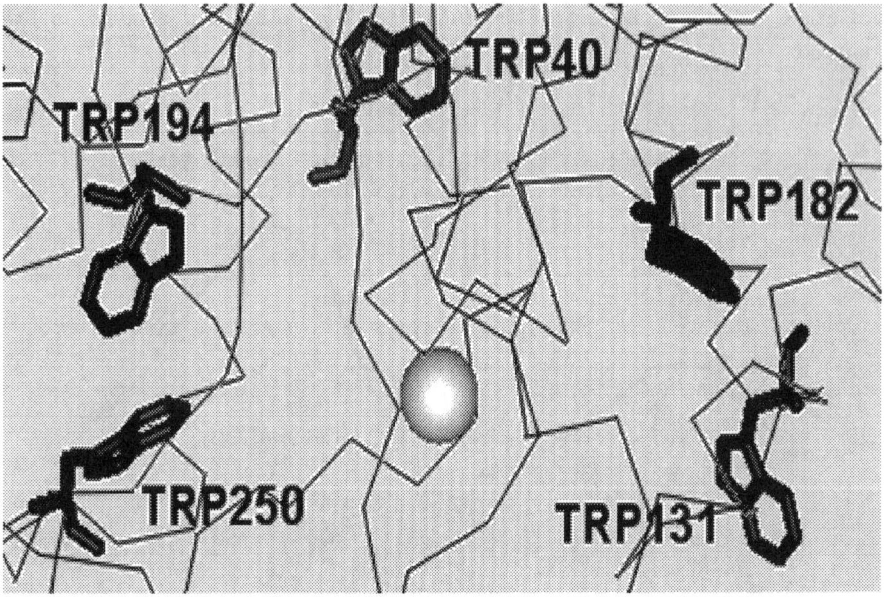

Fig. 3 Structure of the LuxA subunit of luciferase showing the proposed anion binding site (oval) and the relative locations of five of the tryptophan residues.

Acknowledgements

Supported by a grant from the Medical Research Council of Canada (MT4314).

References

1. Li Z, Meighen EA. Fatty acid-enhanced binding of flavin mononucleotide to bacterial luciferase measured by steady-state fluorescence. *Biochem Biophys Res Commun* 1992; **188** 497-502.
2. Li Z, Meighen EA. Tryptophan 250 on the alpha subunit plays an important role in flavin and aldehyde binding to bacterial luciferase. Effects of W-->Y mutations on catalytic function. *Biochemistry* 1995; **34** 15084-15090.
3. Fisher AJ, Raushel FM, Baldwin TO, Rayment I. Three-dimensional structure of bacterial luciferase from *Vibrio harveyi* at 2.4 resolution. *Biochemistry* 1995; **34** 6581-6586.

NATURAL ANIMAL COLORATION CAN BE DETERMINED BY A NON-FLUORESCENT GFP HOMOLOG

KONSTANTIN A. LUKYANOV[1], ARKADY F. FRADKOV[1], NADYA G. GURSKAYA[1], MIKHAIL V. MATZ[1], YULII A. LABAS[2], ALEKSANDR P. SAVITSKY[3], XIAONING ZHAO[4], YU FANG[4], WENYAN TAN[4] AND SERGEY A. LUKYANOV[1*]

[1]Shemiakin and Ovchinnikov Institute of Bioorganic Chemistry RAS, Miklukho-Maklaya 16/10, 117871 Moscow, Russia; [2]Institute of Ecology and Evolution RAS, Leninsky pr. 33, 117071 Moscow, Russia; [3]Institute of Biochemistry RAS, Leninsky pr. 33, 117071 Moscow, Russia; [4]CLONTECH Laboratories, Inc., 1020 East Meadow Circle, Palo Alto, CA 94303-4230, USA

*To whom correspondence should be addressed. Email: luk@ibch.ru

1. INTRODUCTION

It is generally accepted that the enormous variety of colors and fluorescent hues displayed by living organisms are determined by chromoproteins and low-molecular weight pigments. As a rule, chromoproteins typically require a prosthetic group: a small nonpeptide molecule or metal ion, which binds to the protein and is essential for the chromogenic properties of the protein.[1-5] The only known exception to this rule is green fluorescent protein (GFP) from *Aequorea victoria.*[6] In contrast to other naturally occurring fluorescent proteins, GFP s fluorescence is entirely due to an internal interaction between amino acids within the protein; no other cofactors or prosthetic groups are required.[7,8] Recently we described six novel GFP-like fluorescent proteins (FP) from non-bioluminescent *Anthozoa* species.[9] It therefore became clear that GFP-like proteins are not necessarily components of bioluminescent systems but may simply determine fluorescent coloration of animals.

In the present work, we isolated the cDNA for a novel GFP-like non-fluorescent chromoprotein from sea anemone *Anemonia sulcata* and showed that this protein determines non-fluorescent coloration of the tentacles.

2. RESULTS AND DISCUSSION

Using a strategy previously described,[9] cDNA for a novel GFP-like protein was amplified from RNA samples prepared from the tentacle tips of *A. sulcata*. Overall, 20% of the novel protein sequence (named asFP595 according to nomenclature we suggested earlier) is identical to GFP (Fig. 1). When asFP595 is compared with *Anthozoa* FPs the identity level is 37-47%.

After expressing as FP595 in *E. coli*, we measured the absorption and excitation-emission spectra of the purified protein (Fig. 2). With the exception of a slight shoulder at 530 nm, the protein displays a single absorption wavelength maximum, which occurs at 572 nm (ε_{572}=56,200 $M^{-1}cm^{-1}$). Visual inspection of *E. coli* expressing as FP595 shows that the protein confers a strong purple hue to its host. The fluorescence of as FP595 is extremely weak (quantum yield < 0.001); nevertheless, the purple protein can be detected by spectrofluorimetry (emission

```
        10        20        30        40        50
MSKGEELFTGVVPILVELDGDVNGHKFSVSGEGEGDATYGKLTLKFICTTGKLPVP..W GFP
...MASFLKKTMPFKTTIEGTVNGHYFKCTGKGEGNPFEGTQEMKIEVI.EGGPLPFAF asFP595
MRSSKNVI-EF-R--VRM-------E-EIE-E---R-Y--HNTV-LK-T.K-------W drFP585

 60        70        80        90       100       110
PTLVTTFSYGVQCFSRYPDHMKQHDFFKSAMPEGYVQERTIFFKDDGNYKTRAEVKFEGD GFP
HILSTSCMYGSKTFIKYVSGIP..DYFKQSFPEGFTWERTTTYEDGGFLTAHQDTSLDGD asFP595
D---PQFQ----VYV-HPAD--..--K-L------K---VMNF----VV-VT--S--QDG drFP585

120       130       140       150       160       170
TLVNRIELKGIDFKEDGNILGHKLEYNYNSHNVYIMADKQKNGIKVNFKIRHNIEDGSVQL GFP
CLVYKVKILGNNFPADGPVM.QNKAGRWEPATEIVYE..VDGVLRGQSLMALKCPGGRHLT asFP595
-FI----FI-V--S------.-K-TMG--AS--RL-P..R----K-EIHK---LKD-G--L drFP585

180       190        200       210       220       230
 ADHYQQNTPIGDG.PVLLPDNHYLSTQSALSKDPNEKRDHMVLLEFVTAAGITHGMDELYK   GFP
 CHLHTTYRSKKPASALKMPGFHFEDHRIEIMEEVEKGK.CYKQYEAAVGRYCDAAPSKLGHN asFP595
 VEFKSI-MA--...PVQL--YYYV-SKLD-TSHN-DYT.IVE---RTE--HHLFL          drFP585
```

Fig. 1. Sequence comparison of GFP and GFP-like fluorescent proteins. Previously identified[9] red fluorescent protein drFP583 is compared with asFP595: the residues that are identical to the corresponding ones in asFP595 are represented by dashes. Introduced gaps are represented by dots. The residues whose side chains form the interior of the β-can are shaded.

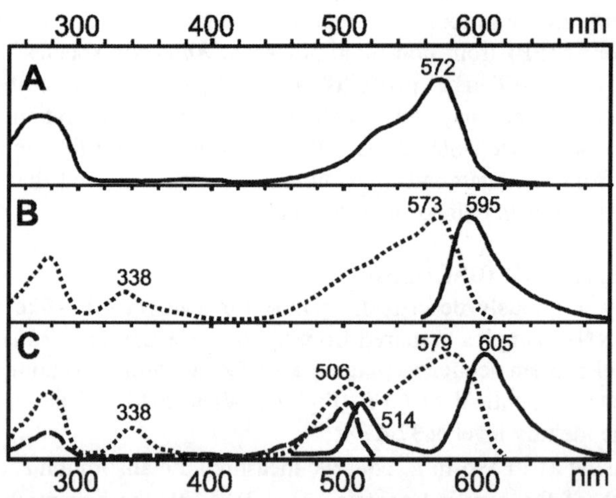

Fig. 2. Spectra of asFP595 and its fluorescent variants. **A**, the absorption spectrum of wild-type asFP595. **B**, the excitation (dotted line) and emission (solid line) spectra of asFP595 mutant A148S. **D**, the excitation-emission spectra of the asFP595 mutant A148S/S165V. Solid line — double-peak fluorescence (excitation at 280nm). Dashed and dotted lines - the excitation spectra for green and red emission, respectively.

λ_{max}=595 nm). When viewed by fluorescence microscopy, the novel protein shows an unexpected feature: although the fluorescence at 595 nm is virtually imperceptible at the start of observation, the emission intensity increases dramatically following a 10—20 sec exposure to green light. This effect is reversible since in the absence of incident green light, the fluorescence capacity slowly decreases to the basal level. Even more surprisingly, asFP595 fluorescence can be quenched by a flash of blue light. This effect strongly depends on intensity of irradiation: the brighter light the more pronounced increasing and quenching of the fluorescence. The spectral properties of recombinant asFP595 closely resemble those observed *in vivo* in the tentacle tips of the corresponding organism, *A. sulcata*. These data suggest that the purple coloration of *A. sulcata* tentacles is determined by asFP595 due to differential light absorbency.

Using site-directed mutagenesis, we attempted to convert asFP595 in the fluorescent state. We paid particular attention to residues surrounding the fluorophore. One of these stood out: residue 148 (numbering based on GFP). In GFP and all *Anthozoa* fluorescent proteins, position 148 is occupied by a polar residue (His in GFP and Ser in *Anthozoa* FPs). The equivalent site in asFP595 is occupied by a nonpolar residue (Ala). Since His-148 is in direct contact with the fluorophore in GFP, it seemed likely that it played a critical role in the protein s fluorescence. To test this hypothesis, we generated an A148S asFP595 point mutant. Indeed, this single substitution increased the quantum yield of red fluorescence by as much as almost two orders of magnitude (Fig. 2B). The quantum yield of red fluorescence for A148S was determined to be 0.012. By means of further random mutagenesis, a brighter variant T70A/A148S was generated (quantum yield = 0.05).

Further mutagenesis was performed on the basis of the mutant A148S. Additional substitutions of amino acids probably adjacent to chromophore (e.g., S68A, W94Y, S165V, E201A,V,L, and H203S) resulted in appearance of an additional green emission peak at 514 or 523 nm, depending on the mutant (Fig. 2C). In these mutants, the nature of the particular amino acid substitution affected the ratio between the green and red emission intensities as well as brightness of the protein but only slightly affected the position of the emission maximums. The congruency of these outcomes may indicate that the green and red fluorescence represent distinct fluorophores in a heterogeneous population of green and red fluorescent protein molecules. This idea is supported by the observation that the red and green emissions can be stimulated independently-excitation of the mutants by 338 nm light produces only red fluorescence, not green.

To evaluate the usefulness of asFP595 for biotechnological applications, we tested its fluorescent variant in eukaryotic expression system. Fusion of protein kinase C β1 subunit (PKCβ) and asFP595 fluorescent mutant T70A/A148S was constructed and expressed in a human cell line. Using fluorescence microscopy, we were able to monitor the real-time translocation of red fluorescence from the cytoplasm to the plasma membrane upon activation of the PKC pathway with phorbol-12-myristate-13-acetate (PMA).

References

1. Fox DL, Pantin CFA. Pigments in the coelenterata. *Biol Rev* 1944; **19:** 121-133.
2. Cheesman DF, Lee WL, Zagalsky PF. Carotenoproteins in invertebrates. *Biol Rev* 1967; **42:** 132-160.
3. Herring PJ. Porphyrin pigmentation in deep-sea medusae. *Nature Lond* 1972; **205:** 103-104.
4. West HH. Pigmentation in the sea anemone *Corynactis californica. Comp Biochem Physiol* 1979; **64B:** 195-200.
5. Wicksten MK. Why are there bright colors in sessile marine invertebrates? *Bull Marine Sci* 1989; **45:** 519-530.
6. Johnson FH, Shimomura O, Saiga Y, Gershman LC, Reynolds GT, Waters JR. Quantum efficiency of *Cypridina* luminescence, with a note on that of Aequorea. *J Cell Comp Physiol* 1962; **60:** 85—104.
7. Cody CW, Prasher DC, Westler WM, Prendergast FG, Ward WW. Chemical structure of the hexapeptide chromophoree of the Aequorea green-fluorescent protein. *Biochemistry* 1993; **32:** 1212—1218.
8. Orm M, Cubitt AB, Kallio K, Gross LA, Tsien RY, Remington SJ. Crystal structure of the Aequorea victoria green fluorescent protein. *Science* 1996; **273:** 1392—1395.
9. Matz MV, Fradkov AF, Labas YA, Savitsky AP, Zaraisky AG, Markelov ML, Lukyanov SA. Fluorescent proteins from nonbioluminescent Anthozoa species. *Nature Biotech* 1999; **17:** 969-973.

MEASUREMENT OF ACYL HOMOSERINE LACTONES WITH MASS SPECTROMETRY

JOHN MAKEMSON[1] AND ANATOL EBERHARD[2]
[1]*Department of Biological Sciences, Florida International University, Miami, FL 33199, USA and*
[2]*Department of Chemistry, Ithaca College, Ithaca, NY 14850, USA*

1. INTRODUCTION

Acyl homoserine lactones (AHL), the major regulatory molecules of bacterial bioluminescence (1), were detected by atmospheric ionization mass spectrometry (AI-MS) as the positively charged pseudomolecular ions. This method takes advantage of the ability of the AI-MS to simultaneously measure many molecules (m/z) non destructively. Thus all extracted organic molecules in a sample can be measured. In luminous and nonluminous bacteria, AHLs of different acyl moieties control a wide variety of genes. The most common AHLs are listed as follows: (MW): Butanoyl, (171); 3-hydroxybutanoyl, (187); Hexanoyl, (199); 3-oxohexanoyl, (213); Octanoyl, (224); 3 oxooctanoyl, (238); Dodecanoyl, (285); 3-oxododecanoyl, (309); Tetradecanoyl, (314).

In the AI-MS, these molecules are resolved by the instrument to have an m/z = MW + 1. In complex mixtures found in biological samples, a variety of peaks are present. We have used a simple extraction procedure on duplicate samples to limit the number of peaks and selectively use the sensitivity of the lactone ring to alkali so that AHLs can be measured over > 5 log range.

2. METHODS

Cultures of *Vibrio fischeri* and *V. harveyi* were grown in Glycerol Marine Broth [0.3% glycerol, 0.5% peptone, 0.1% yeast extract, and 75% artificial sea water (100% is 513 mM NaCl, 25 mM $MgSO_4$, 25 mM $MgCl_2$, 0.75 mM KCl)] adjusted to pH 7.5 (2). Stock cultures were maintained on the same medium with 1.5% agar added.

AHL extraction used acidified ethyl acetate (1 mL glacial acetic acid in 100 mL of ethyl acetate): 3 mL of culture was extracted three times with equal volumes of acidified ethyl acetate, then pooled and evaporated to dryness at 35°C under a gentle stream of nitrogen (3). The residue was dissolved in 1.0 mL acidified ethyl acetate for AI-MS detection. Prior to extraction a duplicate sample was first made alkaline by adding 20 μL of 1 M NaOH. Hydroxide will open the lactone ring so that on subsequent extraction with acidified ethyl acetate, the AHL m/z will be missing. A comparison of m/z peaks and their areas allows specific detection of AHL's.

AI-MS was done in a Finnegan Navigator AI-MS through either direct injection into the nebulizer with the tip at 250°C through a cone voltage of 10kv or through injection of known volumes using the sampling equipment of the attached liquid chromatograph. Direct injection scans were averaged over 30s. Standard AHL used here was 3-oxohexanoyl homoserine lactone (Sigma K-3255 and K-3007), the *V. fischeri* autoinducer dissolved in methanol or acidified ethyl acetate.

3. **RESULTS**

Whether the solvent was methanol or ethyl acetate, the solvent produced two peaks: one at m/z = MW+1 and at 2MW+1. Thus during nebulization, two solvent pseudomolecular ions are produced, the monomer and its dimer. 3-oxohexanoyl-HL (Sigma K-3255) had two peaks, the major one at m/z = 214 and a small peak at m/z = 427. The 427 peak remained when the material was dissolved in methanol saturated with NaCl. The other Sigma product (K-3007) did not have the minor peak at m/z = 427. We believe that the dimer is a product of the preparation of product formed by enolization of the keto reacting with the lactone ring of another molecule. To quantitate the known (Sigma K-3255), we added twice the area at m/z 427 to the area at 214 and plotted that against pmoles of 3-oxohexanoyl-HL (Fig 1). Note in Fig 2 that the background peaks are present with areas around 10^3, thus for this method to be quantitative, we did not count areas at m/z = 214 or 427 less than 10^4.

V. fischeri culture extracted with ethylacetate clearly showed the 214 m/z peak which did not occur if the culture was treated with NaOH prior to ethyl acetate extraction (Fig. 2, middle and bottom panels). Treatment with alkali does not destroy the molecule: we added glacial acetic acid (1 ml) to the aqueous alkali treated sample and re-extracted with ethyl acetate to recover the 214 m/z peak (not shown). *V. harveyi* culture with and without NaOH treatment prior to extraction with ethyl acetate showed the same results as shown in Fig. 2 (middle and bottom panels) but the peak lacking after NaOH treament was m/z = 188 characteristic of 3- hydroxy-butyryl-HL.

While these data do not yet include information on all AHLs, they do provide a basis by which to develop a rather straight forward chemical determination of these important regulatory molecules. We expect that other AHL's to respond identically to NaOH, but they may present other challenges due to the nature of the acyl group.

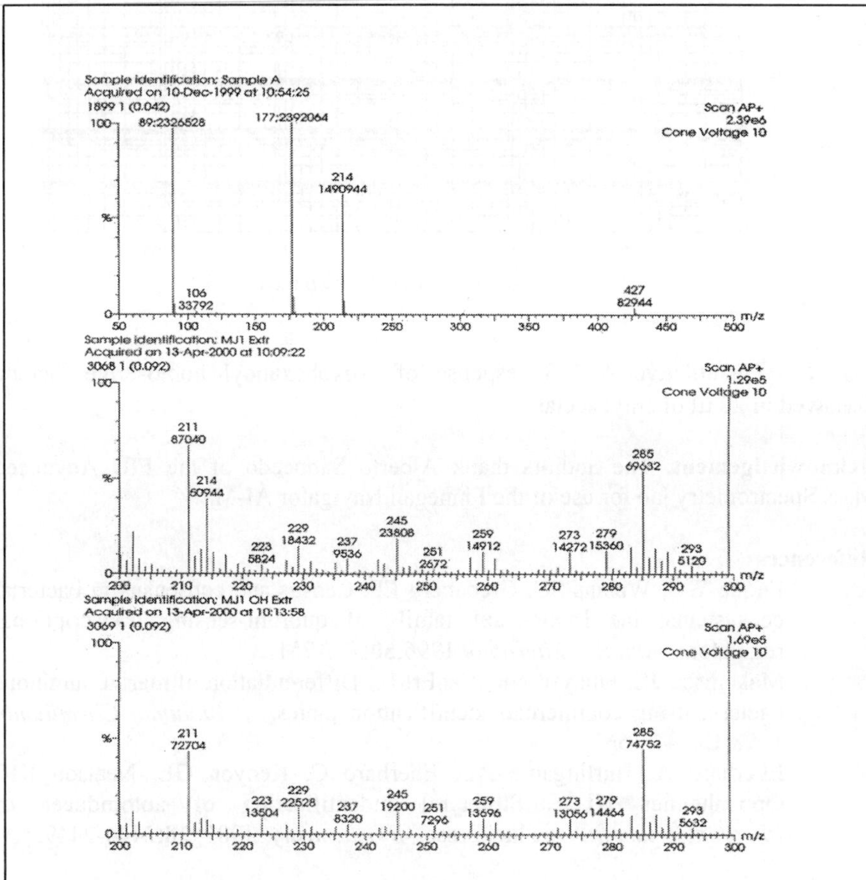

Fig. 1. Atmosphere Ionization Mass Spectra of 3-oxohexanoyl homoserine lactone dissolved in ethyl acetate (top panel), ethyl acetate extract of untreated *V. fischeri* MJ1 culture (middle panel, and ethyl acetate extract of NaOH treated *V. fischeri* MJ1 culture (bottom panel).

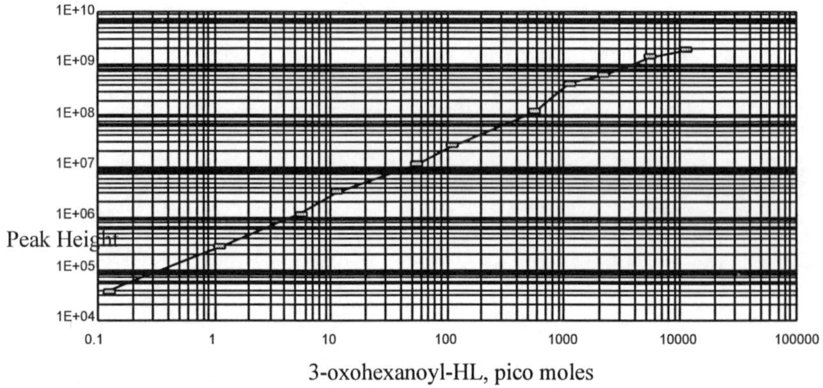

Fig. 2. Quantitative AI-MS response of 3-oxohexanoyl homoserine lactone dissolved in 20 μl of ethyl acetate.

Acknowledgement: The authors thank Alberto Sabucedo of the FIU Advanced Mass Spectrometry lab for use of the Finnegan Navigator AI-MS.

References

1. Fuqua WC, Winans SC, Greenberg EP. Census and consensus in bacterial ecosystems: the LuxR-LuxI family of quorum-sensing transcriptional regulators. *Ann Rev Microbiol* 1996;**50**:727-751.
2. Makemson JC, Fulayfil NR, VanErt L. Differentiation of marine luminous bacteria using commercial identification plates. *J Biolumin Chemilumin* 1998;**13**:147-156.
3. Eberhard A, Burlingame AL, Eberhard C, Kenyon GL, Nealson KH, Oppenheimer NJ. Structural indentification of autoinducer of *Photobacterium fischeri* luciferase. *Biochemistry* 1981; **28**:2444-2449.

OBELIN HYPEREXPRESSION IN *E. coli*, PURIFICATION AND CHARACTERIZATION

SVETLANA V. MARKOVA, EUGENE S. VYSOTSKI, AND JOHN LEE
University of Georgia, Athens, GA 30602, USA and Institute of Biophysics, Russian Academy of Sciences, SB, Krasnoyarsk 660036, Russia
E-mail: jlee@arches.uga.edu

1. INTRODUCTION

It has been demonstrated that the assay of Ca^{2+}-regulated photoproteins is highly sensitive, non-hazardous, and the reaction being rapid and simple. It makes photoproteins attractive for applications as a reporter protein like the enzymes alkaline phosphatase and horseradish peroxidase. The feasibility of using aequorin and obelin as nonradioactive tags is already established in applications in a variety of assay formats for detection of subnanogram amounts of various substances. For commercial application of photoproteins in various assays a high level expression system and a simple procedure for photoprotein purification are very desirable.

Here we describe the hyperexpression of recombinant obelin in *E. coli* cells, a simplified procedure for purification and some properties of the high purity protein.

2. MATERIALS AND METHODS

Reagents: Oligonucleotides were purchased from MWG-Biotech. Competent cells of *E.coli* BL21-Gold and XL1-Blue were from Stratagene. Coelenterazine was from Prolume Ltd.; all other chemicals were the highest purity available.
Construction of obelin expression plasmids. The coding obelin gene sequence was amplified from plasmid pOL101 with previously cloned cDNA[1] using the primers:

3OL-5Nc 5'- ACTccatggCTTCAAAATACGCAGTTA-3'):
2OLG-3Bm 5'- TTCggatccTTAGGGAACTCCGTTGCCAT-3').

An NcoI restriction site (lowercase letters) was introduced into the forward primer and a BamHI into the reverse primer. Underlined regions correspond to areas of homology with the obelin coding sequence. The PCR product was digested with NcoI and BamHI, purified by agarose gel electrophoresis, and inserted into similarly digested, dephosphorylated, and agarose-gel-purified pET19b (Novagen) and pQE60 (Qiagen) expression vectors. The resulting plasmids were named pET-OL8 and pQE-OL18 respectively. Nucleotide sequencing was performed by MWG-Biotech.
Growth of bacterial cultures and induction. *E. coli* BL21-Gold/pET-OL8 was grown on the LB media containing 200 μg/mL ampicillin or 50 μg/mL carbenicillin. For protein production, the recombinant *E. coli* cells were cultivated with vigorous

shaking at 37°C and induced with 1 mM IPTG when the culture reached an OD(600 nm) of 0.5-0.6. After induction, the cultivation was continued for 3 h.

Spectral measurements. Bioluminescence and fluorescence spectra were measured with an SLM spectrofluorometer. The bioluminescence reaction was at 20°C in 1 mM EDTA, 10 mM Tris-HCl pH 7.0, and was initiated by injection of $CaCl_2$. The concentration of free calcium was 0.5 μM in order to provide a constant light level during the spectral scan. Spectra were corrected by reference to the absolute spectral distribution of quinine sulfate, which was also used as the reference for the fluorescence quantum yield. Protein concentration was approximately 10 μg/mL.

3. RESULTS AND DISCUSSION

Previously, it has been reported that apoobelin can be expressed in *E. coli* with pT7-7 vector with relatively good yield.[2] But for some applications (NMR, crystallization trials) such a production level is not enough. Besides, the expression stability of the previously described overproducing strain is not very good. In an attempt to obtain the best expression characteristics of recombinant obelin, other expression systems were tested. The DNA sequence encoding the amino acid residues of obelin was amplified from plasmid with the obelin cDNA and cloned into the pQE60 and pET19b expression vectors. To generate the NcoI restriction site, which is necessary for direct inframe cloning, a single base pair substitution at position nt4 was introduced causing the substitution of Ser to Ala at residue 2.

As judged by SDS-PAGE analysis of whole-cell lysates (Fig. 1) the expression of recombinant apoobelin from the obtained constructions was very different. The production level of recombinant obelin in *E. coli* XL1-Blue with the pQE-OL18 plasmid (Fig. 1A) was still disappointingly low compared to the high level obtained in *E. coli* BL21-Gold with the pET-OL8 plasmid (Fig.1B). To optimize the expression level, the cells were grown during various periods of time after induction. The total amount of apoobelin is increased during the first 3 h after induction (Fig.1); further increasing growing time after induction does not produce a higher amount of protein. This is observed for both *E. coli* XL1-Blue and *E.coli* BL21-Gold cells.

The recombinant obelin was purified in general as previously reported[2] with some modifications. The apoobelin obtained after extraction with 6 M urea and through the first step of purification, was concentrated by ultrafiltration. To fold the protein the concentrated sample containing 6 M urea was diluted with buffer containing 5 mM EDTA, 10 mM DTT, 20 mM Tris-HCl pH 7.0, and coelenterazine. The final concentration of urea in the protein sample after dilution was approximately 0.3 M. Apoobelin was incubated in this solution overnight at 4°C. Then, the charged protein was additionally purified on a Bio-Scale DEAE-10 column (Bio-

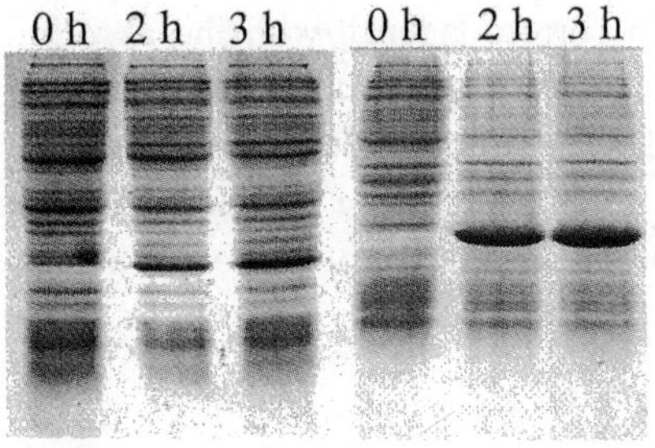

Fig. 1. Coomassie-stained 12.5% SDS-PAGE analysis of obelin expression in *E.coli*. Expression of pQE-OL18 in *E.coli* XL1-Blue (A) and pET-OL8 in *E.coli* BL21-Gold (B) as a function of time. The samples are whole-cell lysates. Time of cell incubation after induction is indicated at the top

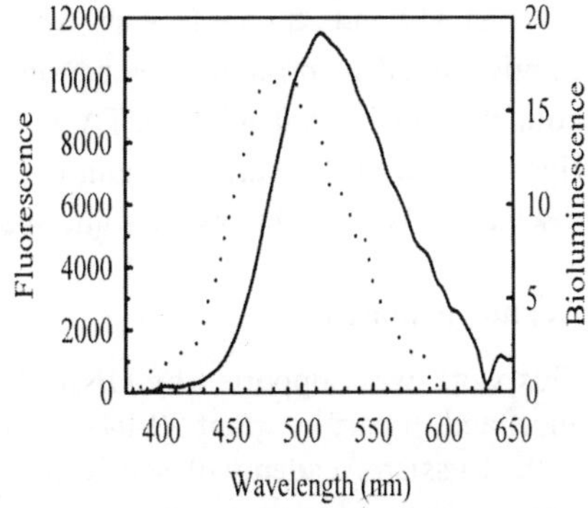

Fig. 2. Bioluminescence emission spectrum of obelin on addition of Ca^{++} (dotted line) and the fluorescence spectrum of the product (solid line).

Rad). The obelin was eluted with a linear NaCl gradient from 0 to 0.35 M in 5 mM EDTA, 20 mM Tris-HCl pH 7.0 at 2 mL/min. This step allows separation of the charged obelin from uncharged protein. The protein was homogeneous according to SDS-PAGE and LC-EIM Spectrometry. N-terminal AA-sequence of the obelin showed that the first Met is digested on producing the recombinant apoobelin in *E. coli* cells. The obelin MW obtained from EIM Spectrometry was in excellent agreement with the AA- sequence. The yield of high-purity obelin was 50-53 and 30-35 mg per liter of LB and isotope labeled defined medium respectively. The recombinant obelin obtained according to the scheme described above was used to produce high quality crystals diffracting to 1.1 Å.

Fig. 2 contains the bioluminescence emission spectrum of obelin and the fluorescence of the product. The result is in significant contrast to that published for the aequorin, where the fluorescence of the product BFP overlaps that of the bioluminescence.[3] The fluorescence of discharged obelin is green with maximum at 510 nm, whereas the blue obelin bioluminescence has a spectral maximum at 478 nm with a shoulder at 400 nm. Photoproteins all have bioluminescence maxima in the range 465-488 nm but a Trp mutant of aequorin[4] and the emission from *Renilla* luciferase,[5] both show a 400-nm shoulder. Spectral studies assign the species responsible for the 400-nm emission to the unionized coelenteramide,[6] the 478-nm the amide anion,[6] and the 510-nm the phenolate monoanion.[7] The quantum yield of the green fluorescence is 0.17 and the uncorrected excitation spectrum for emission at 500 nm has maxima at 353 and 294 nm in correspondence with the absorption maxima 347 and 280 nm (not shown). It should also be noted that coelenteramide is tightly bound to protein after passing the discharged obelin through a BioGel P2 column equilibrated with EDTA removing calcium from the sample. The green fluorescence of discharged obelin is not result of substitution of Ser to Ala since the discharged WT obelin has the same spectral characteristics.

Acknowledgment

This work was supported by ONR N00014-99-1-0414, the Georgia Research Alliance, and in part by grant 99-04-48452 from the Fundamental Research Foundation of the Russian Academy of Sciences.

References

1. Illarionov BA, Markova SV, Bondar VS, Illarionova VA, Vysotski ES. Cloning, expression and nucleotide sequence investigation of apoobelin cDNA from the hydroid *Obelia longissima.* In: Szalay AA, Kricka LJ, Stanley P, eds. *Bioluminescence and Chemiluminescence,* Chichester: Wiley, 1993; 183-185.
2. Illarionov BA, Frank LA, Illarionova VA, Bondar VS, Vysotski ES, Blinks JR. Recombinant obelin: Cloning, and expression of cDNA, purification, and characterization as a calcium indicator. *Methods Enzymol* 2000; **305**: 223-249.
3. Shimomura O, Johnson FH. Calcium binding, quantum yield, and emitting molecule in aequorin bioluminescence. *Nature* 1970; **227**: 1356-1357.
4. Ohmiya Y, Ohashi M, Tsuji FI. Two excited states in aequorin bioluminescence induced by tryptophan modification. *FEBS Lett.* 1992; **301**:197-201.
5. Hori K, Charbonneau H, Hart RC, Cormier MJ. Structure of native *Renilla reniformis* luciferin. *Proc Natl Acad Sci USA* 1977; **74**: 4285-4287.
6. Hori K, Wampler JE, Matthews JC, Cormier MJ. Identification of the product excited states during chemiluminescent and bioluminescent oxidation of *Renilla* (Sea Pansy) luciferin and certain of its analogs. *Biochemistry* 1973; **12**: 4463-4468.
7. Shimomura O, Teranishi K. Light-emitters involved in the luminescence of coelenterazine. *Luminescence* 2000; **15**: 51-58.

MULTIPLE REGULATORY PROTEINS CONTROL INDUCTION OF LUMINESCENCE IN *VIBRIO HARVEYI*

CAROL MIYAMOTO, JAI CHATTERJEE, YIHSING LIN AND EDWARD MEIGHEN

Dept of Biochemistry, McGill University, 3655 Promenade Sir William Osler, Montreal, PQ, Canada, H3G1Y6

E-mail: meighen@med.mcgill.ca

1. INTRODUCTION

The regulation of light emission in the luminescent marine bacterium *Vibrio harveyi* is quite complex. Small autoinducer molecules are secreted by the cells and upon reaching a threshold concentration in the external media interact with the cell resulting in the induction of luminescence at higher cell density (quorum sensing). Moreover, the richness and composition of the media affect the point of induction as well as the maximum level of luminescence. Over the last decade, multiple regulators have been identified by both genetic and biochemical analysis.[1,2] The identification of their specific role and whether they function directly or indirectly in the signal transduction processes controlling light emission will be essential to develop a clear understanding of the control of bioluminescence. Recent analysis has indicated that some of the key lux regulators may affect luminescence indirectly while new regulators that bind directly with the lux operon are being discovered.

2. MATERIALS AND METHODS

Southern blots of genomic DNA were conducted under condition of low stringency at 50¡C with labeled probes. Mobility shift assays were conducted by mixing 0.5 to 1 _g of an extract of *E. coli* expressing the appropriate *V. harveyi* regulatory gene with the *V. harveyi luxR* promoter.

3. RESULTS AND DISCUSSION

3.1 The Quorum Sensing System in *V. harveyi* May be a General Regulatory System

Two different autoinducers are believed to be responsible for the cell density dependence of luminescence in *V. harveyi*; one autoinducer has been identified as N-hydroxybutanoyl homoserine lactone (HBHL) and is synthesized by the LuxLM genes while the structure of a second autoinducer synthesized by the LuxS gene is still unknown. As shown in Fig 1, these interact with the LuxN and LuxQ receptors,

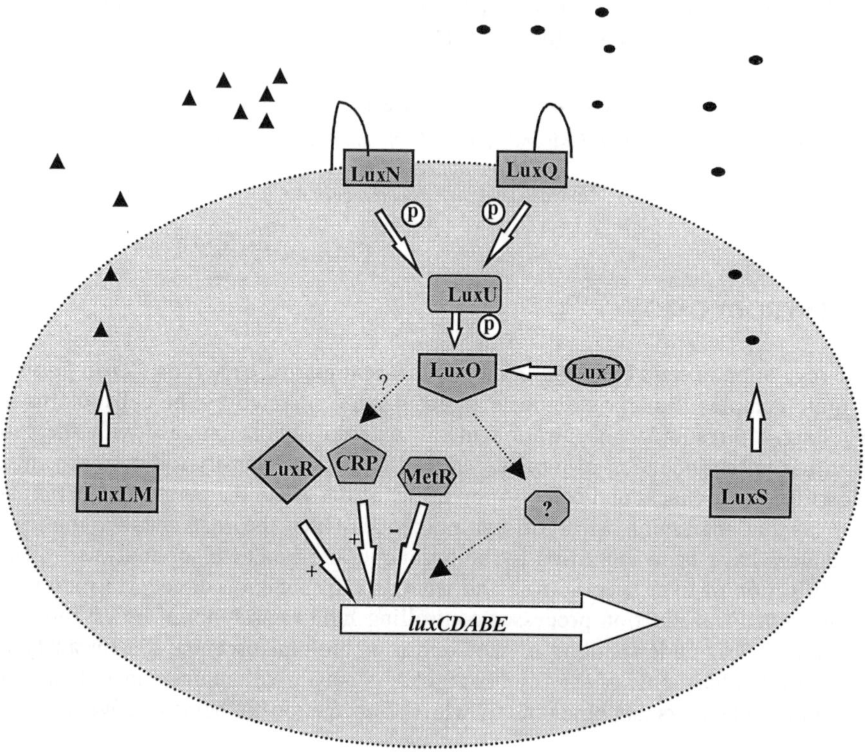

Fig. 1. Regulatory factors affecting luminescence in *V. harveyi*

respectively which transduce the signal through LuxU to LuxO by a phosphorelay system. In its phosphorylated state, LuxO results in inhibition of luminescence while in its absence or in the nonphosphorylated state, luminescence expression is not inhibited.

Recently, the LuxO-LuxU system has been discovered in *V. fischeri*.[3] Gene disruption of luxO-luxU in *V. fischeri* resulted in a higher level and earlier induction of light even though a different well characterized quorum sensing system is already present in this bacteria The luxO gene also appears to a general regulator rather than a lux-specific gene as its disruption affects other properties including survival at high salt concentration (unpublished data). Recently, we have purified a luxO promoter binding protein, designated as LuxT that represses the expression of luxO.[4] A Southern blot shows that *V. harveyi* LuxO and LuxT probes bind to the genomic DNA from three different species of luminescent bacteria (Fig. 2) indicating that the luxO-luxU system may be a general regulatory system in most luminescent marine bacteria including *P. phosphoreum.*

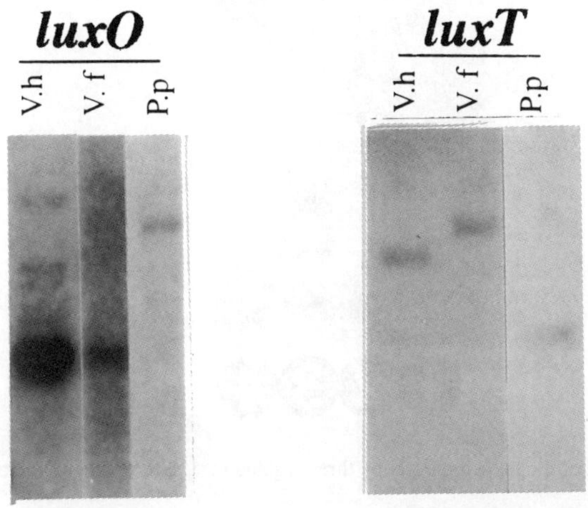

Fig. 2. Southern blots of *V. harveyi, V. fischeri* and *P. phosphoreum* genomic DNA using luxO and luxT probes.

3.2 Direct binding of regulators to the lux operon promoter

Although LuxO is the central regulator controlling cell density luminescence, its binding to the lux operon has not been demonstrated and it may interact indirectly. However evidence now exists for the direct binding of at least three other regulators to the promoters of the lux system (unpublished data).[5] Fig. 3 shows that extracts containing *V. harveyi* LuxR, MetR and CRP all cause a mobility shift of the luxR promoter DNA. Similar shifts are found with the promoter of the *lux* operon containing the structural genes (*lux*CDABE). The relationship of the regulators directly binding to the promoters of the *lux* system and the sensing of the internal and external signals will have to be established before regulation of luminescence is clearly understood.

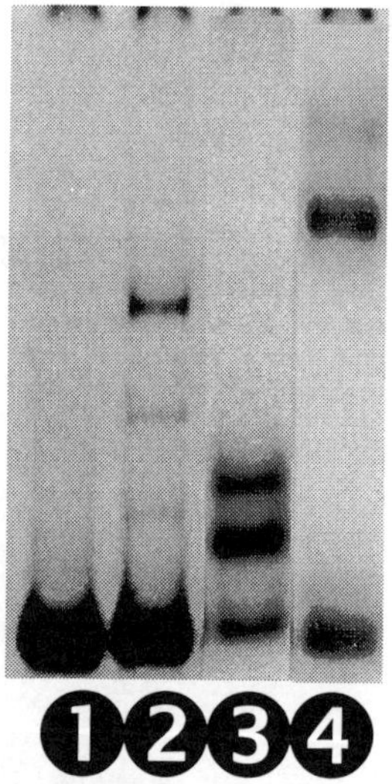

Fig. 3. Mobility shift of the luxR promoter by three regulators. 1. DNA alone, 2. MetR 3. LuxR, 4. CRP

Acknowledgements

Supported by a grant from the Medical Research Council of Canada (MT7672).

References

1. Bassler BL, Wright M, Silverman MR. Multiple signalling systems controlling expression of luminescence in Vibrio harveyi: sequence and function of genes encoding a second sensory pathway. *Mol Microbiol* 1994; **13** 273-286.
2. Bassler BL, Wright M, Silverman MR. Sequence and function of LuxO, a negative regulator of luminescence in Vibrio harveyi. *Mol Microbiol* 1994; **12** 403-412.
3. Miyamoto CM, Lin YH, Meighen EA. Control of bioluminescence in vibrio fischeri by the LuxO signal response regulator. *Mol Microbiol* 2000; **36** 594-607.
4. Lin YH, Miyamoto, CM, Meighen EA Purification and characterization of a luxO promoter binding protein LuxT from V. harveyi Protein Expr Purif 2000; In press.
5. Chatterjee J, Miyamoto CM, Meighen EA. Autoregulation of luxR: the Vibrio harveyi lux-operon activator functions as a repressor. *Mol Microbiol* 1996; **20** 415-425.

SPATIAL AND TEMPORAL VARIABILITY OF BIOLUMINESCENCE POTENTIAL IN COASTAL REGIONS

MARK A. MOLINE AND ERIKA K. HEINE
Biological Sciences Department, California Polytechnic State University, San Luis Obispo, CA 934071, USA
E-mail: mmoline@calpoly.edu

JAMES F. CASE AND CHRISTEN M. HERREN
Marine Science Institute, University of California, Santa Barbara, CA 93106, USA
E-mail: case@lifesci.ucsb.edu

OSCAR SCHOFIELD
Institute of Marine and Coastal Sciences, Rutgers University, New Brunswick, NJ 08901, USA
E-mail: oscar@imcs.rutgers.edu

1. INTRODUCTION

The phenomena of BL in oceanic waters spans the full range of both time and space, from the organismal level of ms flashes on micron scales, to the regional scale of seasonal cycles in BL that can exist on 100-km scales. In order to measure over these scales, appropriate sampling platforms must be utilized. To characterize the time and space variability of *in situ* BL potential in dynamic coastal regions, knowledge of the organisms responsible for the bulk signal and the physical environment which influences the distribution of these organisms needs to be identified. One of the further challenges before addressing the variability in BL is defining the relevant time and space scales for sampling. This study takes advantage of an operational observation network, which is key for sampling episodic events that dominate coastal environments, to make some preliminary BL measurements.

2. SITE DESCRIPTION AND INSTRUMENTATION

The ocean observation network where this study was conducted is located in a dynamic coastal region off New Jersey at the Long-Term Ecosystem Observatory (LEO-15), where frequent upwelling events occur. The lifetime of these transient upwelling centers can last from a few days to a full month depending on the local wind conditions and/or the presence of tropical storms, which are common during the late summer in the Mid-Atlantic Bight. The heart of one of these upwelling centers is instrumented with two autonomous robotic nodes connected to shore via an electrofiber-optic cable that controls and powers moored vertical profilers. A new optical profiler included a newly developed BL bathyphotometer (BBP) that determines total stimulable BL within a defined enclosed volume (500ml). Water is drawn through the chamber at flow rates from 350-400 ml/s by a brushless pump motor, which mechanically excites entrained BL organisms just prior to measurement of emitted light by an integrated PMT module. Water exits past a flow meter. Range of measurable light is from $1E^8$ to $1E^{13}$ photons/s. Hydrodynamic excitation calibration is conducted in the laboratory using cultured dinoflagellates. Unlike traditional BL packages,[1,2] this deployed system allows for continuous vertical coverage. Given the recurrent physical dynamics off the coast of

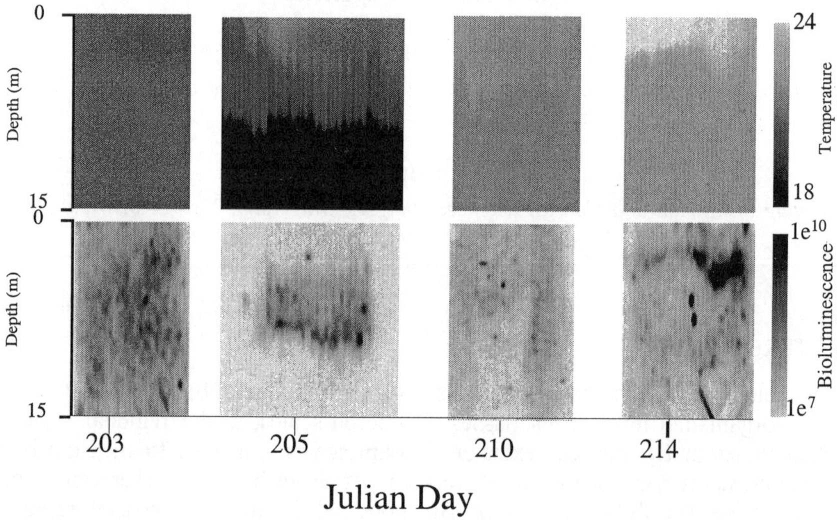

Fig. 1. Time-depth distribution of temperature and BL potential for four night cycles collected by the profiling node. A minimum of 30 profiles for each cycle was collected. The Julian Day for each panel represents the beginning of that day.

New Jersey, this instrumentation provides a powerful opportunity to examine the temporal evolution of inshore BL with relation to other biological, chemical, optical and physical processes. Data was collected continuously each night for two weeks from July 20 to Aug 2, 2000 (Julian Day 202-215). Spatial distribution of BL potential was determined using an identical BBP deployed from a research vessel.

3. RESULTS AND DISCUSSION

BL potential as well as thermal fine structure both showed a full range of vertical distribution from stratified layers to homogeneous distributions over the course of the two-week deployment. The water column was well mixed when the instrument was deployed and remained so until the Julian Day (JD) 204, when an intrusion of cooler water at depth from the shelf pushed up along the coast in response to winds from the southwest, the wind direction responsible for upwelling in the region (Fig.

1). This cold water intrusion ended after three days and was replaced by warm mixed water. From JD 207 until 212, the water column remained mixed and warmed ~1.5 °C. Over the course of the deployment, surface temperatures warmed from 18 to 24 °C.

On JD 202, low BL signals of $1E^9$ photons/350ml/s were associated with the mixed water column and the distribution of bioluminescent organisms extended from the surface to 14 m (Fig. 1). As the cold water intrusion from the shelf intensified, the bioluminescent communities layered and through JD 207 were strongly associated with the thermocline break. With the disappearance of the cold bottom water, there was a 3-4 order of magnitude decrease in BL signal. Warming of the surface layer and stratification in the final 4 days of the experiment was related to the development of another layer of high BL potential associated with the thermocline. There was a homogeneous distribution on the final day of the study that coincides with a decrease in stratification resulting from increased wind stress. Despite the rapid changes in the physical environment and the distribution of BL potential in the water column, there was a consistent repeated pattern in BL signal over the diel cycle; maximum BL signal occurred by 21:00 and remained at that level until 04:30.

In addition to temporal BL studies, another focus of this study was to find, characterize, and describe the dynamics of small-scale optically significant spatial features. Short-term statistical current forecasts were constructed by analyzing shore-based CODAR (high-frequency radar system generating real-time surface current fields), and were used to adjust the position of the research ships looking for bio-optical features in the convergence zones. Convergence zones were caused by an incoming tide hitting a persistent alongshore freshwater jet. CODAR fields suggested convergence zones would be located on the offshore side of the southern portion of our study space (Fig. 2A), depending on tidal cycle. Highly turbid waters with high phytoplankton stocks, on the inshore side of the convergence front, were associated with the southward flowing jet of low saline water. Terrestrial washout in response to heavy rainfall in the region initiated the turbid conditions.

Despite the high turbidity, bioluminescent loads were very low in these low saline waters (Fig. 2B). BL potential was higher below the low salinity waters, and was 3 orders of magnitude higher in the clearer offshore waters that were associated with the red-tide dinoflagellate *Ceratium fusus*. Characterization of this BL front was critically dependent on the short-term forecasts. These techniques will be fundamental in developing effective sampling strategies for BL that require both short and long-term forecasts to prepare and respond to episodic events.

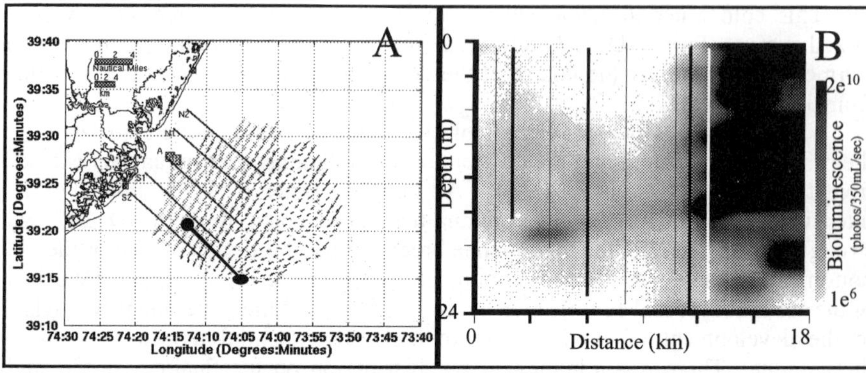

Fig. 2. A) Location of sampling transect (thick line) on JD 211-212 at the LEO-15 site. Superimposed are the surface current velocities and directions for the grid measured by CODAR (Range: 3 cm/s offshore to 20 cm/s onshore). B) Depth distribution of BL potential moving offshore along transect line. Vertical lines indicate profile locations.

Acknowledgements

We wish to thank M. Purcell, C. Von Alt, and C. Johnson for fabrication of the mooring, the Rutgers Marine Field Station and *R/V Arabella* staff for deployment, S. Glenn for the unique ocean observation infrastructure at Rutgers University, and the profiler operators; A. Weidemann, G. Chang, A. Briggs, and D. Peterson. This study was funded by ONR (N00014-00-1-0008).

References

1. Neilson DJ, Latz MI, Case JF. Temporal variability in the vertical structure of bioluminescence in the North Atlantic ocean. *J Geophys Res* 1995; **100**: 6591 — 6603.
2. Widder EA, Case JF, Bernstein SA, MacIntyre S, Lowenstine MR, Cook DP A New Large Volume BL Bathyphotometer with defined turbulence excitation. *Deep-Sea Res* 1993; **40**: 607-627.

SYNTHESES OF THE MECHANISM-BASED INHIBITORS OF COELENTERAZINE BIOLUMINESCENCE

H. NAKAMURA, C. WU, M. KATO, M. KIMURA
Division of Biomodeling, Department of Applied Moleculcar Biosciences, Graduate School of Bioagricultural Sciences, Nagoya University, Nagoya 464-8601, Japan
E-mail: hnak@agr.nagoya-u.ac.jp

S. INOUYE
Yokohama Resear Center, Chisso Co., 5-1 Okawa, Kanazawa-ku, Yokohama 236-8605, Japan.

1. INTRODUCTION

Coelenterazine (**1**) was well known to be involved in bioluminescent reactions such as jellyfishes[1-4], sea pansies[5] and deep-sea shrimps[6]. It distributes widely among bioluminescent marine organisms as well as non-bioluminescent marine organisms[7-9].

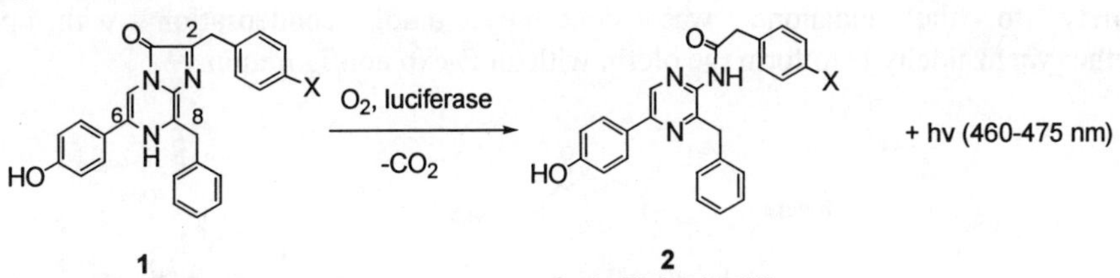

1
Luciferin : X=OH
(Coelenterazine, **1a**)

2
Oxyluciferin : X=OH
(Coelenteramide, **2a**)

Imidazopyrazinones including coelenterazine emit light with or without enzymatic components, defined as bio- and chemiluminescence, respectively, however the efficiency largely increases with a protein component especially in water media. The luminescent reaction was basically accepted as the process of oxidation at C-2 and C-3 position, in which several key peroxy intermediates and/or transition-states (**3-4**) seem to be involved[10,11].

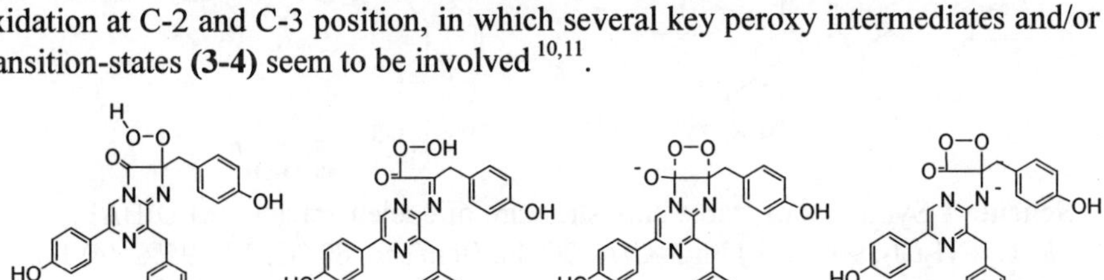

Peroxide (**3a**) Peracid (3b) Dioxetane (**4a**) Dioxetanone (**4b**)

To investigate the detailed process of the enzymatic reactions such as highly efficient chemical excitation step we designed a stable inhibitors based on the reaction mechanism. The synthesized transition-state analogues showed inhibitory effects on luminescence reaction of a recombinant Renilla luciferase[12,13] by which the chirality of the inhibitors was discriminated, suggesting for the first time that the mechanism based inhibitors of bioluminescence might be useful tools to investigate chiral environment of the active sites where an excited molecule forms at high efficiency.

2. SYNTHESIS OF A STABLE CORE MODEL OF COELETERAZINE, DEAZA-COELENTERAZINE.

First our object in this study was the synthesis of a stable core structure of coelenterazine, for which we selected a deaza-analogue (10). The synthesis was started with 3-bromobenzamide (6). Introduction of a p-hydroxyphenyl group and a benzyl group was achieved by Suzuki coupling and ortho-metalation, respectively. After lactonization, two steps transformation provided an acid which was converted to an indanone (8) by a usual manner. Introduction of a second p-hydroxyphenyl moiety to the indanone was done by aldol condensation with p-methoxybenzaldehyde to form the olefin with an E-exo configuration.

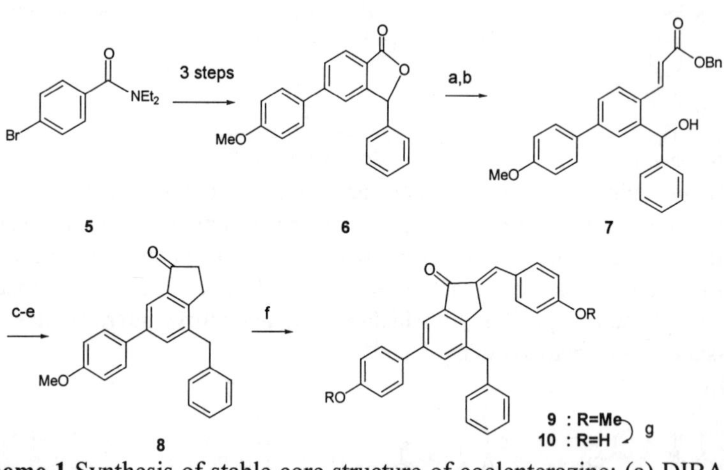

Scheme 1 Synthesis of stable core structure of coelenterazine: (a) DIBAL, -78 °C, CH_2Cl_2, 90%.; (b) Ph_3P-CHCOOBn, Benzene, 80 °C, 1 h, 95%; (c) H_2, Pd/C, EtOH-EtOAc, 23 °C , 36 h, 94%; (d) $SOCl_2$ 80 °C, 3 h; (e) $TiCl_4$, CH_2Cl_2, 12 h, 0 °C to 5 °C, 50% for 2 steps; (f) 4-Methoxybenzaldehyde, 2.5 M NaOH, EtOH, 24 °C, 20 h, 74%; (g) BBr_3, -78 °C to r.t, 48 h, 50%

After deprotection, deaza-coelenterazine analogue **(10)** with an isomeric double bond was obtained. The compound showed moderate inhibitory activity (Ki=10^{-7} M) to a recombinant Renilla luciferase in a competitive manner.

In connection to structural similarity to the model compound, we investigated inhibitory effect of dehydrocoelenterazine on Renilla luciferase. In contrast to **10,** The compound irreversibly inactivated luciferase along with weak luminescence to yield fluorescent inactive protein.

3. SYNTHESES OF THE TRANSITION-STATE MODEL OF COELENTERAZINE BIOLUMINESCENCE.

For the synthesis of the transition-state models having a chiral sp3 carbon at 2 position, the olefin **9** was subjected to hydrogenation with PtO_2 followed by deprotection and separation on a chiral column (DAICEL OC) afforded two enantiomers, (S)-TS-1 **(11-S)** and (R)-TS-1 **(11-R)**. As expected, these are stronger inhibitors than a sp^2 analogue (Ki=4-5x10^{-8} M), however, almost no difference in the activities among the two enantiomers was observed, suggesting the p-hydroxybenzyl group located at similar position in the active site of the enzyme like acetylcholine esterase inhibitors.[14]

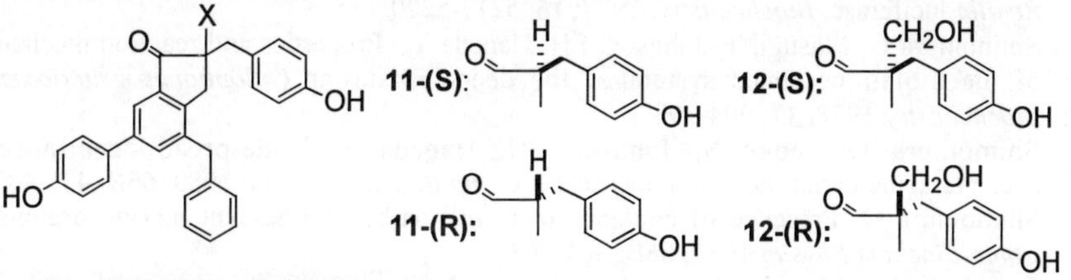

Transition state analogues
TS-1: R=H, TS-2: R=CH_2OH

To add a peroxide moiety to increasing the selective binding, a hydroxymethyl group was introduced at 2 position by an aldol reaction. The enantiomers, **12-(S)** and **12-(R)** were separated by a chiral column and determined their configuration by comparing the CD spectra with those of model compounds. An enantiomer with a R-configuration exhibited the strongest inhibition (Ki=9.3x10^{-9} M) and the large enantio-selectivity was observed (by a factor of about 22).

Bioluminescence reactions of imidazopyrazinones are simple but composed with several transition-states and/or intermediates. Although none of 3D structures of imidazopyrazine-type luciferases has been konwn, the luciferases might stabilize those structures to promote the luminescent reactions with a high efficiency luminescence. It has been reported that coelenterazine 3-enoyl sulfate strongly inhibited Renilla luciferase (Ki=9.1x10^{-9} M). The 3D structure of the strong inhibitors might mimic one of the transition-states involved.

Interestingly, opposite enantio-selectivity of the transition-state analogues **12** to Renilla luciferase was observed in the inhibition of both aequorin regeneration and a catalytic reaction of apoaequorin with benzyl-coelenterazine. The structure of aequorine was solved recently by x-ray analysis.[15] The absolute configuration of the peroxide structure is consistent with one of the inhibitor 12-(S).

Acknowledgements:
We acknowledge to financial support from Grand-in-Aid for Scientific Research on Priority Areas (A) from the Ministry of Education, Sciences, Sports and Culture, Japan and from Naito Foundation. We also thank to JSPS scholarship to C. Wu.

References

1. Shimomura O, Johnson, FH. Calcium binding, quantum yield, and emitting molecule in aequorin bioluminescence. *Nature* 1970; **227**: 1356-1357.
2. Shimomura O. Johnson FH. Chemical nature of the light emitt er in bioluminescence of aequorin. *Tetrahedron Lett.* 1973; **31**: 2963-2966.
3. Shimomura O, Johnson FH. Regeneration of the photoprotein aequorin. *Nature,* 1975; **256**: 236-238.
4. Shimomura O, Johnson FH. Peroxidized coelenterazine, the active group in the photoprotein aequorin. Proc. Natl. Acad. Sci. USA 1978; 75: 2611-2615.
5. Matthews JC, Hori K. Comier, MJ. Substrate and substrate analog binding properties of *Renilla* luciferase, *Biochemistry,* 1977; **16**: 5217-5220.
6. Shimomura O, Masugi, T. Johnson, FH. Haneda Y. Properties and reaction mechanism of the bioluminescence system of the deep-sea shrimp *Oplophorus gracilorostris. Biochemistry* 1978; **17**: 994-998.
7. Shimomura O, Inoue S, Johnson FH. Haneda Y. Widespread occurrance of coelenterazine in marine bioluminescence. *Comp Biochem Physiol* 1980; **65B**: 435-437.
8. Shimomura O. Presence of coelenterazine in non-bioluminescent marine organisms. *Comp Biochem Physiol* 1987; **86B**: 361-363.
9. Campbell AK, Herring PJ. Imidazopyrazine bioluminescence in copepods and other marine organisms. *Mar Biol* 1990; **104**: 219-225.
10. Goto T. Bioluminescence of marine organisms. in "Marine Natural Products, Chemical and Biological Perspective" ed by P. J. Scheuer, New York:Academic Press, 179-222. 1980 Vol. 3.
11. Hori K, Wampler JE, Matthew JC, Cormier MJ. Bioluminescence of *Renilla reniformis.* XIII. Identification of the product excited states during the chemiluminescent and bioluminescent oxidation of *Renilla* (sea pansy) luciferin and certain of its analogs. *Biochemistry*, 1973; **12**: 4463-4468.
12. Inouye S, Shimomura O. The use of *Renilla* luciferase, Oplophorus luciferase, and apoaequorin as bioluminescent reporter protein in the presence of coelenterazine analogues as substrate. *Biochem Biophys Res Commun* 1997; **233**: 349-353.
13. Lorenz WW, Macann RO, Longiaru M, Cormier MJ. Isolation and expression of a cDNA encoding *Renilla reniformis* luciferase *Proc Natl Acad Sci USA* 1991; **88**: 4438-4442.
14. Kawakami Y, Inoue A, Kawai T, Wakita M, Sugimoto H, Hopfinger AJ. The rationale for E2020 as a potent acetylcholinesterase inhibitor. *Bioorg Medicinal Chem* 1996; **4**: 1429-1446.
15. Head JF, Inouye S, Teranishi K, Shimomura O. The crystal structure of the photoprotein aequorin at 2.3 A resolution. *Nature* 2000; **405**: 372 — 376.

CHIMERIC BACTERIAL LUCIFERASES

ROSE SZITTNER, LEO LIN AND EDWARD MEIGHEN

Dept of Biochemistry, McGill University, 3655 Promenade Sir William Osler, Montreal, PQ, Canada, H3G1Y6

E-mail: meighen@med.mcgill.ca

1. INTRODUCTION

Bacterial luciferase contains two nonidentical subunits, α and β, with the α subunit controlling the primary kinetic properties of the enzyme.[1] Based on the crystal structure of the enzyme and comparison to other flavin binding enzymes, it has been proposed that the substrates bind in the carboxyl terminal half of the α subunit of luciferase.[2] Using chimeric α subunits with sequences from two highly divergent luciferases (*Xenorhabdus luminescens* and *Photobacterium phosphoreum*), it was also shown that regions in the carboxyl terminal half were involved in the catalytic function.[3]

Generation of suitable chimeric proteins is often difficult as the interchange of the different sequences can lead to disruption of the physical (e.g. solubility, stability) as well as the chemical properties depending on the specific regions that are exchanged. Replacement of approximately 60 amino acid segments of *X. luminescens* luciferase by the corresponding sequence of *P. phosphoreum* luciferase has previously resulted in chimeras, LuxA1B, LuxA2B and LuxA3B differing by 20 to 25 amino acids from the parental *X. luminescens* luciferase.[3] A central region extending from amino acid 166 to 233 of the α subunit was critical in defining interactions with the aldehyde substrate. In these studies, two additional chimeras, LuxA4B and LuxA5B have been constructed. Partial analyses of their properties have helped define more accurately the region interacting with the aldehyde substrate.

2. MATERIALS AND METHODS

Chimeric luciferases were generated as described previously by insertion of unique restriction sites into *X. luminescens luxA* followed by transfer of the homologous region from *P. phosphoreum luxA* into *X. luminescens luxAB*.[3] Luminescence intensities and decay rates were measured by injection of 50 μM FMNH2 into 0.05 M phosphate, pH 7, containing the appropriate concentration of decanal and the chimeric luciferase.

3. RESULTS AND DISCUSSION

The construction of the chimeric α subunits of luciferases is schematically represented in Fig 1 with the *P. phosphoreum* sequence represented by the slashed

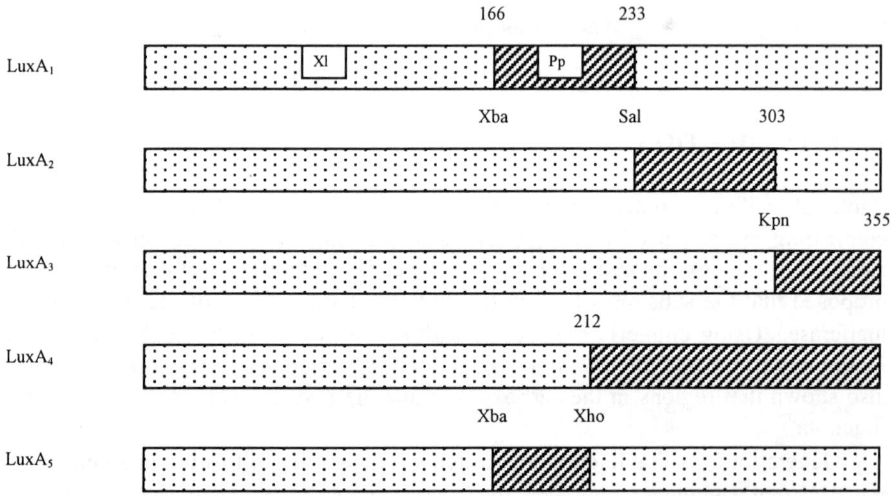

Fig. 1 Chimeric LuxA subunits. Restriction sites and sequence position of inserted Pp LuxA in the Xl LuxA subunit is given for different chimeras.

area inserted into *X. luminescens* LuxA. The two new chimeras, LuxA4B and LuxA5B, contain the Pp sequence extending between residues 166 to 212 and 212 to 355, respectively. These regions were substituted in Xl luxA by using the indicated unique restriction sites of XbaI and XhoI engineered in the plasmid containing the Xl-luxAB genes.

Activity can readily be measured in extracts of *Escherichia coli* containing a plasmid with *lux*A2B or *lux*A3B as reported previously,[3] however the level of activity was much lower for expression of *lux*A1B, *lux*A4B or *lux*A5B. The dependence of the luminescence expression on decanal concentration for LuxA4B and LuxA5B are compared to that for the parental Xl luciferase in Fig. 2. LuxA4B has a dependence on decanal concentration similar to Xl luciferase while LuxA5B binds aldehyde much more weakly as does Pp luciferase and the LuxA1B chimera. As LuxA1B and LuxA5B both contain PP luciferase between residues 166 and 212 in common, it suggests that this region of the sequence is implicated in binding

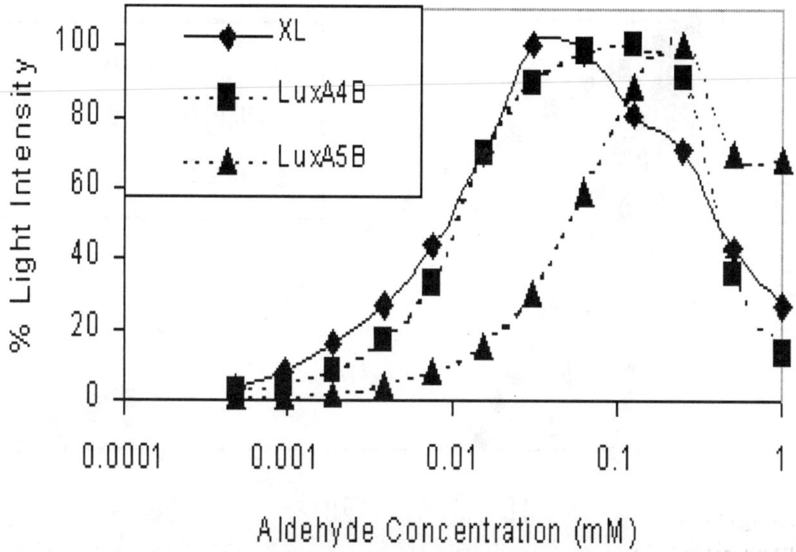

Fig. 2 Dependence of luminescence intensity on decanal concentration.

decanal. This proposal is supported by the fact that the other three chimeras with Xl sequences in this region all have the same dependence on decanal for activity as the parental Xl luciferase. These results indicate that the aldehyde binding regions are encompassed within the secondary structural elements extending between the β5-strand through the α5-helix and β6-strand to the α6 helix of the LuxA subunit of luciferase.[2]

A further comparison of the kinetic properties of the chimeras is given in Fig. 3 with respect to the rate of luminescence decay. The results show that the luminescence of LuxA5B with decanal decays more rapidly than that of Xl luciferase as shown previously for the luminescence decay of LuxA1B and PP luciferase.[3] In contrast the luminescence of LuxA4B decays much more slowly than Xl luciferase in a manner similar to that reported for LuxA3B[3]

These results suggest that both the region implicated in decanal binding and the common PP luciferase region in LuxA3B and LuxA4B between residues 303 and 355 are important in determining the decay rate. Mutation of these regions at single sites is currently underway to determine the role of specific residues.

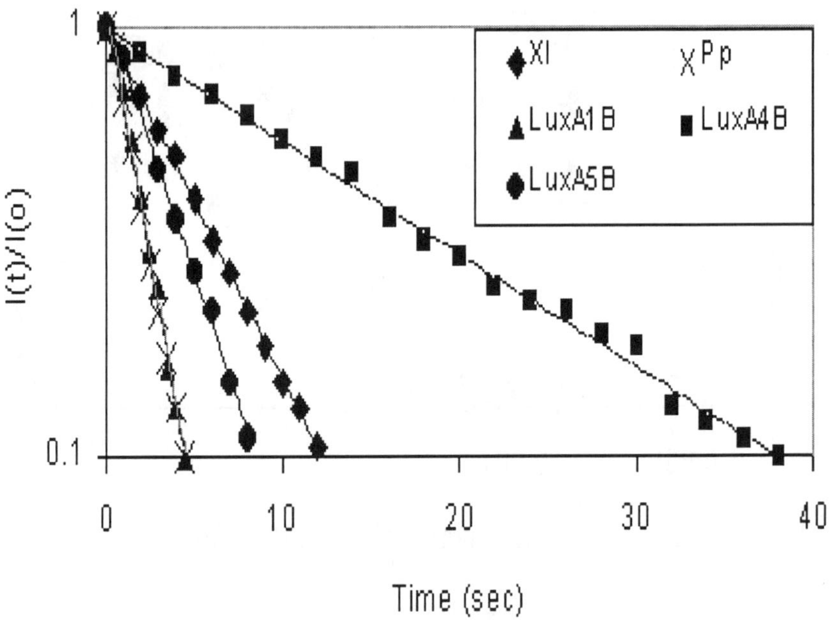

Fig. 3 Exponential decay of luminescence with decanal for parental and chimeric luciferases.

Acknowledgements

Supported by a grant from the Medical Research Council of Canada (MT4314).

References

1. Meighen EA, Bartlet I. Complementation of subunits from different bacterial luciferases. Evidence for the role of the beta subunit in the bioluminescent mechanism. *J Biol Chem* 1980; **255** 11181-11187.
2. Fisher AJ, Thompson TB, Thoden JB, Baldwin TO, Rayment I. The 1.5-Å resolution crystal structure of bacterial luciferase in low salt conditions. *J Biol Chem* 1996; **271** 21956-21968.
3. Valkova N, Szittner R, Meighen EA. Control of luminescence decay and flavin binding by the LuxA carboxyl- terminal regions in chimeric bacterial luciferases. *Biochemistry* 1999; **38** 13820-13828.

OBELIN CRYSTAL STRUCTURE: IMPLICATIONS FOR THE BIOLUMINESCENCE MECHANISM

EUGENE S. VYSOTSKI, ZHI-JIE LIU, LU DENG, JOHN ROSE, B.C. WANG, AND JOHN LEE

University of Georgia, Athens, GA 30602, USA, and Institute of Biophysics, Russian Academy of Sciences (SB), 660036 Krasnoyarsk, Russia.

E-mail: jlee@arches.uga.edu

1. INTRODUCTION

The defining property of calcium-regulated photoproteins is that calcium ion binding triggers bioluminescence through the oxidative decarboxylation of a tightly bound coelenterazine-oxygen moiety.[1] Photoproteins originate mainly from marine bioluminescent coelenterates and include obelin and aequorin, as well as others. All photoproteins show high homology and contain three calcium-binding sites.[2, 3] Some members of the larger family of calcium-binding proteins, for example sarcoplasmic calcium-binding protein, also have only three EF-hands. But the majority of calcium-binding proteins contain an even number, 2 or 4, EF-hands (calmodulin, troponin C, etc.). The structure of a calcium-regulated photoprotein with the coelenterazine-oxygen complex bound is a very desirable goal because it represents a natural stable enzyme intermediate, a snapshot of the bioluminescent pathway.

Recently the crystal structures of both aequorin[4] (resolution 2.3 Å) and obelin[5] (resolution 1.7 Å) have become available, the latter now extended to an exceptional resolution of 1.1 Å. As expected, the overall fold in the two photoproteins is quite similar but the nature of the ligand-oxygen complex in the protein binding site appears to be different.

2. EXPERIMENTAL

The purification and characterization of recombinant obelin have been described elsewhere.[6] Two crystal forms could be obtained, hexagonal[5, 6] and monoclinic (with PEG8000 as precipitant), both containing only one molecule in the asymmetric unit and diffracting to 1.73 and 1.1 Å respectively. Phases for the hexagonal crystal structure were obtained directly from single wavelength anomalous scattering from the eight sulfur atoms in the asymmetric unit. The monoclinic crystal structure was solved by molecular replacement using the hexagonal crystal structure as a search model and refined to 1.1 Å resolution.

3. RESULTS AND DISCUSSION

3.1 Overall structure.

The overall structures of obelin and aequorin as expected are very similar. They are almost entirely helical, highly compact and globular, with radius approximately 25 Å. Their structure contains eight helices, designated helix A — H. As with other calcium-binding proteins, the two photoprotein structures contain two pairs of helix-turn-helix (HTH) motifs, where the second HTH motif here is unable to bind Ca^{++}.

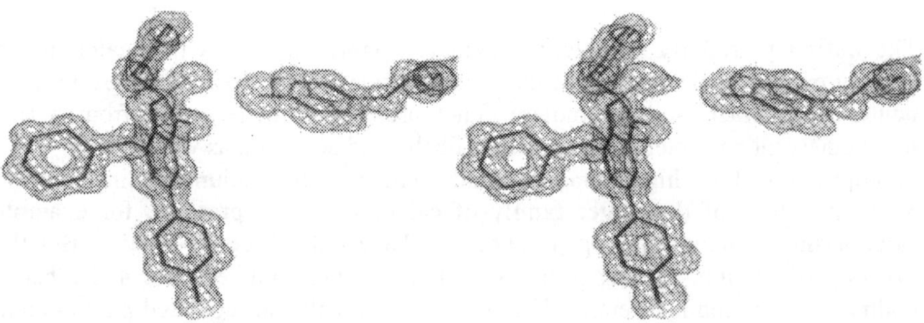

Fig. 1. Electron density distribution around the coelenterazine-oxygen at resolution 1.1 Angstroms. *Left:* obelin. *Right:* After exposure to Ca^{++}.

3.2 Coelenterazine binding site

The coelenterazine resides in an internal cavity surrounded by hydrophobic residues from each of the eight helices. Several hydrophilic residues are also directed into the pocket, in the case of obelin these are His22, Tyr138, His175, and Tyr190, and corresponding residues are seen in the aequorin binding site.

Site-directed mutagenesis has shown that Trp, His, Cys, and the C-terminal Pro, are critical for the bioluminescence activity. In the binding cavity, His22 and His175 lie nearby the coelenterazine and the N atom of His 175 is H-bonded (3.24 Å) to the C3-carbonyl oxygen. It is tempting to suggest that this His assists in the formation of the proposed dioxetane intermediate in the reaction pathway.

3.3 Structure of the coelenterazine-oxygen ligand

Shimomura and Johnson[1] suggested that the coelenterazine was bound to an amino

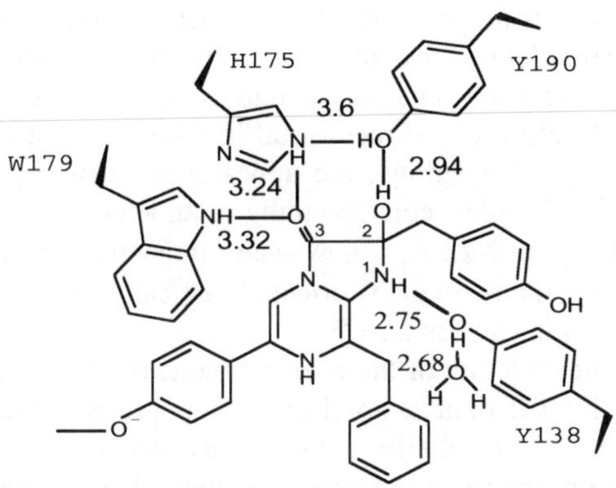

Fig. 2. Coelenterazine hydrate structure proposed for the bound state in the obelin crystal.

acid residue of aequorin through a peroxidic linkage substituted at the C2-position of the imidazopyrazinone ring (Fig. 2). However, for obelin shown here, the electron density distribution (Fig. 1, left) at the C2-position of the coelenterazine molecule clearly corresponds to only one oxygen, not peroxy, with no covalent binding to the protein.[5] Besides, an H-bond is evident between the C2-O and the hydroxyl oxygen of Tyr-190 (2.94 Å) (Fig. 2) leaving no space for a second oxygen

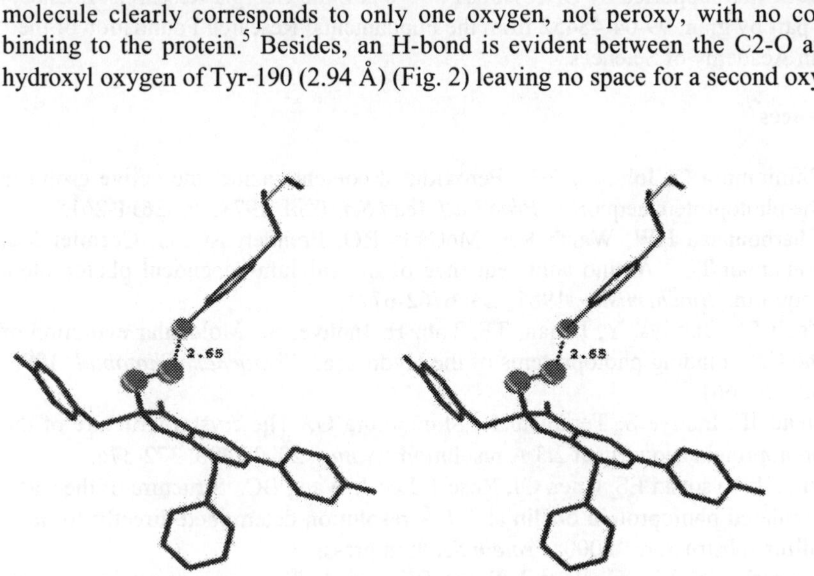

Fig. 3. Stereoview model of the peroxy (balls) substitution on the C2-atom of coelenterazine bound to the crystal of obelin following exposure to low Ca^{++}.

atom. It might be considered as proposed for the aequorin structure, that a weak O-O bond might have been cleaved by the high intensity radiation from the synchrotron. But the same single oxygen is also found using data collected on a rotating anode (not shown). However, if the crystal is exposed to traces of calcium insufficient to cause the crystal to glow, the appearance of increased electron density at this position (Fig. 1, right), corresponding to an additional oxygen and a shift away of the Tyr hydroxyl to 3.22 Å, is now seen. In contrast to the conclusion from the aequorin structure,[4] the peroxide is oriented vertically to the plane (Fig. 3) and poised for dioxetane bridging across C2-C3.

We suggest that the structure of the coelenterazine in obelin could arise by H_2O addition across the N1-C2 double bond during apoprotein charging with coelenterazine (Fig. 2). The obelin crystal has good activity for the bioluminescence reaction on exposure to excess calcium. It is proposed that the first Ca^{++} (out of the three required for the reaction) allows a subtle structural change with rapid dehydration and oxygen addition to the peroxide, followed by dioxetane bridge formation, and the bioluminescence decarboxylation.

Acknowledgements

This work was supported by ONR N00014-99-0414, the Georgia Research Alliance, and in part by grant 99-04-48452 from the Fundamental Research Foundation of the Russian Academy of Sciences.

References

1. Shimomura O, Johnson FH. Peroxidized coelenterazine, the active group in the photoprotein aequorin. *Proc Natl Acad Sci USA* 1978; **75**: 2611-2615.
2. Charbonneau HW, Walsh KA, McCann RO, Prendergast FG, Cormier MJ, Vanaman TC. Amino acid sequence of the calcium-dependent photoprotein aequorin. *Biochemistry* 1985; **24**: 6762-6771.
3. Tsuji FI, Ohmiya, Y, Fagan, TF, Toh, H, Inouye, S. Molecular evolution of the Ca^{2+}-binding photoproteins of the Hydrozoa. *Photochem Photobiol* 1995; **62**: 657-661.
4. Head JF, Inouye S, Teranishi K, Shimomura O. The crystal structure of the photoprotein aequorin at 2.3 Å resolution. *Nature* 2000; **405**: 372-376.
5. Liu ZJ, Vysotski ES, Chen CJ, Rose J, Lee J, Wang BC. Structure of the Ca^{2+}-regulated photoprotein obelin at 1.7 Å resolution determined directly from its sulfur substructure. 2000; *Protein Sci* **9**: in press.
6. Vysotski ES, Liu ZJ, Rose J, Wang BC, Lee J. Preparation and preliminary study of crystals of the calcium-regulated photoprotein obelin from the bioluminescent hydroid *Obelia longissima*. *Acta Crystallogr* 1999; **D55**: 1965-1966.

EFFICIENCY OF APOPHOTOPROTEIN CHARGING DEPENDS STRONGLY ON PROTEIN CONCENTRATION

E.S. VYSOTSKI, L.A. FRANK, AND V.S. BONDAR
Photobiology Lab, Institute of Biophysics, Russian Academy of Sciences, Siberian Brunch, Krasnoyarsk 660036, Russia
E.mail: Eugene.Vysotski@usa.net

1. INTRODUCTION

The Ca^{2+}-regulated photoproteins have been used as reporter proteins both to monitor gene expression in cells and to monitor the process of protein translation in a cell-free system. The application of any gene as a reporter depends on the sensitivity of detection of the product being synthesized, and by the sensitivity of the reporter protein to environmental conditions. This is of a particular importance when, to estimate the amount of a synthesized protein, its enzymatic activity is used. The latter may be affected by numerous environmental factors the influence of which must be taken into account in the use of any reporter enzyme, photoproteins included. In using photoproteins as reporter proteins one must also take into account that the recombinant photoproteins are synthesized as apoproteins. To generate an active photoprotein the apophotoprotein must be charged with synthetic coelenterazine under appropriate environmental conditions.

In this paper we show that the efficiency of apoobelin charging with coelenterazine depends strongly on apoprotein concentration. Also a mathematical model describing the process of coelenterazine binding is suggested.

2. MATERIALS AND METHODS

Chemicals. All chemicals were of the highest grade commercially available. Coelenterazine was purchased from Molecular Probes.

Purification of the recombinant apoobelin was carried out according to procedures described elsewhere.[1] The composition of solutions containing apoprotein samples was changed by running the sample through a BioGel P6 column equilibrated and eluted with 150 mM KCl, 5 mM PIPES pH 7.1 containing either 1 mM $CaCl_2$ or 5 mM EDTA.

Apoobelin charging was performed in 150 mM KCl, 5 mM EDTA, 10 mM DTT, 5 mM PIPES pH 7.1 containing the required concentration of apoprotein and coelenterazine. The coelenterazine was dissolved in cold methanol bubbled with argon. The coelenterazine concentration was estimated using the value 8900 cm^{-1} M^{-1} at 434 nm for the extinction coefficient in methanol.[2]

Light emission was measured with a luminometer equipped with neutral-density filters with different transmission coefficients to extend the dynamic range. The photoprotein luminescence was triggered by injection of 0.2 mL Ca-containing solution (100 mM $CaCl_2$, 150 mM KCl, 5 mM PIPES pH 7.1) into the 0.2 mL of a charging mixture with photoprotein. All measurements of light emission were carried out at 20-23°C.

3. RESULTS AND DISCUSSION

The kinetics of apoobelin charging were studied at 4°C, 22°C, and 37°C. At 4°C, charging progressed slowly, the concentration of active photoprotein reaching a maximum after 4-6 hours of incubation, and remaining constant for at least 24 hrs after that. At 22°C and 37°C, charging was considerably faster, the amount of active obelin reaching a maximum after about one hour, and then declining slowly. The influence of protein (apoobelin) concentration on the efficiency of charging was studied at all three temperatures, in each case with an apoprotein:coelenterazine molar ratio of 1:1. All measurements were made at the incubation times required for maximum light yield. The efficiency of the charging reaction was found to vary strikingly with protein concentration. That is to say, the fraction of apoobelin molecules remaining uncharged at the end of the reaction time decreased as the total protein concentration increased. In Fig. 1, the results of these measurements have been normalized to the maximum light yield obtained at the temperature in question. It will be seen that at each temperature, the fractional yield (the proportion of protein molecules that are charged after the reaction has run its course) is non-linearly related to the apoobelin concentration. The influence on charging yield of the molar ratio of coelenterazine to apoobelin was studied at a single apoobelin concentration (1.3×10^{-7} M), and (because temperature had little influence on dependencies of yield at the molar ratio of 1:1) at a single temperature (22°C). Molar ratios (apoobelin:coelenterazine) of 1:1, 1:5, 1:10, 1:50, and 1:100 were tested. The highest yield of active obelin was obtained in the presence of a 10-fold excess of coelenterazine: in view of this finding, the influence of apoobelin concentration on charging was reexamined at the molar ratio of 1:10, with results similar to those for molar ratio 1:1. The inhibition of charging at high molar ratios is unexplained, but cannot have been the result of the addition of the methanol used to dissolve the coelenterazine, because the amount of methanol added was always the same. A non-linear dependency of charging on apoprotein concentration similar to that shown in Fig. 1 was also observed with recombinant apoaequorin.

For recombinant apoobelin[3], by using a fluorescence method, it was shown that whereas the formation of an active photoprotein from apoprotein, coelentera-

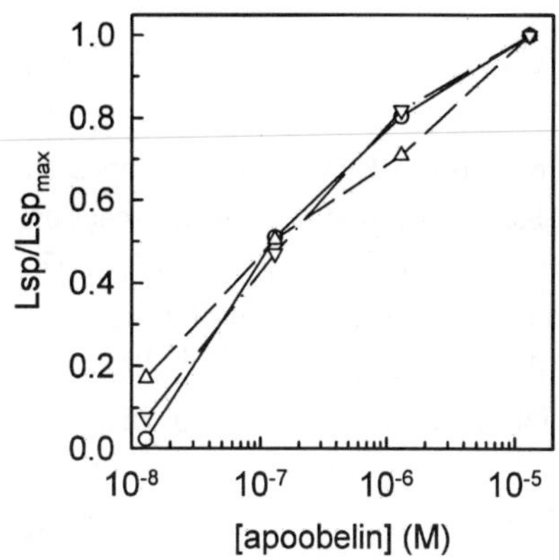

Fig.1. Apoobelin charging at different concentrations of apoprotein at 1:1 "apoobelin-coelenterazine" molar ratio: (o) – 4°C; (Δ) – 22°C; (∇) – 37°C. Lsp - specific luminescent activity of obelin calculated for each protein concentration; Lsp_{max} - maximal specific luminescent activity of obelin for each tested temperature.

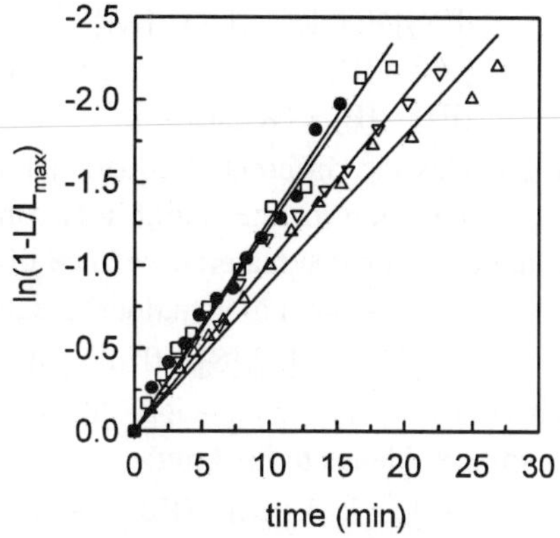

Fig.2. Apoobelin charging kinetics at different coelenterazine concentrations: (Δ) — 1:5; (∇) — 1:10; (□) — 1:50; (•) — 1:100 "apoobelin-coelenterazine" molar ratio. Apoobelin concentration is 1.3×10^{-7} M, temperature – 22°C. L - active photoprotein ($[X_2]$); Lmax - the highest yield of active photoprotein for each concentration of coelenterazine ($[X_2]_{st}$).

zine, and oxygen is slow, the binding of coelenterazine to the apoprotein takes just milliseconds. This suggests that the slow kinetics of charging of recombinant apophotoproteins is most likely due to the conformation transients in a protein taking place on binding of coelenterazine and, perhaps, of the oxygen. With this assumption, the kinetic scheme of the reaction can be presented as follows:

$$E + S \underset{k_{-1}}{\overset{k_1}{\Leftrightarrow}} X_1 \underset{k_{-2}}{\overset{k_2}{\Leftrightarrow}} X_2 \qquad (1)$$

where: E – apophotoprotein; S – coelenterazine; X_1 – intermediate "apophotoprotein-coelenterazine-oxygen"; X_2 – active photoprotein; k_1, k_2, k_{-1}, and k_{-2} – reaction rate constants. If the equilibrium at the first step is much faster than the kinetic steps, the concentration of the X_1 intermediate will be determined by the concentration of free apophotoprotein, coelenterazine, and by an equilibrium constant.[4] (Since oxygen is in excess, and probably is to have no effect on the formation of X_1 intermediate, we do not consider it in the kinetic scheme so far). In this case at $[S]_o >> [E]_o$ and at an insignificant consumption of a substrate, the kinetics are described via the simultaneous solution of the system of equations:

$$d[X_2]/dt = k_2[X_1] - k_{-2}[X_2] \quad (2)$$
$$[X_1] = [E][S]_o/Kd \qquad \text{where } Kd = k_{-1}/k_1 \quad (3)$$
$$[E]_o = [E] + [X_1] + [X_2] \quad (4)$$

that includes the differential equation for the formation of an active photoprotein (X_2), the equation for the equilibrium constant (Kd) at the first step of the reaction, and the equation for material balance of apoprotein (at $[S]_o >> [E]_o$, $[S]=[S]_o$). The solution of this system of equations[4] results in the following expression for X_2:

$$[X_2](t) = (k_2[E]_o[S]_o/((Kd + [S]_o)k_{-2} + k_2[S]_o))\{1-e^{-(k_{-2}+k_2[S]_o/(Kd+[S]_o))t}\} \quad (5)$$

As is evident from the equation, the concentration of active photoprotein increases with time and tends to the limit:

$$[X_2]_{st} = k_2[E]_o[S]_o/((Kd + [S]_o)k_{-2} + k_2[S]_o) \quad (6)$$

Out of the kinetics of active photoprotein formation the elementary rate constants, as well as the equilibrium constant for the first step of the reaction can be calculated. According to equation (7) from the equation of anamorphosis (5) the value of τ^{-1} is determined from the slope of the curve for $\ln(1-[X_2]/[X_2]_{st})$ parameter *vs* t on a semi-logarithmic plot (Fig.2).

$$\ln(1-[X_2]/[X_2]_{st}) = -t/\tau, \quad \text{where } \tau^{-1} = k_{-2} + k_2[S]_o/(Kd+[S]_o) \quad (7)$$

From the dependence of τ^{-1} on the initial concentration of coelenterazine the elementary rate constants $k_{-2} = 0.06\ s^{-1}$, $k_2 = 0.07\ s^{-1}$, and the apparent dissociation constant $Kd = 9.8 \times 10^{-7}$ M were calculated. The low value of the apparent dissociation constant suggests the high stability of the photoprotein complex that is observed experimentally.

Acknowledgments

We thank John Blinks from University of Washington (Friday Harbor Laboratories) for critical comments and discussions at manuscript preparation. This work was supported by Bayer AG (Germany).

References

1. Illarionov BA, Frank LA, Illarionova VA, Bondar VS, Vysotski ES, Blinks JR. Recombinant obelin: Cloning and expression of cDNA, purification, and characterization as a calcium indicator. *Methods Enzymol* 2000; **305**: 223-249.
2. Campbell AK, Patel AK, Razavi ZS, McCapra F. Formation of the Ca^{2+}-activated photoprotein obelin from apo-obelin and mRNA inside human neutrophils. *Biochem J* 1988; **252**: 143-149.
3. Semisotnov GV, Alakhov YB. personal communication.
4. Berezin IV, Varfolomeev SD. Nauka. Moscow: Biokinetica, 1979.

PART 2

FIREFLY BIOLUMINESCENCE

MUTATIONAL STUDIES OF STRINGENTLY CONSERVED FIREFLY LUCIFERASE ACTIVE SITE RESIDUES

BR BRANCHINI, RA MAGYAR, MH MURTIASHAW, NC PORTIER AND M ZIMMER
Department of Chemistry, Connecticut College, New London, Connecticut 06320, USA
E-Mail: brbra@conncoll.edu

1. INTRODUCTION

The crystal structures of firefly luciferase (Luc) without bound substrates and with two bromoform molecules have been reported.[1,2] The beetle luciferases are members of a large and diverse family of enzymes that form adenylates from carboxylic acid substrates. A crystal structure of a second member of this enzyme family—the phenylalanine activating subunit of gramicidin synthetase 1 (PheA) in a ternary complex with Phe, AMP, and Mg ion—also was disclosed.[3] Starting with the crystal structures of Luc and PheA, we used molecular modeling techniques to produce a potential model[4] of the Luc active site with bound substrates (Fig. 1). Although differences are noted, the results of a similar modeling study[5] are in good general agreement with ours.

2. MATERIALS AND METHODS

Previously described[4,6] methods were used to determine bioluminescence activity, corrected bioluminescence emission spectra, and kinetic constants for luciferin (LH_2), Mg-ATP and luciferyl-adenylate (LH_2-AMP). Mutagenesis, plasmid preparation, protein expression and purification of the Luc mutants H245A and T343A were performed as described[4]. For the Luc mutants R218A, R218K, R337A, R337Q and K529A, cloning and protein preparation were carried out according to methods described previously.[6] Fluorescence studies with 6-*p*-toluidino-2-naphthalene sulfonic acid (TNS) were based on a literature method.[7]

3. RESULTS AND DISCUSSION

Our model predicts that His245, Thr343 and Lys529 are near both ATP and the carboxyl group of LH_2, while Arg218 appears to be in close proximity to the LH_2 phenolate at the bottom of the LH_2 binding pocket. These residues are absolutely conserved among the beetle luciferases. Although modeling studies by other groups predict that Arg218 comprises part of the LH_2 binding site, Arg337 is thought instead to form the "base" of the pocket[2] and to be H-bonded to the LH_2

phenolate.[5] Ile and Leu replace Arg337 in two of the twenty Luc proteins. The elevated K_m values of the Arg218 mutants for LH_2 and contrasting near normal values of the Arg337 enzymes (Table 1) support our proposal that Arg218, and not Arg337, is an essential LH_2 binding site residue. It is likely that the greater steric bulk and/or larger polar area of the Arg side-chain accounts for the inability of even the conservatively changed R218K enzyme to bind LH_2 normally. The mutations at position 218 caused greater increases in the K_m values for Mg-ATP than did those at Arg337 and the other conserved residues examined. Our model predicts that Arg218 H-bonds to the side-chain of Ser347 and backbone carbonyl of Ala348. These residues are part of a loop that contains the highly conserved motif II (340YGLTE344 in Luc) in the acyl-adenylate-forming enzyme family thought to be involved in acyl-adenylate formation and/or Mg-ATP binding (Fig. 1). Alteration of the H-bonding interactions between Arg218 and the loop residues may, in turn, cause structural changes that disrupt important binding interactions between nearby motif II and Mg-ATP. Mutation of motif II residue Thr343 does produce an enzyme with decreased catalytic activity and increased K_m values for Mg-ATP and LH_2. Thr343 very probably participates in the catalysis of both the adenylation and oxidation reactions. The location of this residue in our model suggests that Thr343 may also assist in the orientation of the substrates prior to LH_2-AMP formation.

Table 1 Bioluminescence Activity and Kinetic Constants of Luc Enzymes

Enzyme	Relative Specific Activity	K_m(M) Luciferin	K_m(M) Mg-ATP	Relative LH_2- AMP k_{cat}	TNS Fluores. Maxima (nm)	Biolum. Maxima (nm)
Wild-type	100.0	15	160	1.0	425	560
Arg218Lys	33.0	217	690	0.72	430	572
Arg218Ala	3.6	301	6200	0.10	438	595
Arg337Lys	41.0	14	200	0.44	427	595
Arg337Gln	3.3	19	700	0.028	nd	594
His245Ala	26.0	15	240	0.72	425	604
Thr343Ala	1.1	99	857	0.011	427	617
Lys529Ala	0.06	230	1200	0.50	425	562

The greatly diminished catalytic activity and elevated K_m values of K529A for both LH_2 and Mg-ATP provide supportive evidence that Lys529 fixes the position of Mg-ATP and LH_2 at the active site through H-bonding interactions. The catalytic role of Lys529 is mainly in the adenylation reaction as evidenced by the nearly normal rate of catalysis of bioluminescence with the preformed LH_2-AMP.[6] Additionally, our results support the hypothesis[3] that Lys529, an absolutely conserved residue of the adenylate-forming enzyme family, stabilizes a pentavalent transition state of the adenylation step in this enzyme family.

Our model predicts too that the side-chain of His245 is at the opening of the binding pocket near the LH_2 carboxylate group and the γ-phosphate of ATP. We previously demonstrated[4] that His245 was the primary target of singlet oxygen mediated photoinactivation of Luc. However, the steady-state kinetic results with H245A indicate that the side-chain imidazole group has only a minor role in catalysis and is not necessary for productive substrate binding.

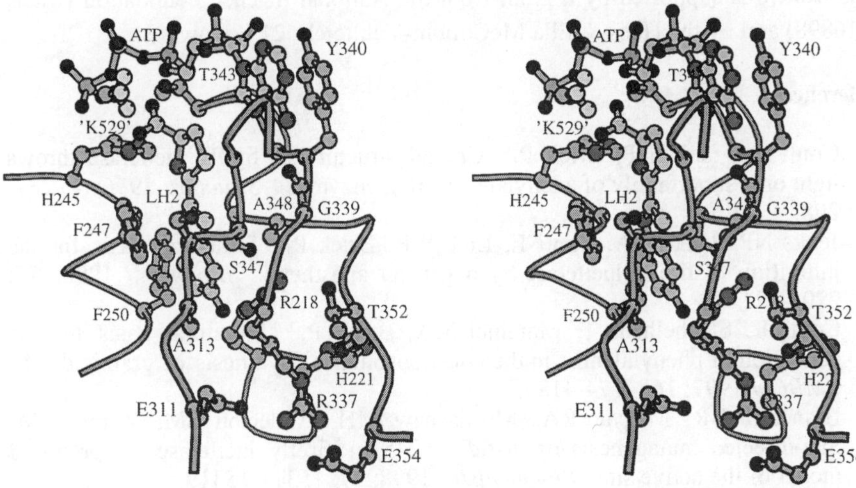

Fig. 1. Stereo diagram showing the interactions predicted by molecular modeling of *P. pyralis* luciferase with substrates luciferin (LH_2) and ATP.

By oxidation of a single LH_2 substrate, beetles emit light with maxima ranging from ~530 to ~630 nm. One proposal to account for color variation requires a Luc base to convert the electronically excited state keto form of oxyluciferin (red emitter) to the corresponding enolate (green emitter). Color modulation also has been attributed to effects on Luc tertiary structure resulting in a change of polarity or rigidity of the emitter-binding site.[8] While mutations of Lys529 have no effect on bioluminescence color, our prior studies suggest that His245 and Thr343 play essential roles in color determination.[6] Presently, we believe that His245 acts as an H-bond donor to the carbonyl oxygen of the keto form of oxyluciferin and that the side-chain imidazole group is held in place by an additional interaction with the Thr343 side-chain. Alternatively, His245 may assist Thr343 in the tautomerization process[8] required for yellow-green light emission. Arg218 appears to be essential to maintaining the polarity and rigidity of the protein environment proximate to the emitter. Using the LH_2 binding site probe TNS[7], differences in the TNS fluorescence emission maxima were found only with the Arg218 mutants (Table 1). These data are consistent with this residue having a key role in maintaining the anisotropic polarity of the emitter site.

Acknowledgements

This work was supported by a grant from the National Science Foundation (MCB 9816898) and by the Hans & Ella McCollum-Vahlteich '21 Endowment.

References

1 Conti E, Franks NP, Brick P. Crystal structure of firefly luciferase throws light on a superfamily of adenylate-forming enzymes. *Structure* 1996; **4**: 287-298.

2 Jenks NP, Jenkins A, Conti E, Lieb WR, Brick P. Structural basis for the inhibition of firefly luciferase by a general anesthetic. *Biophys J* 1998; **75**: 2205-2211.

3. Conti E, Stachelhaus T, Marahiel MA, Brick P. Structural basis for the activation of phenylalanine in the non-ribosomal biosynthesis of gramicidin S. *EMBO J* 1997; **16**: 4174-4183.

4. Branchini BR, Magyar RA, Murtiashaw MH, Anderson SM, Zimmer M. Site-directed mutagenesis of histidine 245 in firefly luciferase: A proposed model of the active site. *Biochemistry* 1998; **37**: 15311-15319.

5. Sandalova TP, Ugarova NN. Model of the active site of firefly luciferase. *Biochemistry (Moscow)* 1999; **64**: 1143-1150.

6. Branchini BR, Murtiashaw MH, Magyar RA, Anderson SM. The role of lysine 529, a conserved residue of the acyl-adenylate-forming enzyme superfamily, in firefly luciferase. *Biochemistry* 2000; **39**: 5433-5440 and references therein.

7. DeLuca M. Hydrophobic nature of the active site of firefly luciferase. *Biochemistry* 1969; **8**: 160-166.

8. McCapra F. Chemical generation of excited states: The basis of chemiluminescence and bioluminescence. *Methods Enzymol* 2000; **305**: 3-47 and references therein.

CENTRAL CONTROL OF LARVAL FIREFLY LUMINESCENCE

ALBERT D. CARLSON
Department of Neurobiology & Behavior, SUNY at Stony Brook, Stony Brook, NY 11794, USA
E-mail: acarlson@notes1.cc.sunysb.edu

1. INTRODUCTION

The larval firefly (*Photuris versicolor*) possesses two tiny lanterns on the sternite of the 8th abdominal segment. These lanterns are innervated by four photomotor (DUM) neurons in the 8th abdominal (lantern) ganglion.[1] The photomotor neurons activate luminescence by release of octopamine onto the lantern photocytes. This study describes experiments that investigate how the central nervous system controls the activity of the photomotor neurons and ultimately luminescence in these insects.

2. METHODS

To expose the photomotor neurons to the drugs, the nerve cord was perfused with collagenase, then the lantern ganglion was ripped open. Acetylcholine in firefly saline was applied to the exposed neurons and the light response was monitored. Light responses could be induced by a burst of electrical stimuli on the abdominal nerve cord between the 1st and 2nd abdominal ganglia delivered by a suction electrode. GABA was then perfused over the desheathed lantern ganglion as stimulation continued.

To further demonstrate central inhibition the nerve cord was stimulated while being perfused with the acetylcholine esterase blocker, eserine. It was necessary to induce glowing to observe inhibitory effects resulting in reduction of the glow intensity. In this situation both excitatory and inhibitory neurons were activated, but it was possible by proper selection of stimulating conditions to preferentially activate either one or the other.

3. RESULTS

The effects of perfused acetylcholine (0.1 and 1 mM) on the desheathed lantern ganglion are shown in Fig. 1. One millimolar acetylcholine induced intense bursting of the photomotor neurons which increased in frequency leading to a high level of luminescence. At the lower concentration of 0.1 mM, acetylcholine induced rhythmic bursts that led to glows of identical intensity.

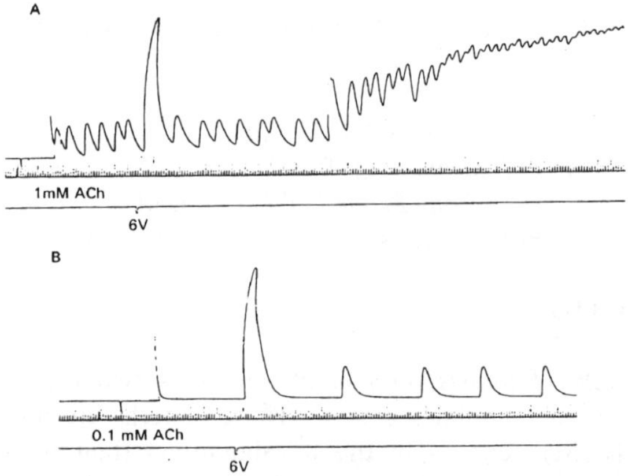

Fig. 1. Effect of acetylcholine on photomotor neurons in a desheathed lantern ganglion. A. 1 mM acetylcholine. B. 0.1 mM acetylcholine. In this and all other figures: top trace: light, middle trace: time base, 1 mark/s, bottom trace: stimulus. See results for further description.

The effect of two concentrations of GABA perfused on the desheathed lantern ganglion during electrical stimulation of the abdominal nerve cord above the 2nd abdominal ganglion is shown in Fig. 2. Note that the 1 mM concentration of GABA completely suppresses electrically induced luminescence (Fig. 2A). A concentration of GABA of 0.1 mM (2B) shows reduction of the light response but incomplete suppression.

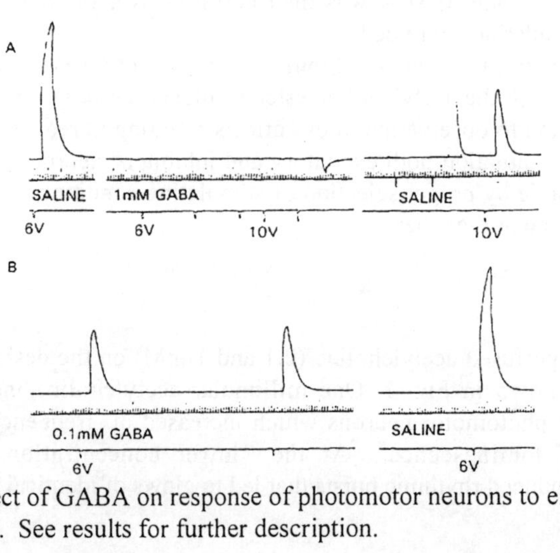

Fig. 2. Effect of GABA on response of photomotor neurons to electrical stimulation. See results for further description.

Electrical stimulation of the abdominal nerve cord of the preparation induced to glow with eserine, results in inhibition as shown in Figs. 3 and 4. Fig. 3A shows intense reduction of luminescence that eventually escapes and begins to glow even during continual stimulation. In Fig. 3B short bursts of stimulation induce transient reductions of luminescence intensity that quickly recover.

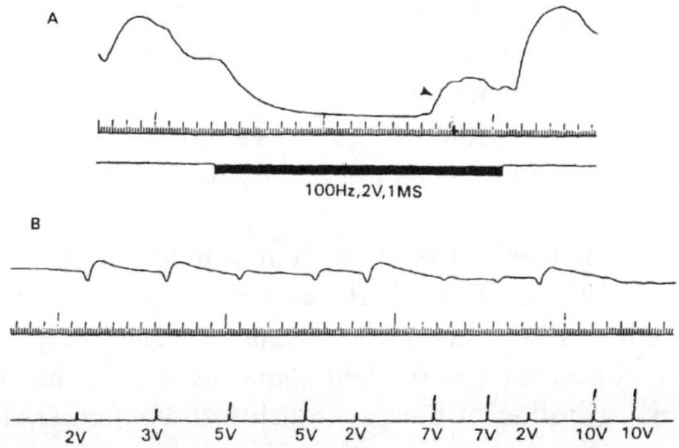

Fig. 3. Inhibition of eserine-induced luminescence during central stimulation. Arrow denotes escape. See results for further description.

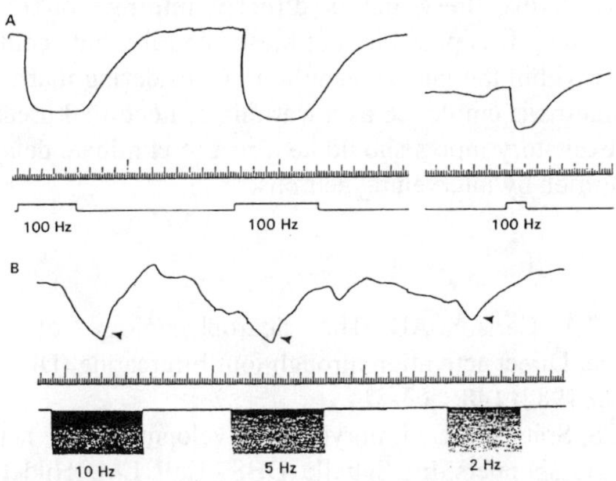

Fig. 4. Inhibition of eserine-induced luminescence during central stimulation. Arrows denote escape. See results for further description.

Strong reduction of luminescence intensity is illustrated in Fig. 4A and B which also show escape from inhibitory stimulation. Fig. 4B demonstrates that inhibition increases as stimulus frequency increases.

4. DISCUSSION

The results of this investigation suggest that the photomotor neurons possess excitatory receptors for acetylcholine and inhibitory receptors for GABA. Acetylcholine perfused on the desheathed lantern ganglion induces pulsed glows that imply that the photomotor neurons are firing in rapid bursts (Fig. 1). Further, the acetylcholine esterase inhibitor, eserine, induces glowing of the lanterns, again pointing to activation of the photomotor neurons (Figs. 3 & 4). The inhibitory transmitter, GABA, perfused on the desheathed lantern ganglion blocks or reduces the luminescence induced by central stimulation (Fig. 2). Further, electrical stimulation of the nerve cord can cause reduction of luminescence in the lantern induced to glow by perfusion of the nerve cord with eserine (Figs. 3 & 4).

No DUM neurons in beetles have been demonstrated to respond to both acetylcholine and GABA. The DUM neurons of the developing locust have been shown to possess both acetylcholine and GABA receptors, not only on their dendritic processes but also on their somas as well.[2] The DUM neurons of the last abdominal ganglion of the cockroach also contain GABA receptors in both areas[3].

Central neural control of the larval firefly lanterns, therefore, is believed to involve both excitatory inputs as well as inhibitory inputs to the photomotor neurons. Most likely these inputs directly impinge on receptors on the photomotor neurons. It is possible that these central inputs could act indirectly via interneurons within the lantern ganglion. Considering that the light reaction of the larval lanterns is employed as a warning to nocturnal predators, it is most likely that the excitatory inputs should be direct to eliminate delay and reduction of synaptic distortion by intervening neurons.

References

1. Christensen TA, Carlson AD. The neurophysiology of larval firefly luminescence: Direct activation through four bifurcating (DUM) neurons. *J Comp Physiol* 1982; **148**: 503-514.
2. Goodman CS, Spitzer NC. Embryonic development of neurotransmitter receptors in grasshoppers. In: Sattelle DBS, Hall LM, Hildebrand JG, eds. *Receptors for Neurotransmitters in Insects,* Amsterdam: Elsevier/North Holland, 1980: 195-209.
3. Dubreil V, Sinakevitch IG, Hue B, Geffard M. Neuritic GABAergic synapses in insect neurosecretory cells. *Neurosci Res* 1994; **19**: 235-240.

A NEUROBEHAVIORAL APPROACH TO FIREFLY FLASH CODE EVOLUTION

JAMES F. CASE
Marine Science Inst., University of California Santa Barbara, Santa Barbara, CA 93106. USA
E-mail:case@lifesci.ucsb.edu

JOHN BUCK
7200 Third Ave., C-20, Sykesville, MD 21784,USA
E-mail:JandEBuck@AOL.COM

1. INTRODUCTION

Buck and Case[1] propose tracing firefly flash code origins by analysis of luminescence-controlling neural circuits, particularly those common to both sexes. Here we report a newly discovered example suggesting that in-flight short-latency flashes are basic elements of the flash code of *Photinus pyralis*. Buck[2] conducted in-flight laboratory analysis of firefly communication behavior by observing patrol and short-latency flashing by male *P. pyralis* flying on a roundabout. In order to provide continuous viewing of flight details and behavioral choices in response to photic stimuli, fireflies of both sexes were flown in an apparatus[3] permitting time correlated radiometric and IR-video recording of flash dialogs and flight details. Long patrol flights by males were obtained with normal flashing periodicity and orientation to LED flashes imitating the female's responses. An in-flight communication mode of the female was discovered and is advanced in support of the possibility of unisex flash code timing elements.

2. FLASHING BEHAVIOR IN FLIGHT

The female of *P. pyralis* is rarely seen flying perhaps because she carries a heavy burden of eggs. In nature she perches in the open and responds to the male's 6 to 7s interval (temperature dependent) patrol flashes with a single flash after about 2s. In flashing she twists her abdomen towards the male. Females flew steadily in the flight apparatus for long periods with a wing-beat frequency approximating the male's 60 Hz. In flight females never flashed spontaneously but responded to an LED flash, or to a male's flash, with two flash types: the newly discovered approximately 300ms latency flash followed by an approximately 2s latency flash (Fig. 1A). The early latency flash was often less bright than the long latency flash. The two flash types almost invariably occurred together. Upon flight termination by landing the 2s responses continued, often without interruption of ongoing responses to a patrol flash test series. The early response ceased until flight resumed, when just as quickly it might appear. The in-flight 2s flash occurred with abdominal aiming as seen in the response of the perched female, but it is delayed by about 100ms in comparison with the flash latency of the perched female. Its kinetics (Fig. 1B) are otherwise indistinguishable from the perched female's 2s flash, suggesting a common neural program for the two. The early flash is not aimed.

The male has two modes of in-flight flashing, the characteristic patrol flash emitted about every 6 — 8s and a male-male interaction in which one male, seeing another male's flash, replies after about 300ms (Fig. 1C) [2]. Its kinetics are indistinguishable from the female's short latency flash, supporting the idea that a common neural program exists in both sexes.

In both sexes the early flash appears to be generated by a defined motor program with eye-specific access because direct microelectrode electrical excitation of the central brain, by-passing the eyes, requires about 450ms as compared with only 300ms for photic excitation.

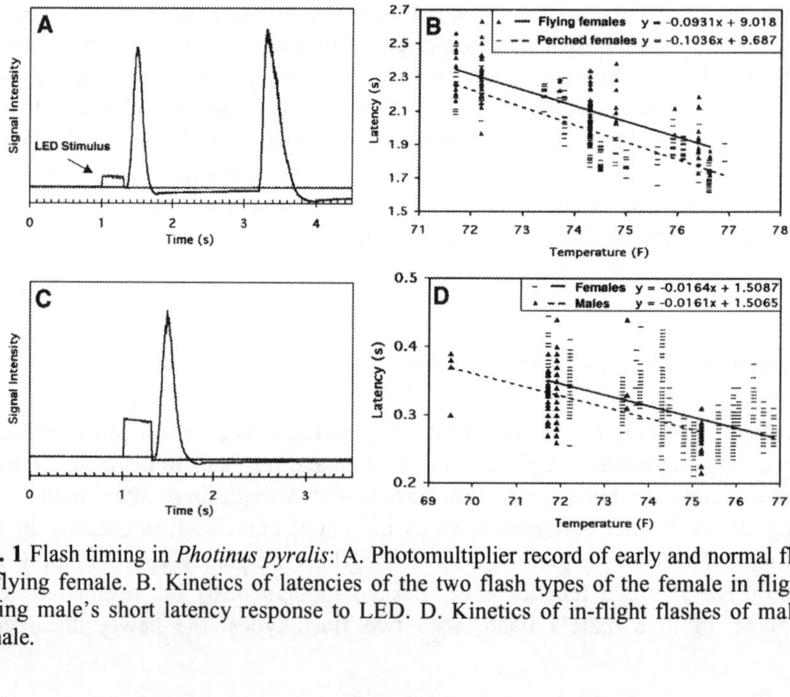

Fig. 1 Flash timing in *Photinus pyralis*: A. Photomultiplier record of early and normal flashes in flying female. B. Kinetics of latencies of the two flash types of the female in flight. C. Flying male's short latency response to LED. D. Kinetics of in-flight flashes of male and female.

3. IS THE SHORT LATENCY FLASH AN ANCESTRAL UNISEX COMMUNICATION MODE?

Fig. 2A is a diagram of the communication code of *P. pyralis* to emphasize the short latency components. Both sexes have in-flight short delay responses and the female's is always followed by a long latency flash and an abdominal aiming movement that is delayed by about 100ms as compared with the perched long latency aimed flash. We believe (Fig. 2B) the in-flight short cycles might have been the timer for communication in the ancestral dialog firefly. Therefore, both sexes are shown communicating with short cycle flashes, even in flight. Since evolution of bats occurred long after insect variety is believed to have been fully established, we presume aerial dialog between the sexes would not have been initially maladaptive. Eventually, as nocturnal predation might have increased, luminescence could have come to serve an aposematic role signaling presence of noxious chemicals, as well known in fireflies today[4]. The propensity of females to fly could have waned, further reducing aerial predation. The primitive code we envision would only promote aggregation of the sexes with pair formation having to depend on other processes. We suggest the signals of males and females would initially be indistinguishable on the basis of timing. Finally, Fig. 2C suggests an obvious way to attain species and sexual specificity of signals by building the short cycle neural circuit into a timer of any required duration by repetitions of the basic timing cycle prior to photomotor output. Theoretically this process of code differentiation has the advantage of allowing flash code changes over evolutionary time by sufficiently gradual timing increments to permit evolutionary change in codes without major timing discontinuities that would likely be maladaptive.

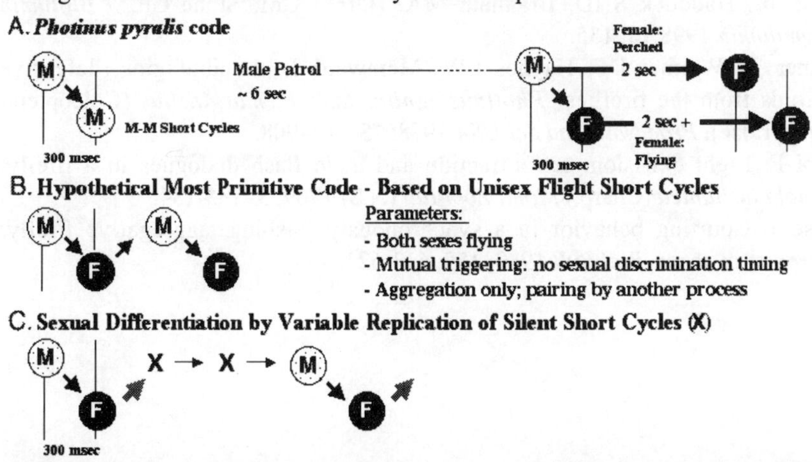

Fig. 2 A scheme involving unisex flight short cycles in firefly code evolution. See text.

4. DISCUSSION

In several firefly species short latency responses are known and are of the same order of magnitude as described here[5]. They have been considered to be reflexes and none have been shown to have the functional dependencies and effects on other flash code elements as found in *P. pyralis*. In *P. pyralis* the reduced latency to visual excitation compared with direct brain electrical excitation argues that the short cycle controller should be classified as a motor program. The experiments described here in which the short latency motor program can be evoked and result in adding to the latency of the 2s response of the female increases the plausibility that the *P. pyralis* flash response timing circuitry might be composed of ganged short latency timers. These we suggest support more complex flash codes with multiple flashes and much longer response latencies than the presumed ancestral 300ms periodicity. Existence of a 3/s flash rate in the SE Asian synchronizing species, *Pteroptyx tener* raises the possibility that a basic 300ms timer element is widespread among fireflies[6].

An alternative explanation of the short latency responses is that they represent a vestige of an alarm response. This seems unlikely since most insect alarm responses have latencies shorter than 100ms.

References

1. Buck J, Case JF. Physiological links in firefly flash code evolution. Submitted
2. Buck J. Unisex flash controls in dialog fireflies. *Biol Bull* 1990; **179**: 87-95.
3. Case JF, Haddock SHD, Branham MA, Herren CM, Stone CF. *J Biolumin Chemilumin* 1998;**13**: 135.
4. Eisner T, Weimer DF, Haynes LW, Meinwald J. Lucibufagins: defensive steroids from the fireflies *Photinus ignitus* and *P. marginellus* (Coleoptera: Lampyridae). *Proc Nat Acad Sci USA* 1978; **75**: 905-908.
5. Papi F. Light emission, sex attraction and male flash dialogues in a firefly, *Luciola lusitanica* (Charp.) *Monit Zool Ital* (N.S) 1969; **3**: 135-184.
6. Case J. Courting behavior in a synchronously flashing, aggregative firefly, *Pteropteryx tener*. *Biol Bull* 1980; **159**: 613-625.

MECHANISMS OF SYNCHRONY IN A COASTAL GEORGIA FIREFLY

JONATHAN COPELAND AND KELSIE FITZGERALD
Department of Biology and Applied Coastal Research Laboratory, Georgia Southern University 30460-8042 USA
E-mail:copeland@gasou.edu

ANDREW MOISEFF
Department of Physiology & Neurobiology, University of Connecticut. 06269-4156
E-mail: moiseff@uconn.edu

1. INTRODUCTION

Most of what is known about the mechanisms of synchrony in fireflies comes from the seminal work of Buck, Buck, Case, and Hanson.[1] *Pteroptyx malaccae* and *Pteroptyx cribellata* were shown to synchronize *via* spontaneously active flash oscillators that showed phase-sensitive flash entrainment to an exogenous pacer flash. During entrainment, the former species' flash oscillator period (interflash interval) could not be changed by the entrainment flash, whereas the latter's oscillator interval was invarient. *Photuris frontalis* is an intermittently synchronous firefly.[2] Although the population of flying flashing males appears to be continuously synchronic, individual fireflies repeatedly flash synchronously for several cycles and then pause for several cycles. Whereas individual Southeast Asian stationary congregational synchronizers must make continual timing adjustments to maintain population synchrony, the non-congregational flying *P. frontalis* has a synchronize-stop-re-synchronize pattern that is a normal part of its flash behavior. These differences prompted our investigation into the mechanisms of synchrony in this *Photuris* firefly.

2. MATERIALS AND METHODS

Photuris frontalis was captured on the grounds of Skidaway Institute of Oceanography (Savannah, GA) and tested in 10 cm petri dishes.

Flashes were recorded using a RCA-1P21 phototube and digitized at 1 kHz sample rate. Entrainment pulses were synthesized and delivered *via* a green LED. To construct phase-response curves, single stimuli were delivered randomly in the firefly's interflash interval. The flashes were either rectangular pulses or natural-like shaped pulses.

3. RESULTS AND DISCUSSION

Flashes were recorded from an isolated caged male (Fig. 1) to demonstrate that *P. frontalis* possessed an endogenous oscillator. In this example, the flash interval was 634 ± 11 msec (N = 89 ; 23.3 °C), a precision similar to that of Southeast Asian synchronizers.[1-2] The flashes are rhythmically continuous (Fig.1). Then, they stop, start again in phase, and once again flash rhythmically and continuously. This is a normal part of *P. frontalis* spontaneous flash behavior that is not normally found in the flash behavior of Southeast Asian synchronizers.[1-2]

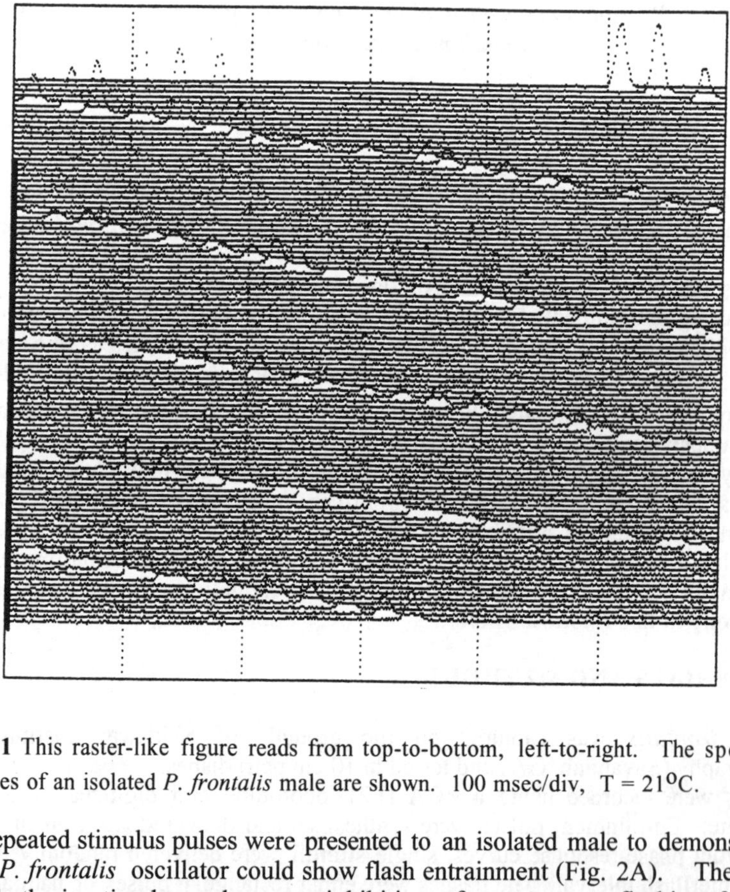

Fig. 1 This raster-like figure reads from top-to-bottom, left-to-right. The spontaneous flashes of an isolated *P. frontalis* male are shown. 100 msec/div, T = 21°C.

Repeated stimulus pulses were presented to an isolated male to demonstrate that the *P. frontalis* oscillator could show flash entrainment (Fig. 2A). The response to rhythmic stimulus pulses was inhibition of spontaneous flashing for several cycles followed by flash entrainment at a relatively constant delay. Unlike entrainment in the Southeast Asian *Pteroptyx* synchronizers,[1] stimulus pulses inhibited spontaneous flashing in *P. frontalis* for several cycles and resulted in a intermittent entrainment (Fig. 2B). Cycle length did not change in *P. frontalis* except at longer stimulus intervals (>700 msec).

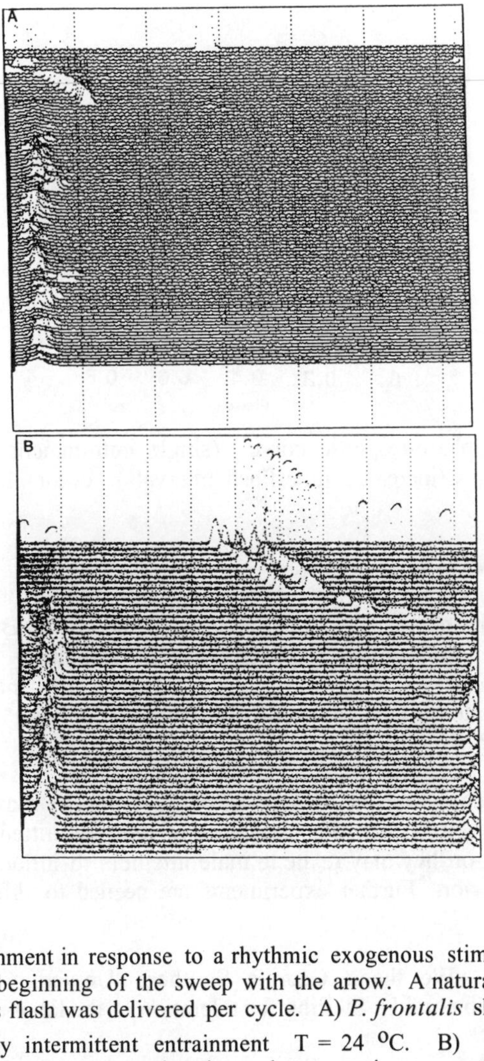

Fig. 2 Flash entrainment in response to a rhythmic exogenous stimulus. The stimulus flash began at the beginning of the sweep with the arrow. A natural-like stimulus was used. One stimulus flash was delivered per cycle. A) *P. frontalis* shows inhibition (11 cycles) followed by intermittent entrainment T = 24 °C. B) *Pteroptyx malaccae* (Malaysia), a Southeast Asian synchronizer, shows continuous entrainment (Copeland and Moiseff, unpub. obs.) T = 26 °C.

To show that the *P. frontalis* flash oscillator was capable of phase-sensitive responses to a single stimulus flash, a single LED pulse was randomly inserted into the interflash interval. Unlike Southeast Asian fireflies that show both phase advance and delay,[1] *P. frontalis* responded mostly with a lengthened/delayed interflash interval (Fig. 3). Sometimes, the duration of the phase delays (in msec) was equal to two or three spontaneous flash cycles.

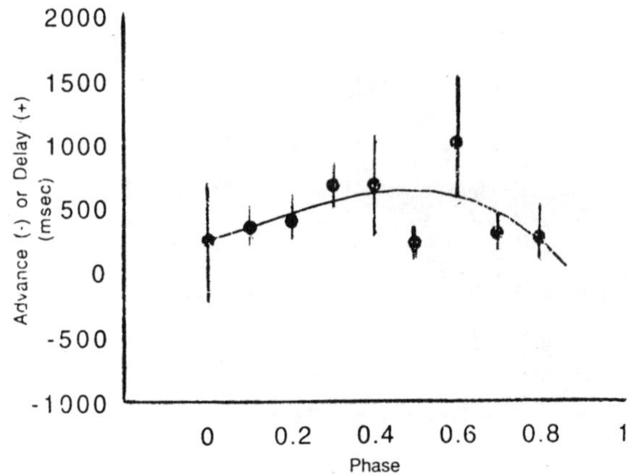

Fig. 3 Summary phase-response curve (single entrainment flash) using four fireflies. Only delays (increased iinterflash intervals) occur after an entrainment pulse.

4. CONCLUSIONS

When tested with the classical flash entrainment paradigm of Buck, Buck, Case, and Hanson, *P. frontalis* shows flash entrainment and many similarities when compared to Southeast Asian synchronizers. However, the species differ in the initial inhibition, intermittent nature of the entrainment, and delay-only phase-response found in *P. frontalis* (Fig. 2-3).

The behavioral meaning of the dynamics of *P. frontalis* flash entrainment is unclear. However, they may be a consequence of the intermittent nature of the *P. frontalis* synchrony, or they may relate to male attempts to influence female choice during mating behavior. Further experiments are needed to differentiate between these possibilities.

Aknowledgements We thank Georgia Southern University and the Noble Foundation for funding and U. Sterling for toleration of the laboratory studies.

References

1. Buck JB. Synchronous flashing of fireflies II. *Q Rev Biol.* 1988; **63**:263-281.
2. Moiseff A, Copeland J. A new type of sychronized flashing in a North American firefly. *J Insect Beh.* 2000 in press.

STABILIZATION OF FIREFLY LUCIFERASE ACTIVITY AGAINST OXIDATION WITH ANTIOXIDANTS

R.H. DAVIES, R.A. VAN WAGENEN AND J. D. ANDRADE

University of Utah, Department of Bioengineering, 2480 Merrill Engineering Building, Salt Lake City, UT, 84112, USA

E-mail: rupert.davies@m.cc.utah.edu

1. INTRODUCTION

Biosensors must be stored at room temperature for extended time periods. Luciferase stability is essential for functional bioluminescent biosensors. Firefly luciferase activity at room temperature can be maintained by lyophilization. However, lyophilized firefly luciferase does lose activity with time. Additionally, the stability of firefly luciferase activity in aqueous solution is important for research and optimization. Some of the activity loss in aqueous and lyophilized states is probably due to oxidative reactions. The oxidative damage to proteins includes scission of the peptide backbone, formation of intra or inter-molecular crosslinks, and modification of several amino acids. This damage leads to changes in net charge, conformation, and/or hydrophobic nature of proteins, reducing enzymatic activity.

Oxidation caused by oxygen free radical species has been shown to significantly decrease the activity of firefly luciferase.[1] Although sulfhydryl groups are not absolutely essential for firefly luciferase activity,[2] the sequential recombinant elimination of sulfhydryl groups reduces firefly luciferase activity.[3] Sulfhydryl groups are particularly susceptible to oxidation as well as other amino acid groups (Fig. 2). Although dithiothreitol (DTT) is generally used to protect sulfhydryl groups in enzymes, naturally occurring antioxidants such as vitamin C (ascorbic acid), vitamin E (a-tocopherol), and glutathione (GSH) have been shown to protect proteins against oxidation. Additionally, these naturally occurring antioxidants have been shown to work in concert in vivo forming an oxidation reduction pathway.[4]

2. METHODS AND MATERIALS

The inhibition of firefly luciferase activity by antioxidants was measured by injecting 50µl of an adenosine triphosphate (ATP) mixture (100µM ATP [Sigma] and 5mM magnesium sulfate [Sigma]) into 200µL of an antioxidant mixture including various concentrations (100mM, 10mM, 1mM, 0.1mM) of antioxidants

(DTT [sigma], GSH [Sigma], Trolox™ [Fluka Chemika], vitamin C [Fisher BioTech]), 0.1 mM D-luciferin (Biosynth AG), 40mM Sucrose (Pfanstiehl), 0.25mM Dextran T40 (Pharmacia), 1mg/ml of bovine serum albumin (BSA, Sigma), and 0.02 mg/ml of recombinant firefly luciferase.[5] The light intensity of the bioluminescence was collected with a Turner TD 20/20 luminometer for 15 seconds. The same antioxidant mixture was lyophilized in two stages using a Virtis 12EL Lyophilizer. The lyophilized mixture was then reconstituted with 200µL of reagent grade water immediately, 1 week, 2 weeks, and 2 months after lyophilization. Storage conditions included room temperature (25 degrees Celsius), in the dark, and in brown glass vials. The firefly luciferase activity after the different time intervals was measure by injecting 50 µl of the same ATP mixture into the reconstituted antioxidant mixture and measuring the light intensity for 15 seconds.

The firefly luciferase activity in the biomimicry antioxidant experiment was measured by injecting an ATP mixture consisting of 10µM ATP, 500µL magnesium sulfate, and 0.1mM luciferin into an antioxidant cocktail, which had been storage at room temperature in micro centrifuge tubes for 40 hours. The antioxidant cocktail consisted of various antioxidants (vitamin C, Trolox, and GSH) at different concentrations (10mM, 1mM, and 0.1mM) in several combinations, 0.5 mg/ml BSA, and 52 µg/ml of firefly luciferase (Promega). The light intensity was measured for 15 seconds with a Turner TD 20/20 luminometer.

3. RESULTS AND DISCUSSION

Fig. 1 shows the inhibition of firefly luciferase activity by various antioxidants at different concentrations. Fig. 2 depicts the results of the biomimic of the naturally occurring oxidation reduction pathway. Fig. 3 depicts the decreased firefly activity with time for various antioxidants. Fig. 4 shows the remaining activity of lyophilized firefly luciferase after two months of storage.

Inhibition of firefly luciferase activity was only observed at concentrations exceeding 10mM, with the exception of Trolox (greater than 1mM). The combinations of Trolox, vitamin C, and GSH did not show stabilization of the firefly luciferase activity greater than GSH alone after aqueous storage of 40 hours. Lyophilized firefly luciferase activity was preserved most effectively with DTT and GSH.

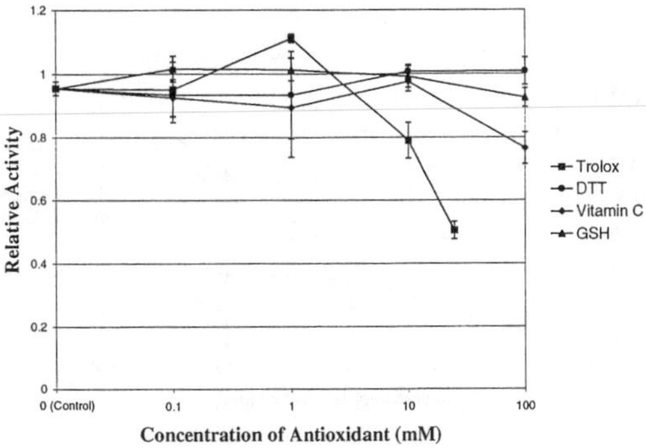

Fig.1 Each point represents the average of three integrated light intensities collected for 15 seconds. The activity is relative to control firefly luciferase cocktail with no added antioxidant or sucrose before lyophilization. The error bars indicate one standard deviation.

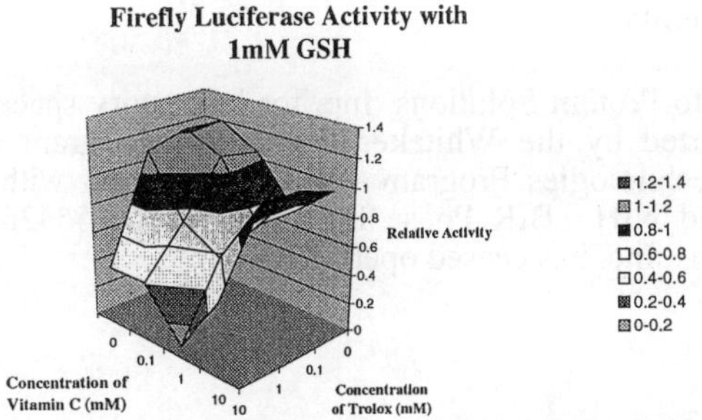

Fig. 2 Each intersection of lines represents a combination of three antioxidants at various concentrations. The activity value is based on the average of three integrated light intensities collected for 15 seconds. The activity is relative to control firefly luciferase cocktail with no added antioxidant after 40 hours of aqueous storage.

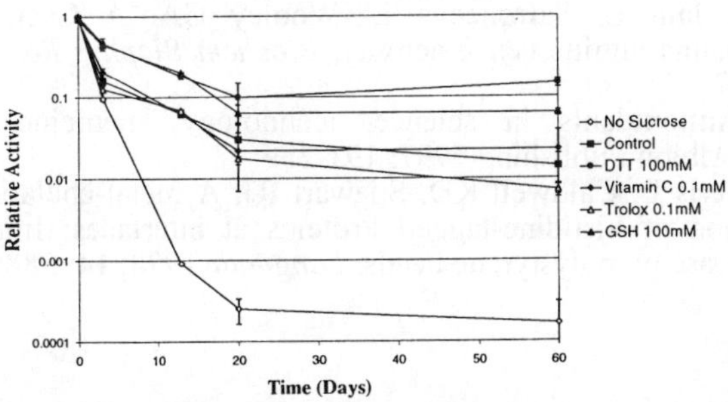

Fig.3 Each point represents the average of three light intensities collected over 15 seconds. The activity is relative to control firefly luciferase cocktail with no added antioxidant or sucrose before lyophilization. The error bars indicate one standard deviation. Only the concentration that maintained the greatest activity for each antioxidant is shown.

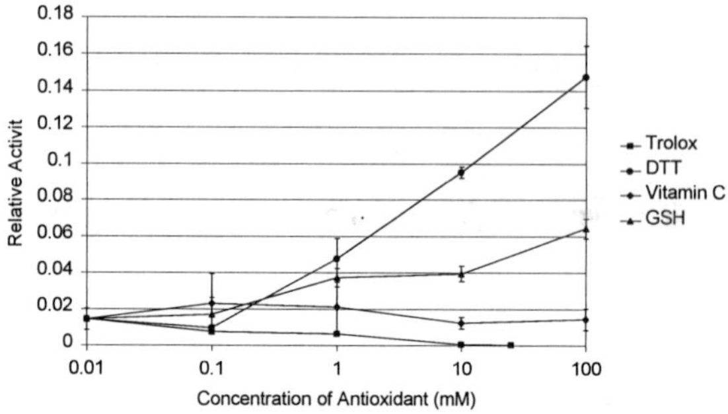

Fig. 4 Each point represents the average of three integrated light intensities collected for 15 seconds. The activity is relative to a control firefly luciferase cocktail with no added antioxidant or sucrose before lyophilization. The error bars indicate one standard deviation.

Acknowledgements

Special thanks to Protein Solutions, Inc. for laboratory space. This work has been partially supported by the Whitaker Foundation (a grant in the Cost Reducing Health Care Technologies Program – a joint program with the National Science Foundation) and NIH SBIR Phase I grant (1R43DK55426-1). Please note that Protein Solutions, Inc. has ceased operations.

References

1. Thompson A, Nigro J, Seliger HH. Efficient Singlet oxygen inactivation of firefly luciferase. *Biochem Biophys Res Commun* 1986; **140**: 888-894
2. Ohmiya Y, Tsuji FI. Mutagenesis of firefly luciferase shows that cysteine residues are not required for bioluminescence activity. *FEBS Lett* 1997; **404**:115-117
3. Kumita JR, Jain L, Safroneeva E, Wooley GA. A Cysteine-free firefly luciferase retains luminescence activity. *Biochem Biophys Res Commun* 2000; **267**: 394-397
4. Scott G. Antioxidants in science, technology, medicine and nutrition. Chichester: Albion Publishing 1997: 191-259
5. Ho C, Limberis L, Caldwell KD, Spewart RJ, A metal-chelating pluronic for immobilization of histidine-tagged Proteins at interfaces: Immobilization of firefly luciferase on polystyrene beads. *Langmuir* 1998; **14**: 3889-3894

GUANIDINIUM CHLORIDE-INDUCED INACTIVATION OF FIREFLY LUCIFERASE AND ITS REACTIVATION

E. I. DEMENTIEVA, L. G. MALOSHENOK AND N. N. UGAROVA
Chemistry Department, Moscow State University, 119899, Moscow, Russia
E-mail: eid@enzyme.chem.msu.ru

1. INTRODUCTION

The problem of protein denaturation is important from both theoretical and practical points of view. The study of denaturation of *Photinus pyralis* firefly luciferase with guanidinium chloride showed that kinetics of enzyme inactivation involves several reversible steps and spontaneous reactivation of the enzyme is possible after the removal of the denaturing agent.[1] It has been shown that subunit interactions play a major role in thermoinactivation of *Luciola mingrelica* firefly luciferase.[2] Mechanism of thermoinactivation involves dissociation of dimers to less stable monomers followed by irreversible inactivation of monomers and their aggregation. The goal of this work was to study inactivation of *L. mingrelica* firefly luciferase under the action of guanidinium chloride (GdmCl) and to examine the possibility of reactivation of the enzyme.

2. MATERIALS AND METHODS

Reagents Recombinant *L. mingrelica* firefly luciferase was isolated from *E. coli* cells, carrying pLR plasmid with the luciferase gene.[3] All the reagents were of analytical grade.
Instrumentation Luciferase activity was assayed on a 1250 LKB-Wallac luminometer. Tryptophan fluorescence spectra were obtained on a Perkin Elmer LS 50B fluorimeter, excitation at 290 nm.

3. RESULTS AND DISCUSSION

Kinetics of GdmCl-induced inactivation of luciferase All the experiments were carried out under the conditions where luciferase activity did not change in the absence of GdmCl at 8°C in the course of the experiment: 0.1 M Na-phosphate buffer solution containing 2 mM EDTA, 0.1 M NaCl, 6.5 mM dithiothreitol, pH 7.8. GdmCl caused gradual inactivation of the enzyme. At low concentrations of GdmCl, lag-period, where no change in activity and tryptophan fluorescence spectra of the luciferase occurred, was observed. The increase in GdmCl concentration from 0.04 to 0.8 M resulted in the decrease in the duration of lag-period from 120 to 0 min. As compared with the total inactivation time (3 days), lag-period was rather short. Conformational changes in the protein structure without changes in the active site,

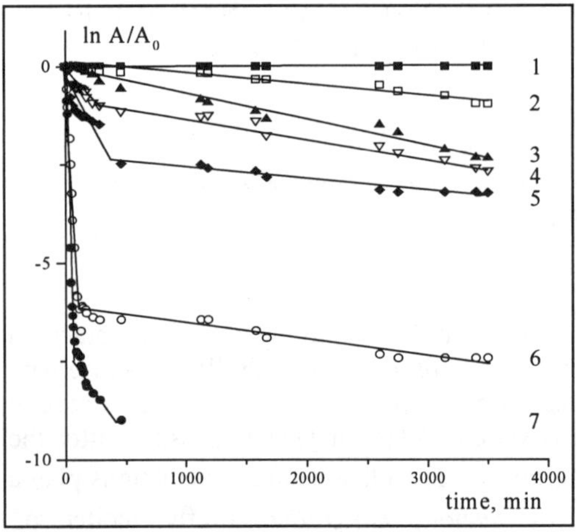

Fig. 1. Kinetic curves of inactivation of *L. mingrelica* firefly luciferase in the presence of GdmCl at 8°C. 2.8 μM luciferase. [GdmCl], M: 0.04 (1); 0.32 (2); 0.5 (3); 0.8 (4); 1 (5); 1.5 (6); 2 (7).

transfer from the more stable form of the luciferase to the less stable one may occur at this step.

At low concentrations of GdmCl (up to 0.5 M) inactivation of the luciferase was described as one exponent. At higher concentrations of GdmCl, kinetics of inactivation corresponded to two exponents (Fig. 1). The kinetic curves on Fig. 1 were similar to those for thermoinactivation of the luciferase at elevated temperatures and are typical of oligomer proteins where the first step is reversible dissociation of a dimer (E_2, scheme 1) into inactive monomer (I) and the second one is irreversible inactivation of a monomer.

<u>Reactivation of luciferase</u> The 100-fold dilution of the solutions of the inactivated luciferase (decrease in GdmCl concentration) resulted in reactivation of the enzyme (Fig. 2). If the inactivation degree corresponded to the 1st exponent (Fig. 1), 100% of luciferase activity was restored. At higher degrees of inactivation (2nd exponent), only partial reactivation occurred. Kinetic curves of reactivation were linearized in $1/a$ – time coordinates (Fig. 3) indicating that reactivation is a 2nd order reaction, *i.e.*, association of inactive monomers to form active dimers. Thus, inactivation of luciferase under the action of guanidinium chloride is described by the following scheme

$$E_2 \underset{k_{ass}}{\overset{k_{diss}}{\rightleftharpoons}} 2I \xrightarrow{k_{in}} 2I' \qquad \text{(scheme 1)}$$

<u>Fluorescent studies of luciferase inactivation</u> *L. mingrelica* firefly luciferase contains single tryptophan residue Trp-419. Changes in tryptophan fluorescence spectra characterize unfolding of protein globule. At inactivation degrees corresponding to the 1st exponent, minor changes in tryptophan fluorescence spectra were observed: shift of fluorescence maximum from 342 to 345 nm and insignificant decrease in fluorescent intensity. At high degrees of inactivation, the spectrum was shifted to 353 nm and fluorescence quantum yield decreased noticeably. Thus, large conformational changes occurred in the latter case: unfolding of the protein and transfer of Trp-419 to aqueous environment. These changes were irreversible. The

changes in the fluorescence spectra correlated completely with the changes in luciferase activity.

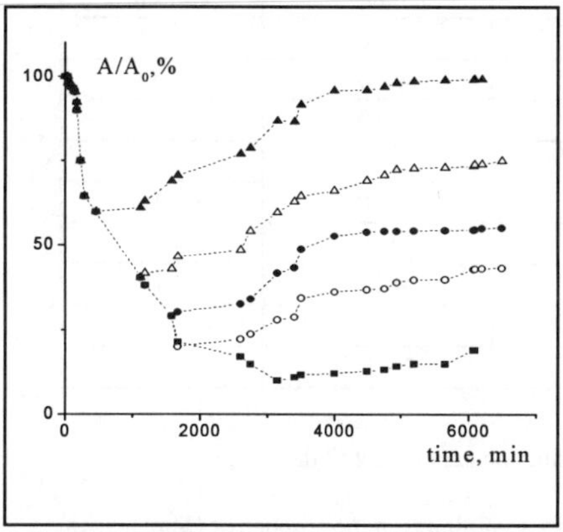

Fig. 2. Reactivation of the luciferase after the removal of GdmCl, 8°C

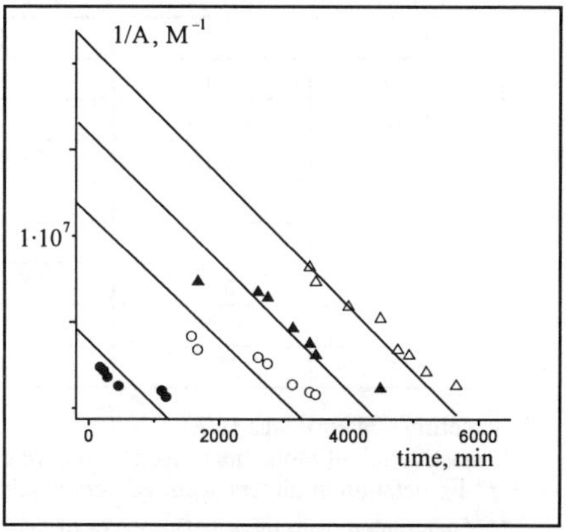

Fig. 3. Kinetics curves of reactivation in 2nd order reaction coordinates

<u>Determination of kinetic constants</u> Kinetic constants for the inactivation of luciferase with guanidinium chloride (scheme 1, Table 1) were determined by the equations for dissociative inactivation of oligomeric enzymes.[4] The value of k_{dis} was calculated from the plot of luciferase activity *vs* time at the initial period of inactivation.

$$2\left(\frac{a}{a_0}\right)-\frac{1}{2}\left(\frac{a}{a_0}\right)^2=\frac{3}{2}-k_{dis}t \qquad \text{(eqn 1)}$$

The constant of association of monomers to dimers (k_{ass}) is equal to the reactivation constant (Fig. 3, $2k_{ass} = \text{tg}\alpha$). At low concentrations of GdmCl, where one exponent was observed, $K_{dis} = k_{dis}/k_{ass}$. At high GdmCl concentrations, where two exponents were present, equilibrium dissociation constant was calculated from the coordinates of the inflection point τ:

$$K_{dis}=4[E]_0\left(\frac{a_0-a_\tau}{a_0a_\tau}\right)^2 \qquad \text{(eqn 2)}$$

The value of inactivation constant (k_{in}) was determined under the conditions when dissociation was practically completed ($t > \tau$):

$$k_{in}=\frac{k_{in,eff}(a_0+a_\tau)}{2(a_0-a_\tau)} \qquad \text{(eqn 3)}$$

where $k_{in,eff}$ is the slope of the curve in semilogarithmic coordinates at high degrees of inactivation.

Table 1. Kinetic parameters of GdmCl-induced inactivation of *L. mingrelica* firefly luciferase at 8°C* (scheme 1)

[GdmCl], M	K_{diss}, M	k_{dis}, min^{-1}	k_{ass}, min^{-1},M^{-1}	k_{in},min^{-1}
0.016	$0.4 \cdot 10^{-8}$	$2.5 \cdot 10^{-4}$	$6 \cdot 10^{4}$	—**
0.04	$0.4 \cdot 10^{-8}$	$2.5 \cdot 10^{-4}$	$6 \cdot 10^{4}$	—**
0.32	$1.0 \cdot 10^{-8}$	$6.2 \cdot 10^{-4}$	$6 \cdot 10^{4}$	—**
0.5	$1.1 \cdot 10^{-8}$	$6.6 \cdot 10^{-4}$	$6 \cdot 10^{4}$	—**
0.8	$0.9 \cdot 10^{-6}$	$1.6 \cdot 10^{-3}$	$1.8 \cdot 10^{3}$	$3.2 \cdot 10^{-4}$
1	$4.9 \cdot 10^{-6}$	$1.5 \cdot 10^{-2}$	$3.0 \cdot 10^{3}$	$2.4 \cdot 10^{-4}$
1.5	$6.5 \cdot 10^{-5}$	$2.6 \cdot 10^{-2}$	$4 \cdot 10^{2}$	$2.8 \cdot 10^{-4}$
2	—***	—***	—***	$\sim 1.4 \cdot 10^{-3}$
0****	$0.38 \cdot 10^{-8}$	$1.1 \cdot 10^{-2}$	$2.9 \cdot 10^{6}$	—**

*experimental error was 10%;
**inactivation of monomers was not observed during the experiment (3 days);
***dissociation of dimers occurred very quickly;
****calculated from the kinetic curves of dissociation of dimers in the absence of GdmCl.[2]

Taking into account kinetic and fluorescence data we propose the general scheme of GdmCl-induced inactivation of the luciferase (scheme 2). It involves the step of conformational changes of the dimer form of the luciferase that occur without change in enzymatic activity (lag-period in the time-course of inactivation):

$$E_2 \rightleftharpoons E'_2 \underset{k_{ass}}{\overset{k_{dis}}{\rightleftharpoons}} I \xrightarrow{k_{in}} I' \qquad \text{(scheme 2)}$$

References

1. Herbst R, Gast K, Seckler R. Folding of firefly (*Photinus pyralis*) luciferase. Aggregation and reactivation of the unfolding intermediates. *Biochemistry* 1998; **37**: 6586–6597.
2. Ugarova NN, Dementieva EI, Lundovskikh IA. Thermoinactivation and reactivation of firefly luciferase. *This volume.*
3. Lundovskikh IA, Leontieva OV, Dementieva EI, Ugarova NN. Recombinant *Luciola mingrelica* firefly luciferase. Folding *in vivo*, purification and properties. In: Roda A, Pazzagli M, Kricka LJ, Stanley PE, eds. *Bioluminescence and Chemiluminescence. Perspectives for the 21 Century.* JohnWiley & Sons: Chichester, 1998: 420–424.
4. Poltorak OM, Chukhrai ES, Torshin IYu. Dissociative thermoinactivation, stability and activity of oligomer enzymes. *Biochemistry* (Moscow) 1998; **63**: 360–369.

LUCIFERASE KNOCK – OUT MUTANTS BY SITE DIRECTED MUTAGENESIS OF THE AMP BINDING SITE

GABRIELA ISPAS, I. FAMELAER, FATIMA ELMALKI AND M. JACOBS
Free University Brussels, Institute of Molecular Biology, Laboratory of Plant Genetics; B-1640 St Genesius Rode, Belgium
E-mail: ifamelae@vub.ac.be

1. INTRODUCTION

Bioluminescence is a subset of chemiluminescence, where the light-producing chemical reaction occurs inside an organism. In one of the major luminescence systems the reaction is catalyzed by specific enzymes called luciferases, and the substrate is generically called luciferin.The firefly luciferase is used in a luciferin-luciferase system that requires ATP as a cofactor and consequently it can be used as a bio-indicator of the presence of energy or "life".

The *Photinus pyralis* luciferase has been intensively studied and the ATP binding domain, the peroxisome tagging sequence as well as some of the amino acids involved in the mechanism of light emission, have been identified. Mutagenesis studies have been reported, leading to a further understanding of the luciferase structure and function. [1-2] Beside the fact that the structure of the firefly luciferase is well known the mechanism of light emission still poses questions. The emission of different light colors, as a function of different pH and /or single amino acid substitutions and the short life time of the reaction make the mechanism of light emission difficult to reveal.

Our study focuses on the AMP binding domain located at the position 194-208 in the *Photinus pyralis* luciferase. By site directed mutagenesis a cassette has been introduced in the AMP binding site of the engineered luciferase and the mutants obtained have been analyzed.

2. STRATEGY OF SITE DIRECTED MUTAGENESIS

The AMP binding domain is a 2-element fingerprint that provides a signature for the STG-rich domain of AMP binding proteins, and it is located in the region of the beta hairpin (PDB - 1ba 3; SWS – P08659) between two beta sheets.

This region is highly conserved among the enzymes acting via an ATP dependent covalent binding of AMP to their substrate.

Photinus pyralis luciferase	ALIMNSSGSTGLPKGVA
Luciola lateralis luciferase	ALIMNSSGSTGLPKGVO
Luciola migrelica luciferase	ALLMNSSGSTGLPKGVR
Hotaria parvula luciferase	ALLMNSSGSTGLPKGVR
Populus tremuloides - CoA ligase	VALPYSSGTTGLPKGVM
Vanilla planifolia - CoA ligase	VALPYSSGTTGLPKGVM

Fig.1 The conserved AMP binding domain in ATP dependent enzymes

The mutagenesis cassettes for the two mutants contain a XhoI site, which is alternatively introduced for the first mutant (M1) and destroyed for the second mutant (M2).

WT	ALIMNS**SG**STGLPKGVA
M1	ALIMNS**LE**STGLPKGVA
M2	ALIMNS**SE**STGLPKGVA

Fig.2 Mutagenesis cassette

In a first aproach two amino acids have been mutated: S199 to L and G200 to E. In this way a XhoI site has been created in the mutant M1. Consequently a knocked-out luciferase has been obtained. In an attempt to assess which of the two mutated amino acids are essential for the switching of the luciferase activity, the S199 has been restored and a second luciferase mutant M2, conserving the mutation G200 to E has been produced. In the second mutant the XhoI site has been destroyed. In the second mutant it has been confirmed that G200 to E is the null mutation responsible for the knock-out of the luciferase activity. The introduced mutations have been confirmed by mapping of the XhoI site and DNA sequencing for both mutants.

3. MEASUREMENT OF LUCIFERASE ACTIVITY IN WILD TYPE AND MUTANTS

The Luciferase Assay System kit supplied by Promega was used for light measurement. No detectable light has been measured for the luciferase mutants. The Promega vectors pSP —Luc$^+$ and pGEM — Luc have been used as expression vectors.

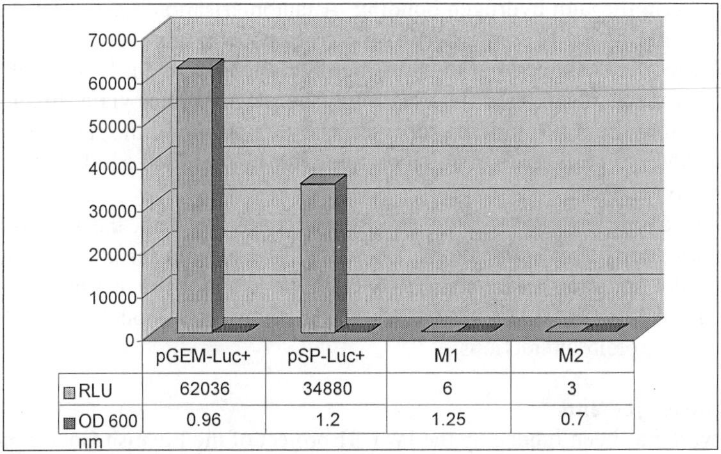

Fig.3 Luciferase activity in mutants and wild type
RLU = relative light units; OD 600 nm = OD of bacteria culture at 600 nm

4. DISCUSSIONS

Studies of mutagenesis located in the AMP binding site of luciferase have been reported by Thompson JF et al.[3] Highly conserved residues including S198, S201, T202 and G203 could be mutated with little effect on the luminescent activity of *Photinus pyralis* luciferase.

The crystal structure of luciferase has been solved and so far two protein entries might be found in PDB, coresponding to luciferase - E.C. 1.13.12.7.[4,5] In both X-ray cristals a gap at the level of S199 and G200 has been found. These gaps might be explained by the high flexibility of the beta hairpins including S199 and G200.

It is difficult to understand why a single amino acid mutation at the level of AMP binding site causes such a dramatic change in luciferase activity. Workers are reporting similar results for GTP binding site. Krenzel U et al [6] have analysed a H – ras p21 mutant containing a point mutation G12 – R. In this case it has been proved that the larger side chain of R interferes with GTP binding and /or hydrolysis. In other cases it has been proved that glycine might have unusual dihedral angles that are forbidden for other amino acids.[7]

For our mutants we can not conclude so far which of the two mechanisms is responsible for the knockout of luciferase activity. Further mutagenesis studies for replacement of the long chain of E with a shorter chain are on the way.

A speculative mechanism for bioluminescence colour determination has been proposed by Branchini BR.[8] Conforming to this model H245, T343, R218 and AMP participate in maintaining a specific conformation of the keto form of excited state oxyluciferin through hydrogen bonding. A tautomerisation reaction of the keto form of oxyluciferin to enol emitter is the source of light emission.

The luciferase knock-out mutants highlight the essential role of G200 in the light emission mechanism. This function might be played either via a direct mechanism conditioning the AMP binding, or indirect by determining the right position for the bound AMP to participate in collaboration with H245, T343 and R218 to the hosting of keto-oxyluciferin.

Due to the versatility of the luciferase reporter system, the knock-out mutants might have different applications. They might be of help for a better understanding of the light emission mechanism; they could be used for the measurement of in vivo missense errors [9] and last but not least as a control in a model system for studying anesthetic-protein interactions. [10,11]

Acknowledgements
This work has been funded by the IWT 81 project of the Flemish Government.

References

1. White PJ, Squirrell DJ, Arnaud P, Lowe CR, Murray JA. Improved thermostability of the North American firefly luciferase: saturation mutagenesis at position 354. *Biochem J* 1996; **319**: 343-350
2. Ohmiya Y, Tsuji FI. Mutagenesis of firefly luciferase shows that cysteine residues are not required for bioluminescence activity. *FEBS Lett* 1997; **404**: 115-117
3. Thompson JF, Geoghegan KF, Lloyd DB, Lanzetti AJ, Magyar RA, Anderson SM, Branchini BR. Mutations of a protease-sensitive region in firefly luciferase alters light emission properties. *J Biol Chem* 1997; **272**: 18766-18771
4. Conti E, Lloyd LF, Akins J, Franks NP, Brick P. Crystallization and preliminary diffraction studies of firefly luciferase from *Photinus pyralis*. *Acta Cryst* 1996; **D52**: 876-878
5. Frank NP, Jenkins A, Conti E, Liels WR, Brick P. Structural basis for the inhibition of firefly luciferase by a general anesthetic.*Biophys J* 1998; **75**: 2205-2208
6. Krengel U, Schlichting I, Scherer A, Schumann R, Frech M, John J, Kabsch W, Pai EF, Wittinghofer A. Three-dimensional structures of H-ras p21 mutants: molecular basis for their inability to function as signal switch molecules. *Cell* 1990; **62**: 539-548
7. Branchini BR, Magyar RA, Murtiashaw MH, Anderson SM, Helgerson LC, Zimmer M. Site-directe mutagenesis of firefly luciferase active site amino acids: a proposed model for bioluminescence color. *Biochemistry* 1999; **38**: 13223-13230
9. Cornut B, Willson RC. Measurement of translation accuracy in *vivo*: missense reporting using inactive enzyme mutants. *Biochimie* 1991; **73**: 1567-1572
10. Franks NP, Jenkins A, Conti E, Lieb WR, Brick P. Structural basis for the inhibition of firefly luciferase by a general anesthetic.*Biophys J* 1998; **75**: 2205-2211
11 Moss GW, Franks NP, Lieb WR. Modulation of general anesthetic sensitivity of a protein: a transition between two forms of firefly luciferase. *Proc Natl Acad Sci USA* 1991: **88:** 134-138.

SPECTRAL TUNING OF BIOLUMINESCENCE EMISSION AND VISUAL SENSITIVITY AMONG BEETLES

ABNER B. LALL
Department of Biology, Howard University
Washington, D.C. 20059, USA
E-mail:alall@fac.howard.edu

ALEXANDRE A. CARVALHO, AND JOHN M. de SOUZA
Instituto de Psicologia, Departamento de Psicologia Experimental,
Universidade de São Paulo, São Paulo, SP, BRASIL
E-mail: johns@usp.br

ETELVINO J. H. BECHARA
Instituto de Quimica, Departamento de Bioquimica, Universidade de São Paulo,
São Paulo, SP, BRASIL
E-mail: ebechara@iq.usp.br

THOMAS W. CRONIN
Department of Biological Science, University of Maryland Baltimore County,
Baltimore, MD, 21250, USA
E-mail: cronin @umbc.edu

1. INTRODUCTION

Von Uexkull[1] showed that in instinctive behavior, the receptor and the effector systems are dovetailed to each other. Such a relationship is described here as tuning between the electroretinographic visual spectral sensitivity $\{S(\lambda)\}$ and bioluminescence (BL) emission spectra among beetles which utilize BL optical signaling between the sexes for mating.[2] The knowledge of the spectral characteristics of vision in fireflies has been pivotal in unraveling the enigma of the presence of different colors of bioluminescence ranging from lime-green to bright yellow among North American fireflies.[3] A mathematical optimization model[4,5] based on signal-to-noise considerations for the detection of BL accounts for the evolutionary selection of the colors of BL among lampyrids. Here we present data summarizing the spectral tuning which exists between BL emission and visual sensitivity among lampyrids[3-7] and elaterids, an example of where co-evolution between sensory and motor systems has occurred.

2. MATERIALS AND METHODS

Detailed descriptions of the collection of different species of beetles,[7] the construction of the optical system, the methods for recording electroretinogram (ERG), the electronics, the calculation of spectral sensitivity curves,[6] *in situ* microspectrophotometric (MSP) absorbance measurements of visual and screening pigments[6,7] from the compound eyes, and measurement of the bioluminescence emission spectra[5] from the lanterns are presented in our earlier papers.

3. RESULTS AND DISCUSSION

Based upon the time of initiation of flashing activity, North American firefly species have been divided into two groups: "twilight-active" species, which are those beginning flashing activity in advance of 30 minutes after sunset, and "night-active" species, beginning after this.[3] In general night-active species (e.g., *Photuris versicolor* and *P. lucicrescens*) emit green BL and the peaks of their BL match their broad green ERG S(λ) functions with peaks around 555 nm. Vision in the green is mediated by P545 rhodopsin in conjunction with a yellow screening pigment absorbing maximally in the blue with a broad peak at 456 nm.[7] The twilight-active fireflies have narrow ERG S(λ) functions shifted toward longer wavelengths such that the shape of the species' yellow BL emission and their species' yellow S(λ) match (Fig.1b, c).[3-7] The presence of species-specific screening pigments with narrow absorption in the blue-green in conjunction with P545 rhodopsin, mediate vision in most twilight-active fireflies (e.g., *Photinus pyralis*). However, the twilight-active *P. scintillans* is the first species to initiate flashing at dusk and it emits a bright yellow flash with its BL peak at 579 nm), the longest lambda among the North American fireflies. In this species, selection has favored P557 rhodopsin[7] in conjunction with an orange colored screening pigment[7] (λ_{max}=525 nm) to mediate vision in the long wavelengths as compared to other night- and twilight-active fireflies in which vision is mediated by P545 rhodopsin.[7]

The S(λ) functions in *Pyrophorus divergens* and *Pyrearinus termitilluminans* have broad peaks in the green with a shoulder in the near-uv (Fig. 2a, b) while a narrow yellow S(λ) function exists in *Fulgeochlizus bruchii* (Fig. 2c). In *P. divergens* and *P. punctatissimus*, vision is mediated by a P540 rhodopsin in conjunction with a screening pigment broadly absorbing in the blue (peak = 463 nm) located in the ventral sector of the eye. In these beetles the BL emission of the dorsal prothoracic lanterns is green (λ_{max} = 535 to 546 nm) and that of the ventral abdominal lantern is green to lime-green (λ_{max} = 543 to 552 nm) in color and would be detected by the insect's visual system with ease. In summary, then, among bioluminescent beetles, vision is shown to be mediated by photopigments (P540 to

P557) in conjunction with screening pigments (λ_{max} = 456-525 nm). These beetles possess superposition compound eyes adapted to night vision.

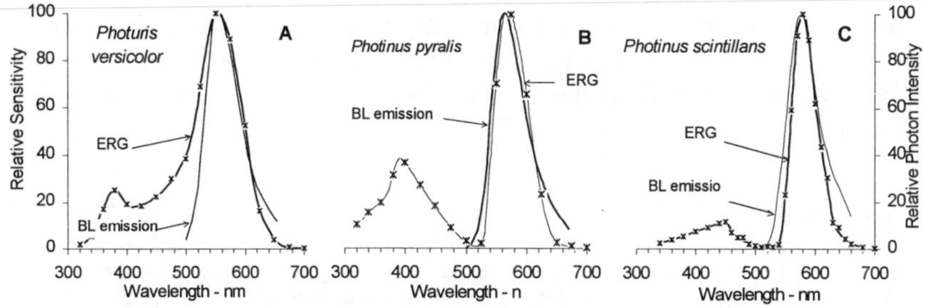

Fig. 1. A comparison of ERG spectral sensitivity and species BL emission for night-active *Photuris versicolor* (λ_{max} = 552 nm)(A) and twilight-active *Photinus pyralis* (λ_{max} = 563 nm)(B) and *P. scintillans* (λ_{max} = 579 nm)(C).

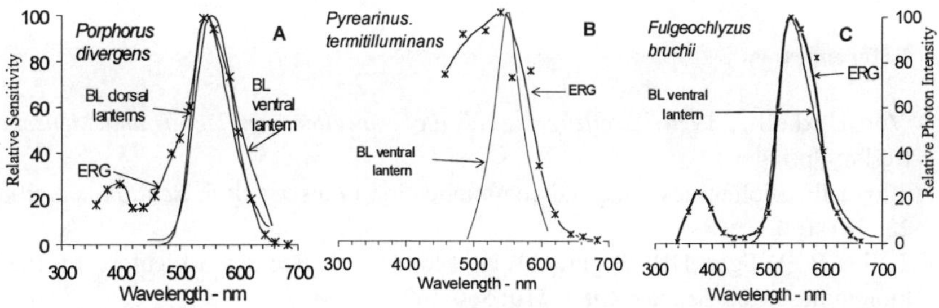

Fig. 2. A comparison of ERG spectral sensitivity and species BL emission for *Pyrophorus divergens* (peaks for dorsal lantern = 546 nm and ventral lantern = 558 nm)(A), *Pyrearinus termitilluminans* (ventral lantern peak = 552 nm)(B) and *Fulgeochlizus bruchii* (ventral lantern peak = 543 nm)(C).

The fireflies inhabit green foliage habitats rich in green photons reflected from the foliage. The selection of rhodopsin maximally absorbing in the green would provide high absolute sensitivity under limited light conditions at night (Sensitivity Hypothesis of Visual Ecology).[5] Therefore the selection of green BL signals would optimize signal detection. Yellow BL is a contrast signal against green foliage background[4,6] and the narrow S(λ) among the twilight-active lampyrids (λ_{max}=560-579 nm lies at wavelengths with minimum amount of photons in the environment (Contrast Hypothesis of Visual Ecology).[3,4] The species-specific screening pigments shift the SS functions mediated by P545 and P557 into long wavelengths (λ_{max} = 555-580 nm) and narrow the S(λ) such that the shapes of species ERG S(λ) and BL match functions among twilight-active firefly species. The species-specific screening pigments act as noise filters to maximize signal-to-noise ratios by absorbing green foliage-reflected sunlight at dusk among twilight-active fireflies and short wavelength space light in the early hours of the night among night-active beetles which possess broad green ERG S(λ) functions. The BL emission and the species visual spectral sensitivity have been selected for mutual tuning among bioluminescent beetle.

Acknowledgments This paper is dedicated to Professor Howard H. Seliger for his recognizing the significance of A. B. Lall's early studies and subsequently nurturing the development of the theme. A. B. Lall is also grateful to Professor Seliger and William H. Biggley for providing the bioluminescence emission spectra of fireflies and ideas in ecology and evolution. C. V. Stevani, M. P. Barros and V. R. Viviani provided bioluminescence emission spectra of the click beetles. Support: NSF grant # IBN-9724028 to TWC, NIH grant # GO6GM08016 to ABL, FAPESP grants to JMS & EJHB, CNPq grants to AAC, JMS & EJHB and FINEP grant to JMS.

References

1. Von Uexkull J. 1934. *Streifzuge durch die Umwelten von Tieren und Menschen.* Berlin: Springer.
2. Lloyd JE. Bioluminescence and communication in insects. *Ann Rev Entomol* 1983; **28**: 131-160.
3. Lall AB, Seliger HH, Biggley WH, Lloyd JE. Ecology of colors of firefly bioluminescence. *Science* 1980; **210**: 560-562.
4. Seliger HH, Lall AB, Lloyd JE, Biggley WH. On the colors of firefly bioluminescence. I. An optimization model. *Photochem Photobiol* 1982; **36**: 673-680.
5. Seliger HH, Lall AB, Lloyd JE, Biggley WH. On the colors of firefly bioluminescence. II. Experimental evidence for the optimization model. *Photochem Photobiol* 1982; **36**: 681-688.
6. Lall AB, Strother GK, Cronin TW, Seliger HH. Modification of spectral sensitivities by the screening pigments in the compound eyes of twilight-active fireflies (Coleoptera: Lampyridae). *J Comp Physiol A* 1988; **162**: 23-33.
7. Cronin TW, Järvilehto M, Weckström M, Lall AB. Tuning of photoreceptor spectral sensitivities in fireflies (Lampyridae). *J Comp Physiol A* 2000; **186**: 1-12.

MATING BEHAVIOR OF A SYNCHRONOUS NORTH AMERICAN FIREFLY

A. MOISEFF

Department of Physiology & Neurobiology, University of Connecticut, 3107 Horsebarn Hill Road, Storrs CT 06269-4156, USA
E-mail:moiseff@uconn.edu

J. COPELAND

Department of Biology and Applied Coastal Research Laboratory, Georgia Southern University, Statesboro GA 30460-8042, USA
E-mail:copeland@gasou.edu

M. F. KUBKE

Department of Zoology, University of Maryland, College Park MD 20742, USA
E-mail:mk145@umail.umd.edu

L. FAUST

11828 Couch Mill Road, Knoxville TN, USA
E-mail:tnlfaust@cs.com

1. INTRODUCTION

Photinus carolinus males produce pulse trains of 5—8flashes with an interflash interval of approximately 500 msec. The pulse trains are repeated once every 10—12 seconds. Groups of *P. carolinus* males produce this pattern synchronously.[1] Synchrony of the flashes within the pulse train also results in a period during which there is not any flash activity (i.e., darkness). We have been studying synchrony as it relates to the mating behavior of this species.

2. MATERIALS AND METHODS

Our study site was located within the boundaries of the Great Smoky Mountains National Park at Elkmont, TN (USA). Fireflies were observed in their natural habitat and/or captured for study in the laboratory.

Behaviors were observed using low—lightvideography (Gen-III night vision system or SONY 0-Lux digital video system). The night vision system was preferred for viewing bioluminescent activity (flashes), whereas the 0—Luxvideo system, combined with infrared illumination, was preferred when viewing the postural behavior of fireflies.

An RCA–1P21 phototube was used to record flashes that were then digitized at 1 kHz. Artificial flash stimuli were synthesized by computer and delivered through a green LED.

Fireflies were placed in 10 cm. petri dishes, small glass terrarium or plastic bags for laboratory experiments.

3. RESULTS AND DISCUSSION

In response to male flashes receptive female *P. carolinus* produced one or two 'doublet' flashes (Fig. 1).

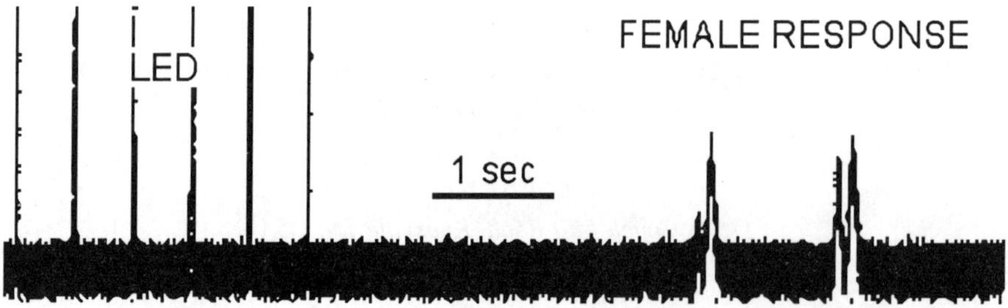

Fig. 1 Response of caged *Photinus carolinus* female. In this photometer recording a caged female (i.e., isolated in a 10 cm petri dish) was stimulated by a green LED emitting a computer-synthesized burst of 6 flashes. The female responded with 2 doublet flashes.

In the field and the laboratory, individual females attracted groups of males that formed "clusters" (Fig. 2). Clusters consisted of males in close proximity (often in contact with) the female as well as males at a greater distance from the female. We defined the former males as "nuclear males" and the latter as "satellite males". During the 1999 field season, we observed an average of 6.7±3.1 males per cluster under natural, field conditions (N=6 clusters).

Nuclear males often form stacked piles (Fig. 2). Males comprising the pile extend and twist their abdomens in an apparent attempt to contact the female. We have also observed males attempting to push into the bottom of the pile. These behaviors may represent forms of male–male competition to mate with a receptive female. Ultimately, a single male couples with the female and the cluster dissolves. The coupling male and female cease all flashing and remain coupled.

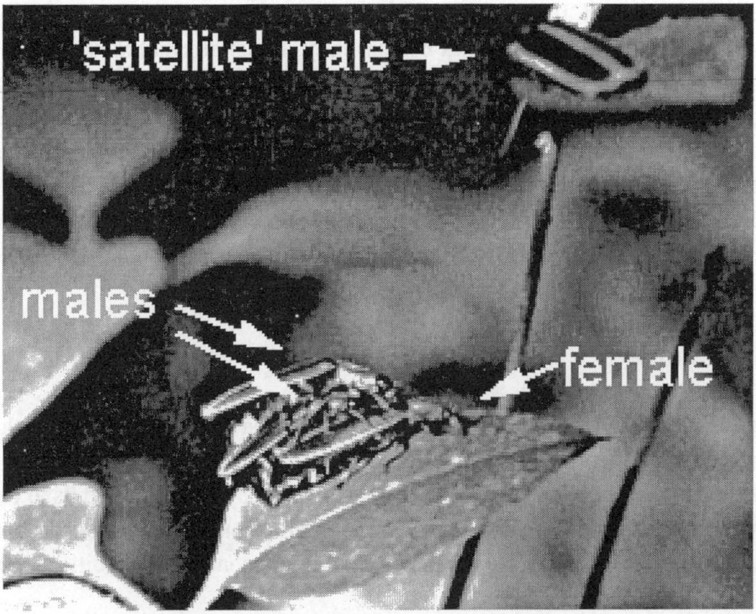

Fig. 2 *Photinus carolinus* mating cluster. Photograph of mating cluster under natural field conditions. From this perspective one of the 'satellite' males is visible.

In experiments designed to determine the basis of cluster formation, an individual receptive female was sealed in an 8 mL glass vial (with rubber lined closure) and placed in a terrarium with 8–10 males. The female responded to the male flashes with doublet flashes (in a manner similar to experiment shown in Fig. 1) triggering the formation of a cluster. Although the males were unable to physically contact the female, they formed what appeared to be natural-like clusters with satellite males surrounding a nucleus of males that formed and reformed piles. This suggested that the female's flash response is sufficient to trigger cluster behavior, at least under experimental conditions where a high concentration of males were in close proximity to a female. This also suggested that female pheromonal cues were not necessary for cluster formation under these conditions. These results were confirmed by using an LED to mimic a female–like response in a terrarium that did not contain any females. When the LED was used to respond to male flashes, males oriented to the LED and formed characteristic clusters.

We have observed at least two forms of male flash patterns associated with clusters: "coincident synchrony" and "serial synchrony". Both of these flash interactions consist of flashes occurring synchronously. In coincident synchrony males tend to terminate their flashes concurrently. In serial synchrony, one or more males tend to continue flashing after other males have ceased flashing. This has the effect of extending the length of the population s flash train. Stimulus—response studies where females were presented with male-like flash trains of different lengths suggest that a female times her doublet-flash response from the final flash of a male-like flash train. Thus, serial synchrony may represent a form of male—male competition insofar as a receptive female will time her response relative to the male that flashes last.

4. CONCLUSION

Mating clusters provide us with an experimental arena for studying and evaluating the relative roles of male-male competition and female-choice in mating behavior, as well as the mechanisms underlying these behaviors. They have enabled us to begin a systematic analysis of the role of synchronous and non—synchronousflashing on *Photinus carolinus* mating behavior under natural and controlled conditions.

ACKNOWLEDGEMENTS

We thank Jennifer Maupin for her assistance in this project and the National Park Service (GSMNP), especially Keith Langden, for permission to work at the Elkmont site and numerous kindnesses. We thank the Great Smoky Mountain Natural History Association, National Geographic Society, Georgia Southern University and the University of Connecticut Research Foundation for their financial support.

References

1. Copeland J, Moiseff A. The occurrence of synchrony in the North American firefly *Photinus carolinus* (Coleoptera: Lampyridae). *J Insect Behav* 1995;**8**:381-394.

PROTEIN ENGINEERING AND MOLECULAR MODELLING OF FIREFLY LUCIFERASE

SE PREBBLE[1]; RL PRICE[1]; B LINGARD[1]; LC TISI[2] AND PJ WHITE[3]

[1]DERA Porton Down, Salisbury, SP4 0JQ, UK, [2]Institute of Biotechnology, Tennis Court Road, Cambridge, CB2 1QT, UK. [3]Babraham Institute, Cambridge, CB2 4AT, UK.

1. INTRODUCTION

Firefly luciferase has great potential as a bioluminescent label, exhibiting greater specificity, sensitivity and localisation than other labels used in immunoassays and nucleic acid detection techniques. However, the inherent lack of stability of the native luciferase enzyme is a factor that must be overcome in order to fully exploit this potential, both for chemical coupling procedures and for its use in the various assay systems. Greater stability at ambient temperature and a generally more robust enzyme would also be useful for reagent production for assays in which firefly luminescence is used to detect ATP/microbial contamination. The native enzyme's K_m (ATP) and thermal stability have been altered using protein engineering, introducing amino acid substitutions by genetic modification. Protein modelling has allowed possible structure/function relationships to be explored.

2. MATERIALS AND METHODS

Starting with the luciferase plasmid pPW6012[1] mutant luciferases were generated by error-prone PCR.[2] Resultant plasmids were transformed into *Escherichia coli* DH5α and initially screened *in vivo* for thermostability. Luminescent colonies expressing more thermostable luciferase than wild-type were selected and cultured, plasmids extracted and sequence information obtained. Plasmids were heat shock transformed into *E.coli* BL21DE3 pLysS, a low protease strain (Promega). Transformants were cultured with appropriate antibiotics to maintain plasmids. Luciferases were purified by the method described in White *et al.*,[1] with the addition of a gel filtration step. Protease inhibitor cocktail tablets (Boehringer Mannheim) were added to the dialysed luciferases before storage at -20°C for later use. Thermostability trials were conducted over a range of temperatures with the enzyme diluted in 10mM HEPES buffer containing two protease inhibitors tablets per 10mL. Luciferase assay reagent (Promega), diluted to 10% in 10mM HEPES, was used to assay luciferase activity. K_m studies were conducted in 20mM Tricine buffer (pH8.0) containing 0.1mM DTT, 0.1mM EDTA, 0.5% bovine serum albumin, 15mM magnesium acetate and 0.1mg/mL luciferin or 1mM ATP for K_m (ATP) and K_m (luciferin), respectively. All bioluminescent measurements were performed using a Biotrace Multilite luminometer. Unless otherwise stated, all reagents were obtained from Sigma-Aldrich (Poole, UK). Protein modelling was conducted with SYBYL software ver. 6.4 (Tripos Associates Ltd., USA) and all energy calculations were carried out using the TRIPOS force-field.[3]. The crystal structure of luciferase complexed with bromoform (pdbcode:1BA3)[4] was used as the starting structure. This structure had fewer disordered regions than the unbound structure (pdbcode:1LCI). Loop building was carried out to construct the disordered region, residues 198-201. Missing sidechains were constructed using SYBYL's BIOPLOYMER tools and the structure was subjected to energy minimisation. Analysis by sidechain conformational searching, packing, interactions, and consideration of hydrophobicity and secondary structural preferences were carried out. A Connolly molecular surface was generated, allowing the identification of a cleft of sufficient shape and size for the potential binding of coenzyme A. An iterative manual and computational approach was used to construct 3' adenine phosphate into the pocket using the DOCK and FLEXIDOCK utilities of SYBYL.

3. RESULTS AND DISCUSSION

Thermostability trials conducted on the various luciferase enzymes at 30°C, 37°C, 40°C and 45°C showed that combining thermostable point mutations had an additive effect on the stability of the enzyme (data not shown). K_m values for both of the substrates were also determined, using glow rather than flash kinetics.

Table 1: K_m (ATP) and (luciferin) values for luciferase under stated buffering conditions

Luciferase Variant	ATP K_m (μM)	Luciferin K_m (μM)
Native luciferase	94.5	200
E354K/A215L	123.8	200
E354K/E270K	53.5	200
T214A/A215L/E354K/I232A/F295L	166.0	200
F14A/L35A/I232A/	100.0	200
T214A/I232A/F295L/E354K/E270K	120.0	200

Of the residues that were modified, T214C/A/N, A215L and F295L were buried within the enzyme. E354K, I232A, F14A, L35A and E270K are all surface accessible to some degree. Each amino acid substitution was modelled using

SYBYL to explore structure/function relationships. For example, T214C, N and A substitutions were all more thermostable than wild-type, with cysteine conferring the highest improvement. Each substituted residue appeared to improve the surface packing of the enzyme. Of the other residues examined; F14A, L35A, A105V, I232A, F295L, H310R, E354K, D234G and S420T, improved packing by reducing the amount of buried hydrophilic surface was the most likely explanation for the observed thermostabilising effects, along with the concomitant displacement of buried water molecules. However, some of the stabilising substitutions (4/10) were also found to have a greater propensity for the secondary structure in which they were found compared with the wild-type residues, which may also have been a contributing factor. Combinations of several thermostable point mutations resulted in luciferases that were incrementally more stable as further mutations were added. The magnitude of the additive effect decreased as more mutations were included. The K_m (ATP) increased simultaneously with additional point mutations, and could reduce the sensitivity of the ATP assay, as more ATP would be needed to produce half-maximal activity of the enzyme. Introducing the E270K mutation, which decreases the K_m (ATP) relative to wild-type, into a multiple thermostable enzyme brought the ATP K_m back towards that of wild-type, thus providing an enzyme that was thermostable and yet retained wild-type sensitivity. However, based on known proteolytic cleavage sites, a lysine at position 270 may increase the enzyme s susceptibility to proteolytic degradation compared with the wild-type enzyme, and may have contributed to its apparent instability. The K_m for luciferin was examined and showed that, under these buffering conditions, the K_m (luciferin) for the wild-type enzyme and examined mutants is ~200μM. Existing reagent formulations in which the luciferin concentration has been optimised will not therefore need to be changed in order to accommodate the variant luciferases luciferin requirements.

The luciferase structure used in the modelling study was complexed with bromoform, located in a pocket bordered by residues D357, E354 and H310, which have all been mutated to improve thermostability. The bromoform pocket measured 8-10 x ~6 Angstroms (wide/deep), with the bromoform removed. A second, adjacent pocket measured ~12 x ~8 Angstroms. Residues F247, Y255, R337 and R218 line the bottom of this pocket. Residues H310 and E354 lie close enough that the 1.4 Angstrom probe used in the Connolly surface generation cannot pass between them, thus they appear to form a bridge separating the cleft into two pockets. However, conformational searching showed many energetically possible conformations for these sidechains. With free space beneath the bridge it is possible that the bridge may move to allow coenzyme A to bind in the cleft. A phosphate probe was used to locate a favourable energy well within the deep pocket, from which construction of the 3 adenine phosphate was begun. From this position, it

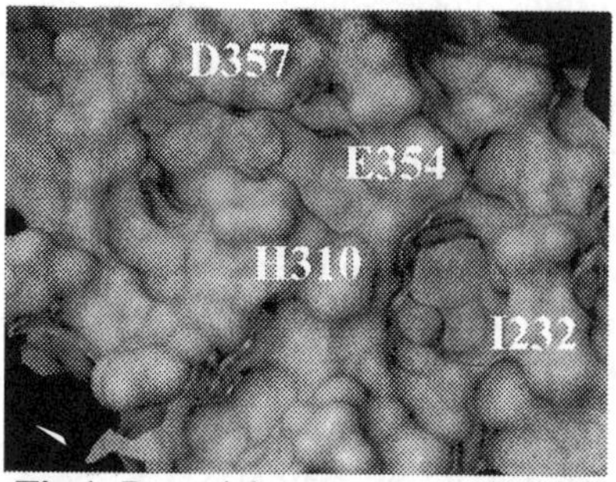

Fig.1: Potential coenzyme A cleft

would be possible for the phosphate moiety to make favourable interactions with Y255, R218 and R337. Conservation of Y255 and R218 is found across 17 species of firefly and click beetle, using a CLUSTAL multiple alignment.[5] Residues R218 and F247 have been implicated in the luciferin binding site[6] and might therefore interact with both the luciferin and coenzyme A sites simultaneously, if CoA could indeed bind here. The pantothenic acid and β-mercaptoethylamine part of CoA were not fitted due to time constraints, but the dimensions of the cleft and regions of polarity within it suggest that this may be possible. The theoretical CoA binding cleft suggested here was supported by experimental data (PJW, data not shown) in which coenzyme A shifted the peak wavelength both *in vivo* and *in vitro* according to its concentration when assayed with the variant luciferase D357Y. This region of the protein had not thus far been examined with regard to coenzyme A binding but a reasonable set of rotamers is conducive to this, and further mutagenesis could therefore be aimed at confirmation or rebuttal of this propostition.

References

1. White PJ, Squirrell DJ, Arnaud P, Lowe CR, Murray JAH. Improved thermostability of the North American firefly luciferase: saturation mutagenesis at position 354. *Biochem J* 1996; **319**: 343-350.
2. Fromant M, Blanquet S, Plateau P. Direct random mutagenesis of gene-sized DNA fragments using polymerase chain reaction. *Anal Biochem* 1995; **224**: 347-353.
3. Clark M, Cramer RD, Van Opdenbosch N. *J Comput Chem* 1989; **10**: 982-1012.
4. Conti E, Franks NP, Brick P. Crystal structure of firefly luciferase throws light on a superfamily of adenylate-forming enzymes. *Structure* 1996; **4**: 287-298
5. Higgins DG, Sharp PM. Fast and sensitive multiple sequence alignments on a microcomputer. *CABIOS* 1989; **5**: 151-153.
6. Branchini BR, Magyar RA, Murtiashaw MH, Anderson SM, Zimmer M. Site-directed mutagenesis of histidine 245 in firefly luciferase: a proposed model of the active site. *Biochemistry* 1998; **37**: 15311-15319.

LUCIFERASE IMAGING REVEALS DISTINCT PATTERNS OF GENE REGULATION IN LIVE BRAIN SLICES

LAURA A SIGWORTH, TIFFANY R. CHANDLER, LUJIAN LIAO, AND MICHAEL E. GEUSZ

Department of Biological Sciences, Bowling Green State University
Bowling Green, OH 43403, USA
E-mail: mgeusz@bgnet.bgsu.edu

1. INTRODUCTION

Bioluminescent reporter genes are commonly used in assays made from cell lysates to determine the level of gene expression at the time of the assay. Similar experiments in live cells are not as common even though they provide continuous measurements of ongoing gene expression. Recently, fluorescent reporter genes based on the green fluorescent protein have gained rapid and widespread popularity for use in live preparations. Despite the relative ease of use and flexibility of these fluorescent assays, bioluminescent reporter genes provide several clear advantages such as a lack of photobleaching and phototoxicity. The considerably lower photon flux of bioluminescence makes it suitable for use in light sensitive tissues, but it also requires more sophisticated equipment for imaging. Also unlike fluorescence, bioluminescence does not require the optics to deliver excitation light into cells and tissues. Nevertheless, the substrates for bioluminescent enzymatic reactions must be delivered to the cells and questions remain about whether their distribution limits or alters luminescent images in any substantial way.

To boost the faint signal from bioluminescent reporter genes, viral enhancer regions have been used to drive expression. In particular, the 5 upstream region of the cytomegalovirus immediate-early gene 1 (*CMV IE-1*) generates constitutive, high level expression. Recently, however, *CMV IE-1* was found to be repressed in liver,[1] selectively expressed in only certain cells of the brain,[2] and regulated through cAMP response elements and pathways involved in viral infection such as cytokines and NfkappaB.[3,4] These properties have raised concerns about reporter genes based on *CMV IE-1* when they are used as controls.

We propose that the pattern of reporter gene expression controlled by viral gene enhancer/promoters is an effective way to evaluate the relative activity of different signal transduction pathways in brain slice cultures. We compared images of gene expression induced through the multiple regulatory elements of *CMV IE-1* and also compared these signals with ones due to the promoter for the immediate-early gene c-*fos* that is used to indicate cell stimulation in the brain. In particular, we were interested in performing long-term imaging lasting days to evaluate control of gene expression by the circadian pacemaker in the mammalian hypothalamus.

2. METHODS

Two lines of transgenic C3H/B6 mice, one containing the promoter/enhancer from human *CMV IE-1* (CMV::luc) upstream from the American firefly luciferase gene and the other with human c-*fos* promoter upstream (fos::luc),[5] were anesthetized with Nembutal and immediately decapitated according to approved animal care guidelines. Brains were removed and placed in chilled artificial cerebrospinal fluid. Trimmed 200 m thick slices were maintained in culture in 1 or 0.1 mM luciferin (as sodium salt, Promega) in medium containing 10mM Hepes, 0.35g/L $NaHCO_3$, B27 supplement (Life Technologies), 100U/ml penicillin and 100mg/ml streptomycin in Dulbeco s Modified Eagle Medium with 10% newborn calf serum. Tissue was transferred to a Millicell-CM membrane (Millipore) pre-wetted with the culture medium and then sealed in a temperature-controlled chamber. Some cultures were also purged with oxygen. Slices were imaged using a 10X objective lens coupled to a liquid nitrogen-cooled, back-thinned CCD camera (CH360, Roper Scientific, 512 x 512 pixels) mounted on a vibration-damping table in a photographic darkroom. Images were analyzed with V^{++} software (Roper). Brightfield images were captured with light from a red light-emitting diode. All preparations containing luciferin were maintained under red light when handled.

3. RESULTS AND DISCUSSION

Brain slices made from CMV::luc mice showed discrete expression in particular cells throughout slices maintained at 30°C or 36°C. Signals were particularly high in cells of the dorsal suprachiasmatic nucleus (SCN) in the hypothalamus (Fig. 1A). The SCN is a major circadian pacemaker and controls daily rhythms in behavior and numerous physiological variables throughout the animal. Signal were often more persistent in the SCN relative to other areas, fading out late during recordings that lasted up to three days. Spontaneous cell excitation in the SCN may initiate this expression through the known Ca^{2+}/cAMP response elements of *CMV IE-1*.

Bioluminescence from brain slices made from fos::luc mice was weaker than with CMV::luc mice and was diffusely distributed across the cells except for high signals in the ependymal cells (Fig. 1B). This result confirmed previous imaging data collected with an intensified CCD camera that also showed expression throughout the SCN and surrounding areas.[5]

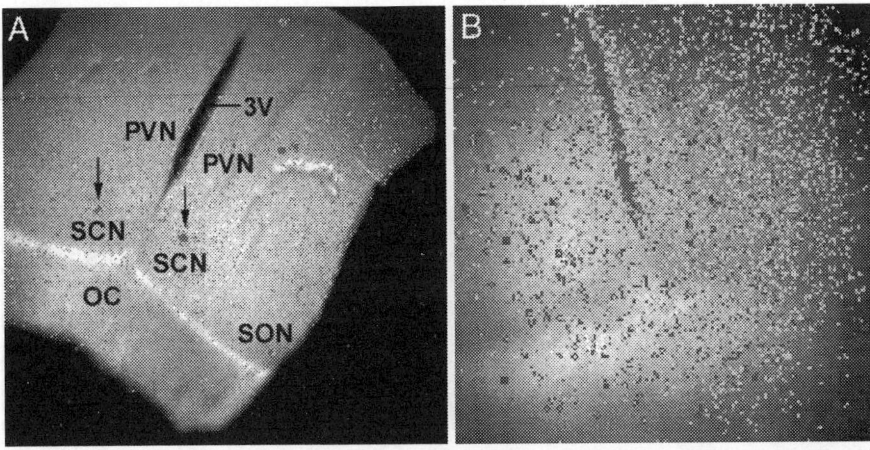

Fig. 1. Bioluminescence overlay (gray) on brightfield images. **A.** CMV::luc brain slice imaged at 30°C showing expression in the dorsal SCN (arrows) and in other nuclei such as the PVN and SON (30-min exposure.) **B.** Fos::luc brain slice imaged at 30°C showing diffuse expression throughout the tissue and high signals around the third ventricle. (1-h exposure.) Similar images of fos::luc slices were obtained at 22°C. SCN: suprachiasmatic nucleus. PVN: paraventricular nucleus of the hypothalamus. SON: supraoptic nucleus. OC: optic chiasm. 3V: third ventricle.

CMV::luc brain slices imaged at room temperature (22°C) and ones that were manipulated excessively before imaging showed a distinctly different pattern of expression from ones at higher temperature. In this case, the capillaries and the ependymal cells lining the third ventricle showed high expression (Fig. 2A). Methylene blue staining of these brain slices after imaging confirmed that expression was in capillaries (Fig. 2B). We interpret the expression pattern observed at room temperature or following mishandling as a response to stress. It is possible that cytokines released in the tissue trigger activation through known regulatory elements that respond to NFkappaB, a transcription factor involved in stress[6].

These results agree with ones from a transgenic mouse containing *CMV IE-1* upstream from green fluorescent protein[2], in which case only some of the cells of the brain expressed. This pattern has the advantage of allowing imaging of individual cells without interference from neighboring cells.

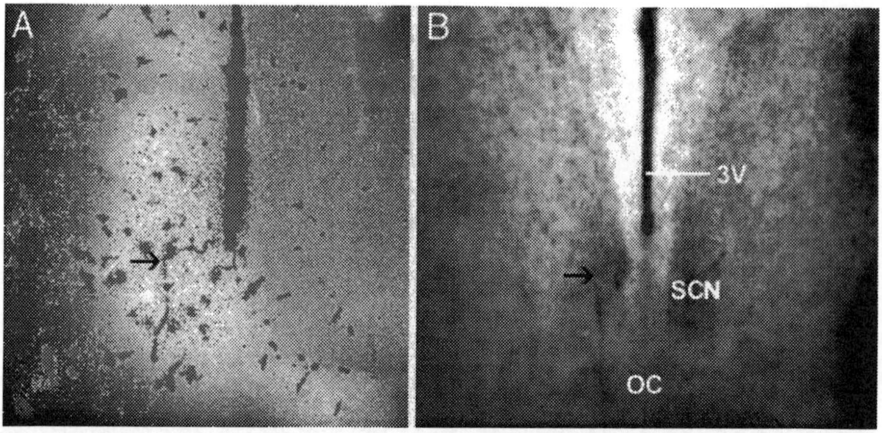

Fig. 2. CMV::luc brain slice imaged at 22°C. **A.** Bioluminescence image as overlay on brightfield image of a brain slice. (30-min exposure.) **B.** After staining with the vital dye methylene blue. Arrow shows the same capillary in both images.

Differences in the expression patterns between temperatures and between reporter genes indicate that luciferin diffusion or localization in the slices does not determine signal patterns. The unique spatial patterns at different temperatures suggests that these preparations may be used to assay various pharmacological treatments that act independently on stress and depolarization. *Supported by NIH grant 1 R21 RR12654-01A1*

References

1. Loser P, Jennings, GS, Strauss M, Sandig V. Reactivation of the previously silenced cytomegalovirus major immediate- early promoter in the mouse liver: involvement of NFkappaB. *J Virol* 1998;**72**,:180-190.
2. van den Pol A.N, Ghosh PK. Selective neuronal expression of green fluorescent protein with cytomegalovirus promoter reveals entire neuronal arbor in transgenic mice. *J Neurosci* 1998;**18**,:10640-10651.
3. Sambucetti LC, Cherrington JM, Wilkinson GW, Mocarski ES. NF-kappa B activation of the cytomegalovirus enhancer is mediated by a viral transactivator and by T cell stimulation. *EMBO J* 1989;**8**: 4251-4258.
4. Speir E, Yu ZX, Ferrans VJ, Huang ES, Epstein SE. Aspirin attenuates cytomegalovirus infectivity and gene expression mediated by cyclooxygenase-2 in coronary artery smooth muscle cells. *Circ Res* 1998;**83**: 210-216.
5. Geusz ME, Fletcher C, Block GD, Straume M, Copeland NG, Jenkins NA, Kay SA, Day RN. Long-term monitoring of circadian rhythms in c-fos gene expression from suprachiasmatic nucleus cultures. *Curr Biol* 1997;**7**, 758-766.
6. Maulik N, Sato M, Price BD, Das DK. An essential role of NFkappaB in tyrosine kinase signaling of p38 MAP kinase regulation of myocardial adaptation to ischemia. *FEBS Lett* 1998; **429**: 365-369.

MUTAGENESIS OF SOLVENT-EXPOSED HYDROPHOBIC RESIDUES IN FIREFLY LUCIFERASE

LAURENCE TISI, CHRISTOPHER LOWE AND JAMES MURRAY

Institute of Biotechnology, University of Cambridge, Tennis Court Road Cambridge, CB2 1QT, UK

1. INTRODUCTION

Firefly luciferase (FFL) is exceptionally hydrophobic in character. Recently, the molecular basis of the hydrophobicity of FFL was clearly demonstrated by its crystal structure.[1] The crystal structure highlights an unusually high number of large hydrophobic residues as being solvent exposed on the surface of the enzyme. For a (supposedly) mono-meric, soluble protein this is noteworthy.

Presented here are some observations of the nature of the solvent exposed residues in FFL as well as data concerning the effect of mutating a sub-set of these amino acids. The primary aim of undertaking this study was to try to identify mutants of firefly luciferase which were less prone to, in particular, thermal inactivation. It was hoped that by reducing the hydrophobic nature of the surface of FFL, that a version of the enzyme could be found with greater stability and a decreased tendency to aggregate. In addition it was hoped that mutants with reduced surface hydrophobicity could be more amenable to crystallography.

2. STRUCTURAL ANALYSIS OF SOLVENT EXPOSED HYDROPHOBIC RESIDUES

The large hydrophobic residues Leu, Ile, Val, Phe, Met & Trp are generally buried from solvent in soluble globular proteins. This can be achieved by sequestering the hydrophobic groups into the hydrophobic cores characteristic of globular domains or by being buried via quarternary structure formation. If a hydrophobic residue is not buried from solvent not only can it destabilise the protein, it can lead it to aggregate. The hydrophobic nature of FFL could therefore explain its instability and its tendency to aggregate.

Accessible surface area calculations (using method of Richmond[2]) using a 1.4 Angstrom radius probe show that there are 31 large hydrophobic residues in luciferase whose sidechains expose over 30 Angstrom^2 to solvent. Interestingly, a large proportion of these amino acids are either completely conserved or at least conservatively conserved. Figure 2a demonstrates just how few of the solvent exposed residues are poorly conserved. The vast majority of the exposed hydrophobic residues occur in loops or turns as indicated in Figure 2b. These results can be largely rationalised on inspection of the crystal structure. Figure 3 demonstrates that a large proportion of the hydrophobic residues lie on the loops making up the interface between the large N-terminal domain (dark) and the small C-terminal domain (light). Thus a large number of the solvent exposed hydrophobic residues are liable to become buried upon the closure of the domains expected during catalysis.[3]

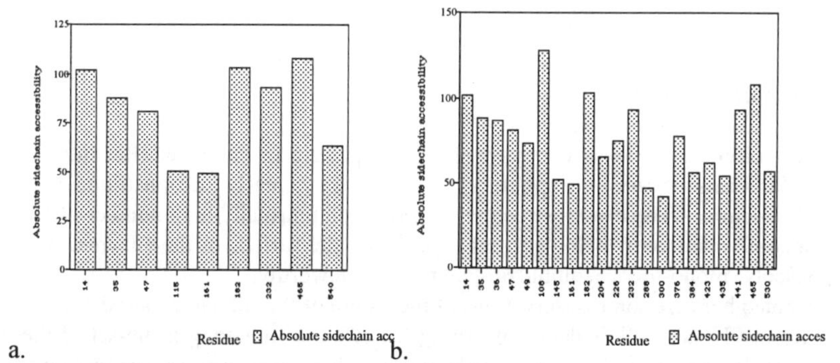

Fig. 1 showing a) the only nine of the 31 solvent exposed hydrophobic residues that are poorly conserved b) hydrophobic residues in neither α-helices nor ß-sheets.

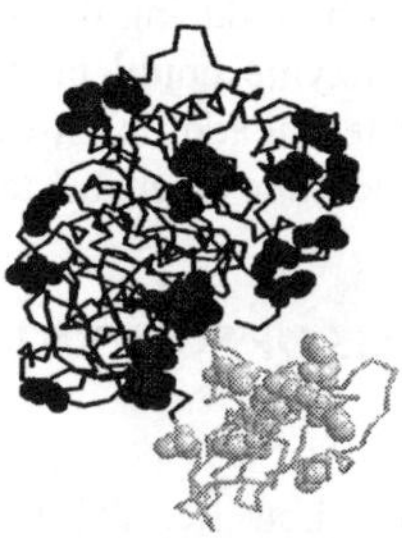

Fig. 2 showing the sidechains of the solvent exposed hydrophobic residues.

3. MUTAGENESIS OF SOLVENT EXPOSED HYDROPHOBIC RESIDUES

A number of the exposed hydrophobic residues were selected to be mutated to alanines. Alanine was chosen as the replacement amino acid as it is significantly less hydrophobic whilst not containing any functional groups which could further complicate interpretation of results. The top dozen most exposed hydrophobic residues include (in descending order) I108, F465, V182, F14, I232, L441, L35, V36, I47, L376, I226 & V49. From the list were selected unconserved amino acids that were not in an α-helix or ß-sheet: F14, L35, I47, V182, I232, and F465. These amino acids were mutated to alanine, with the exception of I47 which is part of a beta-hairpin: by avoiding residues in secondary structures it was hoped to minimise the complications arising from secondary structure propensity effects. In comparison the conserved but solvent exposed amino acids L376 and I226 (which again are not in secondary structure) were also mutated to alanine. Figure 3 shows the location of these in the three dimensional structure of the protein.

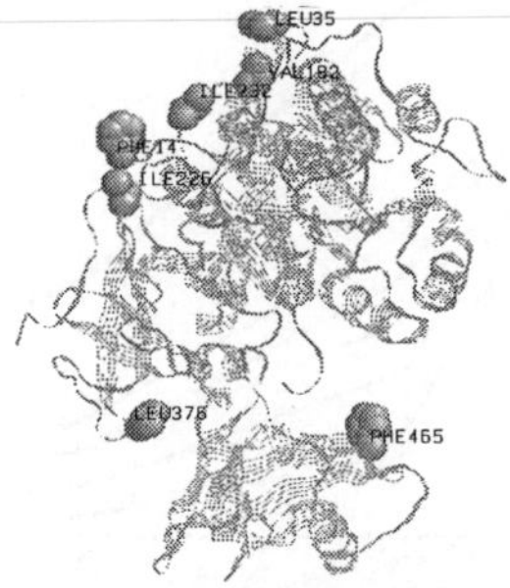

Fig. 3 showing the position of the mutated sidechains

The results of these mutations (and combinations of these mutations) are still being analysed however some of the initial data are interesting. Figure 4 shows that in all cases the mutations have not caused any significant wavelength shift, though the L376A mutant has such low activity that the normalised spectrum is extremely broad. Of more interest were experiments aimed at assessing the thermostability of the different mutants. An experiment was performed to assess the rate of inactivation during incubation at 37°C over time. Figure 5 shows that mutants F14A, I232A, L376A and F465A are more thermostable than wildtype.

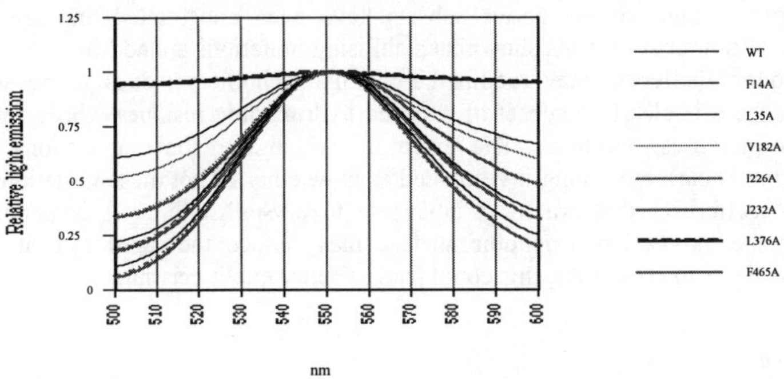

Fig. 4 showing the bioluminescent spectra of the various mutants

4. DISCUSSION

Regarding those five hydrophobic residues which were neither conserved nor in secondary structure (F14, L35, V182, I232, F465), mutating to alanine increased thermostability in three cases, dramatically so for the I232A mutation. There was no effect on the bioluminescent emission spectra however. Mutating the conserved residues I226 and L376 in both cases proved deleterious. The I226 mutant is the least thermostable of all the mutants and the L376 mutant, though thermostable, shows a large decrease in specific activity.

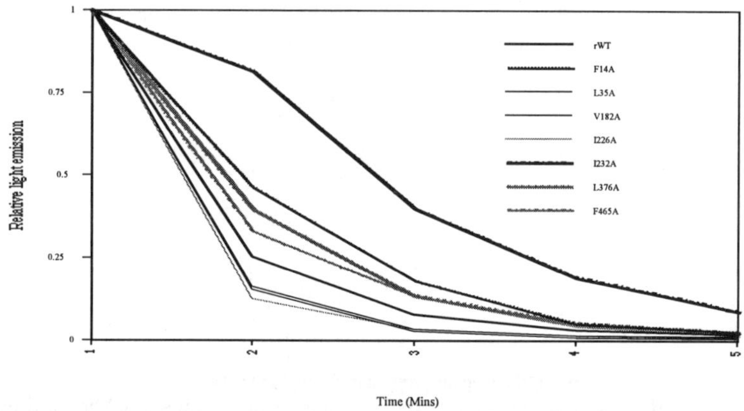

Fig. 5 showing the inactivation of mutants over a period of 5 minutes at 37°C.

least thermostable of all the mutants and the L376 mutant, though thermostable, shows a large decrease in specific activity.

Though the data cannot be claimed to offer a general method for increasing the stability of a protein, it is nonetheless encouraging. In particular for F14, I232 and F465 it will be interesting to introduce hydrophilic amino acids in these positions which may be capable of making favourable electrostatic interactions and hence further increasing stability.

Multiple mutants of the above subsets have been constructed and are being analysed. Preliminary results show that stabilising mutations are additive.

Thus though luciferase may require certain hydrophobic residues to be solvent exposed, there is clearly a subset of exposed hydrophobic residues whose removal certainly increases stability. The extent to which the above mutations effect aggregation is currently being investigated as is whether or not these mutations may generate luciferases that are more amenable to crystallisation. It is hoped that reducing the exposed hydrophobic surface may reduce the rapid rate at which luciferase tends to crystallise, this could lead to better quality crystals.

References

1. Conti E., Franks N.P., Brick P. Crystal structure of firefly luciferase throws light on a superfamily of adenylate-forming enzymes. *Structure* 1996; **4**: 287-298
2. Richmond T.J., Solvent accessible surface-area and excluded volume in proteins : analytical equations for overlapping spheres and implications for the hydrophobic effect *(1984), J. Mol. Biol., **178**, 63-89.*
3. Conti E., Stachelhaus T., Marahiel M.A., Brick P. Structure basis for the activation of phenylalanine in the non-ribosynthesis of gramicidin S. *EMBO* 1997; **16**: 4174-4183

THERMOINACTIVATION AND REACTIVATION OF FIREFLY LUCIFERASE

N. N. UGAROVA, E. I. DEMENTIEVA AND I. A. LUNDOVSKIKH
Chemistry Department, Moscow State University, 119899, Moscow, Russia
E-mail: unn@enzyme.chem.msu.ru

1. INTRODUCTION

Firefly luciferase is an unstable enzyme that readily undergoes inactivation, especially at elevated temperatures. Already in 1982 it was shown that subunit interactions play an important role in the inactivation of *Luciola mingrelica* firefly luciferase.[1] The question about oligomer structure of luciferases remains hitherto unsolved. The literature data are contradictory indicating the existence of monomers or dimers depending on experimental conditions. Early studies of *Photinus pyralis* luciferase revealed the existence of monomers and dimers.[2] Gel-filtration and ultracentrifugation of *P. pyralis* and *Photuris pennsylvanica* firefly luciferases showed that these proteins are monomers that are able to undergo association to dimers, whereas active *Pyrophorus plagiophthalamus* click beetle luciferase exists primarily as a dimer.[3] In the presence of surfactants or guanidinium chloride, only monomers were found. For the inactivated luciferase trimers were also present.[4] The goal of this work was to study subunit interactions in *L. mingrelica* firefly luciferase and their role in thermoinactivation of the enzyme.

2. MATERIALS AND METHODS

Reagents Recombinant *L. mingrelica* firefly luciferase was isolated from *E. coli* cells, carrying pLR plasmid with the luciferase gene.[5] Superdex-200 and molecular weight gel filtration calibration kit were from Amersham Pharmacia Biotech. All the reagents were of analytical grade. All the experiments were carried out in 0.1 M Tris-acetate buffer solution containing 2 mM EDTA, 60 mM $MgSO_4$, 1 mM dithiothreitol, pH 7.8.
Instrumentation Luciferase activity was assayed on a 1250 LKB-Wallac luminometer. Sedimentation of luciferase solution (0.2 mg/mL) was carried out on an analytical ultracentrifuge Spinco, Beckman, model E at 49000 rpm at 10°C.

3. RESULTS AND DISCUSSION

Molecular forms of *L. mingrelica* firefly luciferase We used the recombinant *L. mingrelica* firefly luciferase whose amino acid structure and physicochemical properties are similar to those of the native one.[5] Sedimentation of the active luciferase revealed the existence of monomers (60 kDa) and dimers (112 kDa). For the luciferase incubated for 2 min at 42°C, only monomers were found. After prolonged incubation at 42°C, monomers and high-molecular aggregates, perhaps trimers, were present. Analytical gel filtration of the active luciferase showed the presence of two catalytically active peaks corresponding to the proteins with molecular mass 60±4 and 124±7 kDa. Both of the methods used did not allow us to determine specific activity of monomers and dimers because the activity of the enzyme was almost completely lost during centrifugation and strong dilution occurred during gel filtration. In our case the most appropriate method to study monomer-dimer interactions is the kinetic one.

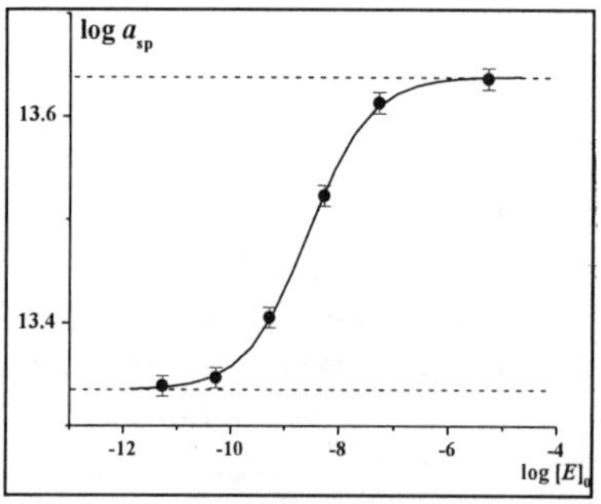

Fig. 1. Plot of equilibrium specific activity of *L. mingrelica* firefly luciferase *vs* enzyme concentration (8°C).

Specific activity of monomers and dimers at 8–22°C At concentrations of the luciferase lower than 10^{-8} M, the decrease in specific activity on dilution was observed. Equilibrium was established in 1 h, after that the activity remained constant during several hours. The plot of equilibrium values of specific activity *vs* luciferase concentration (Fig.1) is typical of oligomeric enzymes when the active particle dissociates to less active subunits. Extrapolation to high enzyme concentrations gives the specific activity of a dimer ($a_{2,sp}$). Specific activity of monomers ($a_{1,sp}$) and dissociation constant were determined by the equation

$$a_{sp} = a_{1,sp} - \frac{2(a_{sp} - a_{2,sp})^2[E]}{(a_{1,sp} - a_{2,sp})K_{dis}} \qquad \text{(eqn 1)}$$

Specific activity of a monomer appeared to be two times lower than that of a dimer, *i.e.*, monomer in a dimer is twice as active as free monomer (Table 1).

Table 1. Kinetic parameters of dimer-monomer system for *L. mingrelica* firefly luciferase at 8–22°C (scheme 1, specific activity was calculated with respect to monomers, experimental error was 10%)

t, °C	K_{dis}, M	Spec. activity of dimer, arb.units/M	Spec. activity of monomer, arb.units/M	k_{dis}, min^{-1}	k_{ass}, M^{-1} min^{-1}	k_{in}, min^{-1}
8	$3.8\cdot10^{-9}$	$4.4\cdot10^{13}$	$2.1\cdot10^{13}$	$1.1\cdot10^{-2}$	$2.9\cdot10^{6}$	$1.4\cdot10^{-4}$
22	$3.0\cdot10^{-8}$	$5.3\cdot10^{13}$	$2.4\cdot10^{13}$	0.1	$3.4\cdot10^{6}$	$6.5\cdot10^{-3}$

Kinetics of luciferase thermoinactivation at 8–22°C Prolonged incubation (about 20 h) of diluted luciferase solutions resulted in inactivation of the enzyme even at 8°C. At concentration higher than 10^{-8} M, where only dimers were present, no inactivation was observed (Fig. 2). The first exponent of thermoinactivation plots characterizes dissociation of active dimers (E_2) into active monomers (E), the second one is inactivation of monomers (Scheme 1).

$$E_2 \underset{k_{ass}}{\overset{k_{diss}}{\rightleftharpoons}} 2E \xrightarrow{k_{in}} 2I \qquad \text{(scheme 1)}$$

Kinetics of luciferase inactivation at elevated temperatures At 30–42°C kinetics of thermoinactivation also involved fast and slow steps (Fig. 3). Position of the inflection point depended on enzyme concentration. At given luciferase concentrations, slopes of the first exponents coincided. Stability of the monomer decreases with the increase in temperature. At elevated temperatures, the monomer activity was lost so quickly that the kinetics curves were better described by the scheme where active dimers (E_2) dissociate to inactive monomers (*I*) (Scheme 2).

$$E_2 \underset{k_{ass}}{\overset{k_{diss}}{\rightleftharpoons}} 2I \xrightarrow{k'_{in}} 2I' \qquad \text{(scheme 2)}$$

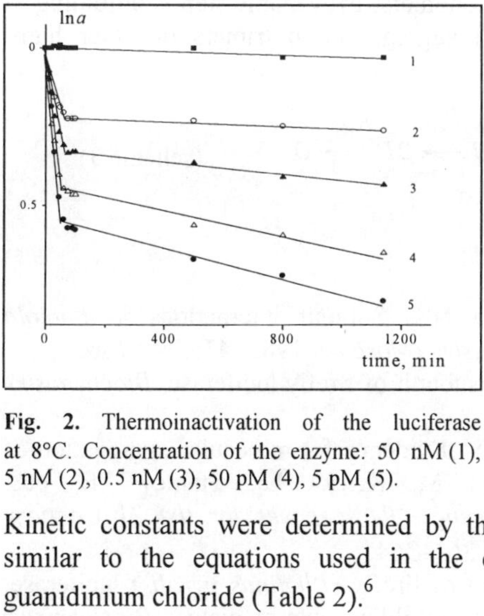

Fig. 2. Thermoinactivation of the luciferase at 8°C. Concentration of the enzyme: 50 nM (1), 5 nM (2), 0.5 nM (3), 50 pM (4), 5 pM (5).

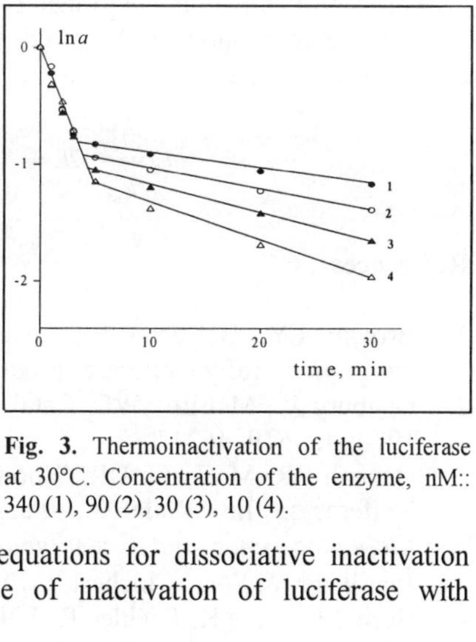

Fig. 3. Thermoinactivation of the luciferase at 30°C. Concentration of the enzyme, nM:: 340 (1), 90 (2), 30 (3), 10 (4).

Kinetic constants were determined by the equations for dissociative inactivation similar to the equations used in the case of inactivation of luciferase with guanidinium chloride (Table 2).[6]

Table 2. Kinetic parameters of inactivation of *L. mingrelica* firefly luciferase at 30–42°C (scheme 2, experimental error was 10%)

t, °C	K_{dis}, M	k_{dis}, min^{-1}	k_{ass}, M^{-1}min^{-1}	k'_{in}, min^{-1}
30	$1.63 \cdot 10^{-7}$	0.22	$1.35 \cdot 10^{6}$	0.02
37	$3.3 \cdot 10^{-7}$	0.40	$1.20 \cdot 10^{6}$	0.03
42	$7.5 \cdot 10^{-7}$	0.70	$0.95 \cdot 10^{6}$	0.06

Reactivation of the luciferase Thermoinactivated luciferase was able to restore spotaneously its catalytical activity after cooling on ice. If the residual activity corresponded to the 1st exponent of the inactivation curve on Fig. 3, 100% of the activity was restored. For higher degrees of inactivation (2nd exponent), only partial reactivation was observed. Kinetic curves of reactivation were linearized in 2nd order

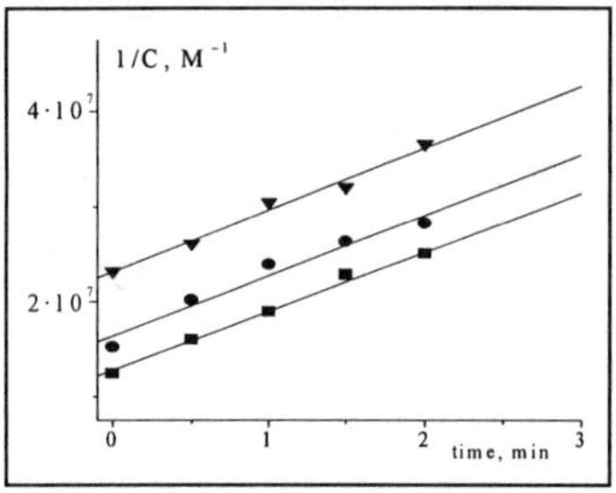

Fig.4. Reactivation of thermoinactivated luciferase

reaction coordinates (Fig. 4) indicating that reactivation is the association of inactive monomers to form active dimers; k_{ass} calculated from the slope of the lines ($3.1 \cdot 10^6$ $M^{-1}min^{-1}$) is equal to the association constant of monomers at low temperatures (Table 1).

<u>General mechanism of thermoinactivation of *L. mingrelica* firefly luciferase</u> At temperatures lower than 22° inactivation of luciferase is described with the first two steps of scheme 3. At elevated temperatures, the first two reversible steps is kinetically seen as dissociation of active dimers to inactive monomers. Irreversible step is unfolding of the inactivated monomers followed with aggregation to trimers or other high-molecular aggregates.

$$E_2 \underset{k_{ass}}{\overset{k_{dis}}{\rightleftharpoons}} 2E \overset{k_{in}}{\rightleftharpoons} 2I \overset{k'_{in}}{\longrightarrow} 2I' \overset{k_{agg}}{\longrightarrow} D \qquad \text{(scheme 3)}$$

References

1. Brovko LY, Belyaeva EI, Ugarova NN. Subunit interactions in *Luciola mingrelica* firefly luciferase. *Biochemistry (Moscow)* 1982; **47:** 760–766.
2. Denburg JL, McElroy WD. Catalytic subunit of firefly luciferase. *Biochemistry* 1970; **9:** 4619–4624.
3. Hannah RR, McCaslin DR, Wood KV. Evidence for molecular aggregation of luciferases. In: Roda A, Pazzagli M, Kricka LJ, Stanley PE, eds. *Bioluminescence and Chemiluminescence. Perspectives for the 21 Century.* JohnWiley & Sons: Chichester, 1998: 361–364.
4. Herbst R, Gast K, Seckler R. Folding of firefly (*Photinus pyralis*) luciferase. Aggregation and reactivation of the unfolding intermediates. *Biochemistry* 1998; **37**: 6586–6597.
5. Lundovskikh IA, Leontieva OV, Dementieva EI, Ugarova NN. Recombinant *Luciola mingrelica* firefly luciferase. Folding *in vivo*, purification and properties. In: Roda A, Pazzagli M, Kricka LJ, Stanley PE, eds. *Bioluminescence and Chemiluminescence. Perspectives for the 21 Century.* JohnWiley & Sons: Chichester, 1998: 420–424.
6. Dementieva EI, Maloshenok LG, Ugarova NN. Guanidinium chloride-induced inactivation of firefly luciferase and its reactivation. *This volume.*

PLAYING WITH LUCIFERIN - NEW RESULTS ON THE LUMINESCENCE OF A WELL-KNOWN MOLECULE

D WEISS,[1] R BECKERT,[1] K LAMM,[1] WJ BAADER,[2] EJH BECHARA[2] AND CV STEVANI[2]

[1] *Inst. fur Org. Chemie - FSU Jena D-07743, Humboldtstr. 10, Jena, Germany*

[2] *Inst. de Quimica - Univ. de Sao Paulo, C. P. 26077, 05599-970, Sao Paulo, Brazil*

1. INTRODUCTION

Terrestrial bioluminescence is most frequent among Coleoptera in the families Lampyridae, Elateridae and Phengodidae, (spectral range ; 536 nm to 620 nm). The bioluminescence colors in these families are governed by tertiary structure of the luciferase, as the luciferin **1** is the same in all these insects. This poses the question of how the same emitter can produce light emission at different wavelengths. The "structural" hypothesis suggests that different colors result from distinct rotamers of excited oxyluciferin **3**, produced by rotation around the benzothiazole-thiazole (2-2') axis[1]. In the "mechanistic" hypothesis, a tautomeric equilibrium in the excited state between the enol-enolate form **4** (yellow/green emitter) and the keto form of the oxyluciferin **3** (red emitter) would modulate the different emission colors[2].

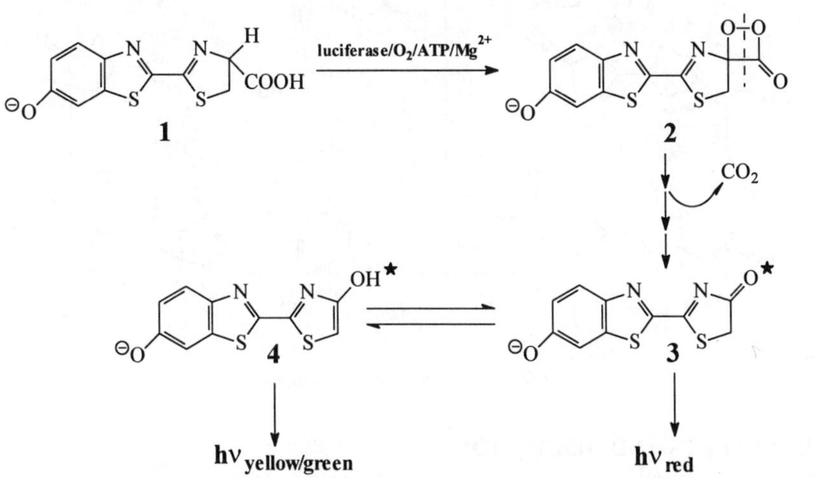

Fig. 1. Firefly bioluminescence reaction.

2. MATERIALS AND METHODS

Intrumentation Chemiluminescence: *SPEX-Fluorolog 1681* (slits: ex 1.25 mm and em 5.00 mm, PMT: 950 V); absorption spectra: *Hitachi U-2000*. Reagents Syntheses of model substances followed literature methods[3]. Starting material **5** demethylated to **6,** then reacted with mercaptoisobutyric acid to yield dimethyloxyluciferin **7**. Methoxyoxyluciferin **8** can be obtained directly from **7** by reaction with mercaptoacetic acid. Dimethylluciferin **9** was synthesized by reaction of **6** with penicillamine in a slightly alkaline ethanol solution. Dicyclohexylcarbodiimide (DCC) allows the conversion of **9** to the pentachlorophenylester **10** (Fig. 2). Methods In a cuvette containing 2.85 mL of phosphate buffer (100 mM, pHs 6.0-10 and 11.6 for **7** and pHs 6.0-10 and 11.6 for **8**) a 150 μL aliquot of a stock solution of **7** or **8** (2.0 mM in ethyl acetate) was added and the spectra were immediately acquired at 25°C (ex 390 and 305 nm, respectively, using right angle detection, for **7** and **8**).

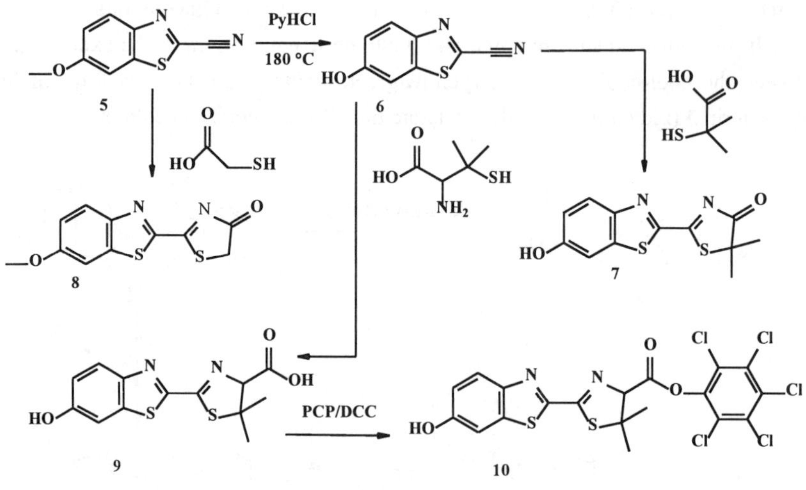

Fig. 2. Syntheses of model-compounds.

3. RESULTS AND DISCUSSION

In compound **7**, the two methyl groups vicinal to the carbonyl group prevent enolization. By raising the pH, one can follow the effect of phenolic proton neutralization on the color of absorbed and emitted light. Fig. 3 shows the pH profile for absorption and emission of **7** (pKa 7.5) in water/phosphate buffer solutions. By raising the pH, the absorptivity of the band at 385 nm decreases and the absorptivity of the band at 486 nm increases. The emission spectra of **7** (ex 390 nm, λ_{max} for the phenolic form) show red fluorescence at pH 6 and yellow/green emission at pH >8. Excitation at 490 nm, where only the phenolate form absorbs, gives a red emission (630 nm) for all pH values studied. However, at pH 6.0 no emission spectrum can be obtained, probably due to the low absorptivity at this wavelength.

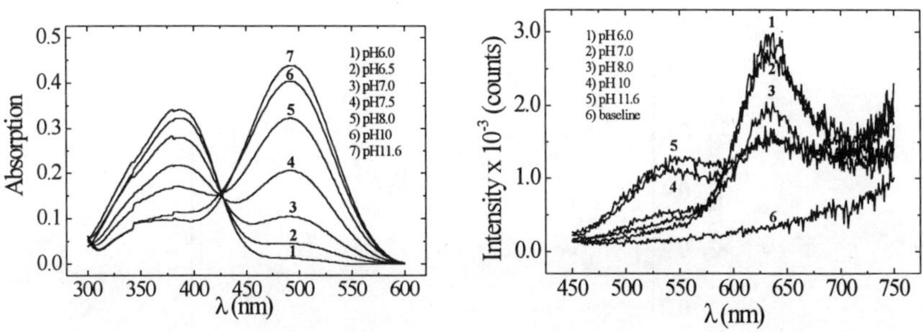

Fig. 3. Absorption and emission profiles of dimethyloxyluciferin **7** with pH.

The fluorescence emission of **7** can be modulated from green (apparently from the deprotonated form) to red (protonated form) (ex 390 nm - absorption max. of the protonated form). On the other hand, compound **8**, with a methyl-protected phenoxy moiety and enolizable hydrogen atoms in α-position to the carbonyl group, may show fluorescence color modulation due to the keto-enol tautomerism in the excited state. Indeed, the absorption spectra reveal the appearance of a long-wavelength absorption band at 440 nm at basic pH values , which should be attributed to the enolate form of **8** (pKa 8.6). Interestingly, the absorption coefficient of the low-wavelength band is also increased at pHs above pH 10. The emission spectra (λ_{exc} = 305 nm) of **7** show a maximum at 425 nm at pH values below 7 and a yellow/green emission (570 nm) at basic pH values, dominant at pH values above 10. These experiments show that the enolate form of oxyluciferin may be responsible for the yellow/green emission of firefly bioluminescence and, that the protonation/deprotonation of the luciferin phenoxide moiety may also lead to color modulation, although results on the pH dependence of the fluorescence emission of 7 cannot be easily rationalized. A very interesting feature of luciferin is the light emission upon treatment with a strong base[4]. To learn more about this process, we synthesized compound **10** (Fig. 2). In organic solvents (DMSO, ethyl acetate) **10** emits light when treated with KOH. Upon treating an aqueous solution of **10** (pH > 8) with H_2O_2, a flash of light can observed, followed by a slow decay of emission intensity. Fig 4 shows the chemiluminescence spectrum of the reaction of **10** with H_2O_2. The observed emission maximum at 600 nm is in the same spectral range as the fluorescence emission obtained from **7** in pH values around 7. Work is in progress to elucidate the fundamental questions regarding the actual mechanism of color modulation in beetle bioluminescence.

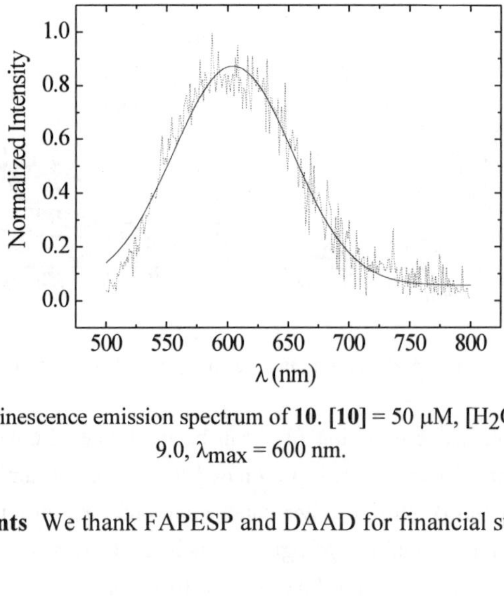

Fig. 4. Chemiluminescence emission spectrum of **10**. **[10]** = 50 μM, $[H_2O_2]$ = 10 mM, pH 9.0, λ_{max} = 600 nm.

Acknowledgements We thank FAPESP and DAAD for financial support.

References

1. McCapra F. Mechanism in chemiluminescence and bioluminescence – unfinished business. In: Hastings JW, Kricka LJ, Stanley PE, editors. *Bioluminescence and Chemiluminescence: Molecular Reporting with Photons.* Chichester: Wiley, 1997: 7-15.
2. Wood K, Lam YA, Seliger HH, McElroy WD. Complementary DNA coding click beetle luciferases can elicit bioluminescence of different colors of light. *Science* 1989;**244**: 700-2.
3. Suzuki N, Sato M, Okada K, Goto T. Studies on firefly bioluminescence-I. *Tetrahedron* 1972;**28**: 4065-74.
4. White EH, Rapaport E, Hopkins TA, Howard H, Seliger H. Chemiluminescence and bioluminescence of firefly luciferin. *J Am Chem Soc* 1969;**91**: 2178-80.

DESIGN AND SELECTION OF FIREFLY LUCIFERASES WITH NOVEL *IN VIVO* AND *IN VITRO* PROPERTIES

T.L. WILLEY, D.J. SQUIRRELL* AND P.J. WHITE
The Babraham Institute, Babraham, Cambridge CB2 4AT, England, UK
**DERA, Porton Down, Salisbury, Wiltshire, SP4 0JQ, England, UK*

1. INTRODUCTION

Firefly luciferase is an efficient bioluminescent reporter in Biotechnology. However, the design of novel and improved firefly luciferases is necessary to keep pace with the development of new applications in the technologies and industries, which use this enzyme. Better performance, both *in vitro* and *in vivo*, will make firefly luciferase a better reagent for use in platform technologies such as high throughput screening. Previously, we have used error prone PCR and a simple *in vivo* screen to identify mutations in firefly luciferase that result in improved thermostability and useful changes to kinetic parameters.[1] To understand further these mutations, and to maximise accessible sequence space, we have used combinatorial cassette mutagenesis[2] (CCM) to probe the informational content of protein sequences in the North American firefly luciferase. Randomisation of just two positions, E354 and D357, in a loop on the surface of the enzyme, yields novel enzymes with very large increases in thermostability and dramatic changes in *in vivo* bioluminescent spectra.

2. MATERIALS AND METHODS

<u>Instrumentation</u> A Biotrace M3 luminometer (Biotrace Ltd, Bridgend, UK) was used to measure RLU signals from extracts of *E. coli* cells. A SPECTRAmax¤ GEMINI microplate Spectrofluorometer (Molecular Devices Corporation, California) was used to measure the spectrum of emitted light from *in vivo* luciferase assays.

<u>Bacterial strains and vectors</u> *E. coli* strain HB101 was used in all cloning experiments and as the host for libraries created using combinatorial cassette mutagenesis. Plasmid pPW601aJ54SpeI/KpnI was used as the template for combinatorial mutagenesis.

<u>Oligonucleotide primers</u> were synthesized by Sigma-Genosys Ltd., Cambridge, UK. The letter s designates either nucleotide g or c at this position.

Lib2A:5 -ctagtgctattctgattacacccnnsggggatnnsaaaccgggcgcggtcggtaaagtggta-3

Lib2B:5 -cactttaccgaccgcgcccggtttsnnatccccsnngggtgtaatcagaat agca-3

<u>Library construction</u> 2μg of each of oligonucleotides Lib2A/B was mixed in a buffer containing 50mM Tris-HCl pH 7.4, 25mM NaCl, and heated to 100°C for 3 min.

This solution was then cooled slowly in a heating block to anneal the complimentary sequences. The annealed oligonucleotides were then ligated into plasmid pPW601aJ54SpeI/KpnI, which had been digested with SpeI and KpnI. Aliquots of the ligation reaction were then used to transform *E. coli* cells using electroporation. Cells transformed this way were plated onto LB agar containing 50µg/mL ampicillin and grown overnight at 37°C. The following day 869 colonies were picked at random from the plates and used to inoculate 1mL of LB containing ampicillin in 96 square-well plates. The plates were then covered and the cells grown overnight with shaking.

In vivo spectral analysis Aliquots (200µL) of cultures of *E. coli* expressing luciferases were centrifuged and the cell pellets resuspended in 150µL of 100mM sodium citrate buffer pH 5.0 containing 0.5mM D-luciferin. Bioluminescent spectra were recorded at room temperature for wavelengths between 500 and 700nm.

In vivo screening Aliquots (50µL) of stationary phase cultures were transferred to two clear plastic round bottom 96 well microtitre plates. One plate was covered and incubated on a heated block for 8 minutes (block surface temperature 45¡C) whilst the other was kept at 37¡C. The *in vivo* luciferase activity in the cells from both plates was then detected and recorded, at room temperature, by adding 50µL of a 100mM sodium citrate buffer pH 5.0 containing 0.5mM D-luciferin to the wells and then transferring the plate to a video imager (Alpha Innotech). The light emitted by the heated and control cultures was integrated over 1 or 2 min and the image recorded on thermal paper. Seventy-nine cultures exhibiting the greatest bioluminescence, as determined by the brightness of the image recorded on paper, were selected for a second round of screening. This time the cultures were incubated for 16 min on the heating block prior to being assayed. Of 55 clones selected from the *in vivo* thermostability screens 25 were chosen for *in vivo* spectral analysis. Fig. 1 shows the *in vivo* spectra for 4 of the mutant luciferases.

Identification of mutations Plasmid DNA was prepared from the 25 clones selected by *in vivo* screening and sequenced using gene specific sequencing primers.

In vitro screening for thermostability Cell free extracts of 25 of the selected clones were prepared by lysis as described in the Promega technical bulletin 101, and the thermostability of the luciferase from each extract was determined in a thermal inactivation experiment. A 50µL aliquot of each extract was placed in an eppendorf tube and incubated in a water bath heated to 45¡C for 4, 9 and 16 min. Table 1. shows the stability, at 45¡C, of 4 of the mutant enzymes.

Effect of growth conditions on the *in vivo* spectrum of emitted light The effect of metabolic status on the spectrum of emitted light from *E. coli* HB101 cells expressing luciferase E354K+D357M was investigated. A 50mL culture of cells was grown to mid log phase in LB medium and then harvested by centrifugation. The cell pellet was resuspended in 1mL of sterile distilled water and 100µl of this suspension was then used to inoculate 5ml of fresh LB, M9 minimal medium + 2mM sodium acetate or M9 minimal medium + 2mM glucose in a sterile 25mL tube. The cultures were grown for a further 120 min after which a 200µL aliquot of cells was removed centrifuged and the spectrum of emitted light recorded (Fig. 2)

3. RESULTS AND DISCUSSION

A total of 25 mutant luciferases were selected from *in vivo* assays. The majority of these luciferases show large changes in the *in vivo* spectrum of emitted light and many show greater contributions from longer wavelengths of light (>580nm). A number of spectra also showed a significant narrowing of bandwidth around a single peak of 610-614nm. DNA sequencing identified mutations responsible for amino acid substitutions at positions 354 and 357 in the primary sequence. Replacements of E354 and D357 with a hydrophobic and an aromatic amino acid respectively e.g. E354I+D357Y give the largest change in the *in vivo* spectrum resulting in a single peak, of narrow bandwidth, around 612nm (Fig. 1). The *in vitro* thermostability of the mutants was analysed and the most thermostable luciferases were found to be those with an aromatic amino acid at position 357 in combination with a large hydrophobic or positively charged amino acid at position 354 (Table 1).

When *E. coli* cells expressing luciferase E354K+D357M were grown on different carbon sources a change in the *in vivo* spectrum of emitted light was observed. Switching from growth in a rich medium (LB) to a defined minimal medium, with either acetate or glucose as the sole carbon source, resulted in shifts to longer wavelengths of emitted light (Fig. 2).

Combinatorial mutagenesis at just two positions within a surface loop on firefly luciferase generates a diverse population of novel enzymes with improved properties. These mutants show large increases in thermostability and have significant changes in *in vivo* spectra. The discrete peak at 612nm observed with the luciferase E354I+D357Y and the changes in spectra seen under different metabolic conditions suggest a ligand effect on the luciferase. Mutagenesis within the surface loop may enhance binding, to the enzyme, of an effector molecule present in the *E. coli* cytoplasm. In summary, knowledge based mutagenesis using CCM is an effective strategy for creating novel luciferases and offers an alternative to evolution based methodologies.[3]

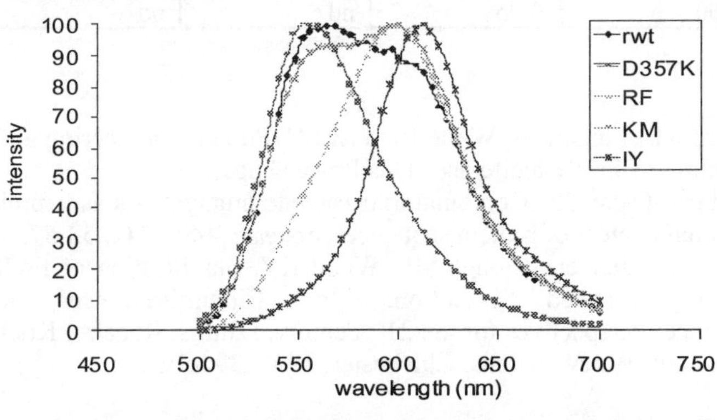

Fig. 1. *In vivo* bioluminescent spectra of r-wt and mutant luciferases

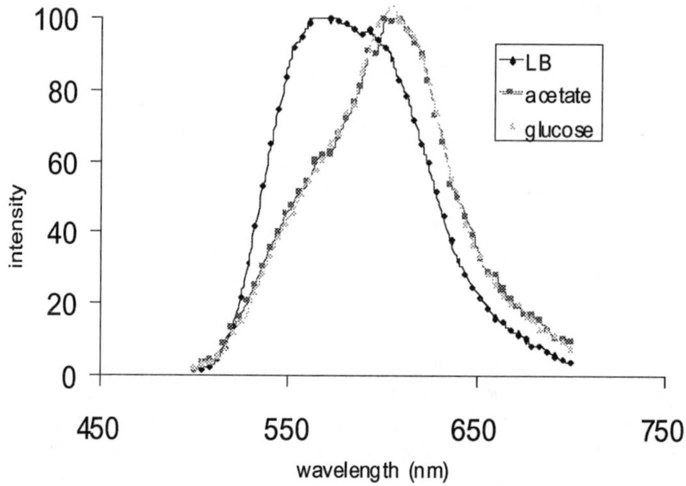

Fig. 2. *In vivo* bioluminescent spectra of luciferase E354K+D357M from *E. coli* cells grown on different carbon sources

Table 1. Stability of luciferases at 45¡C (%remaining activity). Single amino acid codes refer to replacements at positions 354 and 357 respectively.

Luciferase	0min	4min	9min	16min
IY	100	99	85	68
RF	100	89	70	54
KM	100	64	31	12
D357K	100	0.1	nd	nd
r-wt	100	0.05	nd	nd

References

1. Prebble SE, Price RL, Lingard B, White PJ, Tisi LC. Protein engineering and molecular modeling of firefly luciferase. This Proceedings.
2. Sauer RT, Reidhaar-Olson JF. Combinatorial cassette mutagenesis as a probe of the informational content of protein sequences. *Science* 1988; **241**: 53-57.
3. Hall MP, Gruber MG, Jennens-Clough ML, Wood KV. Stabilisation of firefly luciferase using directed evolution. In: Bioluminescence and Chemiluminescence: Perspectives for the 21st Century. Editors, Roda A, Kricka LJ, Stanley, PE. John Wiley & Sons, Chichester, 1999, 392-395.

PART 3
CHEMILUMINESCENCE

MICROWAVE-ASSISTED SYNTHESIS OF CHEMILUMINESCENT ACRIDINIUM SALTS

MACIEJ ADAMCZYK, PHILLIP G. MATTINGLY, YON-YIH CHEN AND JAMES R. FINO

Abbott Laboratories, Diagnostics Division, Department of Chemistry D-09NM, Building AP 20, 100 Abbott Park Road, Abbott Park, Illinois, 60064-6016, USA
E-mail: maciej.adamczyk@abbott.com

1. INTRODUCTION

Replacing the usual N^{10}-methyl substituent on chemiluminescent acridinium salts with the 3-sulfopropyl group brings about a beneficial increase in hydophillicity.[1] That being the case, N^{10}-(3-sulfopropyl)acridinium-9-carboxamides have been used in as the labels of choice in the ARCHITECT® line of immunodiagnostic products and other applications.[2-7] In most cases introduction of the 3-sulfopropyl group proceeds by heating the acridine-9-carboxamide in neat 1,3-propanesultone at 125 °C for 4-5 h. For some sterically demanding compounds, even these harsh conditions were insufficient to effect the reaction. In those cases a more reactive sulfopropyl triflate reagent was required.[1, 8] The conditions were much milder, but the reaction was extremely slow.

In situations were the reactions are sluggish, or require high temperature, dielectric heating in a microwave oven has often offered a convenient means to accelerate the process.[9-11] Our earlier experience in the sulfopropylation of nitrogen heterocycles under the influence of microwave irradiation,[12] encouraged us to explore its usefulness in the context of acridinium-9-carboxamide chemiluminescent labels.

2. MATERIALS AND METHODS

Acridine-9-carboxamide (Scheme 1, **1—6)** (0.45 mmol), 2,6-di-*tert*-butyl-4-methyl-pyridine (3.9 mmol) and 1,3-propane sultone (10 mL, 0.14 mol) were added to a 50 mL round-bottom, two-neck flask equipped with water-cooled condenser, mechanical stirrer, digital recording thermometer (Extech Instruments Inc., model 422315, using an ungrounded type T thermocouple externally grounded to the microwave oven) and inert atmosphere (N_2). A 250-mL Erlenmeyer flask containing 200 mL of ambient temperature water placed next to the reaction flask, opposite the magnetron, was used as a heat sink to moderate the dielectric heating. The reaction mixture was heated on medium-low (2-4 min) using a modified consumer microwave oven (Figure 1, Panasonic, model NN5503A). CAUTION: Avoid exposure to stray microwave radiation.

After cooling to ambient temperature, the reaction was washed with ether (6 _ 40 mL). The residue was refluxed with 6 N HCl (25 mL) for 1h then diluted 1:1 with distilled water. This solution was added to a washed bed (2.5 _ 20 cm) of XAD-16 resin. The column was eluted with water (500 mL) then 60% aq. methanol (750 mL). All of the organic solution was collected and concentrated. This material was further purified by prep HPLC (YMC ODS-AQ, 15 µm, 120Å, 47 _ 300 mm) using a 10–45% acetonitrile/water (both containing 0.1% TFA) gradient over 54 minutes. Fractions containing the pure acridinium salts (**7–12**) were evaporated to dryness. All compounds were greater than 98% pure by analytical HPLC. ESI-MS and ^{1}H/^{13}C NMR were consistent with the assigned structures.

3. RESULTS AND DISCUSSION

The dielectric heating in a microwave oven (1 min, medium power) of a mixture of acridine and 1,3-propane sultone (10 eq,) gave 10-(3-sulfopropyl)acridine in greater than 80% isolated yield.[12] Applying these conditions to acridine-9-carboxamide **4** gave only traces of the desired product **10** after hydrolysis. In the course of optimizing the reaction conditions, we found that three factors influenced the course of the reaction: the ratio of 1,3-propane sultone to acridine; the duration and intensity of the dielectric heating, and presence of an acid scavenger. Portionwise addition of 1,3-propane sultone tended to increase the yield of **10**, but after a point, conversion to **10** stalled. Heating for longer times or at higher power levels led to charring. We considered that the high temperature achieved in the microwave oven might decompose the propane sultone into acidic byproducts that could protonate the acridine-9-carboxamide, rendering it unreactive. Adding a sterically hindered weak base to counteract the acid had the desired effect and led to the optimized conditions listed in the methods section.

The series of acridine-9-carboxamides (Scheme 1, **1–6**) were heated for a total of eight minutes on medium-low power. Under these conditions, the internal reaction temperature rose quickly, fluctuating between 150–200 °C while the microwave oven cycled on and off over the course of the reaction. The acridinium-9-carboxamide products (**7–12**) were isolated in 70–80% yield after extraction of the excess 1,3-propane sultone into ether, acid hydrolysis and chromatographic purification. By way of comparison, the sulfopropylation of **1** under the optimized conditions for conventional heating in the absence of an acid scavenger (5h, 125 °C) gave **7** in 78% isolated yield.[1, 8]

Scheme 1

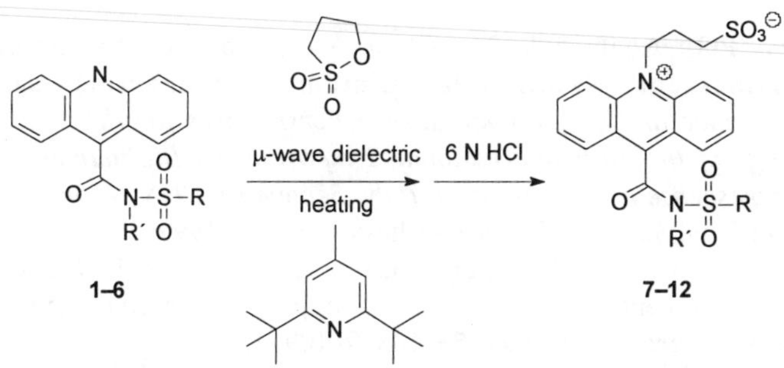

	Substrates		Products	% Yield
1	R = 4-$(CH_3)C_6H_4$- R′ = -$(CH_2)_3CO_2CH_3$	**7**	R = 4-$(CH_3)C_6H_4$- R′ = -$(CH_2)_3CO_2H$	72
2	R= 2-[$C_2H_5O_2C(CH_2)_2$]C_6H_4- R′ = -$(CH_2)_3SO_3CH_2C(CH_3)_3$	**8**	R= 2-[$HO_2C(CH_2)_2$]C_6H_4- R′ = -$(CH_2)_3SO_3H$	77
3	R= 4-[$C_2H_5O_2C(CH_2)_2$]C_6H_4- R′ = -$(CH_2)_3SO_3CH_2C(CH_3)_3$	**9**	R= 4-[$HO_2C(CH_2)_2$]C_6H_4- R′ = -$(CH_2)_3SO_3H$	77
4	R = 4-[$C_2H_5O_2C(CH_2)_3$]C_6H_4- R′ = -$(CH_2)_3SO_3CH_2C(CH_3)_3$	**10**	R = 4-[$HO_2C(CH_2)_3$]C_6H_4- R′ = -$(CH_2)_3SO_3H$	80
5	R = 4-[$C_2H_5O_2C(CH_2)_5$]C_6H_4- R′ = -$(CH_2)_3SO_3CH_2C(CH_3)_3$	**11**	R = 4-[$HO_2C(CH_2)_5$]C_6H_4- R′ = -$(CH_2)_3SO_3H$	72
6	R = 3-C_5H_3N R′ = -$(CH_2)_3CO_2CH_3$	**12**	R = 3-C_5H_3N+$(CH_2)_3SO_3^-$ R′ = -$(CH_2)_3CO_2H$	80

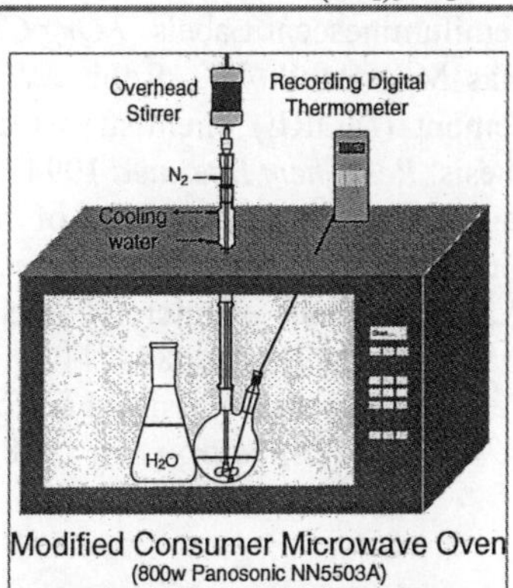

Fig. 1. Microwave Oven Apparatus

References

1. Mattingly PG, Adamczyk M, Chen Y-Y, Pan Y. Sulfopropylated chemiluminescent *N*-sulfonylacridinium-9-carboxamide salts. In *Bioluminescence and Chemiluminescence Perspectives for the 21st Century Proceedings of the 10th International Symposium on Bioluminescence and Chemiluminescence held at Bologna, Italy, September 1998.* Roda A, Pazzagli M, Kricka LJ and Stanley PE, eds. Chichester: Wiley, 1999: 37–40.
2. Adamczyk M, Chen Y-Y, Mattingly PG, Moore JA, Shreder K. Modulation of the chemiluminescent signal from N^{10}-(3-sulfopropyl)-*N*-sulfonylacridinium-9-carboxamides. *Tetrahedron* 1999; **55**: 10899–10914.
3. Adamczyk M, Mattingly PG, Reddy RE. The synthesis of 6β-amino-estradiol and its biotin, acridinium, and fluorescein conjugates. *Steroids* 1998; **63**: 130–134.
4. Adamczyk M, Fishpaugh J, Mattingly PG, Shreder K. Tracermer (TM) signal generators: An arborescent approach to the incorporation of multiple chemiluminescent labels. *Bioorg Med Chem Lett* 1998; **8**: 3595–3598.
5. Adamczyk M, Fishpaugh JR, Mattingly PG. Resin-supported labeling reagents. *Bioorg Med Chem Lett* 1999; **9**: 217–220.
6. Adamczyk M, Mattingly PG, Moore JA, Pan Y. Synthesis of a chemiluminescent acridinium hydroxylamine (AHA) for the direct detection of abasic sites in DNA. *Org Lett* 1999; **1**: 779–781.
7. Adamczyk M, Grote J, Pan Y, Mattingly PG. *O*-(Acridinium)hydroxylamine (AHA): A reagent for the preparation of chemiluminescent acridinium oxime (AO)-steroid conjugates. *Steroids* 2000; **65**: 387–394.
8. Adamczyk M, Chen Y-Y, Mattingly PG, Pan Y. Neopentyl 3-triflyloxypropanesulfonate. A reactive sulfopropylation reagent for the preparation of chemiluminescent Labels. *J Org Chem* 1998; **63**: 5636–5639.
9. Bose AK, Manhas MS, Banik BK, Robb EW. Microwave-induced organic reaction enhancement (MORE) chemistry: technology for rapid, safe and inexpensive synthesis. *Res Chem Intermed* 1994; **20**: 1–11.
10. Mingos MP, Baghurst DR. Applications of microwave dielectric heating effects to synthetic problems in chemistry. *Chem Soc Rev* 1991; **20**: 1–47.
11. Abramovitch RA. Applications of microwave energy in organic chemistry. A review. *Org Prep Proc Int* 1991; **23**: 685–711.
12. Adamczyk M, Rege S. Microwave assisted sulfopropylation of N-heterocycles using 1,3-propane sultone. *Tetrahedron Lett* 1998; **39**: 9587–9588.

TOO MUCH OF A GOOD THING? MODULATING THE SIGNAL OF ACRIDINIUM-9-CARBOXAMIDE SALTS

MACIEJ ADAMCZYK, YON-YIH CHEN, JEFFREY R. FISHPAUGH, PHILLIP G. MATTINGLY, JEFFREY A. MOORE, YOU PAN, KEVIN SHREDER, AND ZHIGUANG YU

Abbott Laboratories, Diagnostics Division, Department of Chemistry D-09NM, Building AP 20, 100 Abbott Park Road, Abbott Park, Illinois, 60064-6016, USA

E-mail: maciej.adamczyk@abbott.com

1. INTRODUCTION

We generally think of chemiluminescent labels in terms of their low detection limit. And much effort has been expended to discover more and more sensitive labels. There are times, however, when the light given off by a chemiluminescent label can be too bright, e.g., when measuring high concentration analytes. *N*-Sulfonylacridinium-9-carboxamide labels can be "dimmed" kinetically[1, 2] to match the sensitivity requirements of such assays by increasing the steric demand of the reactive sulfonamide linker/leaving group. Four labels were prepared (Fig. 1, **1a–d**) with widely different chemiluminescence profiles. Each of these was converted to chemiluminescent tracers for prototypic immunoassays for phenobarbital[3] and phenytoin,[4-7] two analytes that are found in the μM concentration range.

2. MATERIALS AND METHODS

N^{10}-(3-sulfopropyl)-*N*-sulfonylacridinium-9-carboxamide salts (**1a–d**) were synthesized in a three step sequence[2] consisting of acylating a sulfonamide ester with acridine-9-carboxylic acid chloride, sulfopropylating the acridine N^{10} position with 1,3-propanesultone[8] and finally removing the ester protecting group with aqueous HCl. Each was activated to yield **2a–d**, then coupled to phenobarbital derivative **3**[9] and phenytoin derivative **4**[9] in DMF/phosphate buffer (1:1, pH 8). Each tracer (**5a–d**, **6a–d**) was purified by reversed-phase HPLC (C_{18}) to achieve greater than 98% purity. ESI-MS and $^{1}H/^{13}C$ NMR were consistent with the shown structures.

Assay conditions. Purified ovine *anti*-phenobarbital and *anti*-phenytoin IgGs (2.5 μg/mL, pH 9.6) were coated onto 96-well microtiter plates (200 μL/well). The plates were then overcoated with 2% non-fat dry milk (PBS buffer, 1 h). Tracers **5a–d** (250 nM, 145 μL) and **6a–d** (100 nM, 145 μL) were added to their respective antisera and incubated (1 h) with phenobarbital (0–80 μg/mL, 5 μL) or phenytoin

(0–40 μg/mL, 5 μL). Chemiluminescence was triggered by the addition of basic hydrogen peroxide (200 μL, 0.7%; 1% Triton X-100, 0. 0.05% DTPA).

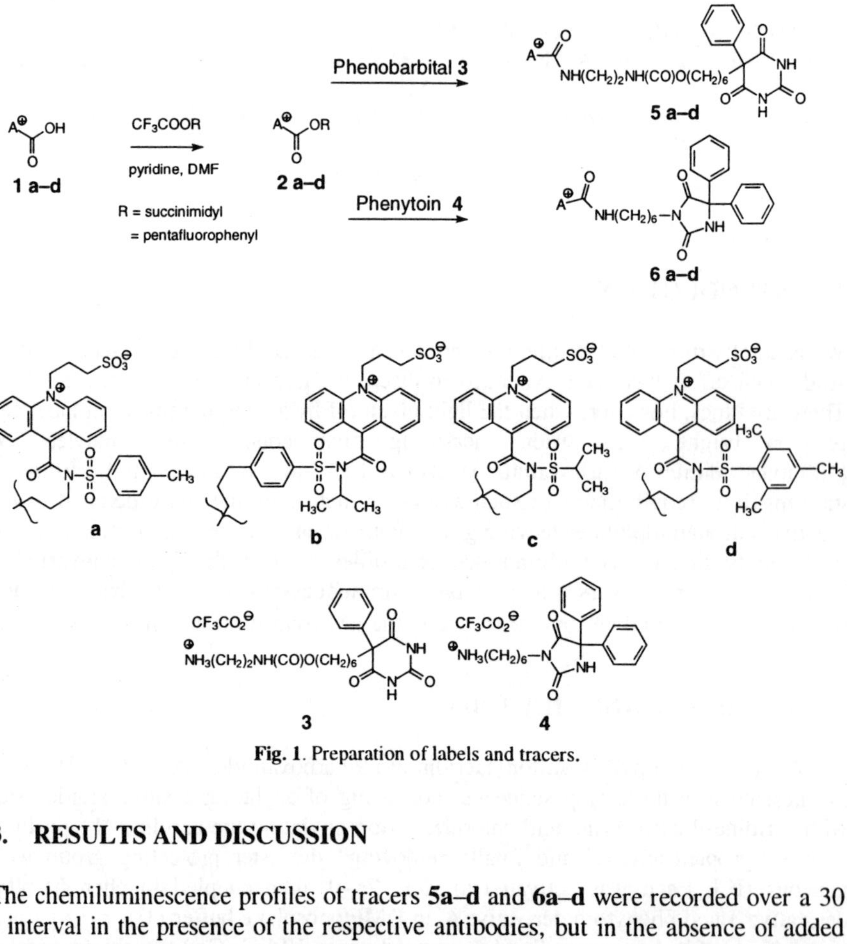

Fig. 1. Preparation of labels and tracers.

3. RESULTS AND DISCUSSION

The chemiluminescence profiles of tracers **5a–d** and **6a–d** were recorded over a 30 s interval in the presence of the respective antibodies, but in the absence of added analyte to produce the curves shown in Fig. 2. The total light output from each tracer was essentially the same, but the rate of light produced by each tracer differed by as much as 20-fold. If a sufficiently long read interval were to be used in assays using these tracers, no difference in the recorded signal would be expected. However, measuring the light produced over a short, fixed interval (0–3 s) serves to kinetically dim the recorded signal. This is illustrated by the dose-response curves shown in Fig. 3. Even the dimmest tracers **5d** and **6d** had reasonably shaped curves that could be used to determine the concentration of phenobarbital and phenytoin in the therapeutic range.

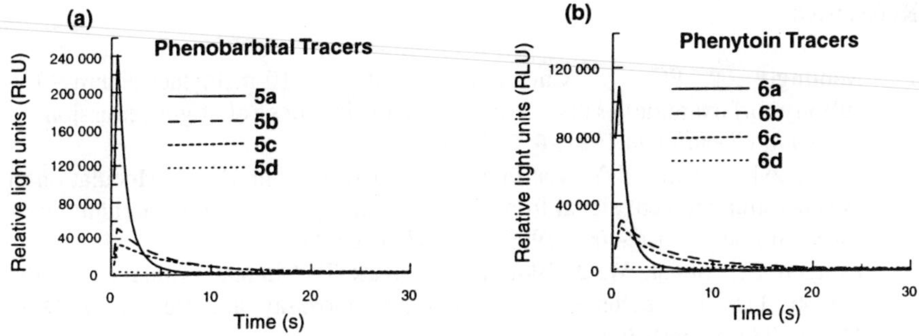

Fig. 2. Chemiluminescence profiles of tracers.

Independent of these results, it was determined that the changes in the sulfonamide substituents in the series (**a–d**) had no effect on the affinity of the tracers for their antibodies, so the results shown in Fig. 3, are due solely to modulation of the chemiluminescent signal.

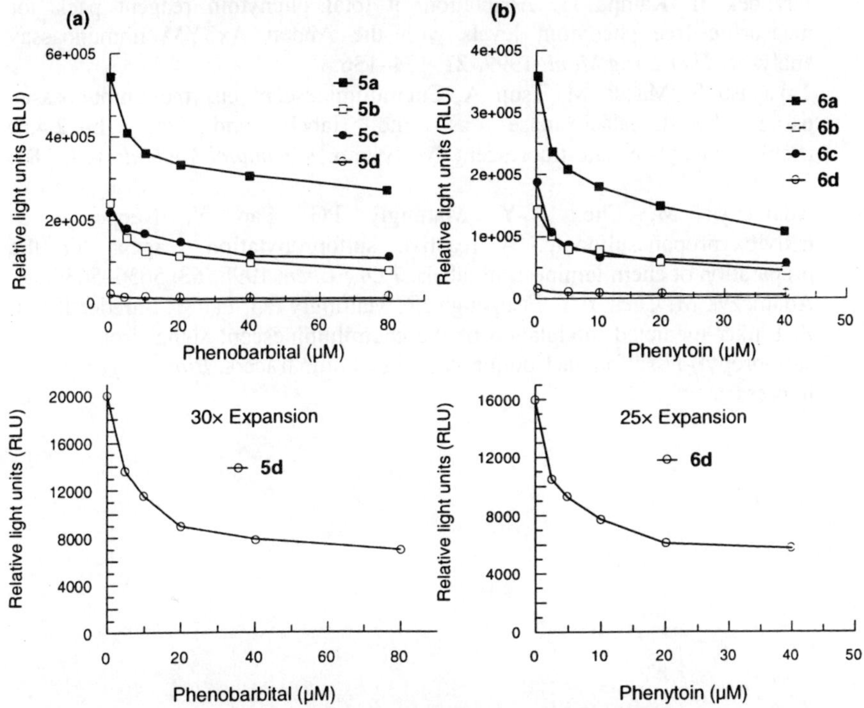

Fig. 3. Dose-response curves for a competitive assay for (a) phenobarbital and (b) phenytoin.

References

1. Mattingly PG. Chemiluminescent 10-methylacridinum-9-(*N*-sulfonylcarboxamide) salts. Synthesis and kinetics of light emission. *J Biolumin Chemilumin* 1991; **6**: 107–114.
2. Adamczyk M, Chen Y-Y, Mattingly PG, Moore JA, Shreder K. Modulation of the chemiluminescent signal from N^{10}-(3-sulfopropyl)-*N*-sulfonylacridinium-9-carboxamides. *Tetrahedron* 1999; **55**: 10899–10914.
3. Oeltgen PR, Shank WAJ, Blouin RA, Clark T. Clinical evaluation of the Abbott TDx fluorescence polarization immunoassay analyzer. *Ther Drug Monit* 1984; **6**: 360–367.
4. Dasgupta A, Datta P, Redlich G, Limmany A. Analytical performance of a new chemiluminescent phenytoin (ACS:180) assay. *Ther Drug Monit* 1997; **19**: 191–194.
5. Dasgupta A, Warner BF, Datta P. Use of alkaline phosphatase to correct the underestimation of fosphenytoin concentration in serum measured by phenytoin immunoassays. *Am J Clin Pathol* 1999; **111**: 557–562.
6. Jarzabek JI, Kampa IS. Adaptation of total phenytoin reagent pack for measuring free phenytoin levels with the Abbott AxSYM immunoassay analyzer. *Ther Drug Monit* 1999; **21**: 134–136.
7. Takayasu S, Maeda M, Tsuji A. Chemiluminescent enzyme immunoassay using beta-D-galactosidase as the label and the bis(2,4,6-trichlorophenyl)oxalate-fluorescent dye system. *J Immunol Methods* 1985; **83**: 317–325.
8. Adamczyk M, Chen Y-Y, Mattingly PG, Pan Y. Neopentyl 3-triflyloxypropanesulfonate. A reactive sulfopropylation reagent for the preparation of chemiluminescent labels. *J Org Chem* 1998; **63**: 5636–5639.
9. Adamczyk M, Chen Y-Y, Fishpaugh JR, Mattingly PG, Pan Y, Shreder K, Yu Z. Linker-mediated modulation of the chemiluminescent signal from N^{10}-(3-sulfopropyl)-*N*-sulfonylacridinium-9-carboxamide tracers. *Bioconjugate Chem* in press.

RECENT ADVANCES IN CHEMILUMINESCENT ENZYME SUBSTRATES

H. AKHAVAN-TAFTI, R. DE SILVA, K. THAKUR, R. A. EICKHOLT, C. CHROVIAN, K. LAUWERS, R. S. HANDLEY, S. SIRIPURAPU, C. ENDRESZL AND A. P. SCHAAP

Lumigen Inc., 24485 W. 10 Mile Rd., Southfield, Michigan 48034 USA

E-mail: hat@lumigen.com

1. INTRODUCTION

Lumigen has developed a family of chemiluminescent enzyme substrates based on the acridan ring nucleus which is used worldwide in immunological and nucleic acid detection. The Lumigen[R] PS reagents are a class of acridan esters and thioesters which produce bright chemiluminescence by reaction with peroxidase and peroxide.[1] Lumigen PS-3 reagent provides extremely sensitive detection of peroxidase-conjugated proteins and nucleic acids in blotting assays. Recently we have also developed a new class of compounds, Lumigen APS, having an acridan ring and an exocyclic enol phosphate group.[2] Reaction of one member, Lumigen APS-5, with alkaline phosphatase (AP) and O2 generates a transient dioxetanone intermediate which undergoes chemiluminescent decomposition. Blue light is emitted from excited N-methylacridone. Enhanced formulations incorporating lucigenin generate a rapid rise to a constant light intensity at room temperature.

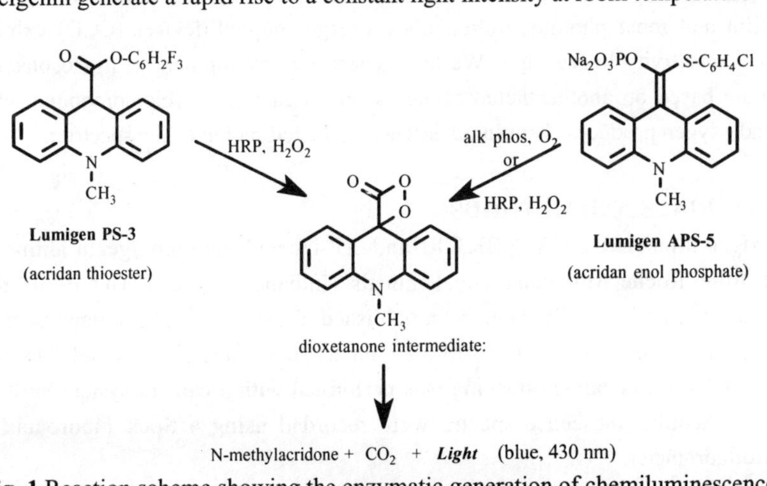

Fig. 1 Reaction scheme showing the enzymatic generation of chemiluminescence from compounds based on an acridan ring nucleus.

We have prepared chemiluminescent labeling compounds incorporating the acridan phosphate structure for use in applications where the signal amplification of enzyme labels is not required.[3] Light emission is produced by a simple chemical reaction as a rapid flash. The chemically triggered reaction involves the sequential reaction of the compound with acid and a strong base in the presence of peroxide. A highly water-soluble compound having a tethered maleimide group was conjugated to BSA. A chemiluminescent assay correlated light intensity with the amount of BSA over six orders of magnitude with a detection limit of 50 amol.[4] Two advantages over other chemiluminescent labels are that labeled proteins or nucleic acids can be separated by electrophoresis without destruction of the label and chemiluminescence can be produced and detected directly in the gel.

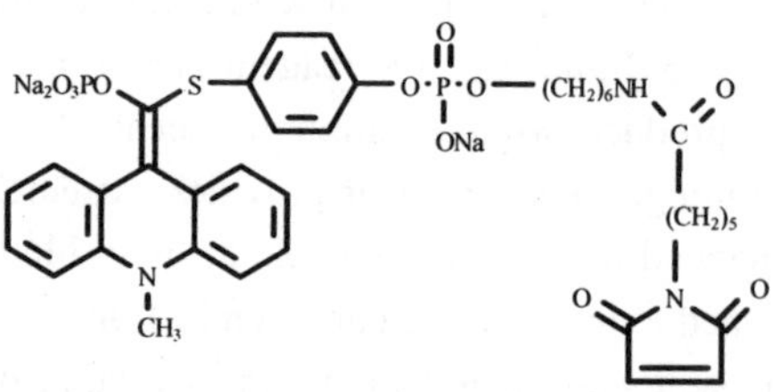

Fig. 2 Structure of chemiluminescent labeling compound.

The emitter in each of the foregoing compounds is N-methylacridone which emits maximally at 430 nm. While blue emission is optimal for detection with x-ray film and most photomultiplier tubes, charge-coupled devices (CCD) exhibit optimal sensitivity to red light. We now report the development of a phosphatase substrate based on another heterocyclic system. Reaction of this compound with AP and oxygen produces chemiluminescence in the red region of the spectrum.

2. MATERIALS AND METHODS

Digoxigenin-labeled DNA (pBR328) and AP-labeled antidigoxigenin antibody were from Roche Molecular Biochemicals (Indianapolis, IN). Details of the synthesis of Lumigen CCD have been published elsewhere.[5,6] Chemiluminescence kinetics measurements were performed on a Turner Designs model TD-20e luminometer. CCD camera imaging was performed with a camera system built in house.[7] Chemiluminescence spectra were recorded using a Spex Fluoromax 3 spectrofluorometer.

3. RESULTS AND DISCUSSION

We have developed a new class of chemiluminescent substrates for AP which generate red chemiluminescence with high efficiency and extended duration. These compounds contain an exocyclic enol phosphate derived from aromatic thioesters of 5,5 -dimethylluciferin. These compounds differ from a previously reported phosphate derivative of firefly luciferin. The latter compound undergoes AP catalyzed removal of a phosphate group from the phenol leaving D-luciferin which is detected by luciferin bioluminescence. The present class of compounds require only one enzyme to generate chemiluminescence. After an evaluation of several compounds in this class, one compound designated Lumigen CCD was selected for further study.

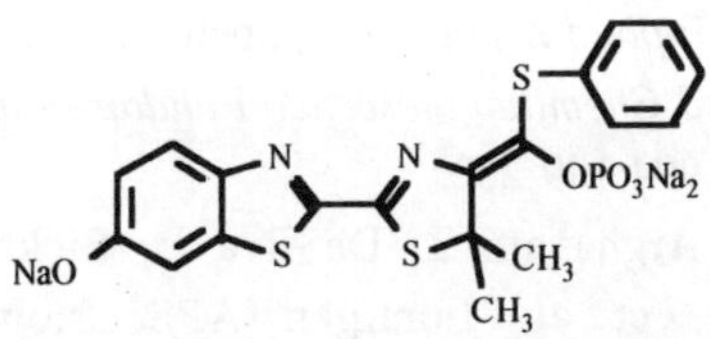

Fig. 3 Structure of red-emitting chemiluminescent substrate Lumigen CCD.

The chemiluminescence spectrum in 0.1 M 221 buffer, pH 9.6 in the presence of AP revealed a broad emission with a maximum at approximately 620 nm, tailing to beyond 700 nm. The chemiluminescence profile from reaction with AP at 25 deg C shows a rapid rise to maximum intensity in ca. 15 min. AP detection sensitivity was assessed by reacting 100 L portions of an enhanced reagent containing 0.33 mM Lumigen CCD, 0.1 % Tween 20, 0.1 mM lucigenin in 0.1 M 221 buffer, pH 9.6 with 10 microL solutions of AP. Room temperature light intensity measured at 10 min correlated with the amount of AP over five orders of magnitude from 0.8 fmol to 8 zmol of enzyme. It should be noted that these results were obtained using an instrument with relatively poor red sensitivity.

Digital imaging of the results of chemiluminescent blotting assays with a CCD camera allows more convenient image acquisition that conventional x-ray film imaging. Data are conveniently stored for postprocessing and analysis. CCD chips with 104-fold dynamic range far exceed the linear range of x-ray film. The potential advantages afforded by detecting the red luminescence with a CCD camera were investigated in a western blot of beta-galactosidase using an AP-labeled antibody and the enhanced reagent described above. The blots produced very intense light emission which allowed imaging of all bands (5-5000 pg of protein) with no substrate incubation. Images could be obtained for many hours.

The enhanced reagent was also used in a dot blot assay of digoxigenin-labeled DNA (pBR328) with AP-labeled antidigoxigenin antibody. Serial dilutions of DNA containing 10-0.01 pg of DNA were spotted on positively charged nylon, probed with AP-labeled antidigoxigenin antibody and detected with the enhanced reagent. Images obtained with the CCD camera revealed the 10 pg-0.03 pg spots within 1 min. All spots including the 0.01 pg spot were detectable within 30 min. Multiple images could be obtained over a 24 h period.

References

1. Akhavan-Tafti H, deSilva R, Arghavani Z, Eickholt RA, Handley RS, Schaap AP. Lumigen PS: New chemiluminescent substrates for the detection of horseradish peroxidase. In: *Bioluminescence and Chemiluminescence Fundamentals and Applied Aspects,* Campbell AK, Kricka LJ, Stanley PE, eds. *Bioluminescence and Chemiluminescence Fundamentals and Applied Aspects*, Chichester: Wiley, 1994:199-202.
2. Akhavan-Tafti H, Arghavani Z, DeSilva R, Eickholt RA, Handley RS, Schoenfelner BA, et al. Lumigen APS: New substrates for the chemiluminescent detection of phosphatase enzymes. In: Hastings JW, Kricka LJ, Stanley PE, eds.*Bioluminescence and Chemiluminescence Molecular Reporting with Photons*, Chichester: Wiley, 1997:311-314.
3. Akhavan-Tafti H, Novel nonenzymatic methods of generating chemiluminescence from acridan alkenes. US Patent 6,017,769. 2000.
4. Akhavan-Tafti H, deSilva R, Sugioka K, Handley R, Schaap AP. Chemiluminescent acridan phosphate labeling compounds for detection in gels. *Luminescence* 2000; in press.
5. Akhavan-Tafti H, Compositions and methods for generating red chemiluminescence. US Patent 5,965,736. 1999.
6. Akhavan-Tafti H, Compositions and methods for generating red chemiluminescence. US Patent 6,045,991. 2000.
7. Schoenfelner BA, Lutz J, Sugioka K, Akhavan-Tafti H, Schaap AP. The Lumigen CCD camera system: An economical cooled CCD camera system for chemiluminescent applications. In: Campbell AK, Kricka LJ, Stanley PE, eds. *Bioluminescence and Chemiluminescence Fundamentals and Applied Aspects*, Chichester: Wiley, 1994:633-636.
8. Miska W, Geiger R. Synthesis and characterization of luciferin derivatives for use in bioluminescence enhanced enzyme immunoassays. *J Clin Chem Clin Biochem* 1987;25:23-30.

ANTIBACTERIAL AND ANTIOXIDANT LEAF FRACTIONS FROM *ACTINIDIA CHINENSIS*

A BASILE AND S SORBO

Dipartimento di Biologia Vegetale, Universit Federico II , Napoli, Italy
E-mail adbasile@unina.it

ML VUOTTO, J SEPE, R MIRANDA, MTL IELPO, I SIMEONE, M MAESTRALE AND L RICCIARDI

Istituto di Patologia Generale e Oncologia, II Universit degli Studi, Napoli, Italy

P DE SOLE

Istituto di Chimica Clinica, Universit Cattolica, Roma, Italy

L FERRARA AND D MONTESANO

Dipartimento di Chimica Farmaceutica, Universit Federico II , Napoli, Italy

1. INTRODUCTION

Plants produce chemicals that are toxic to bacteria and fungi. Inducible defense mechanisms, such as phytoalexins, play a role in plant resistance to pathogens. Phytoalexins belong to a wide range of different substances as flavonoids and isoflavonoids. The antibacterial activity of flavonoids has been reported against microorganisms such as *Proteus vulgaris*, *Staphylococcus aureus* and *Staphylococcus epidermidis*. Flavonoids also have other bioactivities including enzyme inhibition and cytotoxicity. They are known as well for their antioxidant and anti-inflammatory effects. We studied *Actinidia chinensis* Planch. (Actinidiaceae), whose edible fruit (kiwi) is used in human diet, and reported the antimicrobial activity of the acetonic extract in leaves, stems, skin, pulp and seeds. The aim of this study was to evaluate the *in vitro* antibacterial activity of various leaf fractions against Gram+ and Gram- bacteria. The effects on the oxidative burst of human whole blood phagocytes was also evaluated. The metabolic activation of these cells, induced by particulate or soluble stimuli, was easily measured by light emitted during oxidative reactions, amplifying the photon emission with luminol as the chemiluminogenic probe.

2. MATERIALS AND METHODS

2.1. Plant material

Actinidia chinensis C.T. Liang (Actinidiaceae) is a plant from China. The plant material was collected from the Botanical Gardens at Naples University.

2.2. Extract preparation

Leaves (50 g) were air-dried and extracted in acid as reported elsewhere.[1] After filtering and concentrating, the acid aqueous extract fraction was progressively extracted in different solvents (alcohol, ethyl acetate, chloroform).

2.3. Flavon identification

The identification of flavonoids was made by HPLC, according to Games et al.[2] Flavonoid identification was carried out by comparing the retention time with the eluted standards in the same conditions. The standard flavonoids identified in the active fraction were tested for antibacterial activity. The samples were prepared by dissolving 100 mg of the dry residue in 10 mL of sterile physiological Tris buffer (pH 7.4), 0.05 M.

2.4. Antimicrobial assay

Bacterial strains were grown on MH (Mueller-Hinton) agar plates (DIFCO, Detroit, MI) and suspended in MH broth (DIFCO). The MIC values against bacterial strains were performed by MH broth-dilution method using tetracycline hydrochloride (Pharmacia, Milano), benzylpenicillin sodium (Cynamid, Catania) and cefotaxime sodium (Roussel Pharmacia, Milano) as reference antibiotics. Eleven bacterial strains, came from the American Type Culture Collection (ATCC; Rockville, MD, USA).

2.5. Chemiluminescence assay

Modified Krebs-Ringer phosphate medium (KRP*), luminol (Sigma, Chemical Co., St. Louis, MO, USA), zymosan (ZYM, Sigma), opsonized zymosan (OZ) and phorbol myristate acetate (PMA, Sigma) were prepared as reported.[3] The dry residue of the different fractions from *A. chinensis* leaf extract was resuspended at 10 mg/mL in KRP* and diluted in the same medium for CL assays. These were performed using an automatic luminometer (Autolumat LB953, Berthold, Wildbad, Germany). CL activity of WB phagocytes (1.0 μL of whole blood equivalent) was measured in absence or in presence of various concentrations of the fruit extract, without (basal CL) or with 0.5 mg of OZ or 150 nmoles of PMA, in 1.0 mL final volume. To verify interferences between *A. chinensis* extracts and CL backgrounds, control samples containing only luminol and *Actinidia* dilutions were prepared. To exclude any toxic plant effect on cellular performance, we determined the phagocyte viability by the trypan blue exclusion test before and after CL evaluation. The CL responses were measured as percent inhibition over 60 minutes with and without fruit extract.

2.6. Statistical analysis

CL data are expressed as percentages of CL inhibition. The statistical analysis was assessed by Student s *t* test. Values of $p<0.05$ were regarded as significant.

3. RESULTS AND DISCUSSION

Of the fractions, only the ethyl acetate showed antibacterial activity. *P. aeruginosa* and *S. aureus* were more sensitive (MIC = 8μg and 16μg/mL respectively), *S. typhi*, *P. vulgaris* and *P. mirabilis* showed a MIC = 64μg/mL. Finally, *E. faecalis*, *E. cloacae*, *E. coli* (MIC = 128μg/mL) and *K. pneumoniae* (MIC = 256μg /mL) were the least inhibited. In the ethyl acetate extract HPLC identified quercetin, rutin, hesperidin, apigenin and kaempferol. Using standard flavonoids in the antibacterial test, the most active were quercetin (MICs between 16 and 256 μg/mL), rutin (MICs between 32 and 256 μg/mL) and apigenin (MICs between 64 and 128 μg/mL) (Table). Our previous data on the antibacterial activity of the acetonic extract from *A. chinensis* leaves showed that only *S. pyogenes* and *P. aeruginosa* were inhibited by the extract, while the other bacterial strains were resistant at the highest concentration used. Here we confirm the higher sensitivity of *P. aeruginosa*; in addition, the more effective extraction method and purification employed has allowed us to identify the activity even on the other bacterial strains. Extracts from leaves of medicinal plants containing flavonoids have been reported to possess antimicrobial activity. The effects of ethyl acetate leaf extract on CL emission by phagocytes unstimulated (WB) or activated by particulate (ZYM and OZ) or soluble (PMA) stimuli are reported in the Fig. All extracts showed inhibitory effects on leukocyte CL emission, but the most active was ethyl acetate (Figure). A dose-dependent reduction of total light emission was observed with this extract in resting and activated phagocytes. The inhibition of CL emission was greatest when ZYM was the activator. A smaller inhibition occurred with PMA. When the extract acted on OZ-stimulated leukocytes, no significant CL reduction was observed. Resting leukocyte CL emission was also inhibited. Therefore, the antioxidant effect of *A. chinensis* leaf fraction is stimulus-independent, even acting on resting cells, but it is not membrane receptor-mediated. In fact, the inhibitory activity does not appear with OZ-stimulated cells. Because the most active fraction is rich in flavonoids, the biological effects are likely related to these compounds. Hence, *A. chinensis* leaf extracts may benefit medicine and agriculture.

TABLE 1

	EtAc	Ctax	Tet	Penb	Rut	Qu	Ap
Staphylococcus aureus ATCC 13709	16	2	2	0.03	32	16	R
Enterococcus faecalis ATCC 14428	128	R	2	8	R	R	R
Proteus vulgaris ATCC 12454	64	2	R	4	R	R	R
Proteus mirabilis ATCC 7002	64	0.03	32	4	R	R	R
Pseudomonas aeruginosa ATCC 27853	8	16	32	R	64	32	64
Salmonella typhi ATCC 19430	64	0.5	1	4	R	R	R
Escherichia coli ATCC 11229	128	0.1	4	64	64	64	128
Enterobacter aerogenes ATCC 13048	128	R	R	4	128	128	64
Enterobacter cloacae ATCC 10699	128	R	R	4	256	256	64
Escherichia coli ATCC 11229	128	0.1	4	64	64	64	128
Klebsiella pneumoniae ATCC 27736	256	0.1	16	R	64	64	128

MIC values expressed as μg/mL for active fraction, flavonoids and reference antibiotics. **R**=absence of inhibition even at the highest concentration used. **EtAc**: ethyl acetate fraction of the aqueous extract, **Ctax**: cefotaxime sodium, **Tet**: tetracycline hydrochloride, **Penb**: benzyl penicillin sodium, **Qu**: quercetin, **Rut**: rutin, **Ap**: apigenin. Other fractions were inactive, thus the data were not reported.

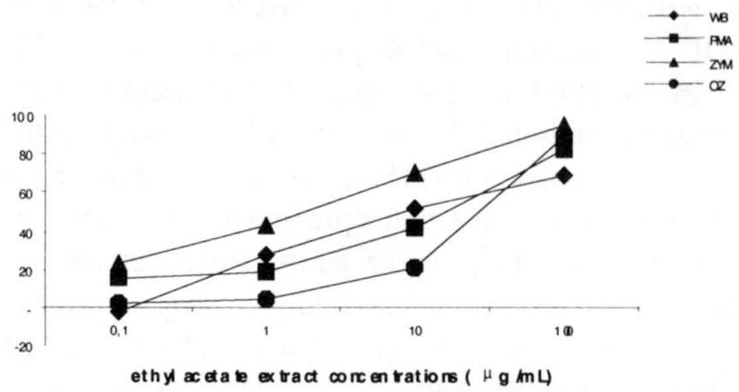

Fig. Influence of ethyl acetate *A: chinensis* extract on CL emission from whole blood phagocytes resting (WB) or activated with soluble (PMA) or particulate (ZYM, OZ) stimuli.

References

1. Basile A, Sorbo S, Giordano S, Ricciardi L, Ferrara S., Montesano D, Castaldo Cobianchi R, Vuotto, M.L. Ferrara L. Antibacterial and allelopathic activity of extract from *Castanea sativa* leaves. *Fitoterapia* 2000, in press.
2. Games DE, Martinez F. Evaluation of the moving belt as an interface for the high-performance liquid chromatographic-mass spectrometric analysis of the flavonoid aglycones. *J Chromatogr* 1989; **474**:1989-2372.
3. Vuotto ML, Basile A, Moscatiello V, De Sole P, Castaldo-Cobianchi R, Laghi E, Ielpo MTL. Antimicrobial and antioxidant activities of *Feijoa sellowiana* fruit. *Int J Antimicrob Agents* 2000; **123**: 197-201.

ANTIMICROBIAL AND ANTIOXIDANT ACTIVITIES IN *FEIJOA SELLOWIANA* SKIN AND PULP

A BASILE

Dipartimento di Biologia Vegetale, Università "Federico II", Napoli, Italy, E-mail adbasile@unina.it

ML VUOTTO, L RICCIARDI, L PEZONE AND D MANCINO

Istituto di Patologia Generale e Oncologia, II Università degli Studi, Napoli, Italy

F SENATORE

Dipartimento di Chimica delle Sostanze Naturali, Università "Federico II", Napoli, Italy

R DE PRISCO

Centro Internazionale di Spettrometria di Massa, Università "Federico II", Napoli, Italy

1. INTRODUCTION

Plant extracts have bioactive compounds that in humans can be important in medicine. Studies have shown that many plant derivatives act as antimicrobial and antioxidant agents.[1] In previous studies we demonstrated that acetonic and aqueous extracts of *Feijoa sellowiana* fruit have bacteriostatic activity and inhibit CL emission from human whole blood (WB) phagocytes and isolated polymorphonuclear leukocytes (PMA), that were activated or not with soluble or phagocytic stimuli. [2,4] *Feijoa* is now grown throughout the Mediterranean area and is widely used for human food. [4] The plant contains large amounts of ascorbic acid, carotenoids, xanthophylls, flavonoids and minerals. [5] This study was designed to evaluate the antibacterial and antioxidant activities in different parts of *Feijoa* fruit (skin and pulp) and compare them to the whole fruit. We measured the *in vitro* antibacterial activity of the lyophilized extracts from *Feijoa* whole, skin and pulp fruit against Gram+ and Gram- bacterial strains by a broth dilution test [6] and its effects on the reactive oxygen species (ROS) production from human peripheral whole blood phagocytes by a chemiluminescence (CL) assay. [7] The CL emission by these cells, unstimulated or activated with particulate or soluble stimuli, was measured by amplifyng the photon emission using luminol, the chemiluminescent probe that mainly measures the superoxide anion production during the oxidative reactions. [7]

2. MATERIALS AND METHODS

Plant material. Ripe *Feijoa* fruits were collected in autumn from the Botanical Gardens of Naples (Italy).
Microorganisms. The following ten bacterial strains were obtained from the American Type Culture Collection (Rockville, MD, USA): *Staphylococcus aureus* and *Streptococcus faecalis*, both Gram+; *Proteus mirabilis*, *P. vulgaris*, *Klebsiella pneumoniae*, *Enterobacter cloacae*, *E. aerogenes*, *Pseudomonas aeruginosa*, *Escherichia coli*, *Salmonella typhi*, all Gram- strains.
Preparation of *Feijoa* extracts. The whole fruit (1000 g), the pulp (500 g) and skin (200 g) were separately blended and exhaustively extracted with acetone. The extracts were concentrated in a small volume under reduced pressure and the residues were fractionated between H_2O and Et_2O. The aqueous layers were lyophilized and analyzed in this study for their activities.
Antimicrobial assay. Bacterial strains were grown on Muller Hinton (MH) agar plates (DIFCO, Detroit, MI) and suspended in MH broth (DIFCO). The MIC values against bacterial strains were performed using the MH broth-dilution method.[6]
Preparation of the extracts for CL assay. Each *Feijoa* fraction was resuspended at 10 mg/mL in modified KRP [7] and diluted in the same medium for CL assay.
Blood. Heparinized peripheral blood was collected from three healthy donors.
Luminol-dependent chemiluminescence assay. Chemiluminescence assays were performed according to De Sole et al.[7], using an automatic luminometer (Autolumat LB953, Berthold, Wildbad, Germany). WB leukocytes were resting or activated with zymosan (ZYM) or zymosan opsonized with autologous serum (OZ).
Data analysis. CL data from the three experiments were calculated as mean ±SEM and expressed as percent activity. Levels of significance were determined using Student's *t* test. Values of $p \leq 0.05$ were regarded as significant.

3. RESULTS

Antibacterial activity. Lyophilized fraction from *Feijoa* pulp showed growth inhibition against *E. aerogenes, E. cloacae* and *P. aeruginosa. S. aureus* and *S. faecalis* were resistant to the extract. Gram- bacteria were more sensitive (Table). Control cultures, containing only sterile physiological Tris buffer, showed no bacterial growth inhibition.
Whole blood CL studies. The CL analysis showed that oxidative metabolism was strongly inhibited in resting leukocytes incubated with *Feijoa* extracts and without stimuli (Figure a). The level of inhibition was greater with skin than pulp or whole fruit extracts. Cells zymosan- or PMA-stimulated showed a significant reduction of CL activity with all the extracts tested and the skin was again the most active (Figure b). When opsonized zymosan was used as cell stimulus, no significant inhibitory effect was found (data not shown).

4. DISCUSSION

Recently, many plants have received great attention as sources of antibiotics and ROS production inhibitors or oxygen free-radical scavengers.[1-5] In fact, they contain pharmacological agents, such as flavonoids and related compounds that can prevent or delay bacteria growth and protect the cells against the potentially deleterious consequences of ROS released from the cells into the extracellular milieu with tissue injury and inflammation. In this study, the lyophilized extract from the pulp was the most active in the inhibition of bacterial growth. The presence of antibacterial activity in this fraction of the fruit could be important in protecting the embryos from pathogen attacks. The set of data concerning CL analysis shows that lyophilized *Feijoa* extract from the skin is the most active, inhibiting, more than other parts of the fruit, native, ZYM- and PMA-, but not OZ-stimulated CL emission. Thus, the activity is not membrane receptor-mediated. The skin is particularly susceptible to free radical damage because of its contact with oxygen and other environmental stimuli; therefore, its antioxidant activity, together with the robust structure of esocarp, could afford fruit protection against degenerative processes especially related to fruit senescence. Various active compounds could be responsible for the observed bioactivities, among those are ascorbic acid and glycosilated flavones.Our findings show that *Feijoa* acts as an antioxidant agent and is bacteriostatic against several enteropathogens. Because ROS could be part of the cellular metabolism of such bacteria, *Feijoa*, counteracting the oxidative stress, may be an important inexpensive clinical agent on gastrointestinal diseases of microbial origin.

TABLE

	F. pulp extract	Cefotax	Tetrac	Penicil.
Escherichia coli	125	0.1	4	64
Salmonella typhi	125	0.5	1	4
Klebsiella pneumoniae	125	0.1	16	R
Enterobacter aerogenes	64	R	4	R
Staphylococcus aureus	R	2	2	0.03
Proteus vulgaris	250	16	2	2.5
Pseudomonas aeruginosa	64	R	R	16
Enterobacter cloacae	64	R	4	R
Proteus mirabilis	250	0.03	32	4
Streptococcus faecalis	R	R	8	2

This table reportes the MIC values expressed as μg/mL for extract from fruit pulp and for reference antibiotics. A comparison is shown between the antibiotic activity of the pulp fruit extract and reference antibiotics. ***R*** = absence of inhibition even at the highest concentration used. ***Cefotax***: cefotaxime sodium; ***Tetrac***: tetracycline hydrocloride; ***Penicil***: benzyl penicillin sodium.

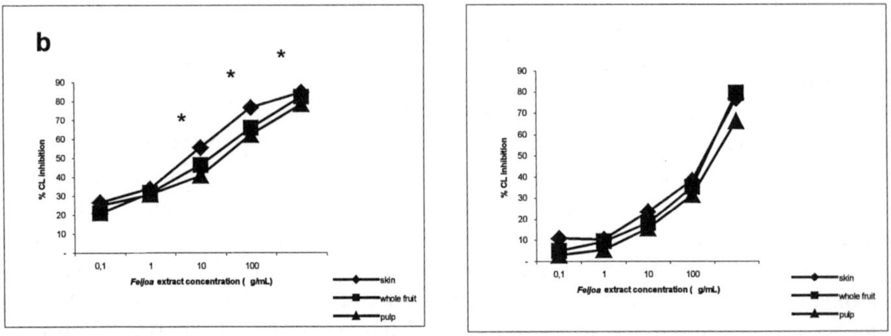

Figure. Peripheral leukocyte CL inhibition by *Feijoa* lyophilized extracts (skin, whole fruit and pulp). Cells were resting (**a**) or activated with PMA (**b**). * p (skin vs. pulp) <0.05

REFERENCES

1. Kandaswami C, Middleton E Jr. Free radical scavenging and antioxidant activity of plant flavonoids. In: Armstrong D, ed. *Free Radicals in Diagnostic medicine*, New York: Plenum Press, 1994: 351-376.
2. Basile A, Vuotto ML, Violante U, Sorbo S, Martone G, Castaldo-Cobianchi R. Antibacterial activity in *Actinidia chinensis, Feijoa sellowiana* and *Aberia caffra. Int J Antimicrob Agents* 1997; **8**:199-203.
3. Ielpo MTL, Moscatiello V, Satriano SMR, Vuotto ML, Basile A. Inhibitory activity on human PMN oxidative metabolism in *Feijoa sellowiana* fruit. *Pharmacol Res* 1997; (Suppl.) **35**:69.
4. Vuotto ML, Basile A, Moscatiello V, De Sole P, Castaldo-Cobianchi R, Laghi E and Ielpo MTL: Antimicrobial and antioxidant activities of *Feijoa sellowiana* fruit. *Int J Antimicrob Agents* 2000; **13**: 197-201
5. Middleton E Jr. Effect of plant flavonoids on immune and inflammatory cell function. In: Manthey and Busling, eds. *Flavonoids in the living system,* New York: Plenum Press, 1998:175-182.
6. Ieven M, Dirk A, Vanden Berghe V, Francis M, Vlietinck A. Lammens E. Screening of higher plants for biological activities. I. Antimicrobial activity. *Planta medica* 1979; **36**: 311-321
7. De Sole P, Fresu R, Frigeri L, Pagliari G, De Simone C, Guerriero C. Effect of adherence to plastic on pheripheral blood monocyte and alveolar macrophage chemiluminescence. *J Biolumin Chemilumin* 1993; **8**:153-158.
8. Heberer M, Ernst M, During M, Algower M, Fisher H. Measurement of chemiluminescence in freshly drawn human blood. *Klin Wochenschr* 1982; **60**:1443-1448.

EFFECT OF SOLVENTS AND POLYMERS ON THE BORONIC ACID ENHANCED PEROXIDASE - LUMINOL - PEROXIDE REACTION

GIOVANNI MAGLIA AND LARRY J KRICKA

Department of Pathology and Laboratory Medicine, 7.103 Founders Pavilion, University of Pennsylvania Medical Center, 3400 Spruce Street, Philadelphia, PA 19104, USA
E-mail: kricka@mail.med.upenn.edu

1. INTRODUCTION

A stable and intense light emission signal is a desirable characteristic of a chemiluminescent assay. Previously, we showed that some hydroxy-polymers stabilized and enhanced light emission of the horseradish peroxidase (HRP) catalyzed chemiluminescent oxidation of luminol enhanced by 4-bromophenyl boronic acid (PBBA)[1,2]. In this study we extend our screening studies to a series of polymers (including non-hydroxy-polymers) and solvents.

2. MATERIALS AND METHODS

Tris[hydroxymethyl]amino-methane (Tris), the sodium salt of luminol (5-amino-2,3-dihydro-1,4-phthalazinedione 3-aminophthalhydrazide) hydrogen peroxide 30% (w/w) solution, horseradish peroxidase type VI-A, dimethyl sulfoxide (DMSO), sec-butanol, isoamyl alcohol, N,N-dimethyl formamide and 2-amino-2-methyl propanol were purchased from Sigma Chemical Company (St Louis, MO 63178, USA). 4-Bromophenyl boronic acid was obtained from Aldrich Chemical Company (Milwaukee, WI 53233, USA). Methanol and ethyl alcohol were obtained from Fisher ChemAlert (New Jersey, 07410, USA). All buffers and solution were prepared using distilled water obtained from Dade International Inc (Newark, DE 19714, USA).

Samples of the following polymers were kindly provided by Tropix (Bedford, MA 01730, USA) - triphenyl phosphonium co-BDMQ 137-83, polyvinyl phthalate, polyvinyl pyrollidone K-90, poly 4-vinyl pyridine, polyvinyl hydrogen phthalate (PUHP EK 5527), 15% C_{12}-dime 85%, Gum Ghatty, Avitex R, DN, 40 and ML (composition; N.J. trade secret), RBQ, Pluronic F-68, TBQ (polyvinyl benzyl tributyl ammonium chloride), poly TBQ, Velvetex OLB-50, Zelec-DP (composition; N.J. trade secret, components include - N-octyl alcohol, isopropyl alcohol, sodium acetate, sodium oleyl sulfate), Lanoquat 1751-A, Polymer 2666 (composition; N.J. trade secret).

We tested the effect of different concentrations of the solvents using an assay mixture comprising Tris buffer solutions of PBBA (20.8 μmol/L 10 uL), HRP (1 nmol/L 10 uL), luminol (1.25 mmol/L)- peroxide (2.7 nmol/L) - solvent (or a dilution of the solvent in Tris buffer)(100 uL). The solutions were pipetted into the wells of a series of white microwell strips. The light emission was then measured for 30 cycles with a pause of 60 seconds at the end of each cycle, using a ML 3000 luminometer (Dynex Laboratories, Chantilly, VA 20151, USA).

In a second series of assays we investigated the effect of polymers on the luminescence reaction. The assay mixture comprised: HRP (0.6 nmol/L, 10 uL), luminol (0.75 mmol/L)-H_2O_2 (1.62 mmol/L)-PBBA (12.5 μmol/L)(100 μL), and the polymer solution (90 μL). The polymers were tested over a final concentration range 0.05 — 4.5 g/L. In both series of experiments the order of addition of components to the microwell was: 1. Luminol-peroxide-(+/- solvent); 2. PBBA; (3. Polymer); 4. HRP.

3. RESULTS

3.1 Screening of solvents

All solvents decreased the light emission from the luminol-PBBA-HRP reaction, but some solvents, for example dioxane, eliminated light emission, even at very low concentration (8%). However all of the solvents tested altered the kinetics of light emission by slowing down the normal rate of light emission. The rank order for inhibition of the light emission was:
dioxane > 2-amino 2-methyl propanol > formamide > DMSO > sec-butanol > methanol > isoamyl alcohol > ethanol > ethylene glycol.

Although, dioxane was the least polar solvent tested and gave the greatest inhibition, the effect of the other solvents was not related to the relative polarities as assessed by the value for the dielectric constant. Thus, DMSO one of the most polar solvents still gave a considerable degree of inhibition. Methanol was chosen as a representative example to test the solvent effects at even lower solvent concentrations, and it was found to decrease light emission and reaction rate over a range of 1.7 to 8.3%.

3.2 Screening of polymers

Seven of the polymers tested both increased and stabilized the light emission from the PBBA enhanced HRP catalyzed luminol-hydrogen peroxide reaction, for example, maximum light emission was achieved approximately 34 minutes (cycle 30) after initiation of the reaction with Polymer 2666 as shown in Fig. 1. The percentage improvement 34 minutes after initiation of the reaction compared to a reaction without polymer was as follows: polyvinyl phthalate (286%), poly 4-vinylpyridine (133%), polyvinyl hydrogen phthalate (250%), Avitex R (240%), Pluronic F-68 (197%), Zelec DP (115%), Polymer 2666 (138%). The effect of polymer concentration on the enhancement was variable, e.g. - for PluronicF-68 no additonal enhancement was observed >0.5 g/L, whereas for polymer 2666 maximum enhancement was observed at 0.05 g/L, but at higher concentrations the polymer was inhibitory.

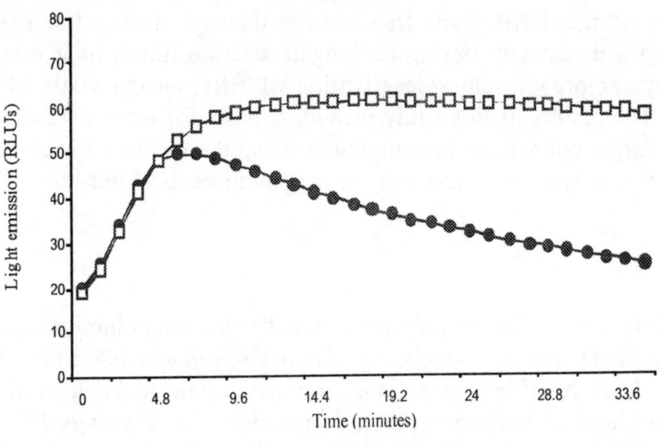

Fig. 1 Effect of polymer 2666 on the 4-bromophenyl boronic acid enhanced HRP-catalyzed oxidation of luminol (RLUs, relative light units)(□ polymer 0.25 g/L, • no polymer)

4. DISCUSSION

We explored the effect of both low MW solvents and high MW polymer moelcules (including non-hydroxy-polymers) on the PBBA enhanced chemiluminecent luminol-HRP reaction. The mechanism of the effect of the different solvents on the reaction does not seem to be due to a specific interaction between particular functional group on the solvent molecule and a component of the chemiluminescent reaction, because all the solvents inhibited the light emission. In previous studies [1,2] enhancement was observed with several polymers with hydroxyl groups. One supposition was a correlation between polymer efficacy and the presence of hydroxy substituents. In this study we tested a diverse range of polymers with varying numbers of non-hydroxy substituents (e.g., polyvinylpyrollidone) and we obtained signal stabilization in all cases, except for the polymers that contained substituted amino groups (e.g., TBQ). The latter may arise via a quenching effect of the amino group on the excited state product by analogy with results from fluorescent reactions[3].

The signal enhancing and stabilizing effect of the polymers may be due to a number of factors. The most plausible is that a polymer molecule provides a hydrophobic pocket for luminol molecules which would prevent external quenching by solvent molecules. This environment would stabilize the excited state aminophthalate product resulting in more molecules reaching the excited state and decaying with emission of light. The diverse chemical composition of the different polymers does not support a specific chemical interaction with components of the chemiluminescent reaction. The stabilizing effect may be a result of protection by the polymer of the HRP from free radical damage during the reaction, thus preserving enzyme activity during prolonged reaction times, or it may simply be that the polymer prevents loss/deactivation of HRP on the walls of the plastic microwell. The results of this study provide a basis for more extensive screening studies with larger collections of compounds using the methods developed for high-throughput massively parallel screening in the pharmaceutical industry.

References

1. Ji X, Kricka LJ. Effect of polymers on enhanced chemiluminescent assays for peroxidase and peroxidase labels. *J Biolumin Chemilumin* 1996; **11**: 303-307.
2. Kricka, L.J., Ji, X. Effects of polymers on the boronic acid enhanced horseradish peroxidase-luminol-hydrogen peroxide reaction. In: Hastings JW, Kricka LJ, Stanley PE. eds. *Bioluminescence and Chemiluminescence: Molecular Reporting with Photons*, Chichester: Wiley, 1997: 473-476.
3. Al-Ani KE, Al-Sabati M. Effect of polar solvents on amine quenching for p-fluorotoluene fluorescence. *J Phys Chem* 1983; **87**: 446-449.

A NEW VISIBLE OSCILLATING CHEMILUMINESCENCE SYSTEM WITH TETRAALKYLAMMONIUM HYDROXIDE

A. SAMADI-MAYBODI AND S.M. OURAD
Department of Chemistry, Faculty of Basic Science, Mazandaran University, Babolsar, Iran.
E-mail: samadi@umcc.ac.ir

1. INTRODUCTION

While a substantial number of chemical oscillators are known,[1] relatively few visible oscillating chemiluminescence reactions in the liquid phase have been described.[2]

In our studies tetraalkylammonium hydroxides (TAAOH), such as tetramethylammonium hydroxide (TMAOH) or tetraethylammonium hydroxide (TEAOH) have been used in the oscillating system of H_2O_2-KSCN-$CuSO_4$- luminol.

In regular oscillations, the oscillating period and amplitude depend strongly on the concentrations of the reaction ingredients. In these studies, both factors were studied with luminol-H_2O_2-KSCN-$CuSO_4$-TAAOH system in batch reactor.

2. EXPERIMENTAL

Stock solutions were prepared from commercially available reagent-grade (Merck) potassium thiocyanate, 30% hydrogen peroxide, tetramethylammonium hydroxide, copper sulphate pentahydrate and luminol without further purification.

The oscillating reaction was monitored with a home-made apparatus equipped with a model BPY47 photocell (Leybold, Huerth, Germany). The apparatus was connected to a personal computer *via* a suitable interface.

In a batch reactor were placed appropriate volumes of potassium thiocyanate, tetramethylammonium hydroxide, hydrogen peroxide and luminol in the reaction vessel. In each series of experiments, only one of the concentrations of the species was varied whilst holding the others constant.

3. RESULTS AND DISCUSSION

The mechanism for this oscillating reaction was established from pervious investigation on the H_2O_2-$CuSO_4$ and H_2O_2-KSCN sub-systems. The model involves 30 kinetic equations and 26 independent variables.

On the basis of work by Orban[3] on the H_2O_2-KSCN-$CuSO_4$ oscillator, it seems likely that one of active species in the system is copper(I) peroxide as an intermediate.

3.1. Tetrametylammonium hydroxide (TMAOH) concentration

To realise the TMAOH effect, in a batch reactor four experiments with final concentrations of 2.1×10^{-2}, 3.6×10^{-2}, 5.1×10^{-2} and 7.7×10^{-2} M tetrametylammonium hydroxide were investigated.

Figure 1 shows increasing the concentration of TMAOH increases the oscillating period, however, increasing concentration of TMAOH decreases the light amplitude. Results indicate the light pulses of TMAOH are narrower than that of tetraethylammonium hydroxide (TEAOH).

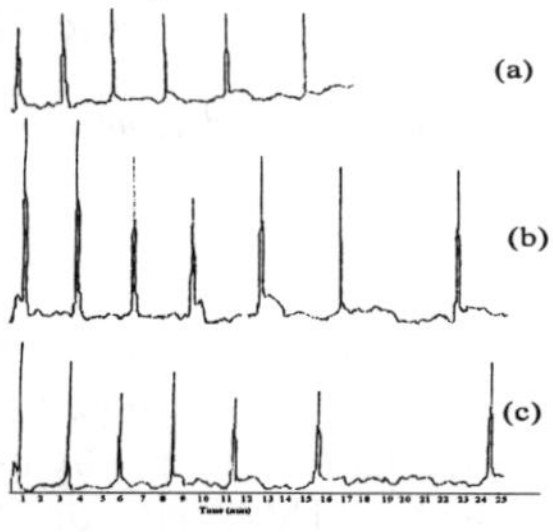

Fig. 1 The effect of TMAOH concentration on the oscillating chemiluminescence. TMAOH concentration: (a) 2.1×10^{-2} M, (b) 3.6×10^{-2} M, (c) 7.7×10^{-2} M. luminol: 1.3×10^{-2} M, H_2O_2:0.3 M, KSCN:6.8×10^{-2} M and Cu(II):3.0×10^{-4} M.

3.2. Luminol Concentration

The effect of luminol has been studied with the final concentration of luminol 2.6×10^{-3}, 5.18×10^{-3}, 7.7×10^{-3}, 1.3×10^{-2}, and 1.5×10^{-2} M.

Figure 2 shows the oscillating pattern of corresponding solutions. It clearly shows the oscillating period and intensity of observing light considerably varied by changing of the luminol concentration. Also increasing the concentration of luminol increases the oscillating period.

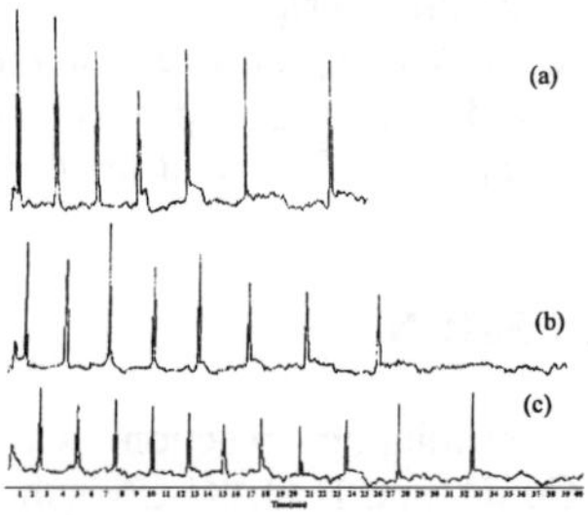

Fig. 2 The effect of luminol concentration on the oscillating chemiluminescence.. Luminol concentration: (a) 1.5×10^{-2} M, (b) 7.7×10^{-3} M and (c) 2.6×10^{-3} M. H_2O_2: 0.3 M, KSCN: 6.8×10^{-2} M, Cu(II) : 3.0×10^{-4} M and TMAOH : 3.6×10^{-2} M.

3.3. Hydrogen peroxide Concentration

The effect of the H_2O_2 has been studied with concentrations of 0.19, 0.30 and 0.58 M.

The corresponding oscillating pattern of those solutions is illustrated in Fig. 3. As can be seen both light-amplitude as well as oscillating period have been greatly influenced by changing concentration of H_2O_2. Fig. 3 illustrates that at the underlying conditions, by increasing the concentration of hydrogen peroxide the oscillating period is decreased appreciably.

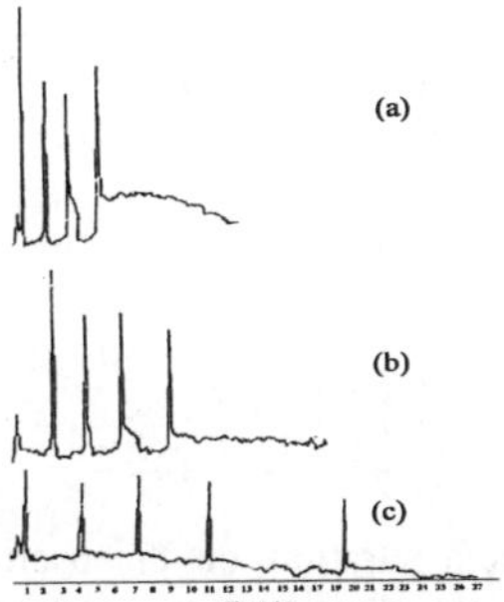

Fig. 3 The effect of H_2O_2 concentration on the oscillating chemiluminescence. H_2O_2 concentration: (a) 0.58 M, (b) 0.30 M and (c) 0.19 M.KSCN: 6.8×10^{-2} M, Cu(II) : 3.0×10^{-4} M, TMAOH:3.6×10^{-2} M and luminol:1.3×10^{-2} M.

3.4. Potassium thiocyanate concentration

The effect of potassium thiocyanate concentration was investigated with different concentrations of 6.2×10^{-2}, 9.2×10^{-2}, 1.2×10^{-1}, 1.5×10^{-1} and 1.8×10^{-1}.

Figure 4 shows the corresponding oscillating pattern of solutions. Variation of KSCN concentration also influences on the oscillating amplitude as well as oscillating period. As can be seen by increasing the concentration of KSCN the number of oscillating is decreased, but the intensity of the emitted light is increased.

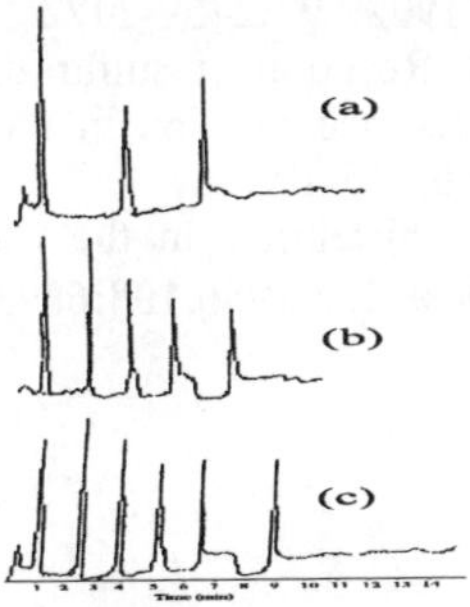

Fig. 4 The effect of KSCN concentration on the oscillating chemiluminescence. KSCN concentration: (a) 1.5×10^{-1} M, (b) 1.2×10^{-1} M, and (c) 6.2×10^{-2} M. Cu(II) : 3.0×10^{-4} M, TMAOH : 3.6×10^{-2} M, luminol : 1.3×10^{-2} M and H_2O_2 : 0.3 M.

3.5. Copper concentration

Five experiments have been carried out with final concentrations of 2.05×10^{-4}, 3.08×10^{-4}, 4.1×10^{-4}, 5.13×10^{-4} and 7.18×10^{-4} M Cu(II).

Figure 5 shows the corresponding oscillating pattern of solutions. This experiment indicates that by increasing of copper concentration the oscillating amplitude is decreased.

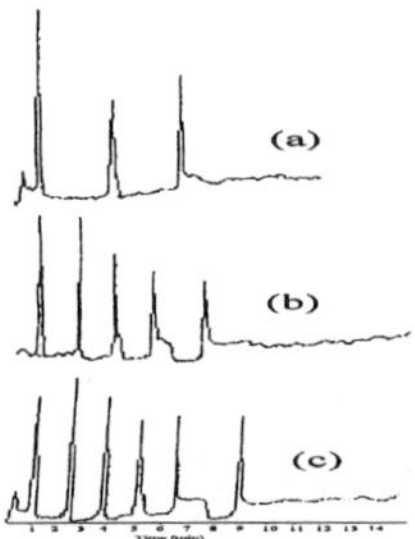

Fig. 5 The effect of copper concentration on the oscillating chemiluminescence. Cu(II) concentration: (a) 1.5×10^{-5} M, (b) 3.08×10^{-4} M, (c) 5.13×10^{-4} M. TMAOH : 3.6×10^{-2} M, luminol : 1.3×10^{-2} M H_2O_2 : 0.3 M and KSCN : 6.8×10^{-2}M.

4. CONCLUSION

The present study introduces a new visible oscillating chemiluminescence using tetramethylammonium hydroxide and indicates how the concentration of components involving in oscillating chemilumenscence has influence on the oscillating period, light amplitude and total time of emitted light. However, results show KSCN, H_2O_2 and Cu(II) have more influence than the others.

References

1. Clarke BL. Stoichiometric network analysis of the oxalate-persulfate-silver oscillators. *J Chem Phys* 1992; **97:**2459-2472.
2. Antipin VA, Parshin GS. Reaction of sulfur dioxide with the dicromate(VI) anion- a new chemiluminescence reaction in an auto-oscillating system. *Dokl Akad Nauk SSSR* 1981; **260:**115-19.
3. Orbàn M. Oscillating and bistability in the Cu(II)-catalyzed reaction between H_2o_2 and KSCN. *J Am Chem Soc* 1986;**108:**6893-6898.

AN ALTERNATIVE TO CHEMILUMINESCENT DETECTION: A FLUOROMETRIC METHOD TO QUANTITATE ACRIDINIUM IN ACRIDINIUM LABELED PROTEINS

DINESH O. SHAH AND VINCE A. SALBILLA

Abbott Laboratories, Abbott Park, IL 60064, USA
Email: Dinesh.Shah@add.ssw.abbott.com

1. INTRODUCTION

Acridinium labeled conjugates are used in various chemiluminescent based immunoassays. 10-methyl-n-tosyl-n-(2-carboxyethyl)-9-acridinium carboxamide trifluoromethyl sulfonate (Ac), Fig. 1, labeled proteins are used as conjugates in the Abbott PRISM® system. The Abbott PRISM® system is a fully automated microparticle-based chemiluminescence immunoassay analyzer (ChLIA). Determining the amount of acridinium in acridinium labeled proteins has historically been done using absorbance techniques. At low concentrations, however, the optical density (OD) method gives variation that affects the precise determination of the incorporation of acridinium to protein. In this situation we need a more sensitive method. The fluorometric method is a good method for this application. Acridinium compounds are known chemiluminescent compounds. These compounds fluoresce when excited in the range 362 – 370 nm and emit in the range 500 – 520 nm. We exploited this characteristic to develop an analytical method to quantitate the amount of acridinium in acridinium labeled proteins. Good correlation between our method and the uv/visible spectroscopic method was observed when several acridinium labeled proteins were compared. However, the fluorometric method shows better sensitivity and reproducibility compared to the uv/visible method at low concentrations. The detection limit for acridinium by our method is ≤50ng/mL. This method shows promise as a potential QC tool to check for purity in acridinium compounds as well as in the routine analysis of acridinium in acridinium labeled proteins.

2. MATERIALS AND METHODS

The process utilizes the Hitachi F-4010 Fluorescence Spectrophotometer as well as the Beckman DU-7 Spectrophotometer. Materials required for the fluorescence procedure consist of a phosphate buffer, the designated samples, and instruments to perform the dilution of the samples. Samples are prepared to a final concentration of 5ug/mL by diluting them with phosphate buffer. The phosphate buffer serves as the blank before the samples are read. Wavelength excitation is set in the range of 362 –

370 nm and emission is set in the range 500 – 520 nm. The method consists of reading the blank prior to running the diluted samples using the Hitachi F-4010. Calculating the Ac/IgG incorporation ratio using fluorescence spectrophotometer readings is a four (4) step process. Given the molecular weights of Ac and IgG, one can determine the incorporation ratio using OD readings and RFU readings.

A. $\frac{\text{RFU of sample} \times 5.0}{\text{RFU of standard (5ug/mL)}}$ = _____ ug/mL of Ac in sample

B. $\frac{\text{A}}{\text{Molecular weight of Ac}^{a}}$ = _____ mM of Ac

C. $\frac{\text{Corrected OD of the sample at 280 nm}^{b}}{224^{c}}$ = _____ mM of IgG

D. Ratio of acridinium / IgG = B/C

a. Sulfo-propyl Ac = 570 g/mol and Methyl Ac = 612.6 g/mol
b. Total OD at 280 nm – 0.25 x OD at 370 nm
c. For IgG: i. OD of 280 nm at 1 mg/mL concentration = 1.40
ii. Molecular weight of IgG = 160,000. Thus, 160 x 1.40 = 224

Materials required for the uv/visible method include phosphate buffer with 0.1% CHAPS, the sample, and instruments for carrying out the dilution steps. Calculation of Ac to IgG ratio using the uv/visible method is as follows:

Acridinium : antibody (protein) ratio = acridinium molarity/antibody molarity

$$\frac{(A_{370} / 14{,}650)}{(\text{corrected } A_{280} / 224{,}000)} = \text{Acridinium : antibody ratio}$$

3. RESULTS AND DISCUSSION

In order to have more precise determinations of acridinium to protein ratios, as discussed above, at low concentrations of conjugate the fluorometric method has an advantage over the uv/visible method (Table 1).

Table 1, below, shows the fluorescence method yields more precise data when compared to optical density reads at wavelength 370 nm. The RFU values show less variation compared to the OD at 370 nm values.

Table 1: Comparison of optical density and relative fluorescence unit data.

Conjugate	OD at 370 nm	RFU
Anti-HBcore conjugate	0.060	260.7
	0.064	260.3
	0.057	260.2
	0.054	260.7

Table 2, below, presents the Ac:IgG ratio determination using both the OD method and the RFU method using four lots of anti-biotin methyl acridinium conjugate.

Table 2: Incorporation ratio determination using the OD and RFU methods.

Conjugate Lot	Ac/IgG input	OD Ac/IgG	RFU** Ac/IgG
1	5:1	1.71	1.69
2	5:1	1.65	1.48
3	6:1	2.25	1.96
4	6:1	2.17	1.75

** RFU = Relative Fluorescence Units.

A serial dilution profile using this method showed greater linearity versus the OD method as well. Fig. 2 shows the results of a 5-fold serial dilution using this method. The Ac stock solution for this experiment was at the usual sample concentration of 5ug/mL. The plot in Figure 1 shows that the RFU method gives a more linear line, indicating greater sensitivity, than the OD method. Finally, two (2) lots of sulfopropyl Ac conjugates showing comparable assay performance also showed similar profiles when evaluated with the fluorometric method. The scan in Fig. 3 shows the profiles of two lots of sulfopropyl Ac run at sample concentration (5 ug/mL). The plot shows the profiles tracking almost identical to each other, indicating that this method of Ac detection is reproducible over several lots.

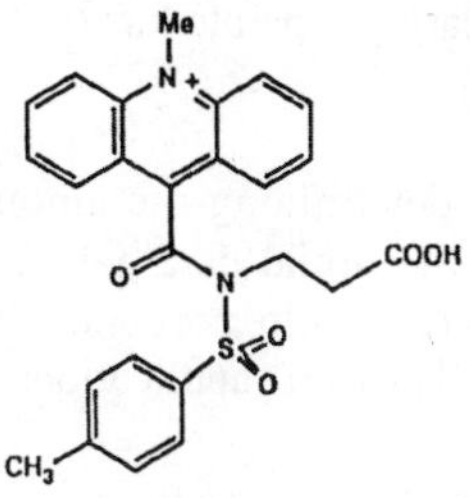

Fig.1 10-Methyl-n-tosyl-n-(2-carboxyethyl)-9-acridinium carboxamide trifluoromethyl sulfonate

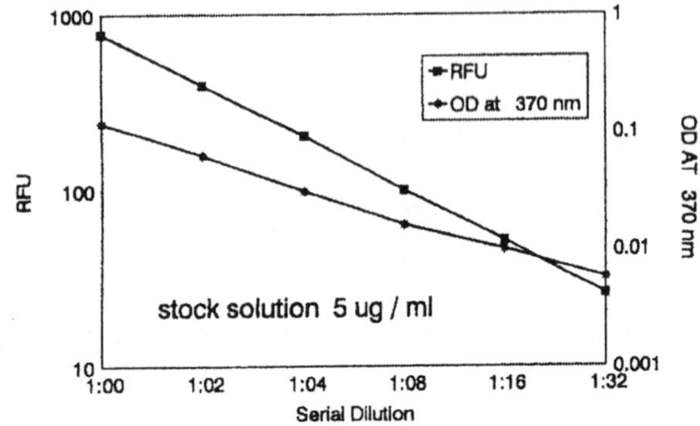

Fig. 2. Plot illustrating OD vs. RFU results from a 5-fold serial dilution of sample at starting concentration of 5 ug/mL.

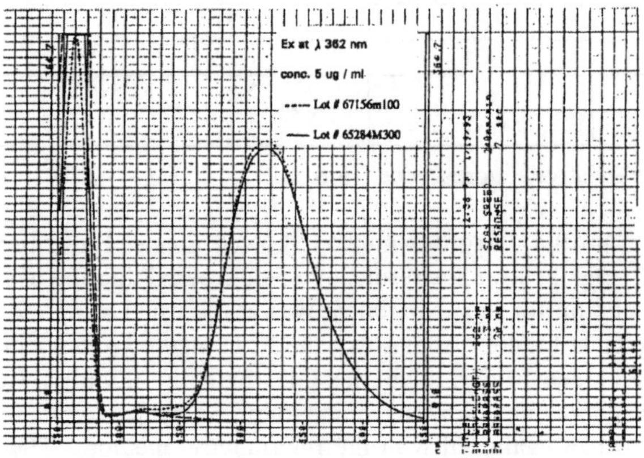

Fig. 3. Fluorescence scan showing good lot to lot reproducibility of the fluorescence method. The scan indicates that this method shows promise as a potential Quality Control (QC) tool when comparing separate lots of Ac labeled proteins.

4. CONCLUSION

The fluorescence method of determining the amount of Ac in Ac labeled proteins appears promising based on the initial data gathered above. This method presents itself as a sensitive and reproducible process. Potential use of this application includes Quality testing as well as R&D applications.

DECOMPOSITION OF "ENVIRONMENTAL HORMONES" BY SINGLET OXYGEN

NOBUTAKA SUZUKI,*[1] IWAO MIZUMOTO,[2] TAKESHI NAGAI,[1] TATEO NOMOTO,[3] HIROKO MATSUYA,[4] BINKOH YODA,[4] KENYA KOZAWA,[5] AND AKIYA KOZAWA[5]
[1]Dept. of Food Science, National Univ. of Fisheries, Shimonoseki 759-6595, Japan
[2]Toyama National Inst. of Maritime Tech., Shin-minato, Toyama 933-0239, Japan
[3]Dept. of Chemistry, Mie Univ., Tsu 514-0008, Japan
[4]Koriyama Women s Univ., Koriyama 963-8503, Japan
[5]ITE/IBA Research Inst., Ichinomiya 491-0806, Japan
suzukin@fish-u.ac.jp

1. INTRODUCTION

"Environmental Hormones", the endocrine disrupter chemicals, have been suspected to disrupt normal hormone actions at extremely dilute concentration when incorporated in living bodies. Therefore, the suspected environmental hormones should be detoxified as fast as possible. Many detoxification methods have been planned and devised using thermolysis, ultraviolet light irradiation, ozone, hydrogen peroxide, titanium oxide/uv, and microorganisms.[1,2] In this study, we describe our successful results using singlet oxygen ($^1O_2$) to treat some representative environmental hormones.

2. EXPERIMENTAL

Chemicals: Rose Bengal (Wako Chem. Co., Osaka, Japan) was used as purchased and also as an immobilized form.[3] Bisphenol A, 4-nonylphenol, DDT, and γ–BHC were purchased from Tokyo Chem. Ind., Tokyo, Japan. PCBs were from Chem. Service Co., U.S.A.

Instrumentation: A metal-halide arc lamp (Wellch Allen, 50 W) was used as an irradiation device. A near-infrared emission spectrometer[4,5] was used for measuring the quenching constant of the environmental hormones against singlet oxygen using a Ge-PIN NIR detector (Applied Detector, COS5020), a monochrometer (JASCO, CT-10S), and a He-Ne gas laser (NEC, GLG5014, 5 mW). Product distribution was analyzed by HPLC (JASCO, 875-UV detector, 880-PU pump, and Finepak SIL C18-5, 46 cm) and LCMS (JEOL Co., JMS-LCmate with HPLC, HP-1100, and Mightysil ODS 2.0 mmID x 15 cm).

Singlet Oxygen Treatment: The apparatus used for the treatment of environmental hormones is almost the same as that used one in shrimp virus inactivation.[6,7] Decomposition of bisphenol A, nonylphenol, DDT, PCBs, and γ-BHC by $^1O_2$

(RB/hν) in MeOH (10^{-5} M) was tested, and the correlation between the decomposition rates and the quenching constants for singlet oxygen was examined.

3. RESULTS AND DISCUSSION

Several representative environmental hormones, such as bisphenol A, 4-nonylphenol, DDT, polychlorinated biphenyls (PCBs), etc., were successfully decomposed by singlet oxygen ($^1O_2$) generated by irradiation of a dye (both as a solution or an immobilized form), such as Rose Bengal, etc., with a metal halide arc lamp (50 W) (light in a shorter wavelength region than 480 nm was eliminated by cut-filters) or an argon laser.

Generation of singlet oxygen was confirmed by the chemiluminescent emission spectrum in the near-infrared (NIR) region (1270 nm) from the reaction mixture as well as by formation and the chemiluminescent degradation of a dioxetane derivative of a chemiluminescent $^1O_2$-probe, provided by Prof. Masakatsu Matsumoto, in the visible region.

The decomposition rates correlate well with their quenching constants (k_q) of $^1O_2$ (at 1270 nm) which were determined by the NIR emission spectrometer constructed in our laboratory. The correlation coefficient between k_q-values and the relative decomposition rates (a reciprocal of the half-lifetime) was 0.9985.

The degradation products of the environmental hormones are under investigation by HPLC, LCMS, and other methods.

Fig. 1 shows the singlet oxygen decomposition of bisphenol A in MeOH; the starting material disappeared in 10 h. Reaction paths were analyzed by HPLC, time curves, and LC-MS of the products.

Half-lifetimes of bisphenol A, nonylphenol, and DDT are 4.7 h, 19.5 h, and 126 h, respectively.

Quenching constants, k_q, for the environmental hormones were measured using an NIR emission spectrometer. The k_q values are plotted against relative reaction rates, the reciprocal of the half-lifetime, and the correlation between them is shown in Fig. 2. They correlate very well as shown in Fig. 2 (correlation factor approximately unity, correlation equation Y = 68.7X -0.17). The correlation equation Y = 68.7X - 0.17, substitute k_q of PCBs, 2.45 x 10^3 for X; the estimated half-lifetime of PCBs can be calculated to be 354 h, which agrees well with the observed half-lifetime, 324 h. From the k_q values for γ-BHC, 2.54 **x** 10^4, the half-lifetime would be estimated to be 162 h.

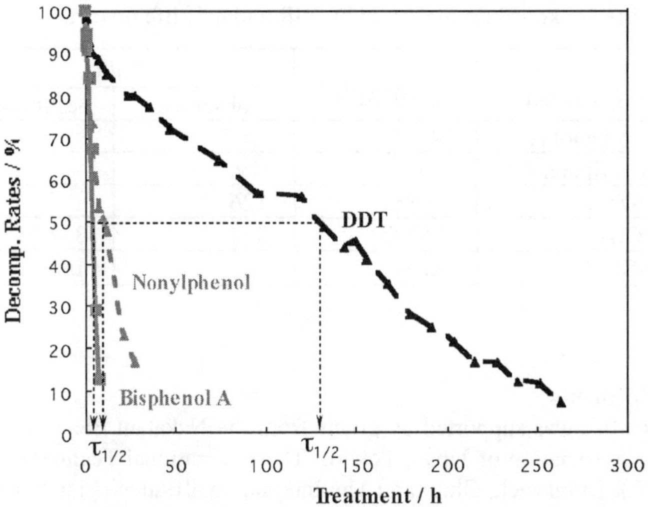

Fig. 1. Decomposition Rates of Environmental Hormones

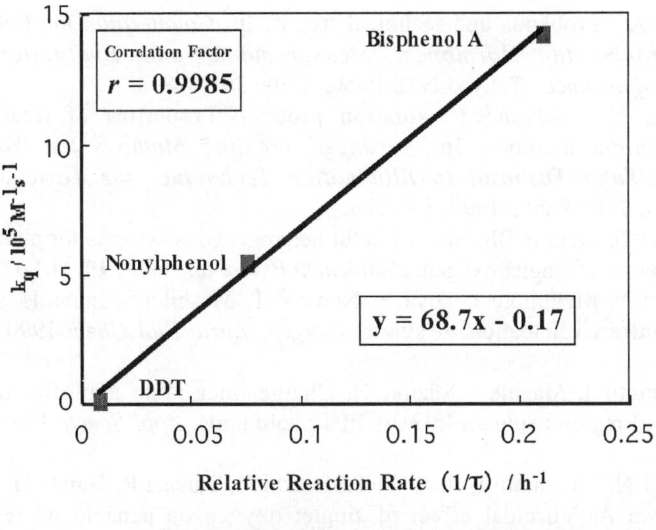

Fig. 2. Correlation between τ and kq

Table 1. kq values measured by NIR and half-life time (τ)

Quencher	$kq/10^5\ M^{-1}s^{-1}$	$\tau_{/2}$/h	
		observed	estimated
Bisphenol A	14.2	4.7	
Nonylphenol	5.44	19.5	
DDT	0.0864	126	
PCBs	0.0245	324	354
γ-BHC	0.254	-----	162

Acknowledgements

Part of this work was supported by grants from the Nakatani Electronic Measuring Technology Association of Japan, Tokyo; The International Technology Exchange Society (ITE), Brunswick, Ohio; and The International Battery Material Association (IBA), Brunswick, Ohio. The authors are sincerely grateful to the donors. We also acknowledge Japan Spectroscopic Co., Tokyo and JEOL Co., Tokyo, for supplying their facilities.

References

1. Miya A. Problems and technical trends. In: *Counterplan for Pollution with Environmental Hormones, Measurements and Evaluation through Correspondence.* Tokyo, NTS Publ., 1999: 391-408.
2. Onari, Y. Advanced oxidation process: Properties of treatments and application methods. In: *Advanced Treating Methods for Water; From WasteWater Disposal to Elimination Techniques for Toxic Substances.* Tokyo, NTS Publ., 1999: 196-226.
3. Nilsson R, Kearns DR. Some Useful heterogeneous systems for photosensitized generation of singlet oxygen. *Photochem Photobiol* 1974; **19**: 181-184.
4. Suzuki N, Mizumoto I, Toya Y, Nomoto T, Mashiko S, Inaba H. steady-state near-infrared detection of singlet oxygen. *Agric Biol Chem* 1990; **54**: 2784-2787.
5. Mizumoto I, Mashiko, Suzuki N. Charge integrating amplifier in the near-infrared region using an InGaAs PIN photodiode. *Appl Spectr* 1993; **47**: 1462-1463.
6. Suzuki N. Mizumoto I, Itami T, Takahashi Y, Tanaka R, Hatate H, Nomoto T, Kozawa A. Viricidal effect of singlet oxygen on penaeid white spot virus (WSSV). *Fisheries Sci* 2000; **66**: 166-168.
7. Matsuura T. Oxygenation Reaction. Tokyo: Maruzen Publ, 1977; p. 244.

CHEMILUMINESCENCE AS A PRODUCT OF THE SPONTANEOUS OXIDATION OF SULFIDE IN SEAWATER

DAVID W. TAPLEY

Dept. of Biology, Salem State College, 352 Lafayette St., Salem, MA 01970, USA
E-mail: dtapley@salemstate.edu

1. INTRODUCTION

The discovery of deep-sea hydrothermal vents stimulated much research into the chemical and biological environment at the vent sites.[1] Research has included investigations into the adaptations of organisms to the toxic effects of sulfide, the chemistry of sulfide oxidation, the biochemistry and physiology of thiotrophic symbioses, and the ecology of vent organisms. One vent organism that has attracted interest is the shrimp *Rimicaris exoculata*, which possesses a pair of light-sensitive organs on its carapace, presumably to see low-level light emissions from the vent plume.[2, 3] Here I present evidence that low-level chemiluminescence during sulfide oxidation can account for light emission at visible wavelengths from the vent plumes.

2. THE CHEMISTRY OF SULFIDE OXIDATION

Most work into the chemistry of sulfide oxidation has examined oxidation rates, rates of appearance of stable end-products, and catalysis by trace metals.[4, 5] Since sulfide is a strong reductant, and readily participates in redox interactions, most workers had assumed a free-radical mechanism for sulfide oxidation,[4] but this has not been demonstrated until recently.[6] Since free-radical reactions are frequently accompanied by chemiluminescence, the production of these reactive chemical species during sulfide oxidation provides a ready mechanism for producing the light that is emitted from the vents.

2.1. Production of Free Radicals During Sulfide Oxidation

In the presence of molecular oxygen and trace metal catalysts, hydrogen sulfide spontaneously oxidizes, with the production of both sulfur- and oxygen-centered free radicals.[6] Figure 1 displays a typical electron paramagnetic resonance (EPR) spectrum obtained when sulfide oxidizes in the presence of the spin-trapping agent dimethylpyrroline-N-oxide (DMPO). The control spectrum shows that no radicals are present in aerated seawater. But when 1 mM sulfide is added, a prominent spectrum appears (Fig. 1B). This spectrum is a composite of at least two spin-trapped signals. Simulation[7] of the signal reveals that it is a composite of the DMPO adducts of the hydroxyl radical ($HO^{\cdot}$) and the sulfite radical ($SO_3^{\cdot -}$) (Fig. 1C). The greater amplitude at the right-hand end of the spectrum is due to accumulation of the trapped adducts during acquisition of the spectrum (about 20 s.)

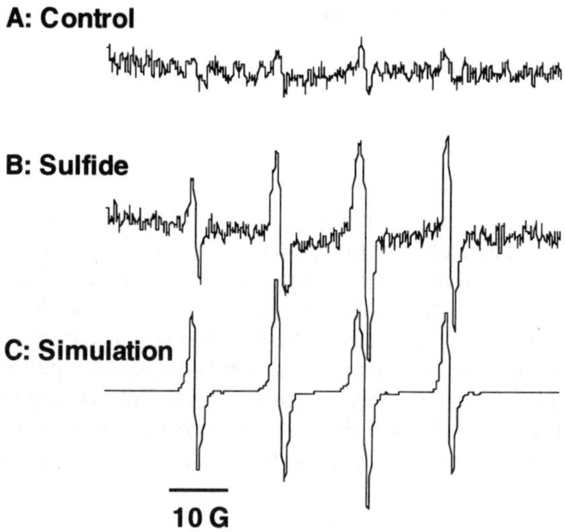

Fig. 1. Electron paramagnetic resonance spectra of sulfide oxidation in artificial seawater using the spin trap DMPO. See text for explanation.

On the basis of these spectra and further experiments, the following reaction mechanism for the initial reactions of sulfide oxidation was proposed:[6]

$$HS^- + O_2 \rightarrow HS^\cdot + O_2^{\cdot -}$$

$$HS^\cdot + O_2 \rightarrow HSO_2^\cdot$$

$$HSO_2^\cdot \leftrightarrow H^+ + SO_2^{\cdot -}$$

$$SO_2^{\cdot -} + O_2 \rightarrow SO_2 + O_2^{\cdot -}$$

$$SO_2 + H_2O \rightarrow H^+ + HSO_3^-$$

$$HSO_3^- + HO^\cdot \rightarrow SO_3^{\cdot -} + H_2O$$

Note that these are not a complete summary of the reaction mechanism; the reactions shown are those that produce the radical species observed in the EPR spin-trapping spectra. All reactions are thermodynamically favorable and proceed with kinetic constants of 10^8 $M^{-1}s^{-1}$ or greater. The complete oxidation of sulfide produces many other products, and it is likely that many side reactions occur simultaneously with those shown here.

2.2. Chemiluminescence during sulfide oxidation

Sulfide oxidation produces weak but measurable chemiluminescence (Fig. 2), as shown by photon-counting experiments conducted using a scintillation counter in the out-of-coincidence mode.[6]

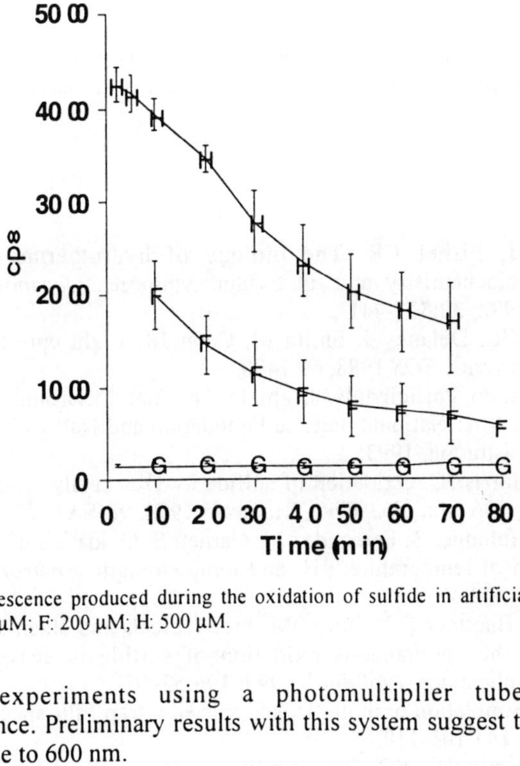

Fig. 2. Chemiluminescence produced during the oxidation of sulfide in artificial seawater. Sulfide concentrations: G: 0 μM; F: 200 μM; H: 500 μM.

More recent experiments using a photomultiplier tube confirm the chemiluminescence. Preliminary results with this system suggest that the emission maximum is close to 600 nm.

3. IMPLICATIONS FOR THE HYDROTHERMAL VENT ENVIRONMENT

The weak chemiluminescence at visible wavelengths observed during sulfide oxidation (Fig. 2) provides a probable mechanism for light production at the deep-sea hydrothermal vents,[2, 3] as suggested by Hastings.[3] Thermal black-body radiation is commonly regarded as a source of visible light in the vent environment,[8] but the conditions necessary for sulfide-dependent chemiluminescence are present at the vent plumes. To determine whether chemiluminescence is a source of light at the vents, in-situ spectra from the vent plumes must be compared with those from sulfide oxidation. Chemiluminescence spectra produced by sulfide oxidation should be determined preferably at the temperature and pressure found at the vents, and the measurement should be made with a low-resolution, high-sensitivity spectrograph of the type under consideration for future vent work.[3]

Acknowledgements

I wish to thank Dr. Garry R. Buettner of The University of Iowa for his expert help with the EPR studies, and Dr. J. Malcolm Shick for his advice throughout this study.

References

1. Childress JJ, Fisher CR. The biology of hydrothermal vent animals: physiology, biochemistry, and autotrophic symbioses. *Oceanogr. Marine Biol. Annu. Rev.* 1992; **30**:337-441.
2. Van Dover CL, Delaney J, Smith M, Cann JR. Light emission at deep-sea hydrothermal vents. *EOS* 1988; **69**:1498.
3. LITE Workshop Participants. Light in Thermal Environments (LITE). In: Woods Hole, MA: National Science Foundation and National Aeronautics and Space Administration, 1993:
4. Chen KY, Morris JC. Oxidation of sulfide by O_2: catalysis and inhibition. *J. Sanitary Eng. Div. Am. Soc. Civil Engineers* 1972; **98(SA1)**:215-227.
5. Millero FJ, Hubinger S, Fernandez M, Garnett S. Oxidation of H_2S in seawater as a function of temperature, pH, and ionic strength. *Environ. Sci. Technol.* 1987; **21**:439-443.
6. Tapley DW, Buettner GR, Shick JM. Free radicals and chemiluminescence as products of the spontaneous oxidation of sulfide in seawater, and their biological implications. *Biol. Bull.* 1999; **196**:52-56.
7. Duling DR. Simulation of multiple isotropic spin-trap EPR spectra. *J. Magnetic Reson.* 1994; **104**:105-110.
8. Pelli DG, Chamberlain SC. The visibility of 350°C black-body radiation by the shrimp *Rimicaris exoculata* and man. *Nature* 1989; **337**:460-461.

A METHOD FOR HIGHLY EFFICIENT CHEMILUMINESCENCE OF IMIDAZOPYRAZINONE IN WATER

K. TERANISHI

Faculty of Bioresources, Mie University, Kamihama 1515, Tsu, Mie 514-8507, Japan
E-mail: teranisi@bio.mie-u.ac.jp

1. INTRODUCTION

2-Methyl-6-(*p*-methoxyphenyl)-[1,2-*a*]pyrazin-3(7*H*)-one (MCLA) is used for the analysis of active oxygen,[1-3] but its light-emitting efficiency and intensity are extremely low in aqueous solvents. Many studies on the enhancement of chemiluminescence have dealt with the regulated medium, such as using a micellar surfactant or cyclomaltooligosaccharide (cyclodextrin).[4-8] It has been shown that the cyclodextrins can enhance the chemiluminescence intensity of MCLA in aqueous solvents.[7,8] Those studies reported that very large amounts of cyclodextrin are needed for the enhancement. For the analysis with chemiluminescence techniques, the analytical system should not affect the subject except to promote luminescence. It is worthwhile to construct a novel chemiluminescence system which promotes greater chemiluminescence activity. We report here that the chemiluminescence of MCLA covalently bound to single cyclodextrin molecule is effectively enhanced in an aqueous solvent.

2. MATERIALS AND METHODS

All chemicals not otherwise mentioned were purchased from Tokyo Kasei Kogyo Co., Ltd. (Tokyo, Japan) or Aldrich Chemical Co. (St. Louis, MO., U.S.A.) in chemically pure grade and were used as such. Cyclodextrins were dried for 12 h at 140 °C under vacuum. Mono-2-*O*-(3-aminopropyl)cyclodextrins[9] and mono-6-amino-6-deoxy-cyclodextrins[10] were prepared according to the published method. Mono-3-amino-3-deoxy-cyclodextrins were prepared by the reaction of mono-2-*O*-tosyl-cyclodextrins[11,12] with aqueous NH_3. 2-Amino-5-(*p*-methoxyphenyl)pyrazine was prepared using a Kishi's method.[13]

Chemiluminescence intensity time curves were obtained using a ALOKA LUMINESCENCE READER BLR-301, and circular dichroism spectra were obtained using a JASCO J-720M spectrometer. Chemiluminescence intensity time curves were obtained as follows: 40 μL of 0.25 mM cyclodextrin-bound MCLA compounds dissolved in distilled water was added to 0.96 mL of 30 mM phosphate buffer (pH 8.0) at 30 °C. The reaction mixture was placed in the photometer and chemiluminescence intensity time curve was obtained at 30 °C. The chemiluminescence quantum efficiencies (ϕ) were determined as relative values to that of luminol as a standard (ϕ= 0.013). Yields of the emitters, amide compounds, generated by the chemiluminescence reaction were obtained follows: an HPLC system was employed for analyzing the luminescence-spent products and the yields of amides were determined by comparison with the corresponding synthesized amides.

3. RESULTS AND DISCUSSION

Syntheses. As shown in scheme 1, mono-aminocyclodextrins and MCLA-COOH, which was prepared from 2-amino-5-(*p*-methoxyphenyl)pyrazine and 2-ketoglutaric acid in EtOH in the presence of H_2 and Pd-C followed by treatment with acetic anhydride in THF, were coupled to afford cyclodextrin-bound MCLA compounds. In order to investigate the influence of distance between MCLA and the cyclodextrins, glycine spacers were introduced between MCLA and the cyclodextrins.

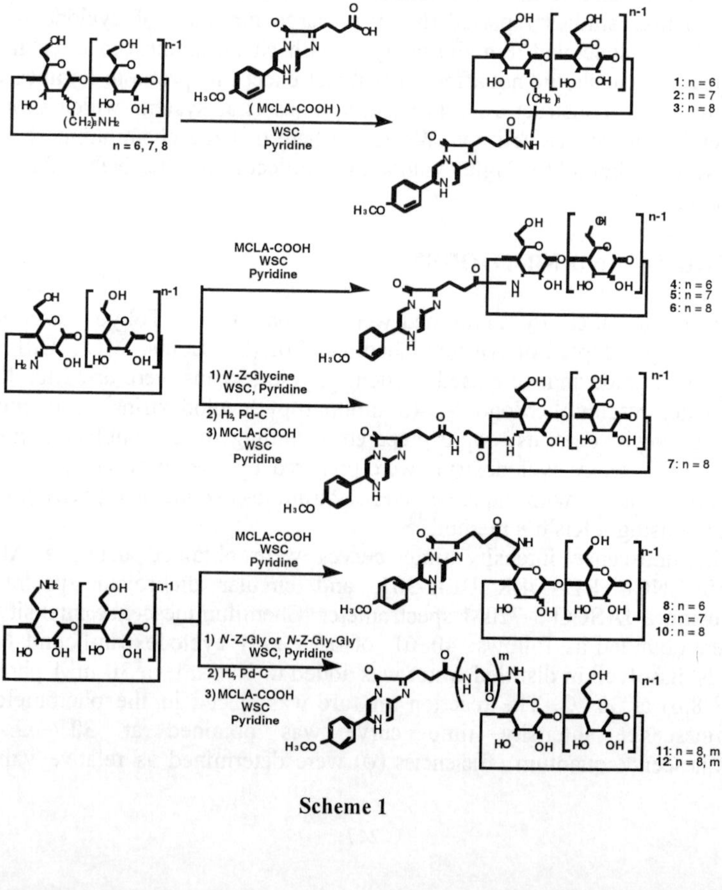

Scheme 1

Chemiluminescence properties. The chemiluminescence efficiency of the oxygen-induced chemiluminescence in phosphate buffer was remarkably dependent on the kind of bound cyclodextrin, the binding site of chromophore to cyclodextrin, and the length of spacer between the chromophore and cyclodextrin as shown in Table 1. When MCLA was covalently attached to C3-carbon of cyclodextrin with a short spacer, namely **6**, its light-emitting intensity showed an up to 44-fold enhancement ($\phi = 0.021$) compared to compound MCLA. ϕ_R and $\phi_S \times \phi_F$ of **6** were 0.41 and 0.052, respectively, and its $\phi_S \times \phi_F$, which was 29-fold as compared to MCLA, was highest in the cyclodextrin-bound MCLA compounds.

Table 1. Chemiluminescent Properties of Cyclodextrin-bound MCLA Compounds

Compound	Φ	Φ_R	$\Phi_S \times \Phi_F$	Compound	Φ	Φ_R	$\Phi_S \times \Phi_F$
MCLA	0.00048	0.27	0.0018	**7**	0.0096	0.36	0.027
1	0.0015	0.16	0.0096	**8**	0.00059	0.39	0.0015
2	0.00048	0.27	0.018	**9**	0.0054	0.49	0.011
3	0.0044	0.14	0.031	**10**	0.0096	0.33	0.029
4	0.00062	0.14	0.0045	**11**	0.010	0.24	0.042
5	0.0036	0.41	0.0087	**12**	0.0058	-	-
6	0.021	0.41	0.052				

The $\Phi_S \times \Phi_F$ is dependent on the molecular state of the light-producing chromophore, amide compound, at singlet excited state generation and generating light. Induced circular dichroism spectra of all the cyclodextrin-bound MCLA compounds exhibited very small bands. It is possible that MCLA chromophores are not included in cyclodextrins, because the MCLA is not sufficiently hydrophobic. Induced circular dichroism spectrum of an amido compound obtained from the chemiluminescence reaction of **6** showed strong positive bands in contrast with those of other amido compounds. This suggests that the light-emitting chromophore of **6** is included in the γ-cyclodextrin cavity. In the chemiluminescence reaction of **6**, after **6** reacts with an oxygen molecule, the 5-(*p*-methoxyphenyl)pyrazine moiety can approach to the γ-cyclodextrin cavity to give the highest $\phi_S \times \phi_F$.

In summary, the light-emitting efficiency of the cyclodextrin-bound MCLA compound in which γ-cyclodextrin was covalently attached with a short spacer showed the high enhancement. This study indicates that the strategy involving covalently attaching a light-producing chromophore onto a cyclodextrin for the enhancement of chemiluminescence is more efficient than the use of an aqueous solution containing very large amounts of cyclodextrin.

Acknowledgements

Author thanks Grant-in-Aid for Scientific Research (No. 12045237 and 11660106) from the Japanese Ministry of Education, Science, and Culture.

References

1.Sawada H, Nakayama M, Detection of linoleic acid hydroperoxide by use of chemiluminescence emitted by cypridina luciferin analogs.*Yukagaku,* 1989; **38**: 103-105.

2.Goto T, Takagi T, Chemiluminescence of a cypridina luciferin analogue, 2-methyl-6-phenyl-3,7-dihydroimidazo[1,2-a]pyrazin-3-one, in the presence of the xanthine-xanthine oxidase system. *Bull Chem Soc Jpn* 1980; **53**: 833-834.

3.Toya Y, Kayano T, Sato K, Goto T, Synthesis and chemiluminescence properties of 6-(4-methoxyphenyl)-2-methylimidazo[1,2-a]-pyrazin-3(7H)-one and 2-methyl-6-(2-naphthyl)-imidazo[1,2-a]pyrazin-3(7H)-one. *Bull Chem Soc Jpn* 1992; **65**: 2475-2479.

4.Grayeski ML, Woolf EJ, Effect of cyclodextrin solutons on chemiluminescence of 10,10'-dimethyl-9,9'biacridinium nitrate. *J Lumin* 1985; **33**: 115-121.

5.Woolf EJ, Grayeski ML, Effect of cyclodextrin solutions on aqueous peroxyoxalate chemiluminescence. *J Lumin* 1987; **39**: 19-27.

6.Karatani H, Effects of cyclodextrin on enhancement for chemiluminescence of the luminol related compound.*Chem Lett* 1986; 377-380.

7.Toya Y, Chemistry of *Vargula* (formerly *Cypridina*) bioluminescence. *Nippon Nogeikagaku Kaishi* 1992; **66**: 742-747.

8.Mitani M, Sasaki S, Koinuma Y, Toya Y, Kosugi M, Enhancement effect of 2,6-*O*-dimethyl-β-cyclodextrin on the chemiluminescent detection of β-D-galactosidase using a cypridina luciferin analog. *Anal Sci* 1995; **11**: 1013-1015.

9.Teranishi K, Tanabe S, Regiospecific alkylation dependent on alkyl chain length of *N*-br,omoalkylphthalimides and an efficient preparation of 2-*O*-aminoalkyl cyclomaltooligosaccharides (cyclodextrins). *ITE Lett Batteries, New Technol Med* 2000; **1**: 53-60.

10.Hamasaki K, Ikeda H, Nakamura A, Ueno A, Toda F, Suzuki I, Osa T, Fluorescent sensors of molecular recognition. modified cyclodextrins capable of exhibiting guest-responsive twisted intramolecular charge transfer fluorescence. *J. Am Chem Soc* 1993; **115**: 5035-5040.

11.Teranishi K, Watanabe K, Hisamatsu M, Yamada T, Efficient regioselective synthesis of mono-2-*O*-sulfonyl-cyclodextrins by the combination of imidazole and molecular sieves. *Carbohydr Chem* 1998; **17**: 489-494.

12.Teranishi K, Tanabe S, Hisamatsu M, Yamada T, Conveniently regioselective mono-2-*O*-sulfonation of cyclomaltooctaose. *Biosci Biotech Biochem* 1998; **62**: 1249-1252.

13.Kishi Y, Tanino H, Goto T, The structure confirmation of the light-emitting moiety of bioluminescet jellyfish aequorea.*Tetrahedron Lett* 1972; 2747-2748.

AB INITIO CALCULATION FOR D-(-)-LUCIFERIN AND ITS INTERMEDIATES IN DIMETHYL SULFOXIDE

NAOHISA WADA* AND KEIICHIRO SAMESHIMA

Department of Life Science, Toyo University, 1-1-1 Itakura –machi, Gunma 374, and Life Science & Material Science Division, Fujitsu Limited, 1-9-3 Nakase, Mihama-ku, Chiba-shi, Chiba 261, JAPAN

*E-mail: *bhwada@itakura.toyo.ac.jp and same@ssd.se.fujitsu.co.jp*

1. INTRODUCTION

Firefly luciferin (Ln) requires a strong base such as t-BuOK in dimethyl sulfoxide (DMSO) to produce the electronically excited oxyluciferin (oxyLn*) in a nonenzymatic reaction.[1] The molecular mechanism of Ln chemiexcitation may consist of two elementary reaction steps. First, the addition of O_2 at the C4 site of Ln occurs only after deprotonation has occurred at the same site to generate linear peroxide. In the next step, the distal end of the peroxide attaches to the carboxylate carbon to form the dioxetane. In order to understand quantitatively the primary process of chemi- and bioluminescence, it is important to clarify the electronic properties of Ln itself in the ground and excited states. Wada et al.[2] applied a molecular orbital (MO) theory to calculate the electronic structure of Ln within the framework of „-electron approximation, and found that the first absorption band of Ln in a basic solution is ascribable to the intramolecular charge-transfer from the benzothiazole to the thiazoline ring,with a large reduction in the amount of pKa in the C6 hydroxyl group. Wada and Shibata[3] observed the spectral changes of Ln in deoxygenated DMSO shows two intermediates of Ln, M440 and M420, using a specifically designed apparatus whereby the O_2 was removed from the DMSO prior to the addition of t-BuOK. Consequently, in this report, we obtained the energy-minimized structures of Ln and its ionized form Ln^{-I} (I=1,2,3) that interact with the solvent by an ab initio calculation. We calculated the optical absorptions for these structures in order to explain qualitatively the observed absorption spectral change of Ln in deoxygenated DMSO.

2. METHOD OF CALCULATIONS

The x-ray diffraction data of Ln [4] was first taken as an initial geometrical configuration so that its energy-minimized structure could be calculated by the semiempirical MNDO-AM1 method in MOPAC2000 (Stewar JJP, Fujitsu Limited, Tokyo, Japan (1999)). Next, using this data as the initial configuration, the energy-minimized Ln structure interacting with the solvent was obtained by an ab initio calculation within the density functional approximation (B3LYP) using the 6-31G* basis set in the Gaussian 98 program (Gaussian Inc., Pittsburgh, USA (1998)) on an SGI workstation. Finally, by the INDO/S method in MOS-F (Fujitsu Limited, Tokyo, Japan (1999)), the optical absorption of Ln was calculated to estimate its observed absorption spectrum.

Although, in accordance with the previous paper, we have no data on the structures of M440 and M420, we calculated the energy-minimized structures and the optical absorptions for Ln^{-I} (I=1,2,3) in the same manner as for the neutral Ln described above. In calculating the optimized geometries of Ln and Ln^{-I} (I=1,2,3), the interaction of these molecules with the solvent was taken into account by the SCRF method.[5] The solvent effect on the optical absorptions of these molecules cannot be considered, however, because the theoretical formulation of the solvent effect is not yet complete in the program package of the INDO/S method.

3. RESULTS AND DISCUSSION

Fig. 1 shows the molecular structures of Ln and Ln^{-I} (I=1,2,3), and the obtained values θ, φ and ψ are listed in Table 1.

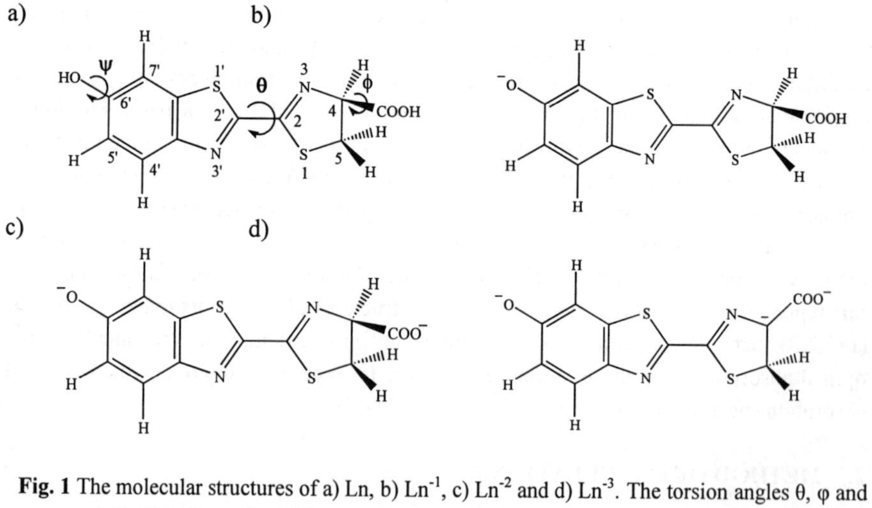

Fig. 1 The molecular structures of a) Ln, b) Ln^{-1}, c) Ln^{-2} and d) Ln^{-3}. The torsion angles θ, φ and ψ are defined as shown in the figure.

In the case of Ln, the torsion angles θ between the benzothiazole and the thiazoline ring are 0°in a vacuum and 0.6°in DMSO. These values correspond well with the experimental value of Ln in the solid phase (3.8°).[4] where both of the hyrdoxyl hydrogen atoms of Ln were involved in hydrogen bonding with the nearest neighbor Ln

Table 1. Torsion angles, θ, φ and ψ, defined as shown in Fig.1, of the energy-minimized Ln and its ionized forms. These values were obtained by the ab initio calculations, where the solvent was assumed to be a continuous medium with the dielectric constant ε and cavity radius a chosen as 45.0 and 5.61□, respectively.

	Ln	Ln^{-1}	Ln^{-2}	Ln^{-3}
θ in a vacuum	0.0°	-2.6°	0.1°	0.0°
in DMSO	0.6°	-2.2°	0.7°	0.7°
φ in a vacuum	53.8°	-14.9°	-81.0°	0.0°
in DMSO	-37.1°	-45.8°	-77.8°	0.0°
ψ □ in a vacuum	0.0°	---	---	---
in DMSO	0.3°	---	---	---

molecules so as to hold the conjugate plane of Ln tightly. Although we have no x-ray diffraction data for Ln^{-I} (I=1,2,3), it can be seen in Table 1 that θ both in a vacuum and in DMSO is almost unchanged with the increments of the ionic valency I. The torsion angles ψ around the C6'-hydroxyl10'O single bond of Ln were calculated as 0° in a vacuum and 0.3° in DMSO. By considering ψ~180° in the solid phase,[4] we found that the C6'-hydroxyl10'O bond tends to fluctuate thermally at room temperature and to rotate around the bond due to a low potential barrier for rotational motion in DMSO.

In contrast with θ and ψ, the torsion angles φ around the C4-carboxyl6Csingle bond were 53.8° in a vacuum and –37.1° in DMSO. Furthermore, these values changed

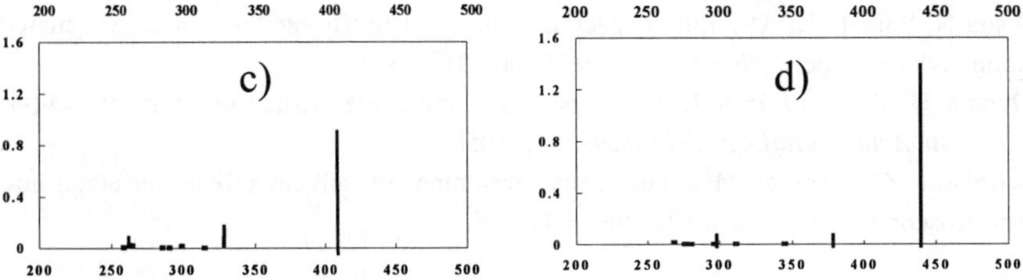

Fig. 2 The optical absorptions of a)Ln, b)Ln^{-1}, c)Ln^{-2} and d)Ln^{-3} as calculated by the INDO/S method, where the ordinate and the abscissa are oscillator strength and wavelength(nm), respectively.

significantly in tandem with the increments of I as shown in Table 1. Based on these energy-minimized structures, optical absorptions were calculated by INDO/S. The results are shown in Fig. 2. The geometries of these molecules are fixed in a vacuum because the INDO/S method is not able to consider the solvent effect at present. Although it is difficult to explain the observed .change of the spectrum of Ln first to that of M440, then to that of M420, according to the calculated optical absorptions of Ln^{-I}(I=1,2,3), each absorption maximum (λmax) of Ln^{-I} (I=1,2,3) was seen to shift as a whole to a wavelength longer than λmax 335 nm of Ln.

One cause of this discrepancy might be the incompleteness of the INDO/S method. Even so, it is necessary to study why the calculated λmax of Ln, 318 nm, is not consistent with the observed value of 335nm. Since the solvent effect on the neutral Ln is much lower than the effect on the ionized Ln^{-I} (I=1,2,3), this inconsistency may be caused by the specificity of the ab initio method itself. An examination of these problems is currently in progress.

Nevertheless, the qualitative conclusions of the present study can be summarized as follows :i)The torsion angle φ is strongly affected by the solvent and also by the ionic valency I in comparison with θ and ψ. ii) The solvent effect causes φ to change greatly in Ln , while causing the largest change of φ in Ln^{-2}, where φ decreases drastically when I is changed from —2 to —3.

References

1.White EH, Steinmetz MG, Miano JD, Wildes PD, Morland R. Chemi- and bioluminescence of firefly luciferin. *J Am Chem Soc* 1980; **80**: 3199-3208.

2.Wada N, Honda M, Suzuki H. Theory of D-(-)-luciferin and dehydroluciferin in acidic or basic solution. *J Phys Soc Jpn* 1985; **54**: 4851-4860.

3.Wada N, Shibata R. Absorption spectral changes of firefly luciferin in deoxygenated dimethyl sulfoxide. *J Phys Soc Jpn* 1997; **66**: 3312-3313.

4.Dennis D, Stanford Jr RH. The crystal and molecular structure of firefly D-(-)-luciferin. *Acta Crystallogr* 1973;**B29**: 1053-1058

5.Karelson MM, Zerner MC. Theoretical treatment of solvent effects on electronic spectroscopy. *J Phys Chem* 1992;**96**: 6949-6957.

PHENOLIC AND INDOLIC RADICAL SCAVENGERS AS LUMINOPHORES

JANINE WENDLER, SABINE HOLST, R. HARDELAND, BEATA K. ZSIZSIK, CHANDANA HALDAR* AND B. POEGGELER

Institute of Zoology and Anthropology, University of Gottingen, Berliner Str. 28,

D-37073, Gottingen, Germany

E-mail: rhardel@gwdg.de

1. INTRODUCTION

Different types of interactions between indolic compounds and free radicals have been described. Several of these possibilities have been studied using melatonin.[1-3] Apart from addition of hydroxyl radicals or formation of neutral indolyl radicals, two reactions have been identified leading to oxidative cleavage of the pyrrole ring: (A) An electron is abstracted from the indole moiety by a highly reactive radical, such as the hydroxyl radical or a cation radical, and the resulting indolyl cation radical combines with a superoxide anion; (B) oxygen of a superoxide anion is transferred to the pyrrole ring under catalysis by the hemoprotein indoleamine 2,3-dioxygenase or, non-enzymatically, by hemin. These two reactions, which presumably involve peroxide intermediates, are relevant for chemiluminescence emitted during melatonin oxidation. Light emission was detected in a xanthine/xanthine oxidase system generating superoxide anions;[4] much stronger luminescence was measured in a hemin-containing alkaline H_2O_2 system generating both superoxide anions and hydroxyl radicals.[5]

This oxidation system proved to be suitable for investigating other free-radical reactions using either metabolites from the melatonin pathway or quinaldic acids and its hydroxylated analogs.[6] In the present study, we extended these experiments to other indolic compounds as well to some phenolic radical scavengers.

2. MATERIALS AND METHODS

The alkaline, hemin-catalyzed H_2O_2 system[5] consisted of glycylglycine/NaOH buffer 0.07 M, pH 8.0, hemin chloride 5 µM, the respective radical scavenger, and 1% H_2O_2. Scavenger concentrations were: resveratrol 50 µM (predissolved at 50 mM in DMSO); sinapinic acid 400 µM (predissolved at 400 mM in DMSO); caffe-

* Visiting scientist to Gottingen 1999; address: Dept. of Zoology, Banaras Hindu University, Varanasi - 221105 INDIA

ic acid 400 μM (predissolved at 1 M in DMSO); curcumin 10 μM (predissolved at 100 mM in DMSO); indole-3-pyruvic acid 400 μM (dissolved in buffer); indole-3-propionic acid 400 μM (dissolved in buffer); 5-methoxytryptophol 400 μM (predissolved at 670 mM in DMSO). The different scavenger concentrations were chosen according to their solubility, avoidance of excess DMSO (which is also a radical scavenger), and avoidance of self-absorption by educts or products. Depending on levels of detectable light, incubation volumes were different: with resveratrol, sinapinic or caffeic acids 2 mL; with curcumin or indole-3-propionic acid 10 mL; with indole-3-pyruvic acid 1 mL. Reactions were started by addition of H_2O_2. Radical reactions leading to chemiluminescence were further analyzed by additions of competing radical scavengers. DMSO was added, at times indicated in Table 1, using final concentrations of 1.27 M for the phenolic, and 0.7 M for the indolic scavengers. Tiron (= 4,5-dihydroxy-1,3-benzene disulfonic acid) was added, at times indicated, in a final concentration of 2 mM. Chemiluminescence was measured by means of a Packard scintillation spectrometer Model 3330, at a temperature of 20°C, operated at settings optimized for determinations of luminescent light.[7]

3. RESULTS AND DISCUSSION

Chemiluminescence was detected during oxidation of all the phenolic and indolic free-radical scavengers tested (Table 1). Light emission rates and kinetics were, however, quite divergent. Among the phenolic substances, resveratrol and curcumin gave highest yields, whereas sinapinic acid was least effective in generating a luminescent emitter. The rate measured for curcumin may even represent underdeterminations with regard to photon fluxes, since the intensely colored compound should absorb a certain proportion of the luminescence. The problem of absorption is even more severe in the case of caffeic acid, from which a pink-colored compound is formed during the initial reaction. It is bleached upon secondary reactions, within a little more than an hour. Addition of a freshly started reaction mix to another one incubated for 1 h resulted in an immediate quenching of luminescence (data not shown). Formation of the pink compound requires hydroxyl radicals, since the presence of DMSO at the start of reaction prevents its appearance. The intermediate, which has not yet been identified, may either represent a benzopyrane-like substance or derive from a quinoid structure. For emitter formation, it seems to be essential. If DMSO is added at the beginning of the reaction, thus blocking production of the pink compound, luminescence is strongly suppressed; however, addition of DMSO at a later time of incubation leads only to moderate inhibition (cf. Table 1).

Table 1. Chemiluminescence emitted during oxidation of radical scavengers and competition experiments with DMSO and Tiron.

Scavenger	Maximal emission [cpm/_mol educt]	Time of maximum [min]	Effect of DMSO: time of addition/ % inhibition after 10 min/60 min	Effect of Tiron: time of addition/ % inhibition after 10 min/60 min
Resveratrol	980,000	60	0 min/ 56/ 69 30 min/ 59/ 64	— /— /— 30 min/ 97/ 99
Sinapinic acid	46,000	160	0 min/ 60/ 46 90 min/ 35/ 68	— /— /— 90 min/ 90/ 98
Caffeic acid	216,000[a]	80[a]	0 min/ 90/ 99 60 min/ 58/ 25	— /— /— 60 min/ 99/ 99
Curcumin	1,540,000[b]	10	0 min/ 54/ 52	0 min/ 90/ 96
Indole-3-pyruvic acid	2,500,000	20	15 min/ 35/ —[c]	20 min/ 98/ 98
Indole-3-propionic acid	186,000[d] 159,000[e]	10[d] 90[e]	0 min/ 46/ 60[f] 30 min/ 19/ 35	0 min/ 97/ >99 30 min/ 81/ 98
5-Methoxy-tryptophol	428,000	10	30 min/ 27/ 33	60 min/ 92/ 91

a: apparent maximum, early emission is quenched by a pink intermediate wich bleaches by time; b: partial autoquench by curcumin; c: no inhibition detectable, even not after two further additions of same amounts of DMSO; d: primary, initial maximum of short duration; e: secondary maximum lasting for about 1 h; f: inhibition by DMSO leads to a prolongation of the reaction (elevated levels last for 3.5 h instead of 2 h in the absence of DMSO).

In competition experiments with the superoxide anion scavenger Tiron and the hydroxyl radical scavenger DMSO, Tiron turned out to strongly suppress luminescence in any of the cases studied. The effect of DMSO remained, however, surprisingly moderate, despite its high concentration. Such amounts of DMSO should suffice for substantially eliminating hydroxyl radicals, and, in comparable studies with kynurenic acid, DMSO was able to almost quantitatively suppress chemiluminescence.[8] The only partial inhibition by DMSO, as observed with the phenolic and indolic scavengers investigated here, indicates multiple reaction mechanisms. One type of reaction seems to be independent of hydroxyl radicals and could involve hemin-mediated oxygen transfer as known from the indolic compound melatonin;[1,2] the second, DMSO-sensitive type involves initial inter-action with a hydroxyl radical followed by combination with a superoxide anion, as known, again, from melatonin.[1,2] The case of indole-3-propionic acid, in which DMSO leads to a partial inhibition, but also to a considerable extension of the luminescent reaction, speaks for two such concurrent mechanisms.

References

1. Hardeland R, Reiter RJ, Poeggeler B, Tan D-X. The significance of the metabolism of the neurohormone melatonin: antioxidative protection and formation of bioactive substances. *Neurosci Biobehav Rev* 1993; **17**: 347-357.
2. Hardeland R, Fuhrberg B. Ubiquitous melatonin – Presence and effects in unicells, plants and animals. *Trends Comp Biochem Physiol* 1996; **2**: 25-45.
3. Hardeland R. Melatonin: Multiple functions in signaling and protection. In: Altmeyer P, Hoffmann K, Stücker M, eds. *Skin Cancer and UV Radiation*, Berlin - Heidelberg, Springer 1997: 186-198.
4. Uemura T, Kadota K. Serotonin- and melatonin-dependent light emission induced by xanthine oxidase. In: Schlossberger HG, Kochen W, Linzen B, Steinhart H, eds. *Progress in Tryptophan and Serotonin Research*, Berlin, Walter de Gruyter 1984: 673-676.
5. Fuhrberg B, Hardeland R. Oxidation of melatonin by free radicals: Catalysis by hemin. In: Hardeland R, ed. *Cell Biological Problems in Chronobiology*, Göttingen, University of Göttingen 1994: 75-81.
6. Hardeland R, Fuhrberg B, Zsizsik BK, Poeggeler B. Chemiluminescence as a tool for monitoring the oxidation of indolic and quinaldic radical scavengers. *Eur J Clin Chem Clin Biochem* 1997; **35**: A108.
7. Hardeland R. Effects of catecholamines on bioluminescence in *Gonyaulax polyedra* (Dinoflagellata). *Comp Biochem Physiol* 1980; **66C**: 53-58.
8. Hardeland R, Zsizsik BK. Kynurenic acid as a free radical scavenger: measurements of educt and product fluorescence and of light emission from an excited intermediate state. In: Hardeland R, ed. *Biological Rhythms and Antioxidative Protection*, Göttingen, Cuvillier 1997: 153-160.

PART 4

APPLICATIONS OF BIOLUMINESCENCE

NEW BIOLUMINESCENT ASSAY OF ALKALINE PHOSPHATASE USING ADENOSINE-3'-PHOSPHATE-5'-PHOSPHOSULFATE AS SUBSTRATE AND LUCIFERIN LUCIFERASE REACTION

H. ARAKAWA, M. SHIOKAWA, O. IMAMURA, A. KOKADO AND M. MAEDA

School of Pharmaceutical Sciences, Showa University, Shinagawa-ku, Tokyo 142-8555, Japan

1. INTRODUCTION

Alkaline phosphatase (ALP) is used as label enzyme in enzyme immunoassay (EIA) and DNA-probe assay. For these assays, the procedure used for the detection of ALP is very important due to the sensitivity and the precision of these assays. Recently, highly sensitive bio- and chemiluminescent method have been developed for detecting ALP labels in EIA and DNA probe assays.[1-4] This paper describes a novel bioluminescent assay of ALP using ATP sulfurylase and luciferin luciferase reaction. The principle of the assay is as follow: adenosine-3'-phosphate-5'-phosphosulfate (APSP) was used as substrate for ALP, which is hydrolyzed enzymatically to produce adenosine-5'-phosphosulfate (APS). APS is converted into ATP by ATP sulfurylase in the presence of pyrophosphate. The produced ATP is detected by a luciferin-luciferase reaction.

The measurable range was 1 zmol to 100 fmol/assay and the detection limit was 10 zmol/assay. The coefficient of variation (CV, n=5) was examined at each point of the standard curve, the mean CV percentage was 4.47 % (n=6). This assay system was applied to enzyme immunoassay of human chorionic gonadotropin (hCG) using ALP as label enzyme. The hCG level in urine could be assayed sensitively and directly by the proposed method.

2. MATERIALS AND METHODS

The bioluminescence intensity was measured by an Aloka Luminescence Reader (Tokyo, Japan). ALP (EC 3.1.3.1) was purchased from Boehringer-Mannheim-Yamanouchi, Co.(Tokyo, Japan). APSP and ATP sulfuryrase (EC2.7.7.4) were purchased from Sigma (St. Louis, Mo). Lucifel-Lu composed of luciferase (EC1.13.12.7) and luciferin were donated by Kikkouman Co., LTD. (Chiba, Japan). hCG, anti hCG antibody and anti hCG-ALP conjugate were donated by Toso Co (Tokyo, Japan). Other chemicals were of analytical reagent grade.

2.1 Procedures

Bioluminescent assay of ALP: A 10 μL ALP sample was mixed with 50 μL of 20 mM APSP solution and incubated for 120 min at 37°C, and then 10 μL of 1 mM pyrophosphate - 0.3 x 10^{-2}U/mL ATP sulfurylase mixtures in 10 mM Hepes buffer (pH 11) containing 1 mM $MgCl_2$ and 0.001% BSA and 100 μL of a luciferin-luciferase solution (Lucifel-Lu) were added successively. The bioluminescent intensity was measured for 10s after a time of 15min following addition of a luciferin-luciferase solution.

Bioluminescent EIA of hCG: hCG standard (100 μL) or a sample solution was added to a microtiter plate coated with anti-hCG antibody. After an incubation of 60 min at 37_, the plate was washed with a saline solution containing 0.1% Tween 20 and 0.05% NaN_3. Next, 100 μL of ALP labeled anti hCG antibody solution was added and the mixture was incubated for 60 min at 37°C. After rewashing again the plate, 100 μ L of a mix solution composed of APSP and Hepes buffer (1:9 of volume) were added and incubated for 2 h at 37°C. Finally, 60 μL of a mix solution composed of ATP sulfrylase, PPi and luciferin/luciferase was added. After incubation of 15 min at 37°C, the luminescent intensity was measured by a MicroLumat LB96P (Berthold).

3. RESULTS AND DISCUSSION

ALP has been widely used as the label enzyme in EIA and DNA probe assay. A sensitive assay of ALP is required for developing better performing assays. At present, it is considered that the chemiluminescent assay of ALP using 1,2 dioxetanes derivative as substrate shows the highest sensitivity (detection limit of 10^{-20} mol). The bioluminescent assay using luciferin derivative has also been reported but with a detection limit inferior to that of the chemiluminescent assay. However, it is possible to develop a sensitive method for ALP by bioluminescence because the luminescence quantum yield in the bioluminescent reaction is higher than that produced in the chemiluminescent reaction. Nyren et al reported a sensitive bioluminescent DNA sequencing method using the enzymes ATP sulfurylase and firefly luciferase.[5, 6] The assay is based on the pyrophosphate (PPi) released during the DNA polymerase reaction, the quantitative conversion of pyrophosphate to ATP by ATP sulfuryrase in the presence of adenosine 5'-phosphosulfate (APS). This method is unique and highly sensitive because this method does not require a separation step by electrophoresis and is determined by a bioluminescence reaction. In this paper, we show an ultrasensitive bioluminescent ALP assay using adenosine-3'-phosphate-5'-phosphosulfate (APSP) as substrate, the sulfurylase for the conversion of the produced APS to ATP in the presence of pyrophosphate and a luciferin-luciferase reaction to detect the generated ATP. The principle for this assay is shown in Fig.1.

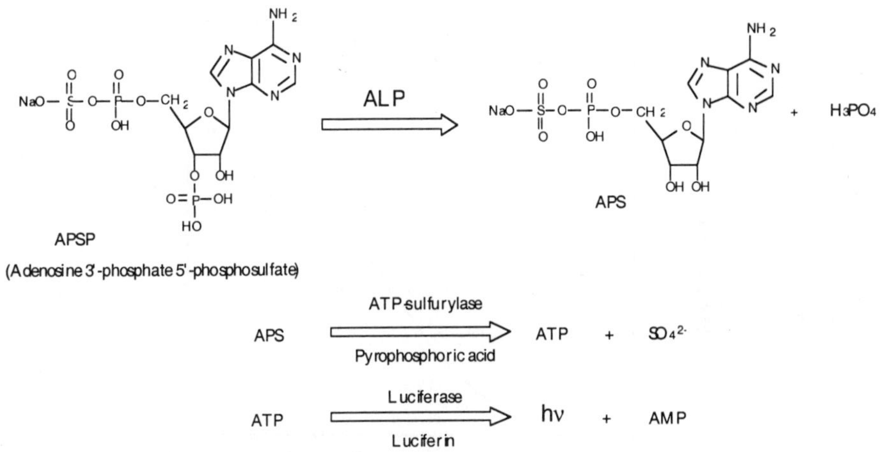

Fig. 1 Principle of the bioluminescent assay for ALP.

3.1 Bioluminescent assay of ALP

The effects of various factors (concentrations of the reagents, pH, reaction time) on bioluminescent assay of ALP were examined to establish the optimal assay conditions as described in Materials and Methods.

A standard curve of ALP was obtained according to the procedure. As shown in Fig.2, a working curve was obtained in the range of 1×10^{-21} mol/assay to 1×10^{-16} mol/assay. The CV percentages (n=6) were 4.47%. The detection limit (blank+2SD) was 1×10^{-20} mol/assay. The sensitivity obtained by the proposed method corresponds to that obtained by a chemiluminescent assay using AMPPD and more sensitive than a bioluminescent assay using luciferin phosphate as substrate.

3.2 Application of bioluminescent assay of ALP in enzyme immunoassay

The new bioluminescent assay for ALP was applied to detect the ALP conjugate in EIA. Immunoassay was carried out according to the method described in the procedure. The measurable range for hCG was 0.001 to 10 mIU/mL. The detection limit was 0.001 mIU/mL (blank+2SD). The sensitivity of this method is about 100 times than that obtained by fluorescence and time resolved fluorescence immunoassay.

In conclusion, a bioluminescent assay of ALP using APSP as substrate has been developed. This assay system of ALP is ultrasensitive. Furthermore, this sensitivity can be improved by the purification of the substrate and the use of enhancement method such as the addition of coenzyme A in the ATP assay. Further studies on an enhancement of sensitivity and the applications of this system are currently in progress in our laboratory.

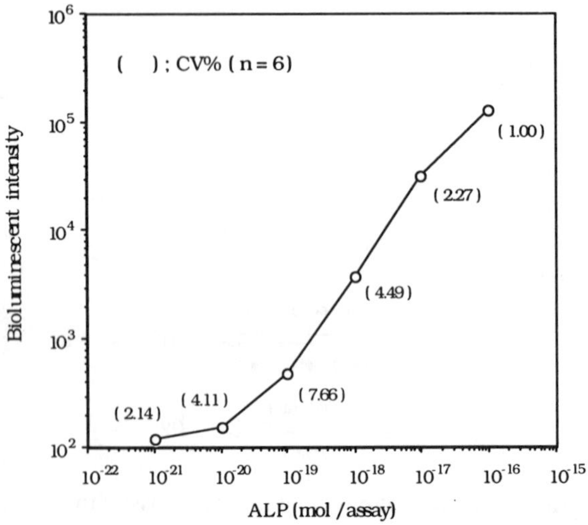

Fig. 2 Standard curve of ALP.

References

1. Bronstein I, Voyta JC, Edwards B. 1,2-Dioxetanes: novel chemiluminescent enzyme substrates. Applications to immunoassay. *Anal Biochem* 1989; **180**: 95-98.
2. Pollard-Knight, Simonds A C, Schaap A, Akhavan H, Brady M. Nonradioactive DNA detection on southern blots by enzymatically triggered chemiluminescence. *Anal Biochem* 1990; **185**: 353-358.
3. Arakawa H, Maeda M, Tsuji A. Chemiluminescent assay of various enzymes using indoxyl derivatives as substrate and its applications to enzyme immunoassay and DNA probe assay. *Anal Biochem* 1991; **199**: 238-42.
4. Miska W, Geiger R. Synthesis and characterization of luciferin derivatives for use in bioluminescence enhanced enzyme immunoassay, new ultrasensitive detection systems for enzyme immunoassays, II. *J. Clin. Chem. Clin. Biochem*, 1987, 25, 23.
5. Ronagh M, Karamohamed S, Pettersson B, Uhlen M Nyren P. Real-time DNA sequencing using detection of pyrophosphate release. *Anal Biochem*, 1996; **242**: 84-89.
6. Nyren P, Pettersson B, Uhlen M. Solid phase DNA minisequencing by an enzymatic luminometric inorganic pyrophosphate detection assay. *Anal Biochem*, 1993; **208**: 171-175.

BIOLUMINESCENCE AS A TOOL TO ASSIST IN THE DEVELOPMENT OF BIOSORBENTS FOR FOOD-BORNE BACTERIAL PATHOGENS

L. BROVKO, W. SUN, F. KHOSRAVI, AND M. GRIFFITHS
Canadian Research Institute for Food Safety, University of Guelph, Guelph, ON N1G 2W1, Canada
E-mail: lbrovko@uoguelph.ca

1. INTRODUCTION

Detection methods for pathogenic bacteria that are currently available are time consuming and not sensitive or rapid enough to meet the requirements of the food industry. Introducing a step to specifically capture bacteria with a biosorbent allows the concentration of the target bacteria in a small volume. This provides an opportunity to avoid enrichment and, at the same time, to decrease the influence of the food matrix on the assay. Several approaches have been used to construct biosorbents for bacterial pathogens, including immunomagnetic separation[1,2] and bacteriophage-based biosorbents.[3] However, the reported capture efficiency was poor.

To improve the performance of these bacterial biosorbents, information regarding both the nature-structure of the biosorbent and conditions of the capture procedure is essential. The methods used at present for the assessment of bacterial capture are time-consuming, laborious and not reliable. It was shown that *in vivo* bioluminescence is a very sensitive and reliable indicator of cell sorption on different surfaces.[4] We propose to use *in vivo* bioluminescence to assess the extent of bacterial capture by the constructed biosorbents.

2. MATERIALS AND METHODS

2.1 Bacterial strains and phages

The strain of *Escherichia coli* O157:H7 #380-94 was obtained from USDA, the strain of *Salmonella enteritidis* and its phage SJ2 were isolated from chicken egg in our laboratory. Both strains were transformed by electroporation with the plasmid pT7 from Dr. E. Meighen, carrying the entire *lux* operon from *Photorhabdus luminescens.*[5] Bacteria were cultured on LB agar (Difco Laboratories, Detroit, MI) supplemented with 50 μg/ml of ampicillin at 37°C, overnight. Bacteriophage was prepared and purified according to Sambrook, et al.[6]

2.2 Construction of antibody- and bacteriophage-based biosorbents.

Affinity purified goat antibodies against *E.coli* O157:H7 (KPL Inc, Gaithersberg, MD) were biotinylated using EZ-Link™ Sulfo-NHS-biotin, Sulfo-LC-NHS-biotin, and Sulfo-LC-LC-NHS-biotin (Pierce, Rockford, IL) according to manufacturer's instructions. For the immobilization of modified antibodies, streptavidin-labelled Dynabeads M-280 (Dynal Inc., Lake Success, NY) were used according to the instructions, supplied by the company.

The purified bacteriophage SJ2 was biotinylated with EZ-Link™ Sulfo-NHS-biotin (1.0-30.0 mg/ml) at room temperature for 30 min, followed by dialysis against 0.85% saline for 24 h. The phage titer (PFU/mL) was examined by the soft agar overlay technique[6] following biotinylation and dialysis. The biotinylated phage was then coated onto streptavidin-labelled Dynabeads M-280 by rocking 100 μL of beads in phage solution (~10^{10} PFU/mL) overnight at room temperature. The immobilized phage was then collected on Magnetic Particle Concentrator Rack (MPCR), washed and resuspended in 100 μL of 0.85% saline.

2.3. Assessment of bacterial capture using *in vivo* bioluminescence

Bacterial cells (1 mL, 10^3-10^8 CFU/mL) were mixed with the prepared biosorbent (20 μL, 10^7-10^8 beads/mL) and incubated at 25°C for 20-30 min on the shaker. The cells captured by the magnetic beads were collected with the MPCR, washed twice and resuspended in LB broth. *In vivo* bioluminescence (RLU) was measured in the supernatant, washing solutions and on the beads using both a tube luminometer BG-P (GEM Biomedical Inc., Hamden, CT) and a luminescence plate reader 1258 Galaxy from BioOrbit (Turku, Finland). The capture efficiency of biosorbent was evaluated by comparing the obtained RLU values with the corresponding calibration curves of RLU versus CFU/mL obtained in a previous experiment.

3. RESULTS AND DISCUSSION

The evaluation of biosorbent performance usually involves enumeration of captured cells by a plate culture technique. However, in this case even if several cells are captured by one bead only one colony will arise on the plate. At the same time, antibody on different beads could react with the same cell, giving a bead cluster that also will give only one colony on the plate. Electron microphotographs of the *E.coli* O157:H7 cells captured by antibody-based biosorbent reveal that both these situations take place.

3.1. Assessment of *E.coli* O157:H7 capture by magnetic immunobiosorbent

Three biosorbents were constructed on the basis of biotinylated antibodies. For all samples investigated it was shown that *in vivo* bioluminescence correlate with intracellular ATP and, thus, with the number of cells in the sample. The percentage of bacteria captured by biosorbents, estimated by *in vivo* bioluminescence depended on the time of bacteria-biosorbent contact, concentration of both cells and magnetic beads and length of spacer between antibody and biotin moiety. Typical results are presented in Table 1.

Table 1. Percentage of *E.coli* O157:H7 captured by commercial immunosorbent and immunosorbents constructed using Sulfo-NHS-biotin (short spacer), Sulfo-NHS-LC biotin (medium spacer) and Sulfo-NHS-LC LC-biotin (long spacer).

Cell density, CFU/mL	Percentage of cells captured on magnetic beads				
	Long spacer	Medium spacer	Short spacer	Anti-*E.coli* O157 beads, Dynal Inc.	Control
3×10^5	70	66	80	76	0
3×10^6	51	49	40	56	2

3.2. Assessment of *Salmonella enteritidis* capturing by phage-based biosorbent

Biotinylation of the phage using different biotin concentrations was performed as described in Materials and Methods. Phage preparation modified by 2.5 mg/mL of biotin retained maximum infectivity. The phage-based biosorbent was constructed using this biotinylated phage. The obtained biosorbent was then employed to capture *S.enteritidis* cells. Non-biotinylated phage adsorbed on the same beads served as a control. The results are presented in Fig. 1.

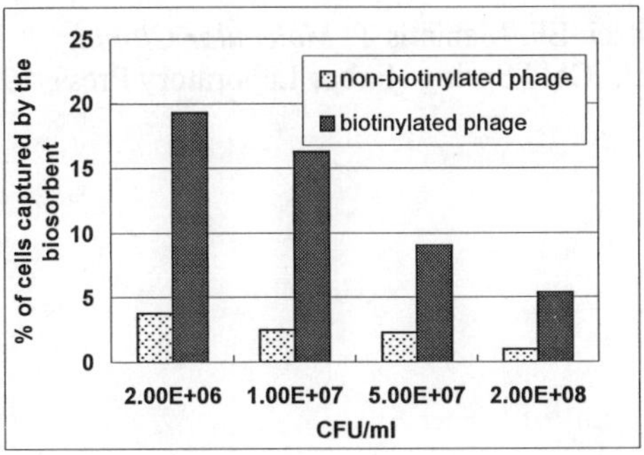

Fig. 1. Percentage of *S.enteritidis* cells captured by the phage-based biosorbents.

4. CONCLUSIONS

The results of present paper confirm that *in vivo* bioluminescence is a reliable indicator of number of captured bacterial cells. The proposed approach could facilitate the development of specific biosorbents and, thus, new rapid methods for the detection of bacterial pathogens.

Acknowledgements

This work was partially supported by Poultry Industry Council, Beef Industry Development Fund and Ontario Cattlemen's Association, Canada

References

1. Prusak-Sochczevski E, Luong JHT. An improved ELISA method for the detection of *Salmonella typhimirium*. *J Appl Bacteriol* 1989; **66**: 127-135.
2. Bennett AR, MacPhee S, Betts RP. The isolation and detection of *Escherichia coli* O157:H7 by use of immunomagnetic separation and immunoassay procedure. *Lett Appl Microbiol* 1996; **22**: 237-243.
3. Bennett AR, Davids FGC, Vlahodimou S, Banks JG, Betts PR. The use of bacteriophage-based system for the separation and concentration of *Salmonella*. *J Appl Microbiol* 1997; **83**: 259-265.
4. Denyer SP, Jassim SAA, Stewart GSAB. *In vivo* bioluminescence for studying the adhesion of bacteria. *Biofouling* 1991; **5**: 125-132.
5. Meighen EA, Szittner RS. Multiple repetitive elements and organization of the *lux* operon of luminescent terrestrial bacteria. *J Bacteriol* 1992; **174**: 5371-5381.
6. Sambrock J, Fritsch EF, Maniatis T. *Molecular Cloning: A Laboratory Manual*, 2nd ed. New York: Cold Spring Harbor Laboratory Press, 1989.

BIOSENSORS FOR THE DETECTION OF HEAVY METALS AND GENOTOXICITY

P. CORBISIER, S. TAGHAVI, A. BOSSUS, E. BRITS, D. VAN DER LELIE AND L. VERSCHAEVE

Vlaamse Instelling voor Technologisch Onderzoek, Vito Boeretang 200, B-2400 Mol, Belgium
E-mail: philippe.corbisier@vito.be

1. INTRODUCTION

The construction of whole cell sensors for the detection of environmentally relevant analytes has received considerable attention. Since most natural promoters are regulated by one or more signals, reporters systems can be placed under the control of such promoters to follow their activity. A number of different micro-organisms have been chosen as cellular hosts taking into account parameters such as their ability to grow in the laboratory, their amenability to genetic manipulation or their environmental relevance. In this paper we will illustrate such bacterial construct that were made to detect several divalent metal ions and genotoxic compounds. In both cases, the bacterial *Vibrio fischeri* luciferase has been preferred as reporter system.

2. RESULTS AND DISCUSSION

2.1. Detection of bioavailable metal ions using metal-lux fusions

Bacterial sensors have been constructed using either *lux* or *luc* reporter systems and reacting to mercury,[1,2] to arsenite and arsenate,[3-5] copper,[6] lead,[7,8] zinc,[6] cadmium,[6,8] thallium,[9] chromate[7] and nickel.[10] The main advantage of such biological systems for the quantification of divalent metal is their ability to detect the fraction of metal that is biologically available for the cellular host containing the metal-lux fusion. The bioavailability of copper[11] or lead[12] in the presence of increasing concentration of metal chelators such as EDTA or CDTA has illustrated this point. We give here another example where a competition between chromate ions and sulphate ions present in the reaction medium is observed. The *chrBΔchrA::luxCDABE* fusion appeared to be specifically induced by chromate ions[7] but the bioluminescence was also dependent upon the sulphate concentration in the medium since chromate is transported into the cell using the sulphate transporter system (Fig.1). Metals ions being complexed, precipitated or competing with other molecule will become less bioavailable and luciferase transcription will be reduced. The exposition of the whole cell sensor to complex matrices is crucial. This is particularly relevant when a solid sample is analysed.[13] A direct contact between the bacterial sensor and the soil particles is recommended since the bacteria are capable to interact with the metal ions that are present on the soil particles and not necessarily dissolved in the medium.

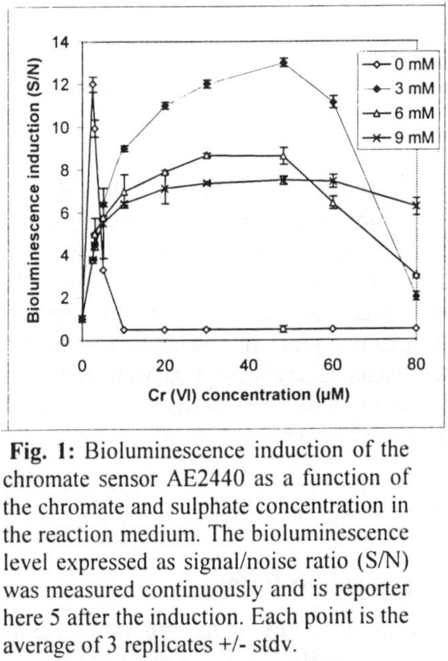

Fig. 1: Bioluminescence induction of the chromate sensor AE2440 as a function of the chromate and sulphate concentration in the reaction medium. The bioluminescence level expressed as signal/noise ratio (S/N) was measured continuously and is reporter here 5 after the induction. Each point is the average of 3 replicates +/- stdv.

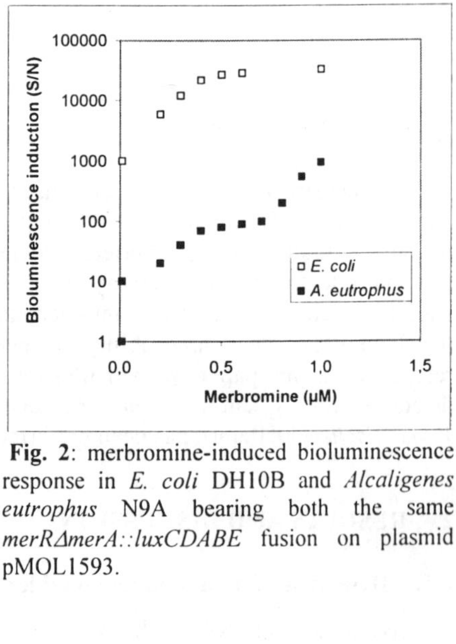

Fig. 2: merbromine-induced bioluminescence response in *E. coli* DH10B and *Alcaligenes eutrophus* N9A bearing both the same *merRΔmerA::luxCDABE* fusion on plasmid pMOL1593.

In this last example, the importance of the cellular host bearing the metal-*lux* fusion is emphasising. A Hg biosensor was constructed by cloning a *merRΔmerA* fragment obtained by PCR using *Ralstonia eutropha* AE104 total DNA as template into a *luxCDABE* promoter expression vector. The plasmid (pMOL1593) was then electroporated in the *E. coli* strain DH10B and transfer by triparental mating into mercury resistant soil bacterium *Ralstonia eutropha* CH34 and AE104, and a mercury sensitive soil bacterium *Alcaligenes eutrophus* N9A (Fig. 2). The *E. coli* strain bearing the pMOL1593 was producing a strong bioluminescent induction in the presence of merbromine. A constitutive bioluminescent level was obtained in *Ralstonia eutropha* CH34 and AE104 probably due to the interference of Hg regulatory system presence in both strains (not shown). However, into the Hg sensitive *Alcaligenes eutrophus* N9A strain, a Hg-inducible signal 1000 x time weaker than in *E. coli* was obtained. Factors such as growth, growth medium composition or outer membrane composition for both bacteria type could influence the bacterial responses.

2.2. Genotoxicity measurement with the VITOTOX on environmental samples

Bacterial assays, like the SOS chromotest, based on the expression of repair genes induced by genotoxic agents are very useful screening tools with high predictive power towards the Ames test results. A new promising screening test such as the Vitotox test developed by VITO[11,14,15] and commercialised by Labsystems Oy, which has the same basic principle as the SOS chromotest test but has the additional advantage of the kinetic evaluation and higher dynamic range, scores very well both on compound consumption and testing time schedule. The Vitotox is a very useful tool for pharmaceutical companies needing an early decision criterion in their product testing. This test has also been successfully applied to environmental samples but improvement of the test is still needed for complex mixtures. A needed extraction procedure with low polarity solvent (water:ethanol, cyclohexane) is recommended when dealing with genotoxic water samples. The solvents must then be removed by evaporation under N_2 pressure to dryness, before being resuspended in DMSO. By performing this pre-treatment of the samples, potential volatile genotoxic molecules may be removed but until now a perfect correlation was observed between Vitotox data obtained using N_2 evaporation to dryness and Ames test data performed after rotavapor evaporation to a small volume.

References

1. Selifonova O, Burlage R, Barkay T, Bioluminescent sensors for the detection of bioavailable Hg (II) in the environment, *Appl Enviro Microbiol* 1993; **59:** 3083-3090.
2. Virta M, Lampinen J, Karp M, A luminescence-based mercury biosensor. *Anal Chem* 1995; **67**: 667-669.
3. Corbisier P, Ji G, Nuyts G, Mergeay M, Silver S, *LuxAB* gene fusions with the arsenic and cadmium resistance operons of *Staphylococcus aureus* plasmid pI258, *FEMS Micro Letters* 1993; **110:** 231-238.
4. Tauriainen S, Karp M, Chang W, Virta M, Recombinant luminescent bacteria for measuring bioavailable arsenite and antimonite, *Appl Environ. Microbiol* 1997; **63:** 4456-4461.

5. Cai J, DuBow MS, Use of a luminescent bacterial biosensor for monitoring and characterisation of arsenic toxicity of chromate copper arsenate, *Biodegradation* 1997; **8:**105-111.
6. Corbisier P, Thiry E, Masolijn A, Diels L, Construction and Development of metal ion biosensors. In: Cambell AK, Kricka LJ, Stanley PE eds. *Bioluminescence and Chemiluminescence: Fundanmentals and Applied Aspects*: John Wiley & Sons, Chichester, 1994:151-155.
7. Corbisier P, van der Lelie D, Borremans B, Provoost A, de Lorenzo V, Brown N, Lloyd J, Hobman J, Csöregi E, Johansson G, Mattiasson B, Whole cell- and protein-based biosensors for the detection of bioavailable heavy metals in environmental samples, *Anal Chim Acta* 1999; **387**: 235-244.
8. Tauriainen S, Karp M, Chang W, Virta M, Luminescent bacterial sensor for cadmium and lead, *Biosen. Bioelectron* 1998; **13:** 931-938.
9. Collard JM, Corbisier P, Diels L, Dong Q, Jeanthon C, Mergeay M, Taghavi S, van der Lelie D, Wilmotte A, Wuertz S Plasmids for heavy metal resistance in *Alcaligenes eutrophus* CH34: mechanisms and applications, *FEMS Microbiol Rev* 1994; **14:** 405-14.
10. Tibazarwa C, Corbisier P, Mench M, Bossus A, Solda P, Mergeay M, Wyns L, and van der Lelie D., A Microbial biosensor to predict bioavailable Nickel in soil and its transfer to plants, *Envir Pollution* 2000 (in press).
11. Van der Lelie D, Verschaeve L, Regniers L, Corbisier P, Use of bacterial tests (The Vitotox genotoxicity test and the BIOMET heavy metal test) to analyze chemicals and environmental samples. In: Persoone G, Janssen C, De Coen W eds. *New microbiotests for routine toxicity screening and biomonitoring*. Kluwer Academic/Plenum, 2000:197-207.
12. Tauriainen, Virta M, Karp M Detecting bioavailable toxic metals and metalloids from natural water samples using luminescent sensor bacteria, *Water Res* 2000; **34:** 2661-2666.
13. Corbisier P, Thiry E, Diels L, Bacterial biosensors for the toxicity assessment of solid wastes, *Env Toxicol Water Qual* 1996; **11**:171-177.
14. van der Lelie D., Regniers L., Borremans B., Provoost A. and Verschaeve L., The VITO-Tox test, a bioluminescent *Salmonella typhimurium* test to measure genotoxicity kinetics, *Mutation Res* 1997; **389**:279-290.
15. Verschaeve L., Van Gompel J., Regniers L., Thilemans L., Vanparys P., van der Lelie D. The VITOTOX genotoxicity and toxicity test for the rapid screening of chemicals, *Environ and Mol Mutagen*; 1999; **33:** 240-248.

DEVELOPMENT OF AN ATP-BASED CHEMOSENSITIVITY ASSAY

I. A. CREE
Department of Pathology, Institute of Ophthalmology, University College London, Bath Street, London, EC1V 9EL.
E-mail: i.cree@ucl.ac.uk

C.M. KURBACHER
Universitats - Frauenklinik, University of Cologne Medical Center, Kerpenerstrasse 34, D-50911Cologne, Germany.
E-mail: christian.kurbacher@medizin.uni-koeln.de

C. SARTORI
DCS Innovative Diagnostik-systeme, Poppenbutteler Chaussee 36, D-22397 Hamburg, Germany.
E-mail: info@dcs-diagnostics.de

P.E. ANDREOTTI
Atlantic Scientific Development Inc., PMB 46, 1511 East Commercial Blvd, Fort Lauderdale, FL 33334, USA
E-mail: pandreotti@csi.com

1. INTRODUCTION

Individual tumors of the same type can vary considerably in their response to chemotherapy, which is the mainstay of treatment for metastatic disease to tissues distant from the tumor origin.[1] This variation has a molecular basis, which is partly dependent on the variable genotype of each tumor, but also environmental factors both within the tissue of origin and at distant sites to which the cells may spread. This heterogeneity of responsiveness means that tumors are essentially unique to the individual and that individualised treatment is necessary. However, individualised therapy has proved far more elusive for tumors than it has for infections, due in part to the difficulty of growing human cells in the same way that bacteria are routinely cultured and tested for antibiotic sensitivity. Nevertheless, the same concepts apply.

The luciferase-luciferin assay for ATP can be used as a very sensitive cytotoxicity assay. We have therefore designed a tumor chemosensitivity assay around this luminescence endpoint and tested it extensively over the last ten years.

2. MATERIALS AND METHODS

The assay method incorporates a short-term cell culture in a non-adherent and serum-free microplate system as previously described.[2,3] Briefly, cells are dissociated from sterile solid tumor biopsies by enzymatic digestion or obtained directly by centrifugation from malignant effusions. The cells are washed in culture medium (DCS Innovative Diagnostik Systeme GmbH, Hamburg, Germany), counted and their viability estimated by trypan blue exclusion. The drugs are prepared for testing at 800% test drug concentration (TDC) and plated in 100 microlitre volumes in the wells of a 96 well plate as illustrated (figure 1). The top and bottom rows of the plate are reserved for maximum inhibitor and no-drug (medium only) controls. Cells are added across the entire plate in a further 100 microlitres/well. The plate is incubated at 37 °C in a 99% humidity environment for six days. ATP is then extracted from the wells by the addition of 50 microlitres of extractant (DCS) to each well. ATP measurement is performed by transferring 50 microlitres from the culture plate to a white microplate (Costar) and adding 50 microlitres of luciferin-luciferase reagent across the plate. The light output is measured 5-20 min later in a luminometer (MPLX, Berthold Detection Systems GmbH, Pforzheim, Germany). The results are transferred to an Excel (Microsoft) spreadsheet, which performs automatic calculation of graphs and other output parameters.

3. RESULTS

The assay has been used with at least 30 different solid tumours and has a high evaluability rate, now around 97% in most tumour types. Experience in the authors' laboratories is estimated at more than 2,500 tumours. Initial studies showed that the degree of heterogeneity observed with different drugs mirrored clinical experience of their efficacy in breast cancer, ovarian cancer and melanoma.

Correlation with clinical outcome in breast and ovarian cancer showed a 70 - 80% sensitivity comparable with microbiological assays of antibiotic sensitivity.[2,4,5] This encouraged us to use the assay to guide treatment for individual patients. The results of a non-randomised phase II clinical trial in recurrent ovarian cancer showed a dramatic difference from a case control group with doubling of progression-free survival from 4.5 months to 9.5 months and an improvement in overall response rate from from 37% to 64%.[6]

The assay has also proved useful of the evaluation of new drugs, and for the design of new drug combinations.[7,8]

4. DISCUSSION

Development and clinical testing of this assay has taken ten years. As a result of the positive phase II study, a phase III clinical trial was launched in recurrent platinum-resistant ovarian cancer, commencing in 1998. This should report in 2001/2, providing a randomised assessment of assay-directed therapy in a very ill group of patients. The assay is now in clinical use in situations where oncologists feel that it can aid patient treatment. This is usually for the selection of regimens thought equally suitable on clinical grounds, or where there is no obvious regimen indicated (e.g. unknown primary cancer, multiple pre-treatment, rare tumours with no trial data on which to base therapy).

If intended to provide information for clinical use, it is important that the assay is performed in accordance with good laboratory and clinical practice guidelines and regulations. Interpretation of the assay results is a skill that requires experience and is expected to benefit from research in progress. In most cases, the pathologist should be involved in obtaining the tissue to avoid situations in which diagnostic information may be lost. It is certainly preferable that the assay is performed in an accredited laboratory run by staff accustomed to providing a clinical laboratory service.

The assay is also providing useful information to investigate the molecular basis of cancer treatment. Evidence from breast tumours suggests that the rapid development of resistance is responsible for the lack of complete correlation, although pharmacological issues may complicate interpretation of the results for some drugs. The information obtained from the assay allows one tumour to provide information on multiple single agents and combinations using an endpoint that approximates well with clinical activity.

References

1. Cree IA, Kurbacher CM. Individualising chemotherapy for solid tumours - is there any alternative? *Anti-Cancer Drugs* 1997; **8**: 541-548.
2. Andreotti PE, Cree IA, Kurbacher CM, *et al.* Chemosensitivity testing of human tumors using a microplate adenosine triphosphate luminescence assay: Clinical correlation for cisplatin resistance of ovarian carcinoma. *Cancer Res* 1995; **55**: 5276-5282.

3. Cree IA. Luminescence-based cell viability testing. In: LaRossa RA, ed. *Bioluminescence Methods and Protocols*, Humana Press, *Methods in Molecular Biology* 1998; **102**: 169 -177.
4. Hunter EM, Sutherland LA, Cree IA, Dewar JA, Preece PE, Andreotti PE. Heterogeneity of chemosensitivity in human breast carcinoma: use of an adenosine triphosphate (ATP) chemiluminescence assay. *Eur J Surg Oncol* 1993; 19: 242-249.
5. Konecny G, Crohns C, Pegram M, *et al.* Correlation of drug response with the ATP tumorchemosensitivity assay in primary FIGO stage III ovarian cancer. *Gynecol Oncol* 2000; **77**: 258-263.
6. Kurbacher CM, Cree IA, Bruckner HW, Mallmann P, Andreotti PE. Use of an ex vivo ATP luminescence assay to direct chemotherapy for recurrent ovarian cancer. *Anti-Cancer Drugs* 1998; **9**: 51-57.
7. Cree IA, Kurbacher CM. ATP based tumour chemosensitivity testing: assisting new agent development. *Anti-Cancer Drugs* 1999; **10**: 431-435.
8. Neale MH, Myatt N, Cree IA, *et al.* Combination chemotherapy for chroroidal melanoma: ex vivo sensitivity to treosulfan with gemcitabine or cytosine arabinoside. *Br J Cancer* 1999; **79**: 1487-1493.

BIOLUMINESCENT METHOD TO CONTROL BACTERICIDAL EFFECT OF BLOOD

V.G. FROUNDJIAN, N.N. UGAROVA, AND L. BROVKO[1]

Department of Chemistry, Lomonosov Moscow State University, Moscow 119899, Russia
E-mail: unn@enzyme.chem.msu.ru

[1]CRIFS, University of Guelph, Guelph, Ontario N1G 2W1, Canada
E-mail: lbrovko@uoguelph.ca

1. INTRODUCTION

The blood of healthy individuals is known to have bactericidal effect due to active immune response. However, in immunocompromised patients bacterial cells can rapidly multiply in blood and induce bacteremia. The bioluminescenct ATP-assay was used to monitor the bactericidal activity of hemolyzed blood against *Escherichia coli* and, thus, evaluate immune status of organism.

2. MATERIALS AND METHODS

Escherichia coli LE 392 was used in all experiments. The bacteria were cultivated in LB broth at 37°C on a shaker (120 RPM). Number of viable bacteria was expressed as Colony Forming Units per mL of sample (CFU/mL) and was evaluated using Plate Count Agar (Serva, Germany) after 56-72 h incubation at 37°C.

Blood samples were received from healthy donors and healthy and health-compromized rats. The group of health-compromized rats consisted of animals after radioactive irradiation and/or surgical procedures, and they initially had bactremia at the levels of $\geq 10^2$ CFU/mL of blood.

For inoculation and hemolysis of blood samples the following procedure was used. One mL of blood sample was mixed with *E. coli* cells to make the cell concentration up to 10^4 CFU/mL. Obtained samples were hemolyzed during 10-60 min in 9-49 volumes of ultrapure water (Milli Q, Millipore, France). After hemolysis osmomolarity of the sample was restored by adding of equal volume of 2×LB broth and samples were incubated additionally for 5 h (37°C, 120 RPM).

For the detection of nonbacterial ATP 0.05% Triton X-100 was used for ATP extraction according to Ugarova et al.[1] Bacterial ATP in blood samples was analyzed according to our previously described method.[2]

ATP concentration was analyzed using Bioluminescent Reagent IMMOLUM[3] (Chemisrty Department, Lomonosov Moscow State University, Russia) and luminometer model 1250 (LKB-Wallac, Finland).

3. RESULTS AND DISCUSSION

Blood samples from donors and healthy and health-compromised rats were inoculated with fresh culture of *E.coli* LE392. After 1-2 h incubation there were no viable bacteria in the blood samples of both donors and healthy rats. On the other hand, there was a substantial growth of bacteria observed in the blood of health-compromised rats. However, these results were obtained only after 56-72 h.

Bioluminescent methods allow measurement of intracellular bacterial ATP and, thus, enumerate bacteria, within several minutes[4,5] and with high sensitivity. However, high levels of non-bacterial ATP in blood samples makes it impossible to reach the same sensitivity for blood samples as is reported for pure bacterial cultures. To remove nonbacterial ATP from blood several methods were used, including treatment with sodium periodate and Triton X-100, incubation in non-physiological conditions, and hemolysis in different conditions. Results are presented in Fig. 1.

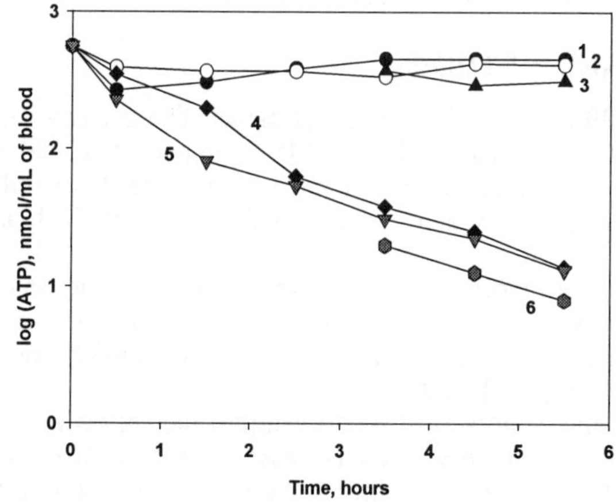

Fig. 1. Time-course of ATP hydrolysis in blood of donor in different conditions: 1 – incubation in LB media: 2 – incubation in LB media supplemented with 0.01% of Triton X-100; 3 – treatment of sample 1 with 1 mM sodium periodate after incubation in LB media: 4 – hemolysis in 24 volumes of water during 30 min at room temperature; 5 – hemolysis in 9 volumes of water during 30 min at 37°C; 6 - treatment of sample 5 with 1 mM sodium periodate in 0.01% Triton X-100.

It was shown that the initial level of non-bacterial ATP in blood could be decreased hundreds times after hemolysis. It is known that hemolysis destroys the blood cells thus decreasing bactericidal effect of blood. We have shown that blood of healthy donor after mild hemolysis in 9 volumes of water totally suppress the growth of *E.coli*, in contrast to the blood hemolyzed in 24-49 volumes of water. Probably this could be explained by the effect of immunoglobulins and complement system, that retain their immunological properties after mild hemolysis. The *E.coli* cells remain viable at the conditions of mild hemolysis.

The optimized conditions of mild hemolysis were used to monitor the growth of *E.coli* cells in the blood samples of healthy and health-compromized rats. The results are presented in Table 1.

Table 1. Time dependence of bacterial and non-bacterial ATP levels in blood of healthy and health-compromised rats after inoculation with *E-.coli*.

Incubation time after inoculation and hemolysis, hours	Blood of healthy rat			Blood of health-compromised rat		
	ATP, nmol/mL		CFU/mL	ATP nmol/mL		CFU/mL
	bac-terial	Non-bacterial		bacterial	Non-bacterial	
0			(0.9±0.3) ×10^4			(4.5±0.6) ×10^4
3	0	0.024±0.011	14±4	0.10±0.004	0.012±0.005	(9.2±0.6) ×10^6
5	0	0.005±0.003	0	0.91±0.01	0.007±0.002	(6.4±0.8) ×10^8

It was shown that after 3-5 hours of incubation of inoculated blood samples there was substantial growth of both number of *E.coli* cells and bacterial ATP in the blood of health-compromized rats. For the healthy rats after 3 h incubation there was no detectable bacterial ATP level observed, and the number of viable *E.coli* cells decreased up to 14±4 CFU/mL.

The obtained results prove that bioluminescent ATP-metry could be used as a rapid alternative to assess the bactericidal effect of blood and, thus, the immune status of organism.

References

1. Ugarova NN; Brovko LY; Trdatyan IY; Rainina EI. Bioluminescent methods of analysis in microbiology. *Appl Biochem Microbiol (Moscow)* 1987; **23**: 11-20.
2. Froundjian VG; Brovko LY; Karabasova MA; Ugarova NN. Bioluminescent assay of the susceptibility of microbial cells to antibiotics in septic blood. *Appl Biochem Microbiol(Moscow)* 1997; **33**: 455-460.

3. Brovko LY; Ugarova NN; Dukhovich AF. Bioluminescent Reagents and "rapid microbiology" methods. In: Stanley PE, Kricka L, eds. *Bioluminescence and Chemiluminescence. Current Status. Proceedings of the 5th International Symposium on Bioluminescence and Chemiluminescence.* Chichester: Wiley 1991: 433-436.
4. Froundjian VG; Brovko LY; Babunova VS; Kartashova VM; Ugarova NN. A bioluminescent assay of total bacterial contamination of fresh milk. *Appl Biochem Microbiol (Moscow)* 1999; **35**; 321-327.
5. Romanova NA; Brovko LY; Ugarova NN. Comparative assessment of methods of intracellular ATP extraction from different types of microorganisms for bioluminescent determination of microbial cells. *Appl Biochem Microbiol (Moscow)* 1997; **33**: 306-311.

RAPID DETECTION OF *STAPHYLOCOCCUS AUREUS* AND COLIFORMS USING FIREFLY LUCIFERASE

S. FUKUDA, I. MASUDA-NISHIMURA AND H. TATSUMI

Research and Development Division, Kikkoman Corporation, 399 Noda, Noda-City, Chiba 278-0037, Japan
E-mail: sfukuda@mail.kikkoman.co.jp

H. IGARASHI

Kokusaigakuin Saitama Junior College, 2-5 Kishiki-Cho, Ohmiya-City, Saitama 330-8548, Japan

S. IGIMI

National Institute of Infectious Disease, 1-23-1 Toyama, Shinjuku-ku, Tokyo 162-8640, Japan

1. INTRODUCTION

The number of reports documenting outbreaks of food-borne illness caused by pathogenic microorganisms is now increasing in all countries. The need for a rapid and easy-to-use detection method for food borne pathogens with high sensitivity led to the development of bioluminescent assays. Recently, we developed a thermostable biotinylated firefly luciferase by genetic engineering.[1] We have used the biotinylated firefly luciferase for a bioluminescent enzyme immunoassay (BLEIA) detecting protein A in *Staphylococcus aureus*. We also investigated a bioluminescent assay using a luciferin derivative, D-luciferin-*O*-beta-galactopyranoside (LuGal), for the detection of beta-galactosidase in coliforms.

2. MATERIALS AND METHODS

2.1. Bioluminescent enzyme immunoassay for *S. aureus*

2.1.1. BLEIA

All reagents for BLEIA were included in CheckLite BH (Kikkoman Co.). Bacterial cultures were prepared as described below and they were heated at 95 °C for 15 min to sterilize the bacteria. To each microtitre well coated with γ-globulins, 40 μL of biotinylated anti-protein A was added. Immediately after that, 40 μL of heat-treated culture or standard solution was added to each well, and they were incubated at room temperature (RT) for 1 hour. After washing, 80 μL of biotinylated luciferase-streptavidin complex was added to each well, and they were incubated at RT for 15 min. After washing, the light intensity was measured as follows: the microtitre wells were broken apart and each well was dropped into a test tube. In the test tube,

80 μL of luminescent substrate was added to the well. Immediately after adding the substrate, the test tube was placed in a luminometer, Lumitester K-200 (Kikkoman Co.), and the light intensity was measured.

2.1.2. Sensitivity test

Protein A (Zymed Laboratories Co.) were diluted serially with 10% (v/v) Blockace (Dainippon Pharmaceutical Co.). Then protein A solutions were applied to the BLEIA and a standard curve was obtained.

The following *S. aureus* strains were used: CCM885; FRI137; FRI326; FRI 361; FRI913; S6; S180; S204; 10; 21; 23; 48; 243; 494; SA123; SA124; SA125; SA126; SA127; SA128; SA130; SA131; Cowan I. Thirty-four strains of *S. aureus*, which were isolated from healthy individuals, were also used.

Fresh cultures of *S. aureus* strain 494 were washed and diluted to *ca* 1000 cfu/mL in saline supplemented with 0.1% Bacto Peptone (Difco Laboratories). Cultures were prepared by inoculating 0.1 mL of the bacterial suspension to 2 mL of Nutrient Broth No.2 (*ca* 50 cfu/mL) and incubating at 37 °C for 2 to 7 h. Bacterial counting was performed by 10-fold serial dilution of the cultures in saline supplemented with 0.1% Bacto Peptone and surface plating (0.1 mL) on Plate Count Agar (Eiken Chemical Co.), followed by incubation at 37 °C for 1 day.

For assay of 57 *S. aureus* strains, the preparation of cultures (*ca* 50 CFU/mL) was the same as described above except that a modified medium, which was Nutrient Broth No.2 supplemented with 0.5% LiCl and 0.5% glycine, was used in addition to Nutrient Broth No.2, and the incubation time was 7 h.

2.2. Bioluminescent assay for coliforms

2.2.1. Bioluminescent assay

The bioluminescent assay using LuGal was performed according to Masuda-Nishimura *et al.*[2] In brief, 40 μL of beta-galactosidase solutions or bacterial cultures permeabilized by benzalkonium chloride (Sigma Chemical Co.) were added to 200 μL of assay reagent containing 8.75 mg/L of LuGal and 0.16 g/L of LUC-H (a surfactant-resistant recombinant firefly luciferase, Kikkoman Co.). After incubating them for 10 min at RT, the light intensity was measured using Lumitester K-200.

2.2.2. Sensitivity test

Following 26 strains in 12 genera of coliforms were used: *Citrobacter freundii* JCM1657, *Budvicia aquqtica* JCM3902, *Eschrichia coli* ATCC9637, *E. coli* ATCC25922, *E. coli* ATCC35218, *E. coli* IAM12119, *E. hermannii* JCM1473, *Enterobacter aerogenes* IAM12348, *Ent. cloacae* IAM12349, *Ent. cloacae* IFO3320, *Ent. sakasakii* IAM12660, *Ewingella americana* JCM5911, *Klebsiella planticola* IAM1133, *K. pneumoniae* NISL4308, *Kluyvera ascorbata* JCM1681,

Moellerella wisconsensis JCM5895, *Pantoa aggromerans* IAM12659, *Rahnella aquatilis* JCM1681, *Serratia ficaria* IAM13540, *Ser. fonticola* IAM13541, *Ser. liquefaciens* IFO12979, *Ser. marcescens* ATCC13880, *Ser. odourifera* IAM13542, *Ser. plymuthica* IAM13543, *Ser. rubidaea* IAM13545, *Yersinia intermedia* JCM7579. These coliforms were seeded in 2 mL of C medium[2] at several tens to several hundreds cfu ml^{-1} and incubated at 37 °C for 7 or 24 h. These cultures were used for bioluminescent assay, and the luminescence that was 1.5-fold or higher than noise, which was obtained using the C medium without seeding coliforms, was classified as positive. As the control experiment, the bacteria were plated on desoxycholate agar medium (Eiken Chemical Co.) and Chromocult agar medium (Merck).

3. RESULTS AND DISCUSSION

3.1. Bioluminescent enzyme immunoassay for *S. aureus*

The detection limit of protein A of the BLEIA was 1 pg/mL (Welch s test, n=4, $P<0.05$) and the quantitative range was 1 to 1000 pg/mL. This is remarkably more sensitive than colorimetric ELISAs for protein A[3, 4] and the amperometric immunosensor method.[5] Cultures of *S. aureus* strain 494 with 4 h incubation showed significant light intensity and bacterial counts reached 7 x 10^3 cfu/mL (Fig. 1).

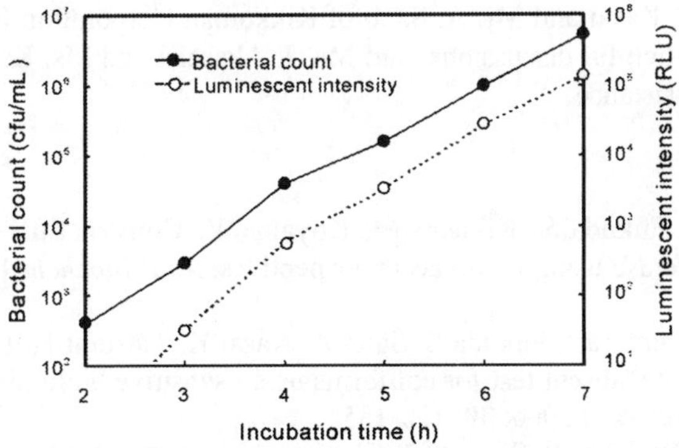

Fig. 1 Growth and reactivity in the BLEIA of *S. aureus* strain 494.

In the sensitivity test, all 57 strains were detected in the modified medium after 7 h incubation, while 56 out of 57 strains were detected in Nutrient Broth No.2. Conventional surface plating methods need incubation for 24 to 48 h. This BLEIA is much faster than conventional methods.

3.2. Bioluminescent assay for coliforms

The detection limit of beta-galactosidase in our method was 2 x 10^{-7} U/mL

(ANOVA, n=5, $P<0.05$), which was converted to 3×10^{-20} mol per assay. The detection limit of the assay using the fluorescent substrate, 4-methylumbelliferyl-beta-D- galactopyranoside, was 6×10^{-18} mol per assay. Therefore, it was shown that this method using LuGal was approximately 50-fold more sensitive than the method using a fluorescent substrate.

Using 26 strains of coliforms, the results were compared between this established method using LuGal and two current plating methods (Table 1). The number of strains detected as positive by the LuGal test after 7 h is comparable to those in the desoxycholate medium after 24 h and Chromocult medium after 22 h. To confirm the specificity of the LuGal method, the test was performed using the 22 non-coliform strains. As a result all strains were negative.[2] The results of the sensitivity and specificity tests showed that the LuGal method allows a more rapid detection than the current methods.

Table 1. Sensitivity test of 26 coliform strains

Methods	Incubation time (h)	Positive strain	Negative strain	Unclear
LuGal	7	17	9	0
LuGal	24	20	6	0
Desoxycholate	24	13	8	5
Chromocult	22	19	7	0

Acknowledgements

We thank Dr. K. Kasai and Mr. A. Sano of Kikkoman Corporation for synthesizing LuGal and their helpful discussions, and Ms. J. Akasaka and Ms. K. Fukushima for their technical assistance.

References

1. Tatsumi H, Fukuda S, Kikuchi M, Koyama Y. Construction of biotinylated Firefly luciferase using biotin acceptor peptides. *Anal Biochem* 1996; **243**: 176-180.
2. Masuda-Nishimura I, Fukuda S, Sano A, Kasai K, Tatsumi H. Development of a rapid positive/absent test for coliforms using sensitive bioluminescence assay. *Lett Appl Microbiol* 2000; **30**: 130-135.
3. Larsson A, Wejaker P, Sjoquist J. Chicken anti-protein A for the detection and capturing of protein A from *Staphylococcus aureus* in the presence or absence of mammalian IgG. *Hybridoma* 1992; **11**: 239-243.
4. Godfrey MAJ, Kwasowski P, Clift R, Marks V. A sensitive enzyme-linked immunosorbent assay (ELISA) for the detection of staphylococcal protein A (SpA) present as a trace contaminant of murine immunoglobulins purified on immobilized protein A. *J Immunol Methods* 1992; **149**: 21-27.
5. Mirhabibollahi B, Brook JL, Kroll RG. An improved amperometric immunosensor for the detection and enumeration of protein A-bearing *Staphylococcus aureus*. *Lett Appl Microbiol* 1990; **11**: 119-122.

RECOMBINANT FIREFLY LUCIFERASES AND THEIR APPLICATION FOR THE DETECTION OF MICROORGANISMS

S. FUKUDA, S. MURAKAMI AND H. TATSUMI

Research and Development Division, Kikkoman Corporation, 399 Noda, Noda-city, Chiba 278-0037, Japan

E-mail: sfukuda@mail.kikkoman.co.jp

1. INTRODUCTION

Firefly luciferase has been used to detect low numbers of microorganisms by monitoring ATP. The low thermal stability, low surfactant tolerance and low stability against chemical modifications have restricted the enzyme from further practical applications. For the analysis of microbial biomass and coliforms, surfactant is used to extract ATP and β-galactosidase respectively from microorganisms. But surfactants cause significant inactivation of luciferase during extraction. For highly sensitive enzyme immunoassays (EIAs), firefly luciferase could be an attractive marker. However, the enzyme is inactivated during chemical modifications, and it has not been possible to obtain labeled luciferase with intact activity.

Luciferase cDNAs have been cloned,[1, 2] which enabled us to improve the characteristics of luciferase by genetic manipulations. In this manuscript, we describe the applications of various recombinant luciferases (thermostable,[3, 4] surfactant tolerant,[5] color[6] and biotinylated[7, 8] derivatives), which have been constructed and characterized in our laboratory for the detection of microorganisms.

2. MATERIALS AND METHODS

2.1. Construction of thermostable mutants

Plasmids containing cDNA for *L. cruciata* luciferase (LcL) were treated in a solution containing 0.8 M hydroxylamine, followed by transformation to *E. coli* JM101. Colonies grown on Luria-Bertani agar plates at 37¡C for 12 h were transferred to nitrocellulose filters. The filters were put on new agar plates, incubated at 60 ¡C for 30 min, soaked with 0.5 mM luciferin in 100 mM sodium citrate buffer, pH 5, and then exposed by contact to X-ray film for 1h. Some thermostable mutants were obtained.[3] With further experiments *L. lateralis* luciferase with T217L (LlL-217L) was constructed as the most thermostable one.[4]

2.2. Construction of surfactant tolerant mutant

We assayed the residual activity of thermostable mutants from our collections in the presence of 0.1% benzalkonium chloride (BAC). We found that LlL-217L showed the highest tolerance and E490K in LcL contributed to the enhancement of tolerance. Then we constructed LlL with both T217L and E490K (LUC-H, Fig. 1) and confirmed that the double mutant is the most tolerable to BAC (Fig. 2).[5]

2.3. Construciton of biotinylated luciferases

Biotinylated luciferase bL248 was constructed by genetic fusion of the thermostable luciferase and a biotin acceptor peptide.[7] It consists of LlL-217L with the last Met replaced by Gly-Ser, and the natural biotin acceptor peptide BCCP-87 (the carboxy terminal 87 residues of a biotin carboxyl carrier protein, a subunit of acetyl-CoA carboxylase from *E. coli*) at the C-terminal (Fig. 1).

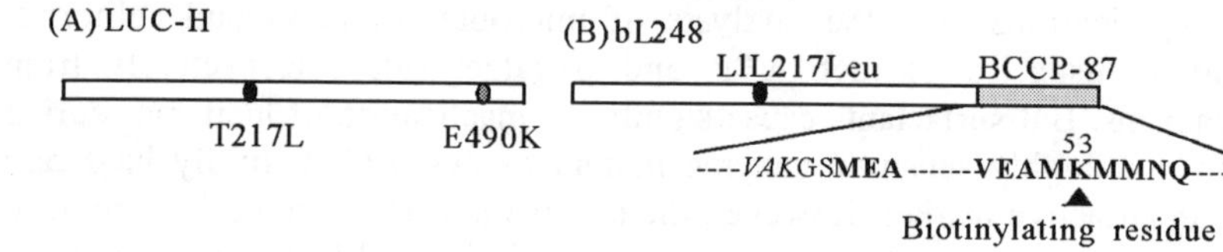

Fig. 1 Structure of LUC-H and bL248

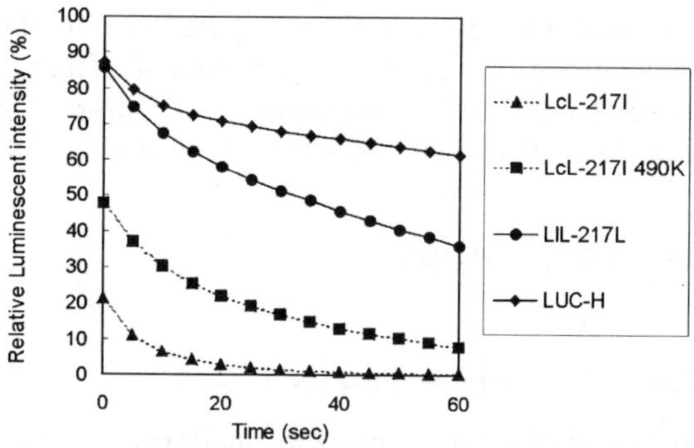

Fig. 2 Surfactant tolerance of mutant luciferases

3. RESULTS AND DISCUSSION

3.1. Microbial biomass assay

Over night cultures of *E. coli* ATCC 25922, *S. aureus* ATCC 25923, *Pseudomonas aeruginosa* ATCC 27853 and *Enterococcus faecalis* ATCC29212 were diluted 100 times with sterile water. One hundred μL of the bacterial solution was mixed with 100 μL of 0.25% BAC. After 20 s extraction, 100 μL of luminescence reagent containing luciferase, luciferin and Mg^{2+} was added and the luminescence was measured for 20 sec. Compared to LcL contained in our former kit, the surfactant tolerant mutant LUC-H yielded 190, 110, 88 and 49 times higher luminescence with *E.coli*, *S. aureus*, *P. aeruginosa* and *E. faecalis*, respectively.[5] LUC-H enabled us to use high concentrations of BAC, which enhanced the efficiency of ATP extraction. LUC-H combined with ATP eliminating enzymes[9] increased the sensitivities of these microbial biomass assays.

3.2. Bioluminescent assay of coliforms

The detection of β-galactosidase from bacteria grown in a selective medium for gram-negative bacteria demonstrates the presence of coliforms. We constructed the assay system for coliforms using a luciferin derivative, D-luciferin-*O*-β-galactopyranoside (LuGal) and the surfactant tolerant luciferase, LUC-H.[10] The system was able to detect β-galactosidase at 50-fold more sensitivity than the fluorescent substrate, 4-methyl-umbelliferyl-β-D-galactopyranoside. Twenty six coliforms were seeded in 2 mL of medium at several tens to several hundreds cfu/mL and incubated at 37 ¡C for 7 h. The number of strains detected as positive by the LuGal test after 7 h is comparable to those detected desoxycholate medium (Eiken Chemical Co.) after 24 h and Chromocult medium (Merck) after 22 h.

β-Glucuronidase is an indicator for *E. coli*. It would be possible to construct a rapid detection system of *E. coli* with a novel luciferin derivative, D-luciferin-*O*-β-glucuropyranoside.

3.3. Bioluminescent EIA of *S. aureus*

Almost all strains of *S. aureus* produce protein A. We constructed an EIA system to detect protein A using bL248 as a label.[8, 11] The detection limit of protein A was 1 pg/mL and the quantitative range was 1 to 1000 pg/mL. It is remarkably more sensitive than colorimetric EIA for protein A.[12] All 57 strains (initial inoculation is several tens cfu per mL) were detected after 7 h incubation. This EIA is also much faster than conventional surface plating methods (24 to 48 h). It would be possible to construct rapid detection systems for any food-borne pathogens using specific antibodies and bL248.

3.4. Possible application of color mutants

Using biotinylated luciferase with yellow-green emission (maximum wave length of light is 559 nm) and a mutant biotinylated luciferase with red emission (607 nm), Okuma *et al.*[13] succeeded in measuring pepsinogen I and pepsinogen II simultaneously. We might be able to construct simultaneous detection system for two kinds of food-born pathogens in the future.

References

1. De Wet J R, Wood K V, Helinski D R, DeLuca M. Cloning of firefly luciferase cDNA and the expression of active luciferase in *Escherichia coli*. *Proc Natl Acad Sci USA* 1985; **82**: 7870-7873.
2. Masuda T, Tatsumi H, Nakano E. Cloning and sequence analysis of cDNA for luciferase of a Japanese firefly, *Luciola cruciata*. *Gene* 1989; **77**: 265-270.
3. Kajiyama N, Nakano E. Thermostabilization of firefly luciferase by a single amino acid substitution at position 217. *Biochemistry* 1993; **32**: 13795-13799.
4. Kajiyama N, Nakano E. Enhancement of thermostability of firefly luciferase from *Luciola lateralis* by a single amino acid substitution. *Biosci Biotech Biochem* 1994; **58**: 1170-1171.
5. Hattori N, Murakami S. Luciferase and method for assaying intracellular ATP by using the same. Patent publication WO99/33997.
6. Kajiyama N, Nakano E. Isolation and characterization of mutants of firefly luciferase which produce different colors of light. *Protein Eng* 1991; **4**: 691-693.
7. Tatsumi H, Fukuda S, Kikuchi M, Koyama Y. Construction of biotinylated firefly luciferases using biotin acceptor peptides. *Anal Biochem* 1996; **243**: 176-180.
8. Fukuda S, Tatsumi H, Maeda M. Bioluminescent enzyme immunoassay with biotinylated firefly luciferase. *J Clinic Ligand Assay* 1998; **21**: 358-362.
9. Sakakibara T, Murakami S, Hattori N, Nakajima M, Imai K. Enzymatic treatment to eliminate the extracellular ATP for improving the detectability of bacterial intracellular ATP. *Anal Biochem* 1997; **250**: 157-161.
10. Masuda-Nishimura I, Fukuda S, Sano A, Kasai K, Tatsumi H. Development of a rapid positive/absent test for coliforms using sensitive bioluminescence assay. *Lett Appl Microbio* 2000; **30**: 130-135.
11. Fukuda S, Tatsumi H, Igarashi H, Igimi S. Rapid detection of *Staphylococcus aureus* using bioluminescent enzyme immunoassay. *Lett Appl Microbiol* 2000; **31**:134-138.
12. Godfrey MAJ, Kwasowski P, Clift R, Marks V. A sensitive enzyme-linked immunosorbent assay (ELISA) for the detection of staphylococcal protein A (SpA) present as a trace contaminant of murine immunoglobulins purified on immobilized protein A. *J Immunol Methods* 1992; **149**: 21-27.
13. Ohkuma H, Abe K, Kosaka Y, Maeda M. Simultaneous assay of pepsinogen I and pepsinogen II in serum by bioluminescent enzyme immunoassay using two kinds of *Luciola lateralis* luciferase. *Anal Chim Acta* 1999; **395**: 265-272.

EVOLUTION OF A THERMOSTABLE LUCIFERASE FOR APPLICATION IN ATP ASSAYS

R.R. HANNAH, M.P. HALL, M.T. BECK AND K.V. WOOD
Promega Corporation, 2800 Woods Hollow Road, Madison, WI 53711, USA
E-mail: rhannah@promega.com

1. INTRODUCTION

Firefly luciferase from *Photuris pennsylvanica* (*LucPpe2^{m}*) was evolved under select conditions to produce a thermostable enzyme with specific characteristics that make it an ideal candidate for use in homogeneous ATP detection systems. Initial selection was carried out under increasing temperature increments to produce a thermostable luciferase[1]. The thermostable enzyme was evolved further for resistance to L-luciferin, an inhibitor to luciferase. Some of the more commonly used biochemical parameters such as pH optimum, K_m as well as stability of enzyme activity at different temperatures were compared with firefly luciferase from *Photinus pyralis* (*LucPpy*). The latter is used extensively in ATP assays but has low stability *in vitro* and is sensitive to its environment. The results from this paper demonstrates how the evolved thermostable firefly luciferase, *LucPpe2^{m}* is better adapted for the development of two different homogeneous ATP assays. In both cases *LucPpe2^{m}* produces a more robust and homogeneous ATP assay.

2. MATERIALS AND METHODS

<u>Firefly luciferases</u>: The thermostable *LucPpe2^{m}* was developed through a program of directed evolution[1] and was purified[2] as described earlier. *LucPpy* was purchased from Promega Corp.

<u>Detection of bacteria and yeast</u>: An overnight culture of E. coli (JM 109) diluted in Tris buffered saline was filtered onto grid filters and then incubated on LB agar for 7hr at 37°C. Dried Baker s yeast was diluted into water and then filtered. The yeast and bacteria captured on filters were detected by two methods. The first method of detection is a standard protocol where the filters were sprayed with an ATP extraction agent, allowed to dry, and then sprayed with a detection agent. The released ATP was imaged with a CCD camera. The second method of detection utilizes the newly evolved *LucPpe2^{m}* luciferase. The filters were sprayed once with a single homogeneous reagent that combines lysis and detection components. Released ATP was then imaged with a CCD camera.

<u>Detection of Viable Cells in Culture Medium:</u> Varying concentrations of cells (100 — 100,000) were plated in 100µL of culture medium in a 96-well plate. The standard method used to detect viable cells is a two-step assay. First 50µL of a reagent was added to lyse cells and inhibit ATPase activity. The plate was mixed for 2 min. and then 50µL of an ATP detection reagent was added. The plate was mixed again for 2

min. followed by measurement in a plate reading luminometer. This standard protocol was compared to a protocol that incorporated the newly evolved *LucPpe2*^{m} into a single homogeneous reagent resulting in a single addition assay. A 100μL of the reagent was added to cells in 100μL and was mixed on a shaker for 1 min. Readings were taken immediately in a microplate luminometer. Readings for both assays were taken at intervals over a 4-hour period to demonstrate decay of the reaction over time.

3. RESULTS AND DISCUSSION

The table below compares some properties of *LucPpe2*^{m} and *LucPpy*. These parameters illustrate why *LucPpe2*^{m} is superior to *LucPpy* in homogeneous assays. *LucPpe2*^{m} has greatly increased stability even at 70°C. Although stability at this high temperature is not necessary for ATP assay function, it has been observed (data not

Table1. A comparison of properties between *LucPpe2*^{m} and *LucPpy*

	LucPpy	*LucPpe2*^{m}
pH optima	7.8	8.0
Luciferin — K_m at pH optima	72μM	0.57μM
Saturation for luciferin	470μM	20μM
Half-life at 65°C	< 1min.	> 1 day
Inhibition by L-luciferin (ratio of D-Luc (100) to D- & L-Luc (50-50)	18.6	3.8
Half-life of enzyme in L/L reagent pH 8.0	11.7min @ 37°C	8.3hr @ 65°C
Half-life of enzyme in L/L reagent pH 6.0	2.3min @ 37°C	16hr @65°C
Signal Stability in L/L reagent over 3 min	3% remaining	100%

shown) that thermostable luciferases are more resistant to a variety of chemicals that comprise homogeneous ATP assays. For example D-luciferin in aqueous solutions as in the L/L (luciferase/luciferin) mixture, will racemerize to L-luciferin, an inhibitor of the luciferase reaction. This results in decreased stability of the L/L reagent. Since *LucPpe2*^{m} is more resistant to L-luciferin its L/L reagent is more stable at both pH 6.0 and 8.0. Additionally *LucPpe2*^{m} has greater signal stability in comparison to *LucPpy*, which results in a more robust assay. *LucPpe2*^{m} has a lower K_m for luciferin and therefore is more economical for assay development. It is evident from this table that some of the characteristics of *LucPpe2*^{m} make it a potential candidate for use in developing reagents for ATP assays.

Figs. 1 and 2 show images of ATP released from yeast and bacterial cells using the standard 2-step reagent and the homogeneous reagent that utilizes *LucPpe2*^{m}. These images clearly show that the homogeneous reagent outperforms the standard reagent for detection of yeast and bacteria. Since the sensitivity is vastly improved when *LucPpe2*^{m} is incorporated into the reagent it is possible to detect

detect bacteria more rapidly than the standard test as can be seen in Fig 2c where a 10sec exposure is adequate.

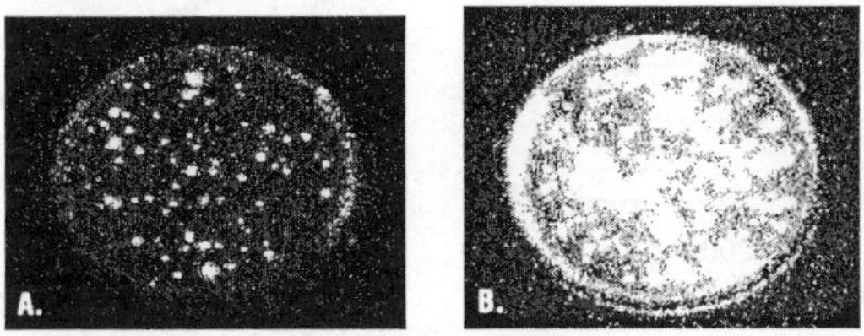

Fig. 1a Image of yeast cells with the standard reagent utilizing *LucPpy* and (**1b**) the one step reagent prepared with *LucPpe2^m*.

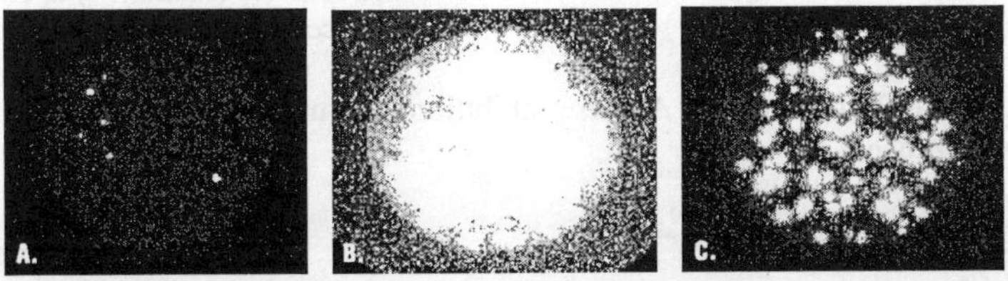

Fig. 2 Image of bacterial cells (**a**) with standard reagent after 2 min exposure (**b**) with one-step reagent using *LucPpe2^m* after a 2 min exposure and (**c**) same as (b) with a 10sec exposure

Similarly, in a different application *LucPpe2^m* was shown to be superior to *LucPpy* such as determination of cell viability. This assay is also based on the detection of ATP an indicator of the functional integrity of living cells. Once again we have been able to demonstrate the versatility of *LucPpe2^m* when used to replace *LucPpy* in the cell viability reagent and develop a single addition reagent. Fig. 3 illustrates that 10-fold brighter luminescence is achieved when the one step reagent that incorporates *LucPpe2^m* is used instead of the standard two-step reagent. The single reagent accomplishes cell lysis, ATPase inhibition and activity assay all with one addition. The allows the development of a user friendly assay for detection of viable cells and results in shorter set up to read times.

During cells lysis there is release of enzymatic activities such as ATPases. If ATPase activity is present, it can degrade ATP and make it difficult to accurately quantitate viable cell number. Fig. 4 shows that in the presence of an ATPase inhibitor it is possible to maintain constant levels of cellular ATP over time. The experiment was carried out by removing aliquots of cell lysate incubated in the absence or presence of inhibitor at various time points. These aliquots were then assayed immediately for ATP level with an L/L reagent. In absence of the ATPase inhibitor the level of ATP decreases 5-fold in 3 hrs. However, medium containing spiked ATP and the lysate with ATPase inhibitor show no loss of ATP over that same period of time. This example illustrates that ATPase activity has been effectively inhibited. It also demonstrates that we have been able to effectively incorporate lysis

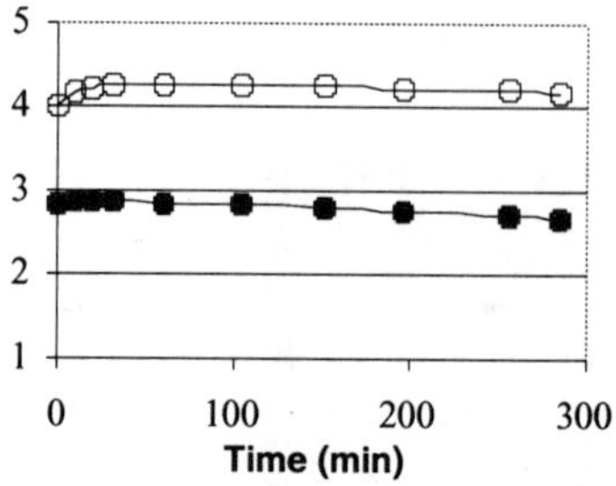

Fig. 3 A comparison of the 1-step reagent using *LucPpe2^m* (o-o) to the 2-step reagent using *LucPpy* (•-•).

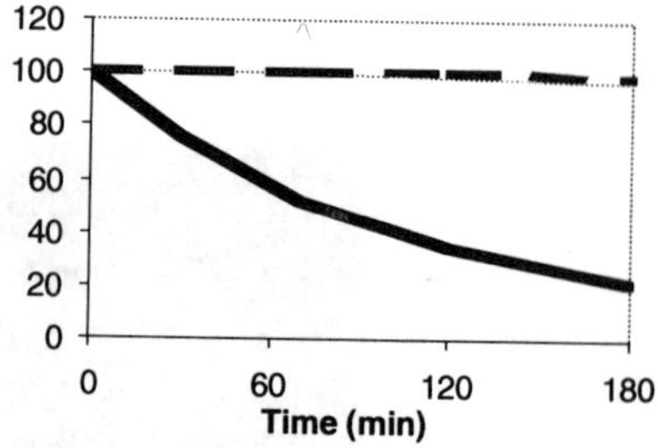

Fig. 4 The effect of ATPase inhibitor on maintaining ATP level (----) similar to the spiked ATP in medium (---) in comparison to absence of inhibitor in cell lysate (_)

(extraction of ATP from cells), ATPase inhibition and measurement of total ATP pool in a one step process.

4. CONCLUSIONS

Through directed evolution of firefly luciferase from *Photuris pennsylvanica* under select conditions a thermostable mutant *LucPpe2^m* was evolved with characteristics that made it resistant to adverse environmental factors. This has made it more adaptable to the environment necessary for developing a homogeneous ATP assay and allows development of single addition reagents. We have observed greater stability of the working reagent and sensitivity over a similar log range as *LucPpy* along with a stable and bright luminescent output during direct comparative studies (data not shown). In brief firefly luciferase from *Photuris pennsylvanica* has great potential for applications requiring detection of ATP and can be used in the preparation of more stable as well as homogeneous single step reagents.

References

1. Hall MP et al. Stabilization of firefly luciferase using directed evolution. In: Roda A, Pazzagli M, Kricka LJ, Stanley PE, eds. *Bioluminescence and Chemiluminescence: Perspectives for the 21st Century*. Chichester: Wiley, 1998:391 -5
2. Kutuzova GD, Hannah RR and Wood KV. Bioluminescence color variation and kinetic behavior relationships among beetle luciferases. In: Hastings JW, Kricka LJ, Stanley PE, eds. *Bioluminescence and Chemiluminescence: Molecular Reporting with Photons*. Chichester: Wiley, 1996:248 -52.

DETECTION OF SURFACE ANTIGENS ON TUMOR CELLS BY IMMUNOLUMINESCENCE AND ENZYMATIC ANALYSIS

FRIEDEMANN LAUBE
Martin-Luther-University Halle-Wittenberg, Institute of Physiological Chemistry, Hollystr. 1, D-06097 Halle, Germany
E-mail: friedemann.laube@medizin.uni-halle.de

1. INTRODUCTION

Intercellular adhesion molecule 1 (ICAM-1/CD54), specific CD44 isoforms, urokinase-type plasminogen activator and its receptor (uPA/uPAR) and amino-peptidase N (APN/CD13) are functionally important cell surface proteins involved in invasive and metastatic behaviour of tumor cells. This contribution describes the application of immunoluminescence to the detection of the above mentioned cell surface antigens. The immunoluminescent technique is an advantageous method compared to immunofluorescent labeling.[1] Additionally, enzymatic assays were used to measure the activity of uPA and APN in cell suspensions and monolayers. Although the enzyme activity is not a feature that results from the immunologic response it was possible to characterize membrane-bound proteases by suitable enzyme activity assays. Tests of cell extracts or tissue homogenates by ELISA can not differentiate between intracellular and extracellular antigen localization, however, the immunological technique for intact cells described here enables us to detect the existence and extent of proteins on the cell surface. Additionally, enzyme activity assays of vital tumor cells result in values for the specific proteolytic capacity that can be attributed to the tumor cell itself. Intracellular tumor enzymes or enzymes associated with or secreted from surrounding non-tumor cells did not interfere this enzyme activity compared with those detected in tissue homogenates.

2. MATERIALS AND METHODS

Cell lines. The human melanoma cell line IGR 1 was supplied by W. Wohlrab (Halle, Germany) originally from C. Aubert.[2] The human renal cell carcinoma (RCC) cell lines 49RC 43STR and 75RC 2STR were from B. Göhring and originally provided by G. Mickisch (Rotterdam, The Netherlands).
Antibodies. All monoclonal antibodies (mAbs) were of the IgG 1 type: anti-CD44(v5), (VFF-8); anti-CD44(v6), (VFF-18); anti-CD44(v7), (VFF-9); polyclonal Ab: goat-anti-CD44(v3-v10), (Bender MedSystems, Austria); anti-uPA (IM14L), anti-uPA (IM13L), anti-uPAR (GR35), anti-CD54 (15.2), Calbiochem-Novabiochem, Germany); anti-CD54 (HA58); anti-CD13 (WM15), Pharmingen,

Germany); anti-CD13 (SJ1D1), Coulter-Immunotech, France); conjugates: anti-mouse-IgG (Fab)-HRP and anti-sheep/goat-IgG (Fab)-HRP (Boehringer Mannheim, Germany). Normal mouse-IgG 1 (Sigma) and normal goat-IgG (Boehringer Ingelheim, Germany) were used for blocking and negative controls.
Reagents. Luminol, phenylmethylsulfonyl fluoride (PMSF), (Boehringer Mannheim, Germany), 6-hydroxybenzothiazole, plasminogen (from human plasma, P-5661), o-phenanthroline, 5,5′-dithio-bis(2-nitrobenzoic acid, (DTNB) (Sigma), Z-Gly-Gly-Arg-AMC, D-Val-Phe-Lys-CMK (Calbiochem-Novabiochem, Germany), Z-Lys-SBzl, H-Ala-pNA (Bachem, Germany).
Cell labeling and luminescent measurement. Procedure was performed as described recently.[1]
uPA-activity. **a)** endogenous membrane-bound uPA activity was assessed as described[3] with modifications: the reaction was started by adding 100 µL of a cell suspension ($2x10^6$ cells) to 1 mL substrate solution (Z-Gly-Gly-Arg-AMC, final concentration 150 µmol/L in PBS) with or without inhibitor. The excitation /emission wavelengths were set at 380/450 nm. The fluorescence of 7-amino-4-methylcoumarin (AMC) released was monitored continuously over a period of 5 min. Values for the maximal velocity (Δ/min) were used for calculations. **b)** an indirect assay for uPA activity was performed by release of plasmin from plasminogen (PG) after preincubation of cells with PG (final concentration: 0,5 U/mL, 10 min, 37°C) using the same procedure as described under a). **c)** the colorimetric assay with Z-Lys-SBzl was performed according to Green and Shaw.[4] Final concentration of the substrate: 250 µmol/L in PBS, pH 7,5. The procedure was the same as described under a) but Z-Lys-SBzl and DTNB were used instead of AMC-substrate. The absorbance at 412 nm of supernatants was measured after 5 min at 25°C followed by centrifugation.
APN-activity. Tests were performed with intact cells freshly harvested and washed thoroughly. After 45 min preincubation of cells ($2x10^6$/0,5 mL PBS, pH 7,4) in the presence or absence of effectors (inhibitors: 25°C; Abs: 4°C) 0,5 mL substrate solution was added (final concentration of Ala-pNA: 1,5 mmol/L in PBS). Cells were allowed to react 20-30 min at 37°C with gentle shaking. The amount of p-nitroaniline formed was measured in the supernatant at 405 nm.
Instrumentation. Luminescent detection was performed with the luminometer Lumat LB 9501 (Berthold, Germany), the fluorescent measurement with the spectrophotometer F-2000 (Hitachi, Japan) and the colorimetric assay using the Zeiss spectrophotometer UV-VIS 3.01.
Cell viability was assessed by exclusion of trypan blue.

3. RESULTS AND DISCUSSION

The immunophenotyping of the cell lines IGR 1, 49RC 43STR and 75RC 2STR with the Abs to CD44 variant sequences was described before.[1] The melanoma cell line IGR 1 exhibits a strong reaction with anti-CD44(v5) whereas only one

RCC cell line shows a positive reaction with anti CD44(v6). Numerous investigations have shown that the expression of specific CD44 splice variants in malignant tumors and tumor cell lines is associated with tumor progression and metastasis formation. In several tumors enhanced expression of epitopes especially those encoded by the exons v5 and/or v6 was observed. The presence of these specific sequences could be advantageous for these cells, e.g. in the process of metastasis into certain organs as these sequences confer the ability to attach to hyaluronat-rich tissues (targeting).[5] Detailed results from the immunologic phenotyping are summarized in Table 1.

Table 1. Immunophenotyping of tumor cells by immunoluminescence. Signal intensities (RLU)

Antigen / (Ab)	49RC 43STR	75RC 2STR	IGR 1
CD44(v5)	<3000	<3000	30.612
CD44(v6)	<3000	31.235	<3000
CD44(v7)	<1000	<1000	<3000
CD54 (15.2)	4743	6.390.000	357.045
CD54 (HA58)	<3000	6.754.510	342.725
uPA (IM13)	18.541	11.831	4.682
uPA (IM14)	27.660	14.591	17.682
uPAR	38.385	63.180	17.890
APN (SJ1D1)	620.712	<3000	148.775
APN (WM15)	532.230	4.140	14.870

Unlike 75RC 2STR the cell line 49RC 43STR expressed high levels of APN and uPA. Contrary to that the cell line 75RC 2STR expressed high levels of CD54 (ICAM-1) and CD44(v6), whereas 49RC 43STR showed a low level of CD54 and no distinct light signal with anti-CD44(v6). The melanoma cell line IGR 1 exhibited the adhesion proteins CD44(v5) and ICAM-1 as well as both proteases.

Table 2. Direct and indirect uPA enzyme activity assay

Cell line	uPA activity (%) (control)	Inhibition (%)		uPA + plasmin activity (%)	Inhibition (%)	
		PMSF	CMK		PMSF	CMK
49RC 43	100,0	33,5	61,8	165,5	n.d.	58,4
75RC 2	100,0	35,4	100,0	111,0	n.d.	100,0
IGR 1	100,0	60,3	92,4	308,7	26,8	84,5

Because both uPA and plasmin are serine proteases with similar specificity the fluorogenic substrate Z-Gly-Gly-Arg-AMC was used for the detection of both proteases. It is noteworthy that the activity measured after preincubation with plasminogen was up to threefold higher (IGR 1) than the directly tested uPA activity (control). The activator activity as well as the plasmin activity were inhibited in a concentration dependent manner by PMSF and the CMK inhibitor (data not shown). Values of inhibition in Table 2 are caused by an inhibitor concentration of 50 μmol/L.

Besides the analysis of cell suspensions it was advantageous to use intact cell monolayers grown in Petri dishes (diameter: 35 mm). Using this procedure the uPA activity of the cell line 49RC 43STR was inhibited by the anti-catalytic uPA-Ab (IM13L) and anti-uPAR in a concentration dependent manner (Ab: 5-20 μg/mL). uPA activity was almost completely inhibited by IM13L and partly inhibited by anti-uPAR. Activity and inhibition assays with Z-Lys-SBzl as substrate and the above mentioned inhibitors confirmed the results from Table 2.

APN activity was tested using o-phenanthroline and EDTA as inhibitors (0,5-2,0 mmol/L). o-Phenanthroline was the more effective inhibitor and with the highest concentration used it gave an inhibitory effect of 80-90%. Both anti-CD13 Abs used for the immunophenotyping demonstrated only a low inhibitory effect (Ab: 20 μg/mL → max. 15% inhibition).

Experiments described here revealed that the combination of immunologic and enzyme activity analysis is a suitable approach for the functional characterization of tumor cells.

References

1. Laube F. Cell surface expression of CD44: Detection on tumor cells by immunoluminescence compared to fluorescence analysis. In: Roda A, Pazzagli M, Kricka LJ, Stanley PE, eds. *Bioluminescence and Chemiluminescence: Perspectives for the 21st Century*, Chichester (UK): J. Wiley & Sons, 1999: 126-129.
2. Aubert C, Rouge´ F, Galindo JR. Tumorigenicity of human malignant melanocytes in nude mice in relation to their differentiation in vitro. *J Nat Cancer Inst* 1980; **64**: 1029-1040.
3. Zimmerman M, Quigley JP, Ashe B, Dorn C, Goldfarb R, Troll W. Direct fluorescent assay of urokinase and plasminogen activators of normal and malignant cells: Kinetics and inhibitor profiles. *Proc Natl Acad Sci USA* 1978; **75**: 750-753.
4. Green GDJ, Shaw E. Thiobenzyl benzyloxycarbonyl-L-lysinate, substrate for a sensitive colorimetric assay for trypsin-like enzymes. *Anal Biochem* 1979; **93**: 223-226.
5. Penno MB, Hart JC. CD44, hyaluronan and lung cancer. In: Mousa SA, ed. *Cell adhesion molecules and matrix proteins: Role in health and diseases*, Berlin-Heidelberg-New York: Springer-Verlag, 1998: 175-190.

SOLUBILITY OF A LUCIFERASE-GFP FUSION PROTEIN IN BACTERIAL AND MAMMALIAN CELLS

G. M. NELSON AND A. ESCHER

Center for Molecular Biology and Gene Therapy and Department of Microbiology and Molecular Genetics, 11085 Campus Street, Loma Linda University, Loma Linda, CA 92350, USA

E-mail: aescher@som.llu.edu

1. INTRODUCTION

Protein folding has been studied extensively *in vitro*, using mostly model substrates with molecular weights less than 55-kDa. Sequence analysis of several complete genomes has revealed that eukaryotes generally synthesize larger proteins when compared to bacteria.[1] It is therefore likely that eukaryotic cells have acquired mechanisms for folding large proteins with higher efficiency than bacterial cells. We have constructed a gene fusion, *lux-gfp,* consisting of a *luxab* bacterial luciferase gene fusion,[2] and the green fluorescent protein cDNA.[3] The protein encoded by this construct has a molecular weight of 104-kDa, a much larger substrate than used in most protein folding studies. Previous work demonstrated that the chaperonins GroES and GroEL could rescue the temperature sensitive Fab9 luciferase *in vivo* using *Escherichia coli* as the model organism,[4] and that the fusion of the *gfp* cDNA to the *fab9* gene permitted *in vivo* monitoring of protein folding in *E. coli.*[5] In this work, we investigate the folding behavior of this large multidomain protein in bacteria and mammalian cells using GFP fluorescence and differential centrifugation.

2. MATERIALS AND METHODS

Plasmid Constructions: Plasmid pACGN1 carries a gene that confers resistance to chloramphenicol and contains a gene fusion of the bacterial luciferase *fab9* gene and the green fluorescent protein (GFP) gene from the jellyfish *Aequorea victoria* under the transcriptional control of the tetracycline resistance gene promoter. Plasmid p3.1FG contains the *luciferase-gfp* gene cloned into the mammalian expression vector pcDNA3.1(+) (Invitrogen, San Diego, CA, USA) and under transcriptional control of the human cytomegalovirus (CMV) immediate-early promoter/enhancer. Plasmids pGroEL and pGroESL, cloned into plasmid pBluescript SK (Stratagene, La Jolla, CA, USA) carry an ampicillin resistance gene and the *groE* genes under transcriptional control of the *groE* promoters.

Visualization of GFP fluorescence in *Escherichia coli* and COS-7 cells: *E. coli* DH5α were visualized with ultraviolet fluorescent microscopy under 1000X magnification. COS-7 cells were grown on glass coverslips 24 hour prior to transfection. Coverslips were removed from plates and placed on a microscope slide that had a small drop of mounting solution containing DAPI, a DNA staining agent (Vectashield, Vector Laboratories, Burlingame, CA, USA). Digital photographs were captured with a cooled CCD camera and imaged in black and white with MetaMorph software (Universal Imaging Corp, West Chester, PA, USA). Images were acquired for 2500 ms.

Differential Centrifugation and Immunoblot analysis of *Escherichia coli* and COS-7 cells: DH5α cell lysates were separated into soluble and insoluble fractions as described previously.[6] Proteins separated by SDS-PAGE using 12% gels were transferred to PVDF membrane (Immobilon-P, Millipore, Bedford, MA, USA) and subsequently reacted with polyclonal rabbit anti-LuxAB antibodies (dilution of 1:15,000). Antibodies were detected with horseradish peroxidase conjugated secondary anti-rabbit IgG antibody (Sigma, St.Louis, MO, USA, dilution of 1:30,000) and reacted with enhanced chemiluminescent substrate (Pierce, Rockford, IL, USA).

COS-7 cells were washed 1X with cold DPBS 48 hours after transfection and removed by scrapping with a rubber policeman into 1 mL of DPBS. Cell pellets were kept on ice until resuspended and lysed as previously reported.[7] Lysates were clarified by room temperature centrifugation at 15,500 X *g* for 15 minutes. The cell lysates were separated into soluble and insoluble fractions as previously described.[6] Western blots were reacted with polyclonal anti-GFP (Molecular Probes, Eugene, OR, USA) at a dilution of 1:3000.

3. RESULTS AND DISCUSSION

3.1. Excess chaperonins GroESL assist the folding of LUX-GFP in *Escherichia coli*, but are not required in COS-7 cells

The introduction of the LUX-GFP protein into bacterial cells via plasmid transformation demonstrated the requirement of overexpressed chaperonin genes from *groESL* for increased protein activity. *E. coli* cells transformed with the plasmid pACGN1 containing the *lux-gfp* gene had a very low basal level of GFP fluorescence (Fig.1A). Luciferase activity was low, but detectable (data not shown). In contrast, cells cotransformed with the *lux-gfp* and *groESL* genes showed significant and dramatic increases in levels of GFP activity (Fig.1B). The pattern of GFP fluorescence revealed that the LUX-GFP protein was soluble throughout the cell, but also formed compact, bright focused areas of fluorescence.

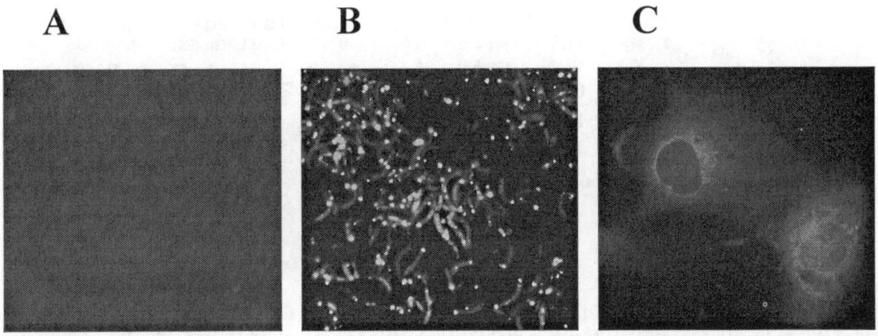

Fig. 1 GFP fluorescence activity at 37°C in *E. coli* (A and B) and COS-7 cells (C). Cells were transformed with (A) plasmid pACGN1 and vector only, (B) plasmid pACGN1 and pGroESL, and (C) transfected with p3.1FG.

To determine if there were differences in the state of the LUX-GFP fusion protein between bacteria and mammalian cells, COS-7 cells were transfected with the *lux-gfp* gene construct p3.1FG (Fig. 1C). In contrast with *E. coli*, COS-7 cells did not show focused fluorescent aggregates. The pattern of fluorescence was diffuse throughout the cell, with no obvious subcellular localization, indicating that COS-7 cells were able to fold the LUX-GFP fusion protein without excess chaperonins. Cotransfection of the *groESL* genes with the *lux-gfp* gene did not change the pattern or increased the intensity of the fluorescence (data not shown).

3.2. Mammalian cells fold the LUX-GFP fusion protein with higher efficiency than bacteria

Cell extracts were prepared from *E. coli* and COS-7 cells that contained various combinations of chaperonin and *lux-gfp* genes. The extracts were separated into soluble and insoluble fractions followed by western blotting against either luciferase or GFP proteins.

E. coli cells expressing the *lux-gfp* gene appeared to contain mostly low levels of soluble LUX-GFP protein (Fig.2 ,left). When the *groEL* gene was cotransformed with the *lux-gfp* gene, the amounts of LUX-GFP protein within the cell increased, but without a concomitant increase in GFP fluorescence or luciferase activity (data not shown). However, when the *groESL* genes were cotransformed with the *lux-gfp* gene, there was an increase in the soluble and insoluble protein amounts, and there was a significant increase in GFP fluorescence and luciferase activity.

Direct comparison between *E. coli* and COS-7 cells revealed that the mammalian cell extracts did not contain any detectable LUX-GFP aggregates, indicating that the cells were capable of folding the large multidomain protein without the assistance of excess chaperonins as in *E. coli* (Fig.2, right). Overexpression of the *groEL* and *groESL* genes in COS-7 cells with the *lux-gfp*

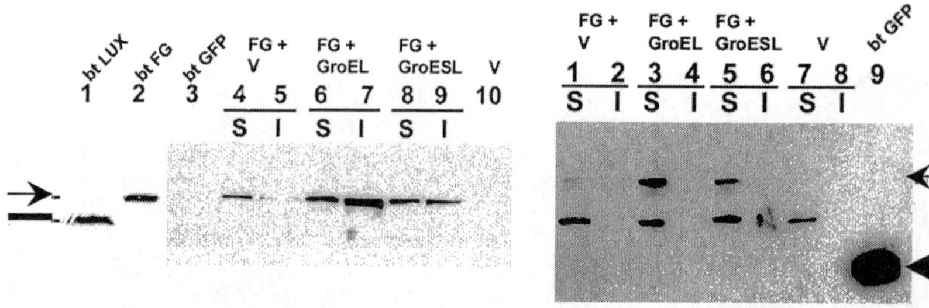

Fig. 2 Solubility of the LUX-GFP fusion protein in *E.coli* (left) and COS-7 cells (right). Anti-luciferase or anti-GFP antibodies were used for immunoblot analysis of fractionated lysates of cells expressing the *lux-gfp* gene. *E. coli* and COS-7 cells were cotransformed or cotransfected with the *lux-gfp* gene, or the *lux-gfp* gene with the *groEL* gene or the *lux-gfp* gene with the *groESL* genes or the vector only. S, soluble fraction; I, insoluble fraction; bt, bacterial synthesized protein. The arrow indicates the position of the LUX-GFP fusion protein; the line indicates the position of the luciferase and the thick arrow indicates the position of the GFP protein.

gene construct resulted in an increased amount of the LUX-GFP protein. Coimmunoprecipitation of GroEL and LUX-GFP demonstrated that the LUX-GFP protein does not interact with GroEL in mammalian cells at a detectable level (data not shown). These results support the hypothesis that mammalian cells have developed mechanisms that allow the cell machinery to efficiently and productively fold large multidomain proteins.

References

1. Netzer WJ, Hartl FU. Recombination of protein domains facilitated by co-translational folding in eukaryotes. *Nature* 1997; **388:** 343-349.
2. Escher A, O'Kane DJ, Lee J, Szalay AA. Bacterial luciferase alpha beta fusion protein is fully active as a monomer and highly sensitive in vivo to elevated temperature. *Proc Natl Acad Sci USA* 1989; **86:** 6528-6532.
3. Zolotukhin S, Potter M, Hauswirth WW, Guy J, Muzyczka N. A "humanized" green fluorescent protein cDNA adapted for high-level expression in mammalian cells. *J Virol* 1996; **70:** 4646-4654.
4. Escher A, Szalay AA. GroE-mediated folding of bacterial luciferases in vivo. *Mol.Gen.Genet.* 1993; **238:** 65-73.
5. Nelson GM, Escher A. Visualizing the chaperonin-mediated folding of a luciferase-GFP fusion protein. In: Roda, A, Pazzagli, M, Kricka, LJ, and Stanley PE, ed. *Bioluminescence and Chemiluminescence: Perspectives for the 21st Century,* Chichester: John Wiley & Sons, 1999: 425-428..
6. Siemering KR, Golbik R, Sever R, Haseloff J. Mutations that suppress the thermosensitivity of green fluorescent protein. *Curr Biol* 1996; **6:** 1653-1663.
7. Beckmann RP, Mizzen LE, Welch WJ. Interaction of Hsp 70 with newly synthesized proteins: implications for protein folding and assembly. *Science* 1990; **248:** 850-854.

RAPID DETECTION OF FOOD-BORNE BACTERIA USING BACTERIOPHAGE AND AK BIOLUMINESCENCE

RL PRICE[1], MJ MURPHY[1], DJ SQUIRRELL[1] AND MF SANDERS[2]

[1]*Detection, DERA Porton Down, Salisbury, Wiltshire, SP4 0JQ, UK*

[2]*MAFF Central Science Laboratory, Sand Hutton, York, YO4 1LZ, UK*

1. INTRODUCTION

Food-borne bacterial pathogens in ready-to-eat chilled foods pose a significant health risk. Conventional microbiological testing takes several days to identify a positive sample. In short shelf-life foods, this means that the food is already on sale by the time test results are obtained. Food testing that provided results within one working day would allow manufacturers to release foods to supermarkets knowing that they were contamination-free, avoiding the need for product re-calls should a positive result occur. Presented here is a more rapid method (5h) that detects and identifies food-borne bacteria by combining immunomagnetic separation, phage-mediated specific lysis and detection using AK bioluminescence.

Adenylate kinase (AK) is an essential metabolic enzyme that catalyses the equilibrium reaction: $Mg^{++}.ATP + AMP \rightleftharpoons Mg^{++}.ADP + ADP$. By adding excess purified ADP the reaction can be driven to the left to produce ATP, which can then be detected with the standard firefly luciferase reaction. Thus, once broken open, the cell being detected provides its own enzyme label. We previously showed that combining an AK bioluminescent endpoint with specific lysis by bacteriophage[1,2] can enable the (specific) detection of fewer than 10^3 bacteria. The method was adapted for the detection of food-borne bacteria by including a bacterial capture step using immunomagnetic beads. Target bacteria were captured from food onto paramagnetic beads coated with antibodies against the target bacteria, immobilised and washed to remove food particles.

Light generated in the reaction can be measured using standard luminometers, but can also be detected using high sensitivity Polaroid® film. Steps towards automation have been taken by adapting a commercially available immunomagnetic bead ELISA machine, the EiaFoss Analyser.

2. MATERIALS AND METHODS

Ready-cooked pilau rice samples, (purchased from supermarkets) were spiked with 10^4 *Listeria monocytogenes* strain 004 (determined by plate counts). Control samples were not spiked. 25g rice samples were stomached for 60 sec with 225mL buffered peptone water using a Labsystems Stomacher 400. The slurry was then incubated for 2h at 20°C. 1mL aliquots were removed to sterile Eppendorfs and 20μL of ListerTest anti-*Listeria* immunobeads 20x concentrate (Vicam, Watertown, MA, USA) were added to each. The beads and slurry were incubated for 15 min at room temperature with gentle mixing. Samples were then washed twice with 1mL tryptose broth and resuspended in 1mL tryptose broth. Half of the samples were diluted 1/10 in broth before all were incubated for 1h at 37°C. Samples were then infected with phage 92431 and incubated for 1h, before being tested for their extracellular AK content. 100μL samples were added to 100μL 15mM magnesium acetate. 0.3mM purified ADP was then added (100μL) and incubated for exactly 5min. The addition of 100μL bioluminescence reagent ("RB1L", Celsis-Lumac, Cambridge) produced light which was then read using a Biotrace Multilite luminometer. Unspiked samples were also tested for *Listeria* contamination by conventional microbiological methods using selective enrichment and plating. This took 7 days.

A non-luminometer based test was developed in which the endpoint AK phage assay was recorded with instant photographic film instead of a luminometer. This was based on the CamLight system from Perkin Elmer and used 20,000 ASA speed Polaroid film. The sample tray, which used individual microwells, was adapted to prevent light cross-talk between tubes by isolating wells with black silicon sealant. Coating the lid and well insides with silver-coloured paint and foil enhanced the light signal. AK phage assays using *Escherichia coli* 10243 and phage 10360 were performed; 100μl *E.coli* samples in broth containing 10^3 phage were incubated at 37°C for 1h before the addition of 50μl 15mM magnesium acetate and 50μL 0.3mM ADP. Samples were incubated for exactly 35min at room temperature before 50μL bioluminescence reagent was added. Light produced was then recorded using 20,000 ASA Polaroid® film, exposed to the microwells for 5min. Automation of the endpoint bioluminescent assay was investigated using the EiaFoss Analyser (Foss Electric, Denmark), a commercial instrument that performs fluorescent immunomagnetic bead sandwich ELISAs. Experiments were performed in which the EiaFoss assay's fluorescent endpoint was substituted for a standard AK assay to detect *Salmonella typhimurium*. Software modifications are currently under way to enable the entire AK phage assay to be performed on the instrument.

3. RESULTS AND DISCUSSION

When the AK phage assay was applied to spiked and control ready-to-eat pilau rice samples, the luminescent signal from the spiked sample was at least double that obtained from the control (Fig. 1). This demonstrated that the AK Phage method was able to detect *Listeria monocytogenes* in a fairly complex food matrix. Conventional food microbiology showed that no *Listeria* strains were present in the unspiked, control samples. The rapid AK bead and phage method took 5h to complete, compared with 7 days for the standard method.

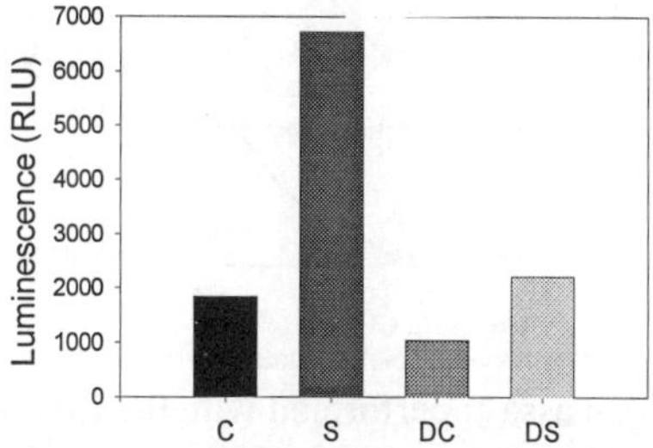

Fig. 1. AK Phage assay on control (C) and spiked (S, Listeria) pilau rice samples. Some samples had been diluted 10-fold to facilitate bead washing (D).

Light produced can be measured using conventional luminometers, or can be detected with high sensitivity instant photographic film. AK Phage assays in which the light generated was recorded using instant photographic film were also successful, easily detecting $2.5x10^3$ *E.coli* cells (Fig. 2). Very faint light spots were also visible for microwells containing a mean of 250 *E.coli* cells.

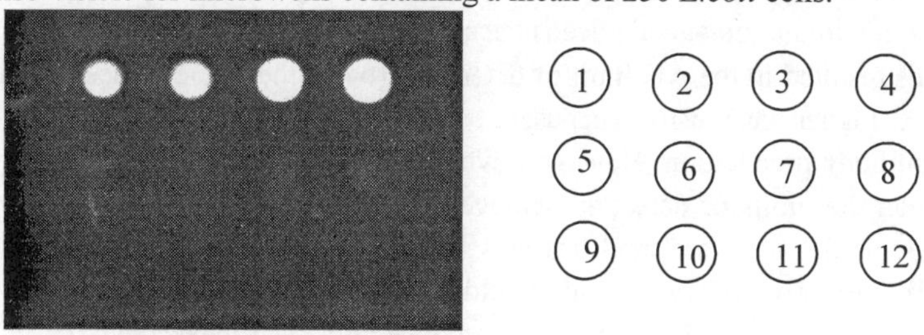

Fig. 2. Detection of *E.coli* by AK phage assay, measured with instant photographic film. [Key: 1,2 = $2.5x10^3$ *E.coli* + phage; 3,4 = $2.5x10^4$ *E.coli* + phage; 5,6 = 25 *E.coli* + phage; 7,8 = 250 *E.coli* + phage; 9 = Broth + phage; 10 = Broth only; 11,12 = $2.5x10^4$ *E.coli* without phage]

Control samples containing $2.5x10^4$ *E.coli* without phage did not produce light, showing that the lysis observed by the AK reaction was due to the phage infection. A simple, low-cost, assay format such as this, which provides a hard copy of the results following analysis, could have utility in "low-tech" applications with non-scientific operators as numerical interpretation is not required; presence of light signifies presence of target bacteria.

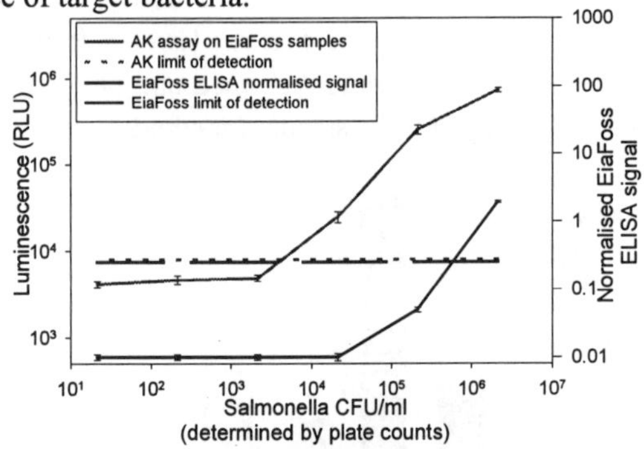

Fig. 3. *Salmonella typhimurium* assays performed with the EiaFoss Analyser.

Assay automation is desirable for higher throughput food screening. The first step towards automation was based on the EiaFoss Analyser. *Salmonella tyhimurium* was assayed to determine whether the detection limits achieved using a fluorescence-based ELISA could be improved by substituting with an AK endpoint (Fig. 3). AK assays using a chemical extractant were performed on samples removed from the machine following the wash step and addition of the secondary antibody. It was not possible to perform the assay with just the bound material captured onto the immunomagnetic beads because of the way in which the EiaFoss operates. This resulted in the AK limit of detection, (twice the highest background signal), being higher than usual because of the extraneous AK content of the secondary antibody preparation. However, even with this constraint the AK assay improved upon the limits of detection achieved with the standard ELISA by two orders of magnitude. Software modifications to alter the times at which reagents are added are currently underway, and should enable the entire AK phage assay, with initial immunomagnetic bead separation, to be automatically executed.

References

1. Blasco R, Murphy MJ, Sanders MF, Squirrell DJ. Specific assays for bacteria using phage mediated release of adenylate kinase. *Journal of Applied Microbiology* 1998, 84; 661-666
2. Murphy MJ, Squirrell DJ, Sanders MF, Blasco R. The use of adenylate kinase for the detection and identification of low numbers of microorganisms. In: Hastings JW, Kricka LJ, Stanley PE, *Bioluminescence and Chemiluminescence; Molecular Reporting with Photons.* 1998: 319-322.

BIOLUMINESCENT ASSAY OF ATP LEAKAGE FROM YEAST CELLS AFTER PHOTODYNAMIC DAMAGE

N. A. ROMANOVA, EV. V. POMETUN, L. YU. BROVKO,
N. N. UGAROVA AND A. P. SAVITSKY

Chemistry Department, Lomonosov Moscow State University, 119899, Moscow, Russia
E-mail: unn@enzyme.chem.msu.ru

1. INTRODUCTION

Photodynamic treatment of cells of any origin generally requires a photosensitizer accumulated in cells and activated by visible light.[1] The mechanisms of photodynamic damage of mammalian cells are widely studied,[2] and photodynamic therapy of cancer is under clinical trials in different countries.[3] Much less information is available on photodynamic damage of bacterial cells. Photoactivated porphyrins[4,5] and phthalocyanines[6,7] are recognized as antimicrobial drugs.[8] However, the detailed mechanism of photodynamic damage of bacterial cells is not studied yet. The primary step of the antimicrobial action of photosensitizers should include damage of outer cell membrane where photosensitizer is adsorbed. The loss of membrane integrity should lead to ATP leakage from damaged cells; thus, highly sensitive bioluminescent method can be used for measurement of ATP leakage. On the other hand, photodynamic treatment of microbial cells might be used as an alternative method to release intracellular ATP in so-called "rapid microbiology", that is based on the ATP measurement using the firefly luciferin–luciferase system. Bioluminescent ATP-assay of microorganisms is not specific. At the same time different photosensitizers may specifically accumulate in cell membranes depending on pH, charge of photosensitizer molecule and cell membrane.[6,7] Moreover, some photosensitizers[4,6] selectively accumulate in Gram-positive or Gram-negative bacteria. The aim of this work was to study the ATP leakage from the yeast cells after photodynamic treatment and elucidate the possibility of using this method in the bioluminescent ATP assay of microorganisms.

2. MATERIALS AND METHODS

Instrumentation Bioluminescence was measured on a microluminometer, Model 3550 (New Horizons Diagnostics, USA). Home-made installation including 350 Wt tungsten lamp, condenser, water filter, long pass red filter OS-14 (LOMO, Russia, cut-off 580 nm) and optic fibre was used for illumination of samples in microtiter well.

Reagents All reagents were of analytical grade. Dimegin (2,4-di(α-metoxyethyl)deuteroporphyrin was kindly donated by Dr. G. V. Ponomarev. ATP-reagent containing immobilised firefly luciferase, luciferin, $MgCl_2$ and buffer components was from Chemistry Department, Moscow State University (Russia).

Methods The *Saccharomyces cerevisiae* yeast culture was grown in nutrition media for 10–12 h at 37°C, and washed twice with buffer solution. Each cell suspension was divided into three samples, put into microtiter wells, 10–100 μg/mL dimegin was added to two samples, the third was used as a blank. All samples were incubated in darkness for 0–115 min. Only one of the samples containing dimegin was illuminated by red (filter OS-14) or white light with power density 0.02–0.13 Wt/cm^2 for 3–15 min. Two other samples were used as controls. Concentration of intercellular ATP was measured in all three samples by adding 50 μL of the cell suspension to 150 μL of ATP-reagent. The concentration of leaked ATP ($[ATP]_{out}$) was determined as a difference of the ATP concentration of illuminated sample and controls. The concentration of total ATP ($[ATP]_{total}$) was measured after treatment of 0.1 mL of cell suspension with 0.9 mL of DMSO.

3. RESULTS AND DISCUSSION

Illumination of cell suspension by red or white light (doses 6.6 and 19.8 J per sample) had no effect on $[ATP]_{total}$ in the absence of dimegin. Dark incubation of the sample containing dimegin did not induce ATP leakage whereas the sample containing dimegin demonstrated the marked increase in $[ATP]_{out}$ after illumination with red or white light. The ATP leakage depended both on light dose, durability of illumination and photosensitizer concentration. At dimegin concentration in the range 10–100 μg/L and at the time of illumination from 30 s up to 2 min, the ATP leakage reached 10–50% of total ATP pool.

To optimize the conditions of illumination, we have studied the effect of different factors (time of dark incubation, dose and power of light, spectra of light, phase of cell growth) on the $[ATP]_{out}$. At dark incubation time 0–20 min in the presence of dimegin, no the ATP leakage was observed after illumination. The $[ATP]_{out}$ increased linearly with time till 60 min of dark incubation, kept constant till 80 min, and then decreased slowly. It can be explained by the fact that it takes time to achieve equilibrium $[dimegin]_{solution} \Leftrightarrow [dimegin]_{cell\ membrane}$, and only absorbed dimegin molecules cause photodynamic effect. The $[ATP]_{out}$ increased with the growth of dimegin concentration and reached the maximal value at 50 μg/mL of dimegin. The same photodynamic effect was observed at illumination with red (λ > 580 nm) or white (λ > 400 nm) light. It is known that the concentration of intracellular ATP decreases a little for the cells in stationary phase in comparison with that of logarithmic phase. But we did not found marked difference in $[ATP]_{out}$ for both types of cells after photodynamic damage.

Under optimal conditions (50 μg/mL dimegin, dark incubation time 60 min) we studied the effect of light dose and power density on the $[ATP]_{out}$ (Fig.1). The same dose was achieved in two ways: (1) at fixed power density (0.10 Wt/cm^2), the duration of illumination was varied from 3 till 15 min (Fig.1, line A); (2) at fixed illumination time (3 min) the power density was grown from 0.03 till 0.22 Wt/cm^2 (Fig.1, curve B). When the dose was not above 4.5 J per sample, the $[ATP]_{out}$ showed linear dependence on the dose, and lines (A) and (B) coincided. At doses above 4.5 J per sample the $[ATP]_{out}$ increased exponentially when the dose was reached by the power increase. This can be explained by the fact that at high power of light, the cells have no time to switch on its repair mechanisms and to reconstruct partial integrity of outer membrane, whereas at lower power density and more prolonged illumination this can be effected.

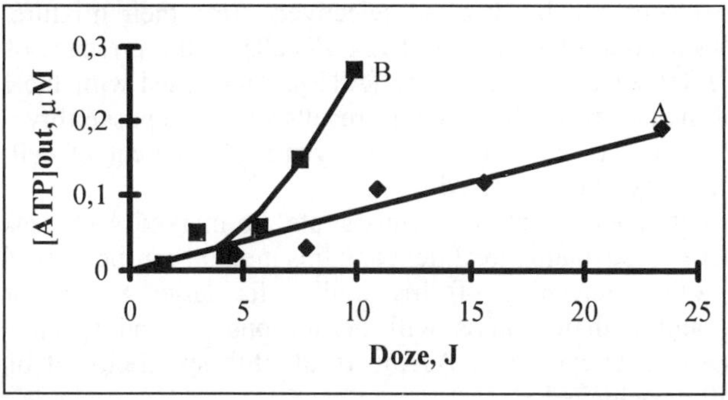

Fig. 1. Concentration of $(ATP)_{out}$ as a function of red light dose: A) at fixed power intensity (0.10 Wt/cm^2) and varied illumination time (3–15 min); B) at fixed illumination time (3 min) and varied power intensity (0.03–0.22 Wt/cm^2)

To investigate the possibility of using photodynamic treatment of the cells in "rapid microbiology" we obtained the dependence of the $[ATP]_{out}$ on the cell content in the initial sample (Fig.2), that is described by linear equation $y = 0.0131\ x$ ($R^2 = 0.96$), where x is the content of the yeast cells in the initial sample, and y is $[ATP]_{out}$ in the measuring cuvette of the luminometer. In parallel for the same samples, we obtained the dependence of the $[ATP]_{total}$ on the cell content measuring the ATP concentration in the cell solution treated preliminary with 10-times excess of DMSO. In this case this dependency was described by the equation $y = 0.011\ x - 0.0076$ ($R^2 = 0.99$).

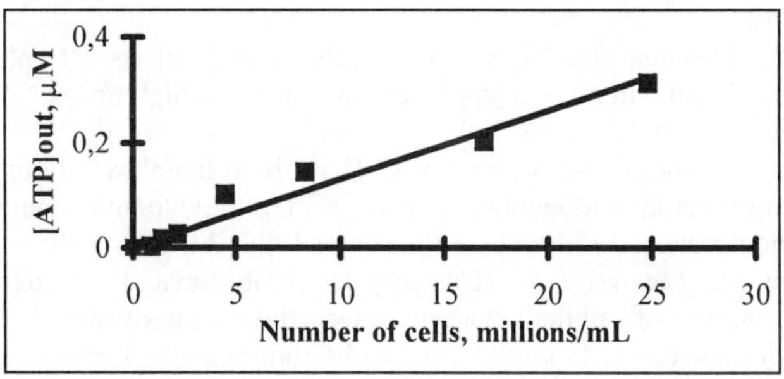

Fig. 2. Concentration of $(ATP)_{out}$ as a function of yeast cell content in the initial sample. Dose of light was 7.8 J per sample, exposition 5 min.

The used ATP reagent had the ATP detection limit 1 nmol/L. This corresponds to the cell content 76,000 cells/mL for the $[ATP]_{out}$ data shown in Fig.2, and 780,000 cells/mL for the $[ATP]_{total}$ data. Thus, the ATP release by photodynamic damage permitted us to increase the sensitivity of the cell detection more than 10 times in comparison with ATP extraction by DMSO.

It is known that Gram-negative *E. coli* cells are resistant to photodynamic damage in the presence of photosensitizers.[6,8] We proposed that *Saccharomyces cerevisiae* yeast cells can be detected selectively from their mixture with *E. coli* cells. It was shown that illumination of *E. coli* cells in the presence of dimegin did not cause the ATP leakage, whereas the $[ATP]_{out}$ increased with the light dose for the mixture of the two types of cells. Our results show the possibility to distinguish Gram-negative and yeast cells using photodynamic treatment of cell mixture and bioluminescent assay of the released ATP.

Thus, the data obtained clearly demonstrate the marked ATP leakage from the yeast cells after photodynamic treatment and this method can be used for the release of microbial ATP. Screening of this ability for large groups of porphyrins, porphyrazines and phthalocyanins will permit one to find specific effectors for different classes of bacterial cells. As the result, the specificity of bioluminescent ATP assay will be achieved.

Acknowledgements
This work was supported by NATO Science Programme (C.L.G.977173).

References

1. Bayley H, Gasparro F, Edelson R. *Photoactivatable Drugs*. TIPS-April, 1987; 138–143.
2. Phillips D. Chemical mechanisms in photodynamic therapy with phthalocyanines. *Prog Reaction Kinetics* 1997; **22**: 175–300.
3. Dougherty T, Marcus S. Photodynamic therapy. *Eur J Cancer* 1992; **28A**: 1734–1742.
4. Malik Z, Hanania J, Nitzan Y. Bacteriocidal effects of photoactivated porphyrins – an alternative approach to antimicrobial drugs. *J Photochem Photobiol* 1990; **5**: 281–293.
5. Fraikin G, Strakhovskaya M, Rubin A. The role of membrane bound porphyrin type compounds as endogenous sensitizer in photodynamic damage to yeast plasma membranes. *J Photochem Photobiol* 1996; **34**: 129–135.
6. Bancirova M, Lasovsky J, Kolarova H, Frebortova J, Medkova J. The chemiexcitation of phthalocyanines for the photodynamic therapy by phthalhydrazide on the Gram-positive and Gram-negative bacteria. In: Roda A, Pazzagli M, Kricka LJ, Stanley PE, eds. *Bioluminescence and Chemiluminescence. Perspectives for the 21st Century,* New York: John Wiley & Sons, 1999: 567–570.
7. Millson Ch, Wilson M, MacRobert A, Thurrell W, Mikvy P, Davies C, Bown S. Lethal photosensitisation of *Helicobacter* species. *Proc SPIE* 1994; **2325**: 174–180.
8. Wainwright M. Photodynamic antimicrobial chemotherapy (PACT) *J Antimicrob Chemother* 1998; **42**: 13–28.

OPTIMIZING BIOSENSOR DESIGN WITH COMPUTER MODELING: A CASE STUDY INVOLVING CREATINE

[1,2]J.W. SMITH, [1]R.H. DAVIES, [1]J.D. ANDRADE, AND [1,2]R.A. VAN WAGENEN

[1]Department of Bioengineering, University of Utah, Salt Lake City, UT, USA

[2]Protein Solutions, Inc., Salt Lake City, Utah, USA

E-mail:jws@xmission.com

1. INTRODUCTION

A bioluminescence-based biosensor with the ability to measure several metabolites at once would be a great asset to research as well as to clinical and preventative medicine. One of the advantages of bioluminescence is its sensitivity and ability to measure analytes in the micromolar (μM) range. However, to do so, an analytical instrument such as a photomultiplier tube (PMT) based luminometer which can accurately measure light intensity, is typically required. This approach can be too expensive for a home-based monitor; therefore, a hand-held CCD based luminometer is a desirable alternative. Additionally, a CCD has the advantage of being able to easily measure multiple analytes at once compared to a PMT-based luminomter, which is generally limited to only one.

Computer modeling is a useful tool for evaluating the feasibility of various biochemical pathways for measuring a specific metabolite of interest. The parameters of the computer model are governed by the desired detection limits of the metabolite of interest, the reaction kinetics, and the measuring device (in this case a CCD based luminometer). The purpose of this investigation was to apply a computer model of the ATP firefly luciferase reaction, present a case study of a homogeneous, ATP depletion assay for creatine (Cr), and discuss the use of computer modeling for optimizing biosensor design and development.

2. FIREFLY LUCIFERASE PLATFORM REACTION

All computer modeling was performed using Gepasi, a program for the simulation and optimization of biochemical reactions.[1] This investigation specifically investigated reactions coupled to firefly luciferase. Thus, the first step was to model the firefly luciferase (FFL) reaction and validate the model results with experimental data. The FFL reaction is shown below (rxn(1)):

$$ATP + luciferin + O_2 \xleftrightarrow{FFL} AMP + oxyluciferin + CO_2 + PP_i + light \quad (1)$$

The multi-step reaction scheme and kinetic constants used in the computer model were based on Gandelman et al. (1993).[2]

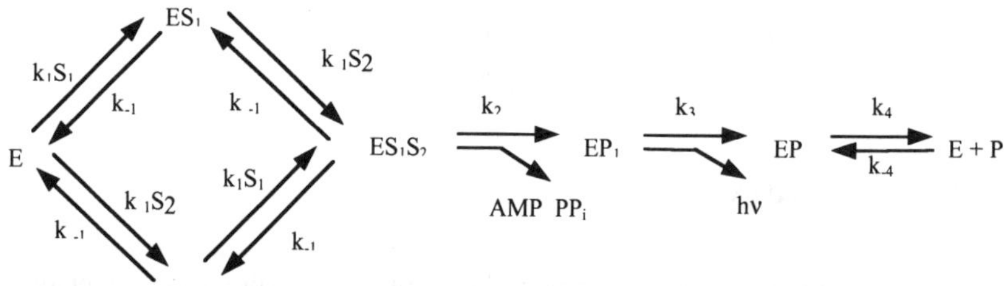

The above reaction scheme was modeled with an initial luciferin (LH_2) concentration of 30 μM, luciferase concentration of 0.7 μM, and ATP concentration ranging from 0.3 μM to 3.3 mM. To validate the computer model, experiments were performed using concentrations presented above. A Turner 20/20 luminometer was used to measure the light output as a function of time. The model and experimental results are shown in Figures 1a and 1b.

3. COUPLED REACTION SCHEMES

There are two basic types of sensor reactions: ATP depletion and ATP production. This investigation focused on ATP depletion assays (rxn(2)):

$$substrate + ATP \xleftrightarrow{E1} product + ADP \quad (2)$$

An example of an ATP depletion assay is the conversion of Cr to creatine phosphate (CrP) via creatine kinase (CK) (E.C. 2.7.3.2). To quantify Cr using bioluminescence, rxn(2) is coupled to the firefly luciferase reaction (rxn(1)). The difference between the ATP consumed by the firefly luciferase reaction alone and the ATP consumed by the coupled reactions is proportional to substrate concentration. This difference can be measured from the light output.

As an example for biosensor design, the creatine assay was modeled to determine feasibility and optimal conditions. The maximum velocity and Michaelis-Menten constants used in the model for the CK reaction were for rabbit muscle and were based on those previously determined by Morrison and James.[3]

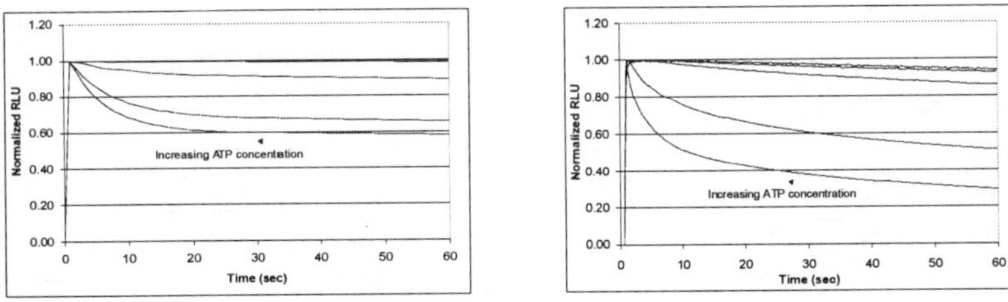

(a) **(b)**

Fig 1. Results of the (a) computer model and (b) experiments for the firefly luciferase reaction. Luciferin concentration = 30 μM, luciferase concentration = 0.7 μM, and ATP concentrations of 0.3 μM, 3 μM, 30 μM, .3 mM and 3.33 mM.

4. RESULTS

The computer model was used to optimize the results of the assay by varying the luciferin concentration between 0.01 and 1 mM, the luciferase concentration between 3×10^{-4} and 3×10^{-3} mM, and the ATP concentration between 0.1 μM and 1 mM. The results of the creatine model at varying ATP concentrations and at creatine concentrations between 0 and 300 μM are included in Figures 2a-d.

5. DISCUSSION

The design parameters for a biosensor are based on the required resolution and sensitivity of the assay. For example, a biosensor that measures creatine must be capable of detecting creatine concentrations between 0 and 300 μM with a sensitivity of 10 μM and a precision of ± 5%. To achieve these results, both the reaction kinetics and instrumentation limitations must be considered.

One issue with the CCD is the fact that it is approximately 100-1000x less sensitive than a typical PMT based luminometer and generally requires an integration of light intensity. As a result, it is desirable to maximize the light intensity while maintaining adequate sensitivity. To increase the light intensity, ATP and luciferase concentration can be increased. However, as ATP and luciferase concentrations are increased, the sensitivity of the assay (ability to detect differences between creatine concentrations) decreases. Potential solutions are to manipulate cofactors such as Mg^{2+} or to use completely different reaction schemes.

A hand-held CCD analyzer for a multi-channel, multi-analyte biosensor offers unique challenges. Future work will investigate the relationship between the results of the computer model and the CCD detector output. Ultimately, this will help to maximize the applicability of the computer model and will reduce the need for PMT-based detectors.

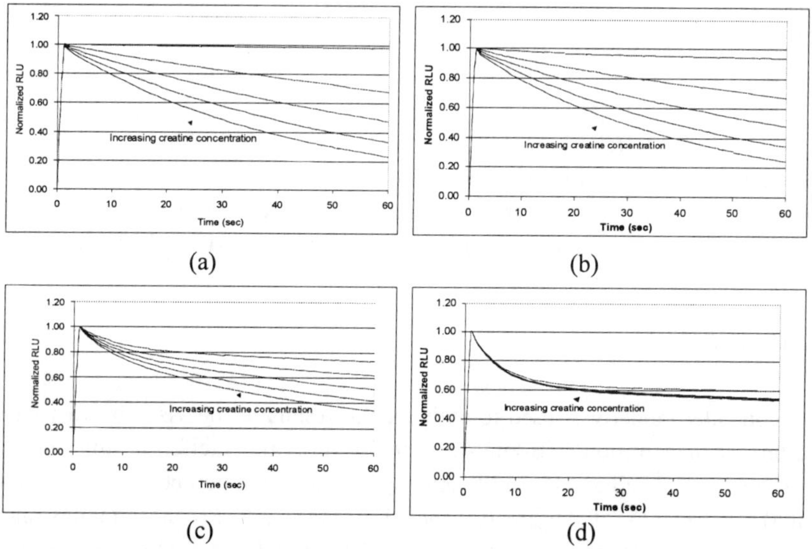

Fig 2. Results of computer model of creatine assay with creatine concentrations of 0, 75 μM, 150 μM, 225 μM, and 300 μM; luciferase concentration = 0.7 μM; luciferin concentration = 10 μM; and ATP concentrations of (a) 0.001 mM, (b) 0.01 mM, (c) 0.1 mM, and (d) 1.0 mM.

Acknowledgements

Financial support was provided by NIH SBIR Grant # 1R43DK5542601.

References

1. Mendes P. GEPASI: a software package for modeling the dynamics, steady states and control of biochemical and other systems. *Comput Applic Biosci* 1993; **9**: 563-571.
2. Gandelman OA, Brovko LYu, Polenova TE, Ugarova NN. Non-steady state kinetics in bioluinescent firefly luciferase system. *Bioluminescence and Chemiluminescence*, eds., Szalay AA, Kricka LJ, Stanley P. New York: John Wiley & Sons, 1993: 79-83.
3. Morrison JF, James E. The mechanism of the reaction catalyzed by adenosine triposphate-creatine phosphotransferase *Biochem J* 1965; **97**: 37-52.

BIOMATE: DEVELOPMENT OF CUSTOM-DESIGNED BIOLUMINESCENT SENSORS FOR TOXICITY TESTING

SIOUXSIE WILES[1,2], ANDREW WHITELEY[1], JIM PHILP[2] AND MARK BAILEY[1*]

[1]*Institute of Virology and Environmental Microbiology, Mansfield Road Oxford, OX1 3SR, England*

[2]*Department of Biological Sciences, Napier University, Edinburgh EH10 5DT, Scotland*

[*]*E-mail: mbj@ceh.ac.uk*

1. INTRODUCTION

Bioluminescent microbial biosensors have become increasingly used in toxicity sensing due to their apparent ease of use and potential to rapidly report on toxic events as evidenced by the widespread implementation of such systems as the Microtox[¤] field testing assay. At present however, disadvantages exist in using traditional marine based organisms (eg *Photobacterium* sp.) which tend to comprise the majority of commercially available assays: high salt dependence, limited sensing range (tolerance) to toxicants and physiological limitations as potential *in situ* sensors in wastewater treatment plants. These problems are exacerbated in highly polluted systems where either the sensors are too sensitive to pollutants, potentially giving a false impression of toxicity, or the luminescence is reduced to extremely low levels thereby reducing the sensing range. Efforts to overcome these problems have focused primarily on the use of plasmid-borne bioluminescent phenotypes in laboratory-derived hosts such as *E. coli* and *Pseudomonas* sp.[1-3] Although significant advances have been made using these types of sensors, the fact remains that they are not derived from the process under study and hence may have inherent physiological drawbacks to accurate sensing in treatment remediation.

This communication constitutes a preliminary report on the development of a panel of novel bacterial biosensors based on organisms native to a highly toxic phenolic-processing wastewater treatment plant.[4] The sensors accurately predict toxicity of the major toxicants responsible for modulating microbial community activity (and hence process efficacy) within the system.

2. MATERIALS AND METHODS

Bacterial strains: A panel of bioluminescent sensors were developed by chromosomally marking a number of environmental Pseudomonad isolates using a mini-Tn*5* construct containing the *luxCDABE* operon from *Photorhabdus luminescens*, as well as a kanamycin resistance gene.[5]

Toxicity measurements: Biosensor strains were grown to stationary phase in LB medium, supplemented with 50❄g/mL kanamycin, at 28…Cin a rotary shaker. Toxicity measurements were carried out in 96 well microtiter plates. Plates were incubated for five minutes at room temperature, after which luminescence was measured on a microtiter plate luminometer (Lucy1, Anthos Labtec Instruments, Austria). Luminescence values are expressed as a percentage of the luminescence of a control well containing cells incubated with distilled water or LB, depending on the assay being performed. Assays were performed in triplicate on three replicate cultures. Pure phenol toxicity assays were carried out using 1.5x stock concentrations of phenol made up in LB medium. Doubling dilutions of phenol (100❄L) were added to 50❄L of cells to give final toxicant concentrations in the range 0-800mg/L. Environmental toxicity testing was carried out using samples taken from an undisclosed industrial vitox-based processing system (150❄L of sample added to 100❄L of cells).

3. RESULTS AND DISCUSSION

A panel of novel biosensors was developed based on the expression of the full lux operon, chromosomally located in a variety of Pseudomonad strains isolated from the polluted site. Inclusion of the genes for substrate production and recycling negated the need for substrate addition and hence aided the development of a rapid and reproducible toxicity assay. Toxicity evaluations confirmed good negative linear correlations of sensor luminescence with phenol concentration (Fig.1). In addition, each strain showed differential tolerance to phenol toxicity with EC_{50} values ranging from 500-800mg/L phenol. This fulfilled the prediction that natural isolates would provide strains that could be implemented as biosensors for a wide range of acute toxicities.

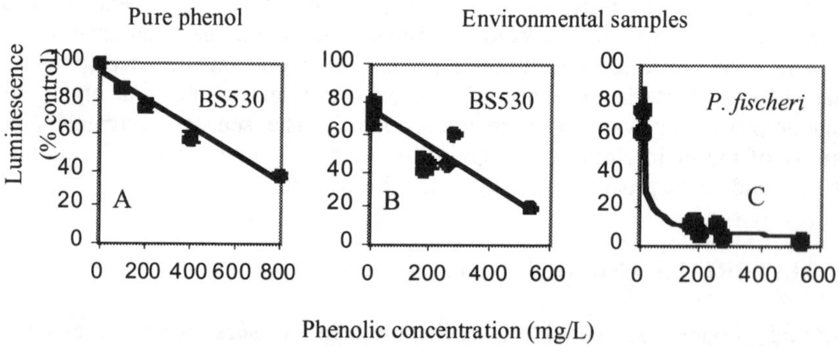

Fig 1. Luminescent response of biosensors to pure phenol (A) and response to different phenolic concentrations in polluted bioreactor samples (B/C).

Environmental toxicity measures performed using wastewater samples with varying levels of phenolic load indicated that the luminescence response was linearly related to phenolic concentration in complex cocktails (Fig.1) and as a result the sensors could be used in highly toxic compartments. This was in direct contrast to *Photobacterium fischeri* which demonstrated a non-linear response to wastewater samples and had a limited sensing range, with luminescence output negligible in samples whose phenolic concentration was higher than 150mg/L (Fig.1).

Temporal analysis of the various compartments of the industrial wastewater treatment system under normal operating parameters allowed the determination of an operational range for each sensor in a given compartment. This is presented as the mean luminescence obtained during temporal analysis of a particular compartment (0) within the range of one standard deviation. Any sample can be analysed to determine whether it falls within a sensor s operational range (and hence give an indication of the sample s toxicity) using the following equation:

$$\textbf{Relative toxicity } (R_{tox}) = L_x - L_m$$

where L_x is the % luminescence obtained for the sample of interest and L_m is the mean % luminescence obtained during temporal analysis of the compartment from which the sample is taken. If the calculated value for R_{tox} is negative, this represents an increase in toxicity whilst a positive value represents a decrease. Within the standard deviation, fluctuations in toxicity are considered within the sensors operational range. However, values outside the standard deviation can be considered to represent significant changes in the toxicity of the system which lie outside the tolerance of a particular sensor and may be indicative of a pollution event. The temporal responses of sensor BS530::LuxKm to samples from the pre-treatment (toxic) reservoir obtained during the field trial analyses (normal toxicity) are given in Figure 2, illustrating the reliability of a sensor based on a natural isolate indigenous to the bioreactor.

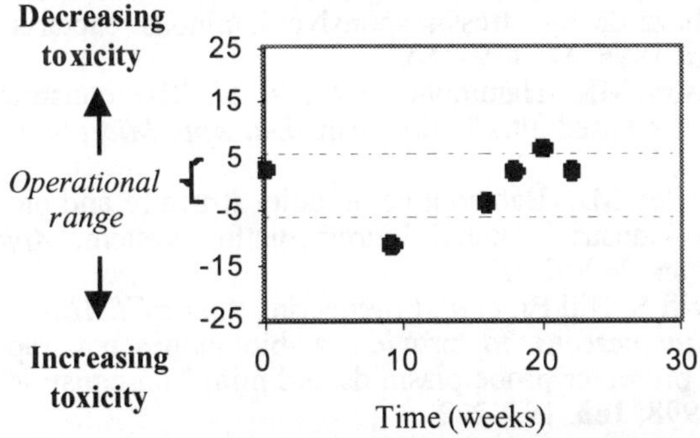

Fig 2. Temporal evaluation of biosensor BS530::LuxKm: response to samples from the pre-treatment (toxic) reservoir of the operating bioreactor plant.

The relationship between microbial activity, assessed by rRNA probing,[4] and sensor luminescence to the toxic loads in the same environmental samples revealed a significantly positive correlation between the two parameters (Fig.3). This confirmed that the sensors accurately predicted the microbial activity within the samples and would therefore accurately predict the toxic effects of pollutant loading on the microbial activity within the wastewater treatment system. This provides a reliable method of sensing acute toxic shock and its associated effects on microbially driven processes.

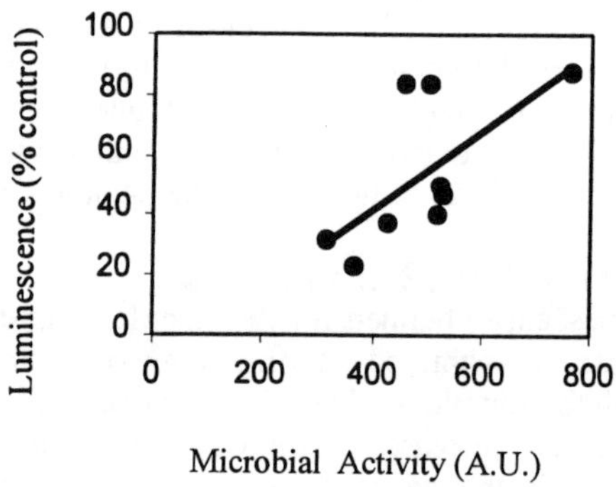

Fig 3. Correlation between biosensor activity (luminescence) and microbial activity to samples from the studied wastewater treatment plant.

References

1. Van Dyk TK, Majarian WR, Konstantinov KB, *et al.* Rapid and sensitive pollution detection by induction of heat shock gene-bioluminescence gene fusions. *Appl Env Microbiol* 1994; **60**: 1414–1420.
2. Belkin S, Van Dyk TK, Vollmer AC, *et al.* Monitoring subtoxic environmental hazards by stress-responsive luminous bacteria. *Environ Toxicol Wat Qual* 1996; **11**: 179–185.
3. Prest AG, Winson MK, Hammond JRM, *et al.* The construction and application of a lux-based nitrate biosensor. *Lett Appl Microbiol* 1997; **24**: 355–360.4.
4. Whiteley AS, Bailey MJ. Bacterial community structure and physiological state within an industrial phenol bioremediation system. *App Environ Microbiol* 2000; **66**: 2400-2407.
5. Winson MK, Swift S, Hill PJ, *et al.* Engineering the *luxCDABE* genes from *Photorhabdus luminescens* to provide a bioluminescent reporter for constitutive and promoter probe plasmids and mini-Tn*5* constructs. *FEMS Microbiol Lett* 1998; **163**: 193–202.

PART 5

APPLICATIONS OF CHEMILUMINESCENCE

CHEMILUMINESCENT ASSAY OF ALKALINE PHOSPHATASE USING DIHYDROXYACETONE PHOSPHATE AS SUBSTRATE WITH LUCIGENIN

AMANE KOKADO, HIDETOSHI ARAKAWA AND MASAKO MAEDA

School of Pharmaceutical Sciences, Showa University, 1-5-8 Hatanodai, Shinagawa-Ku, Tokyo 142-8555, Japan

1. INTRODUCTION

Chemiluminescent (CL) and bioluminescent (BL) assays are attractive analytical methods in the field of clinical chemistry because of their high sensitivity, rapidity of reaction and wide dynamic range. Highly sensitive CL methods and BL methods have become available for detecting alkaline phosphatase (ALP). Lucigenin produces strong chemiluminescence with hydrogen peroxide or reducing agents. Previously, we have reported that many compounds having an α-hydroxycarbonyl moiety in the side chain reacts with lucigenin to generate chemiluminescence.[1, 2] Dihydroxyacetone (DHA) gives the most intense chemiluminescence in the reaction of lucigenin with various reducing compounds. We devised the CL assays of ALP using glycerol 3-phosphate or $NADP^+$ as substrate and glycerol dehydrogenase (GDH) as the coupled enzyme.[2] The principle of the method as follows: ALP causes hydrolysis of $NADP^+$ to NAD^+ that is used as cofactor of GDH in the presence of glycerol 3-phosphate. The DHA produced by GDH reaction was measured by CL assay using lucigenin. Though this method was sensitive, was not simple. Therefore, we studied to improve the assay.

We report the development of detection method for ALP using dihydroxyacetone phosphate (DHAP) as a new substrate and lucigenin as CL reagent. And this CL assay was applied to enzyme immunoassay using ALP as a label. The CL mechanism of lucigenin with DHA also investigated by ESR spectrometry using spin trapping methods.

2. MATERIALS AND METHODS

Materials Lucigenin (*N*, *N*'-dimethyl-9, 9'-biacridinium dinitirate) were purchased from Tokyokasei Industrial Co. Ltd. (Tokyo, Japan). Alkaline phosphatase (ALP, EC 3.1.3.1) was from Roche Molecular Biochemicals (Germany). Dihydroxyacetone phosphate (DHAP), dihydroxyacetone phosphate dimethyl ketal (DHAP-ketal), 17β-estradiol and 17β-estradiol 3-carboxymethyl ether were from Sigma Chemical Co. (St. Louis, MO). Anti-17β-estradiol antibody was from Cosmo

Bio Co., Ltd. (Tokyo, Japan). Other chemicals used were of analytical reagent grade.

ALP assay To each well of microtiter plate, 25 μL of ALP solution and 25 μL of 0.1 or 0.2 mmol/L of DHAP or DHAP-ketal in 0.1 mol/L Tris-HCl buffer (pH9.5) containing 0.1 mmol/L $MgCl_2$ were added and incubated for 1h at 37 °C with mixing. Then, adding 10 μL of 1.0 mol/L H_2SO_4 stopped the enzymatic reaction. After addition of each 50 μL of 1.0 mol/L KOH and 8 x 10^{-4} % lucigenin containing 0.01 % Triton X-100, the CL intensity was measured by Berthold-MicroLumat LB96P (measurement time 5s, delay time 10s).

Determination of superoxide anion[3, 4] Superoxide anion (O_2^-) was determined by measurement of the 5-hydroperoxy-2, 2-dimethyl-1-pyrolinyloxy (DMPO-OOH) adduct using spin trapping method (JEOL JES RE1X spectrometer).

Enzyme immunoassay for 17β-estradiol To each well of microtiter plate coated with anti-rabbit IgG, 50 μL of 17β-estradiol (E_2) standard or sample solution was added, and then, 25 μL of anti-E_2 antiserum was added. After incubation for 1 h at 37 °C with mixing, 25 μL of ALP labeled E_2 was added, and incubated for 30 min. The well was washed three times, by adding 100 μL of 0.05 mmol/L DHAP, enzyme activity bound in each well was determined as described above.

3. RESULT AND DISCUSSION

New substrates, DHAP and DHAP-ketal for ALP, were examined by CL assay with lucigenin (Scheme). These substrates were hydrolyzed enzymatically, produced DHA or DHA-ketal, respectively. Furthermore, DHA-ketal was transformed into DHA by adding H_2SO_4 (Method B). For CL emission with lucigenin, same behavior of kinetics was observed (data not shown); the CL intensity was increase until 40-60s, decreased gradually and disappeared at 180s. The detailed comparisons of Method A and Method B were carried out for detection method of ALP with CL assay.

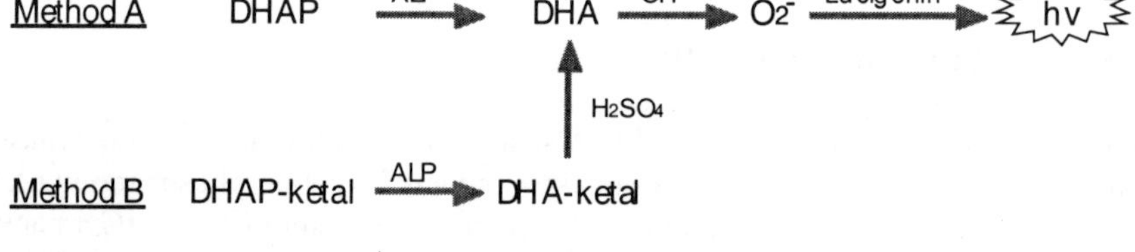

Scheme CL mechanism of the proposed method using DHAP and DHAP-ketal as substrate.

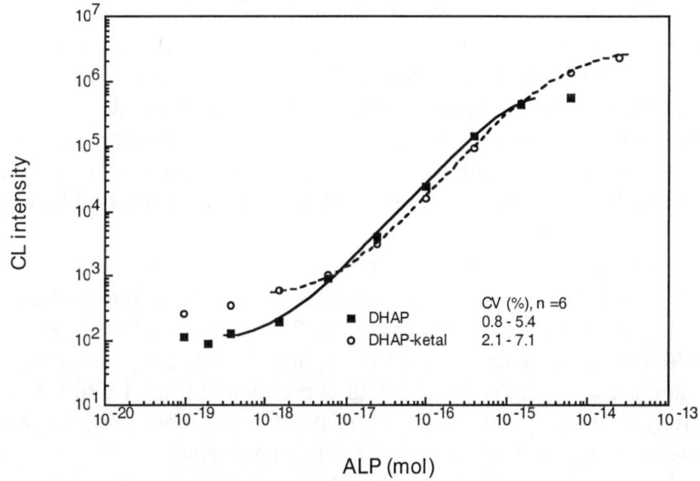

Fig. 1 Standard curves of ALP and precision of intra-assay (n = 6) using DHAP (Method A) or DHAP-ketal (Method B) as substrates.

The optimum assay conditions (pH, incubation time and concentration of reagent) were examined for the ALP assay. As shown in Fig. 1, the measurable range of ALP by Method A and Method B was 2 x 10^{-14} - 6 x 10^{-11} mol/L and 6 x 10^{-14} - 1 x 10^{-9} mol/L at the optimum assay conditions. And the detection limit (2x standard deviation) was 3.8 x 10^{-19} and 1.5 x 10^{-18} moles of ALP, respectively. ALP assay by using Method A was more sensitive and had a wider range than that of Method B.

The CL mechanism of lucigenin was studied by ESR spectroscopy using DMPO as spin trapping reagent. The spectrum of DHA in alkaline solution (KOH) showed the signal pattern of the DMPO-OOH adducts of O_2^-. Furthermore, addition

Table 1 The relations between O_2^- and CL intensity with lucigenin in various compounds having α-hydroxycarbonyl group.

	CL intensity [(a)]	O_2^- intensity [(b)]
Cortisol	1297	1.09
DHA	3597	1.44
p-Hydroxy acetophenone	8537	1.63
DHA dimer	11645	2.49

Each 50 μL (a) or 100 μL (b) of 0.1 mmol/L various compounds, 1 mol/L KOH and 8 x 10^{-4} % lucigenin were mixed. And the CL intensity was measured by proposed method (a) and O_2^- intensity was measured by ESR spectrometry (b).

of SOD under these conditions almost eliminated the adduct signal. Table 1 shows O_2^- intensity of various compounds having α-hydroxycarbonyl group. It was assumed that their O_2^- intensity was correlated to CL intensity. Kubo et. al. also reported that O_2^- was generated from various reducing compounds such as uric acid, dihydroxyacetone, cortisone, and glucose in NaOH solution.[5] Therefore, we speculated as shown in Scheme: an α-hydroxycarbonyl group of DHA reacts with oxygen to yield O_2^- in alkaline medium. And then lucigenin react with O_2^- and emit light.

The proposed ALP assay was applied for enzyme immunoassay of 17β-estradiol (E_2). ALP labeled E_2 was prepared from activated 17β-estradiol 3-carboxymethyl ether. In the competitive immunoassay detected using Method A, E_2 could be determined in the range 15 - 4000 pg/mL. The detection limit was 3.4 fmol per assay, and the accuracy at the standard curve ranged from 4.0 to 8.5 % (n = 6). The sensitivity of the proposed method was about 4-fold more sensitive than that of the colorimetric assay for ALP using 4-nitrophenyl phosphate.

Acknowledgements

This work was supported in part by a grant in aid for the Scientific Research from the Ministry of Education, Japan, and by the Sasagawa Scientific Research Grant from the Japan Science Society.

References

1. Kokado A, Tsuji A, Maeda M. Chemiluminescent assay of alkaline phosphatase using cortisol 21-phosphate as substrate and its application to enzyme immunoassays. *Anal Chim Acta* 1997; **337**: 335-340.
2. Kitamura M, Maeda M, Tsuji A. A New highly sensitive chemiluminescent assay of alkaline phosphatase using lucigenin and its application to enzyme immunoassay. *J Biolumin Chemilumin* 1995; **10**: 1-7.
3. Kanemitsu (Hosaka) M, Arakawa H, Yoda R, Maeda M. Chemiluminescent determination of lucigenin using thioureid derivatives. *Anal Chim Acta* 2000; **403**: 125-130.
4. Ueno I, Kohno M, Haraikawa K, Hirono I. Interaction between quercetin and superoxide radicals. Reduction of the quercetin mutagenicity. *J Pharm Dyn* 1984; 7: 798-803.
5. Kubo H, Saitoh M, Mochii Y, et. al. Chemiluminescence of reducing agents based on the luminol reaction. *BNSKAK (J Anal Chem Japan)* 1998; **47**: 599-603.

RESPONSE OF AN ENHANCED CHEMILUMINESCENCE ASSAY TO SELECTED SUBSTANCES

JAMES C. MC HILLIS AND B. RIPPEY

School of Environmental Studies, University of Ulster, Coleraine, Co. Londonderry, Northern Ireland, BT52 1SA
E-mail: JCM.Hillis@ulst.ac.uk

S. P. FITZGERALD

Randox Laboratories Ltd. Diamond Road, Crumlin, Co. Antrim, Northern Ireland, BT29 4QY
E-mail: mail@randox.com

1. INTRODUCTION

There are difficulties in regulating effluent disposal and water quality. There is a need for rapid and economic assays in the field of ecotoxicology. Wharfe and Tinsley state that rapid assays may be included in toxicity based consents for waste waters.[1] The regulatory bodies seek comprehensive testing, of the aquatic ecosystem and effluent inputs, at reduced cost. Full chemical characterisation on all wastewaters is not economically practical and mass biological screening is expensive and unpalatable to many. It is now also widely recognised that neither chemical characterisation nor whole organism toxicity testing are satisfactory alone.[2]

Aquanox® is a chemiluminescence based water quality monitoring unit, developed for the rapid toxicological screening of water samples. The components of the reaction are; horseradish peroxidase (HRP), luminol, hydrogen peroxide and p-iodophenol.[3] luminol absorbs energy released by the chemical reaction between the enhancer (p-iodophenol), the oxidant (hydrogen peroxide) and the enzyme (HRP) to form an excited intermediate product, the aminophthalate ion. This excited intermediate relaxes back to its ground state emitting the excess energy as a photon. The reaction is a free radical reaction (the oxidation of luminol) between hydrogen peroxide, the oxidant, and luminol, in the presence of the enzyme HRP and enhancer p-iodophenol. Inhibition of the reaction causes a reduction in light output. This reduction in light output is measured by a handheld luminometer. Expected inhibitors are free radical scavengers (antioxidants that remove oxygen from the water), enzyme inhibitors, reactive species and species that increase Biological Oxygen Demand (BOD) and Chemical Oxygen Demand (COD).

2. MATERIALS AND METHODS

All glassware was subjected to a rigorous cleaning and contamination control procedure. All solutions were made up on the day of use and all volumetric glassware was grade A.

The chemiluminescence assay was carried out as per Randox method, revised 06/01/97.[3] The reagent kit contains borate buffer pH 8.5, signal reagent containing luminol, p-iodophenol and hydrogen peroxide, and the enzyme reagent containing horseradish peroxidase. The reference is set by adding 1000 μL of Milli Q water and 100 μL of signal reagent with mixing. 20 μL of enzyme reagent is then added with mixing, the luminometer started at the same time and the reaction cuvette quickly inserted into the luminometer. After 4 min. the luminometer sets this level of light output as the maximum possible for the current system and stores it as such. All samples are then measured with reference to this value. Sample testing follows the same procedure apart from the addition of 800 μL of milli Q water and 200 μL of sample, in the place of the 1000 μL of Milli Q water for the reference and the % reduction in light output noted after 4 minutes. Relative Standard Deviation (RSD) was calculated as (standard deviation/mean) s/x X 100, for 6 replicate measurements at the same concentration that Limit of Quantitation (LOQ) was measured. Limit of Detection (LOD) was set as the lowest point on the linear range. LOQ was calculated as the concentration corresponding to the % inhibition level of 3s + Av. Value for 6 replicate measurements near the expected limit of quantitation (Level used, see table).

3. RESULTS AND DISCUSSION

The overall aim of the investigation is to develop chemiluminescence inhibition further as a rapid toxicology assay. The aim of this study is to explore the response of the enhanced chemiluminescence system to a range of selected toxicants. For each compound: linear range, (LOD), (LOQ) and (RSD) have been performed. As certain chemicals inhibit the reaction and therefore reduce the light output, the reduction in light output in comparison with Milli-Q water can be used as a measure of contaminant concentration.

Microtox® is an analogous bio-luminescence system utilising the marine bacterium *Vibrio fischeri*. The volume of information available gives credence to the Microtox® test and this study aims to offer a similar database for the Aquanox® test. Aquanox® is a younger technique and is less well documented. However, the application of chemi-luminescence is a popular field and there is a corresponding growth in the volume of literature on the area.[4-11]

Table 1. Response of the Aquanox® system to a selection of compounds

Compound	Linear Range mgL^{-1} Bottom	Linear Range mgL^{-1} TOP	R^2	Level Used For LOQ/RSD mgL^{-1}	RSD %	LOQ mgL^{-1}	LOD mg L^{-1}
Ammonia	35	17500	0.89	350	3.31	559	35
Bromoxynil	N/A	N/A	N/A	132	N/A	N/A	>132
Cadmium	0.1	150	0.97	0.1	6.66	21.52	0.1
Ethanol	100	797600	0.97	1000	5.67	1280	100
Hydroquinone	0.01	0.99	0.94	0.01	2.18	0.15	0.01
Lead	1	175	0.81	4	0	14.68	1
Mecoprop	N/A	N/A	N/A	31.8	N/A	N/A	>31.8
Phenol	0.05	25	0.88	1	4.21	0.75	0.05
Potassium Cyanide	0.05	0.46	0.94	0.1	2.45	0.14	0.05
Potassium Permanganate	0.01	9.95	0.88	0.5	6.4	0.15	0.01
Salcylic acid	98.6	986	0.95	200	3.45	237.70	98.6
Thiosemicarbazide	1	10	0.89	2	9.06	1.54	1

Table 1 shows the range of responses to the selection of compounds. The system was most sensitive to cadmium, hydroquinone, phenol, potassium cyanide and potassium permanganate. The system was less sensitive to ammonia, and ethanol and insensitive to salcylic acid. Lead and thiosemicarbazide gave LODs of 1 mgL^{-1}. Aquanox® was insensitive to both mecoprop and bromoxynil at the respective limits of solubility in water. There was a strong relationship between concentration and % inhibition for each compound. The R^2 values for every compound were statistically significant.[12] Correlations were found to be particularly high for cadmium (R^2 0.97), ethanol (R^2 0.97), hydroquinone (R^2 0.94), potassium cyanide (R^2 0.94) and salcylic acid (R^2 0.95). % Inhibition response for each compound was directly proportional to concentration over the given range.

The results show a varied response for the range of compounds. However, Aquanox® could be successfully used to measure the concentrations of cadmium, hydroquinone, phenol, potassium cyanide, potassium permanganate, lead and thiosemicarbazide. The system could also be applied as a screen for the afore mentioned chemicals, to cheaply, reliably and quickly detect the chemicals, their concentrations and changes in concentration. The Aquanox® system is intended to be a holistic toxicity screening tool. However, it is important to know how it responds to single chemicals in solution. Chemicals to be added to the list include; 4-chloro-1-napthol, 2-chlorophenol, 3-chlorophenol, 4-chlorophenol, chromium (II) in solution. Chemicals to be added to the list include; 4-chloro-1-napthol, 2-chlorophenol, 3-chlorophenol, 4-chlorophenol, chromium (II) chloride, 3,5-dichlorophenol, thiouracil, uric acid butyric acid, ferulic acid, manganese (II) chloride, gallic acid and eugenol. The next stage of this investigation will be the response of the system to binary and tertiary mixtures of the chemicals listed.

Acknowledgements
The Department of Education for Northern Ireland (DENI), The University of Ulster (UU), Randox Laboratories and The Royal Society of Chemistry (RSC).

References

1. Wharfe J, Tinsley D. The toxicity-based concent and the wider application of direct toxicity assessment to protect aquatic life. *J CIWEM* 1995; 9 :526-530.
2. Grothe DR, Dickson KL, Reed-Judkiins DK. Whole effluent toxicity testing: an evaluation of methods and prediction of receiving system impacts. Pensacola, Florida: SETAC, 1996.
3. Aquanox® operator's manual. Crumlin: Randox, 1996.
4. Billings C, Lane M, Watson A, Whitehead TP, Thorpe G. A rapid and simple chemiluminescent assay for water quality monitoring. *Anaysis* 1994; 22 :27-30.
5. Campbell AK. *Chemiluminescence, Principles and Applications in Biology and Medicine.* Cambridge: VCH, 1988.
6. Coulet PR, Blum LJ. Bioluminescence/chemiluminescence based sensors. *Trends Anal Chem* 1992; 11 2:57-61.
7. Fearnside D, Caffoor I. The detection of pollution-some novel approaches. *Novel Approaches in Pollution Detection* 1998.
8. Johnson TB, Long ER. Rapid toxicity assessment of sedimnents from esturine ecosystems: A new tandem in vitro testing approach. *Environ Toxicol Chem* 1998; 17 6:1099-1106.
9. Johnson I, Harman M. Review of the enhanced chemiluminescence (ECL) test. Bristol: SNIFFER (Scotland & Northern Ireland Forum For Environmental Research) The Environment Agency, 1997.
10. Kricka LJ, Thorpe GHG. Chemiluminescent and bioluminescent methods in analytical chemistry. *Analyst* 1983; 108 : 1274-1296.
11. Thorpe GHG, Sawcer KE, Yeomans WE. Enhanced chemiluminescence as an indicator of water quality. *J CIWEM* 1999; 13 : 170-176.
12. Fowler J, Cohen L, Jarvis P. *Practical Statistics for Field Biology.* New York: John Wiley, 1999.

A CHEMILUMINESCENT WHOLE CELL BIOSENSOR FOR ASSESSING ESTROGENIC ACTIVITY

P. PASINI, G. GENTILOMI, M. GUARDIGLI, M. BARALDINI, M. MUSIANI AND A. RODA

Department of Pharmaceutical Sciences, University of Bologna, Via Belmeloro 6, 40126 Bologna, Italy

E-mail: pasini@alma.unibo.it

1. INTRODUCTION

Many of the widely-used natural compounds and manmade chemicals that are continuously released into the environment interact with the endocrine system and alter steroid-hormone-receptor-mediated physiological processes, causing adverse effects on the health of humans and wildlife. Such endocrine disrupting compounds (EDCs) include phytoestrogens, certain pesticides and herbicides, plasticizers, surfactants breakdown products, some polychlorinated biphenyls and synthetic estrogens used as drugs. These substances are very toxic even when present in small quantities in food, water and air, they persist in the environment, can accumulate in living organisms along the entire food chain and concentrate in target organs and body fat.[1]

The study of the environmental distribution of EDCs and their actual impact on humans and wildlife requires the evaluation of their potential toxicity for living organisms by means of suitable bioassays at the genetic and molecular level.[2,3] Biological tests, based on genetically transformed cells which express estrogen receptors, have been developed to study compounds with estrogenic activity.[4-7] Yeast cells are often used because they are easily engineered and cultured and have rapid growth rates, thus providing a very robust model suitable for wide-scale screening. β-Galactosidase coding sequence is commonly used as a reporter gene and the enzyme is generally detected by colorimetric methods which are time-consuming and offer poor sensitivity, with inadequate detection limits.

In the present work we proposed chemiluminescence (CL) as a valid alternative to conventional detection methods, due to its higher detectability and rapidity. A luminescent whole cell biosensor based on recombinant yeast cells was developed and optimized for environmental monitoring of estrogen-like compounds.

2. MATERIALS AND METHODS

A recombinant yeast strain (*Saccharomyces cerevisiae*) in which the DNA sequence of the human estrogen receptor was stably integrated into the main chromosome was provided with permission by Prof. John Sumpter (Brunel University, U.K.).[4] The yeast cells also contained an expression plasmid carrying estrogen-responsive sequences and the reporter gene *lac-Z*, encoding the enzyme β-galactosidase. Upon binding an active ligand, the estrogen-occupied receptor interacted with the estrogen-responsive elements to modulate gene transcription. This caused expression of the reporter gene *lac-Z* and the β-galactosidase produced was secreted into the medium and detected by a chemiluminescent 1,2 dioxetane-β-D-galactopyranoside substrate (Galacto-Light Plus, PE-Tropix, Bedford, MA).

Cell culture was performed as previously described.[4] Briefly, 4 mL growth medium were inoculated with one cell colony and incubated at 28°C for 24 h on an orbital shaker until an absorbance at 620 nm of 1.0 was reached. The assay medium was prepared by seeding 50 mL fresh growth medium with 2 mL yeast from the 24-h yeast culture.

The assay was performed in triplicate in 96-well microtiter plates. Ten μL ethanol standard solutions of the natural estrogen 17β-estradiol (17β-E2) (1000-10 pmol/L final concentrations in the microtiter plate wells) were incubated with 200 μL assay medium containing approximately 1 x 10^6 yeast cells per mL. After a 24-h incubation at 28°C, 20 μL assay medium were added to 70 μL CL substrate and incubated at room temperature for 1 h. One hundred μL light emission accelerator were then added and the CL signal was measured using a photomultiplier-tube-based microplate luminometer (Luminoskan Ascent, Labsystems, Helsinki, Finland).

3. RESULTS AND DISCUSSION

The CL reaction showed a glow-type kinetics, with the light emission reaching the maximum intensity in a few minutes and the steady-state lasting for at least 30 minutes, thus facilitating the measurement procedure.

The CL method was compared with the conventional colorimetric one based on the use of a chromogenic chlorophenolred-β-D-galactopyranoside substrate. Using the CL substrate, a detection limit of 10 pmol/L 17β-E2 (corresponding to 2 fmoles/well) was achieved after incubating the recombinant yeast cells with the analyte for 24 h (Fig. 1b). Positive signals could not be detected with the chromogenic substrate after 24 h (data not shown) and a 72-h incubation was required for the colorimetric method to reach the 10 pmol/L 17β-E2 detection limit (Fig. 1a). This is probably due to the less efficient enzyme-catalyzed transformation of the chromogenic substrate compared to that of the chemiluminescent one.

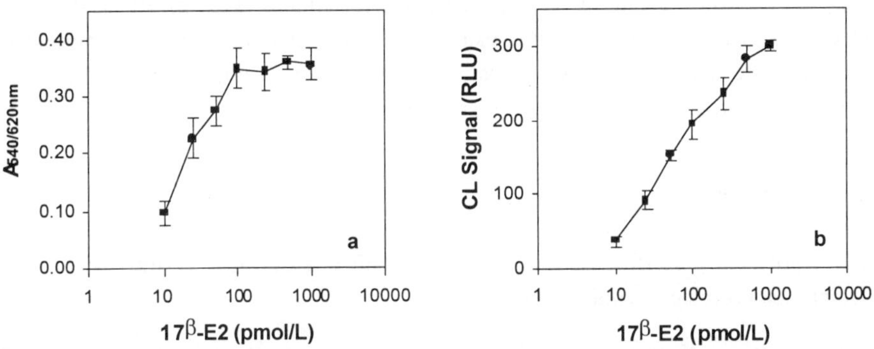

Fig. 1 Calibration graphs achieved with colorimetric detection (a) after a 72-h incubation of the recombinant yeast cells with 17β-E2 and with chemiluminescent detection (b) after a 24-h incubation.

The CL method also exhibited a wider dynamic range (2 decades vs. 1 decade), thus allowing the direct analysis of concentrated samples. In the colorimetric system the coloured reaction product accumulated over time. As a consequence, the reaction saturation occurred in the presence of high 17β-E2 concentrations during the long incubation time required to detect the lower 17β-E2 concentrations.

The reproducibility of the CL method was assessed intra-assay and inter-assay by measuring the response of the cells to replicate dilutions of 17β-E2. The response over the concentration range examined (1000 to 10 pmol/L) was quite reproducible, with values of the coefficients of variation ranging 12 to 20 %.

After using the natural estrogen 17β-E2 for the optimization and evaluation of the analytical performance of the CL method, known environmental estrogens were tested. Genistein (a phytoestrogen), bisphenol-A (a plasticizer) and diethylstilbestrol (a synthetic estrogen) were chosen as representatives of the main classes of EDCs. All the examined substances were able to induce the maximum cell response, but at different concentrations; diethylstilbestrol possessed an estrogenic activity similar to that of 17β-E2, while genistein and bisphenol-A were about four orders of magnitude less potent. These results are consistent with those reported in the literature[4] and could be explained on the basis of different affinities for the human estrogen receptors.

The CL method proved to be rapid, sensitive, robust, inexpensive, easy-to-use and easily automatable, thus being suitable for wide-scale screening. These features make the chemiluminescent biosensor a very useful tool for the first-level monitoring of environmental and biological samples. Only the positive samples will then be tested by more accurate analytical techniques (such as HPLC-MS) to detect and quantify the individual estrogen-like compounds. The CL biosensor could also be exploited to screen large numbers of molecules and find out the estrogenic activity of new endocrine disrupting compounds.

Achnowledgements

This work was partially supported by the Ministero dell'Università e della Ricerca Scientifica e Tecnologica (MURST).

References

1. Natural and anthropogenic environmental oestrogens: the scientific basis for risk assessment. *Pure Appl Chem* 1998; **70**: 1617–1865.
2. *Trends in Analytical Chemistry* 1997; **16**: 544-619.
3. Soto AM, Sonnenschein C, Chung KL, Fernandez MF, Olea N, Serrano FO. The E-screen assay as a tool to identify estrogens: an update on estrogenic environmental pollutants. *Environ Health Perspect* 1995; **103**: 113-122.
4. Routledge EJ, Sumpter JP. Estrogenic activity of surfactants and some of their degradation products assessed using a recombinant yeast screen. *Environ Toxicol Chem* 1996; **15**: 241–248.
5. Arnold SF, Robinson MK, Notides AC, Guillette Jr LJ, McLachlan JA. A yeast estrogen screen for examining the relative exposure of cells to natural and xenoestrogens. *Environ Health Perspect* 1996; **104**: 544–548.
6. Klein KO, Baron J, Colli MJ, McDonnell DP, Cutler JB, Jr. Estrogen levels in childhood determined by an ultrasensitive recombinant cell bioassay. *J Clin Invest* 1994; **94**: 2475-2480.
7. Balaguer P, Francois F, Comunale F, Fenet H, Boussioux A-M, Pons M, Nicolas J-C, Casellas C. Reporter cell lines to study the estrogenic effects of xenoestrogens. *Sci Total Environ* 1999; **233**: 47–56.

PART 6

IMMUNOASSAY AND DNA PROBE ASSAYS

CARBONYL-REACTIVE CHEMILUMINESCENT LABELS: ACRIDINIUM HYDROXYLAMINES (AHA)

MACIEJ ADAMCZYK, JONATHAN GROTE, PHILLIP G. MATTINGLY, JEFFREY A. MOORE, AND YOU PAN
Abbott Laboratories, Diagnostics Division, Department of Chemistry D-09NM, Building AP 20, 100 Abbott Park Road, Abbott Park, Illinois, 60064-6016, USA
E-mail: maciej.adamczyk@abbott.com

1. INTRODUCTION

Electrophilic acridinium-9-carboxamide chemiluminescent labels, principally carboxylic acid active esters[1, 2] or maleimides,[3] have been used to derivatize nucleophilic amines and thiols, respectively. Rarely, acridinium-9-carboxamide labels bearing a nucleophilic amine have been used.[4, 5] In those cases, of course, the complementary carboxylic acid must be activated on the substrate. This doesn t present a problem unless the substrate also bears an amine or it is difficult to introduce the carboxylic acid functionality. For instance, a simple substrate such as an amino acid would require protection of the amine before activating the carboxylic acid. Protection schemes are not practical though if the substrate is as complex and multifunctional as a protein. Further, many substrates lack a carboxylic acid, e.g., clinically significant steroids, yet have an electrophilic carbonyl functional group.

O-Alkylhydroxylamines differ in reactivity from the analogous *N*-alkylamines; they can be coupled to carboxylic acids under acidic conditions where their amine counterparts are less reactive,[6, 7] and they form stable oximes with carbonyl groups, whereas the corresponding imines are easily hydrolyzed. It was therefore of interest to us, to prepare an ***a***cridinium-9-carboxamide ***h***ydroxyl***a***mine label **3** (AHA) with the versatile hydroxylamine functionality and test its applicability for labeling carbonyl-containing substrates.

2. MATERIALS AND METHODS

Acridinium-9-carboxamide active ester **1**[1] in DMF was coupled to the imidate-protected aminooxyalkylamine **2**, then hydrolyzed in 3 N HCl to afford AHA **3** (94%) in greater than 98% purity by analytical HPLC (C_{18}). ESI-MS and $^{1}H/^{13}C$ NMR analyses were consistent with the proposed structure.[8]

Coupling to oxo-steroids.[9] The oxo-steroid (Table 1, **4–8**)(0.05 mmol) in methanol (1.5 mL) containing excess sodium acetate was heated at reflux for 24 h with AHA **3** (0.05 mmol), then purified and analyzed by reversed-phase HPLC (C_{18}). The reaction of aldehydic steroids[10] (**9**, **10**) was carried out at ambient temperature over 6 h.

Labeling abasic sites on DNA.[8] DNA containing abasic sites[11] (25 μg/mL, pH 6.8) was heated (2 h, 37 °C) with AHA **3** (317 μM). The labeled DNA was isolated by ethanol precipitation. Chemiluminescence was initiated on aliquots (10 ng) by the addition of basic hydrogen peroxide (200 μL, 0.7%; 1% Triton X-100, 0. 0.05% DTPA).

Scheme 1

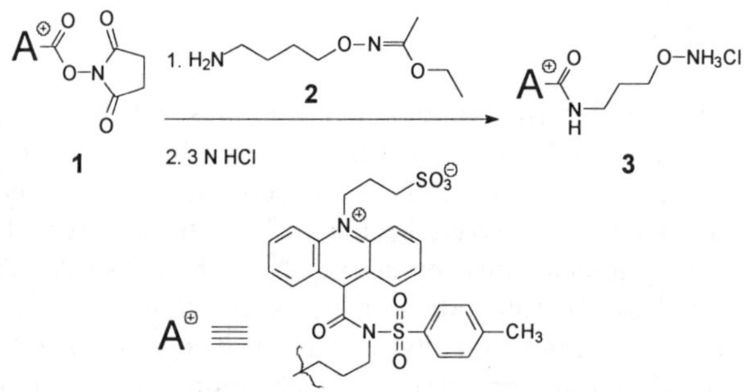

3. RESULTS AND DISCUSSION

Table 1 lists the range of oxo-steroids (**4–10**) labeled with AHA **3**. In general the yields were good, and the procedure was much more convenient than alternative routes that proceed through the carboxymethyloxime (CMO) derivatives of the steroids. Quite surprisingly, the dioxo-steroids (cortisol **7** and progesterone **8**) reacted only at the 3-position to give the mono-oximes. It the preparation of CMOs of these compounds, it is necessary to activate the 3-position with pyrrolidine to achieve selectivity.[12] The cardiac glycoside dialdehydes (**9** and **10**) gave mixtures of mono- and bis-oximes with only one equivalent of **3**, but could be driven to the bis-AHA-oximes in good yield using 200 mol% of the reagent.

Table 1. Labeling of oxo-steroids

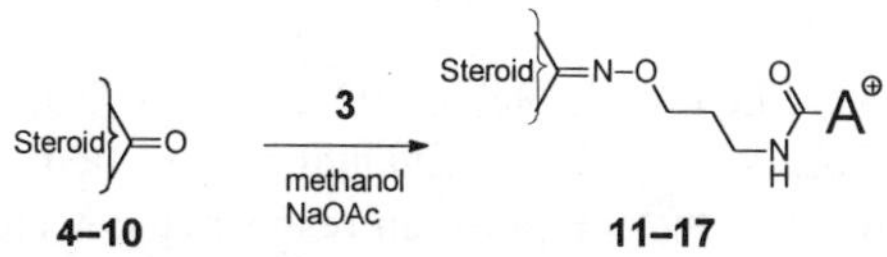

	Oxo-steroid	Oxime	% Yield
4	6-oxoestradiol	**11**	53
5	estrone	**12**	68
6	norethindrone	**13**	64
7	cortisol	**14**	37
8	progesterone	**15**	61
9	digitoxin-dialdehyde	**16**	64[a]
10	digoxin-dialdehyde	**17**	50[a]

[a]The bis-oxime is formed with 200 mol% of **3.**

Abasic sites on calf thymus DNA were generated in a time dependent fashion by acid treatment. The open aldehydic sites reacted with an excess of AHA **3**. Fig. 1 shows the correlation between the number of lesions and the chemiluminescent signal generated from the AHA that was incorporated. Based on this experiment, using this sample size (10 ng), 1 abasic site per 10^7 nucleotides could be detected. This should be a convenient and sensitive method to assess the extent of abasic lesions in untreated, native DNA samples.

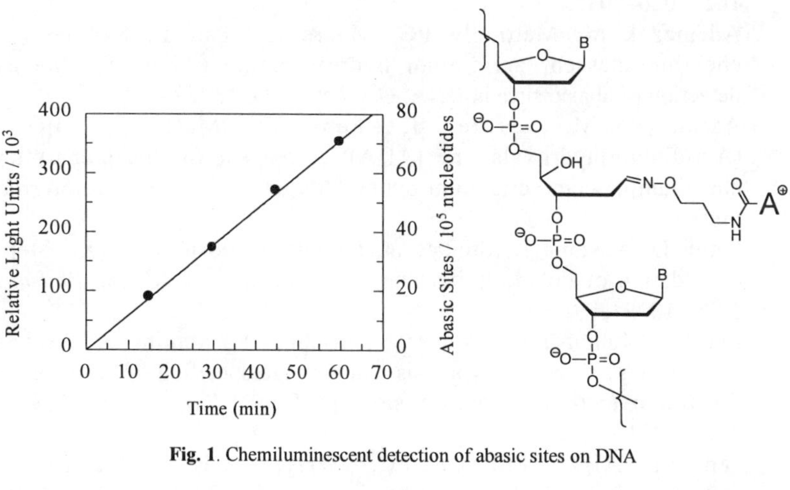

Fig. 1. Chemiluminescent detection of abasic sites on DNA

References

1. Adamczyk M, Chen Y-Y, Gebler JC, Johnson DD, Mattingly PG, Moore JA, Reddy RE, Wu J, Yu Z. Evaluation of chemiluminescent estradiol conjugates using a surface plasmon resonance detector. *Steroids* 2000; **65**: 295–303.
2. Adamczyk M, Chen Y-Y, Fishpaugh JR, Mattingly PG, Pan Y, Shreder K, Yu Z. Linker-mediated modulation of the chemiluminescent signal from N^{10}-(3-sulfopropyl)-*N*-sulfonylacridinium-9-carboxamide tracers. *Bioconjugate Chem* in press.
3. Adamczyk M, Mattingly PG, Moore JA, Rege SD. Thiol-specific acridinium reagents for labeling proteins and nucleic acids. This volume.
4. Adamczyk M, Mattingly PG, Reddy RE. The synthesis of 6β-amino-estradiol and its biotin, acridinium, and fluorescein conjugates. *Steroids* 1998; **63**: 130–134.
5. Adamczyk M, Mattingly PG, Reddy RE. An efficient stereoselective synthesis of 6α-aminoestradiol: preparation of estradiol fluorescent probes. *Steroids* 1997; **62**: 462–467.
6. Adamczyk M, Mattingly PG, Shreder K, Yu Z. Surface plasmon resonance (SPR) as a tool for antibody conjugate analysis. *Bioconjugate Chem* 1999; **10**: 1032–1037.
7. Miller MJ, Mattingly PG, Morrison MA, Kerwin JF, Jr. The synthesis of β–lactams from substituted hydroxamic acids. *J Amer Chem Soc* 1980; **102**: 7026–7032.
8. Adamczyk M, Mattingly PG, Moore JA, Pan Y. Synthesis of a chemiluminescent acridinium hydroxylamine (AHA) for the direct detection of abasic sites in DNA. *Org Lett* 1999; **1**: 779–781.
9. Adamczyk M, Grote J, Pan Y, Mattingly PG. *O*-(Acridinium)hydroxylamine (AHA): A reagent for the preparation of chemiluminescent acridinium oxime (AO)-steroid conjugates. *Steroids* in press.
10. Satoh D, Aoyama K. Studies on digitalis glycosides. XXXI. Stepwise degradation of polydigitoixosides of cardenolides. *Chem Pharm Bull* 1970; **18**: 94–99.
11. Ide H, Akamatsu K, Kimura Y, Michiue K, Makino K, Asaeda A, Takamori Y, Kubo K. Synthesis and damage specificity of a novel probe for the detection of abasic sites in DNA. *Biochemistry* 1993; **32**: 8276–8283.
12. Janoski AH, Shulman FC, Wright GE. Selective 3-(*O*-carboxymethyl)oxime formation in steroidal 3,20-diones for hapten immunospecificity. *Steroids* 1974; **23**: 49–64.

THIOL-SPECIFIC ACRIDINIUM REAGENTS FOR LABELING PROTEINS AND NUCLEIC ACIDS

MACIEJ ADAMCZYK, PHILLIP G. MATTINGLY, JEFFREY A. MOORE, AND SUSHIL D. REGE

Abbott Laboratories, Diagnostics Division, Department of Chemistry D-09NM, Building AP 20, 100 Abbott Park Road, Abbott Park, Illinois, 60064-6016, USA

E-mail: maciej.adamczyk@abbott.com

1. INTRODUCTION

Maleimide-containing derivatives of biotin and a variety of fluorescent labels are commonly used in thiol specific conjugation reactions. In contrast, maleimide derivatives of acridinium-9-carboxamide labels have not been reported. Such labels are desirable due to the increased sensitivity of chemiluminescent conjugates relative to the alternative biotin or fluorescent analogs.

The propensity of acridinium salts to add nucleophiles at the 9-position is well known. The addition of hydrogen peroxide anion to the 9-position initiates the chemiluminescent reaction, while hydroxide adds to form an inactive pseudobase form of the label. The list of other reactive nucleophiles is extensive; some have been reported in methods to stabilize analytes labeled with acridinium salts.[1, 2] Nucleophilic thiols were particularly effective.

It was with some doubt that we initiated studies to prepare acridinium-9-carboxamide labels containing an additional thiol-reactive maleimide group. However, addition to the 9-position should be reversible whereas reaction with the maleimide should not. In this report five acridinium derivatives were prepared and the relative reactivity of the maleimide group and the 9-position of the acridinium nucleus were evaluated.

2. MATERIALS AND METHODS

*Maleimido-acridinium-9-carboxamides (*Fig. 1, **1–5**). The precursor acridinium-9-carboxamide active esters[3] (0.22 mmol) were coupled to the requisite aminoalkylmaleimide[4, 5] (0.26 mmol) in DMF/phosphate buffer (3 mL, 1:2, pH 8) for 2 h. The reaction solution was then evaporated to dryness, and the residue was purified by reversed-phase HPLC (C_{18}, 30% aq. acetonitrile containing 0.01% TFA). The fractions containing the product were combined and lyophilized. All compounds were greater than 99% pure by analytical HPLC. ESI-MS and $^{1}H/^{13}C$ NMR were consistent with the assigned structures.

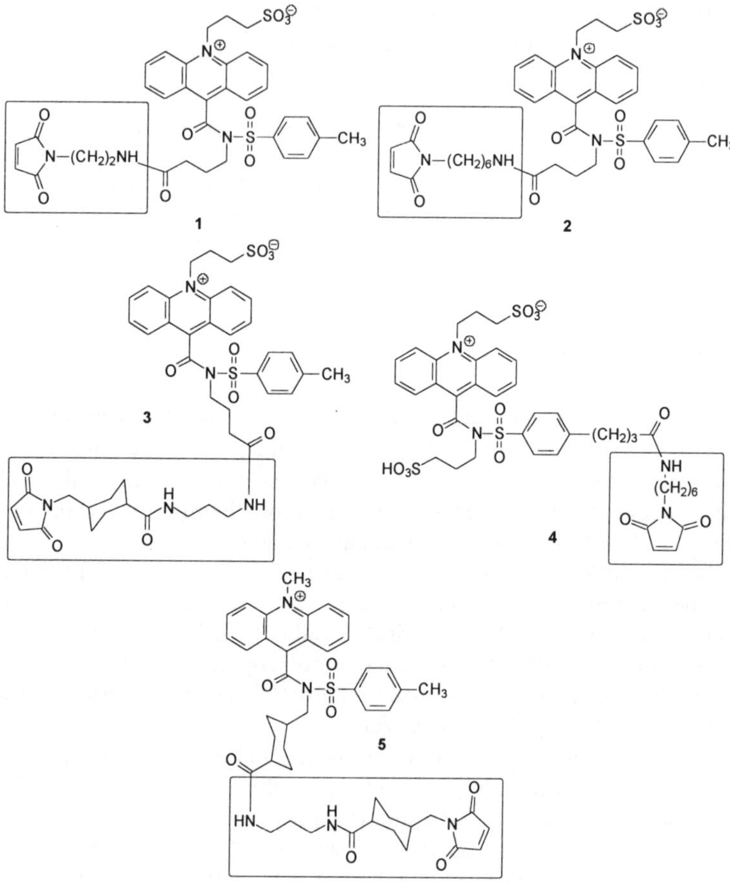

Fig. 1. Maleimido-acridinium-9-carboxamide labels

Addition of N-acetyl-L-cysteine. Acridinium-9-carboxamides (**1–5**) (1.2 μmol) and *N*-acetyl-L-cysteine (1.2 μmol) were dissolved in buffered aq. acetonitrile (50% 100 mM sodium phosphate, pH 6.7, 1.7 mL) and equilibrated for 3 h. The materials were purified and analyzed by reversed-phase HPLC.

Conjugation to proteins. Acridinium-9-carboxamide **1** (35 μM) and bovine hemoglobin (7.46 μM thiol content) were coupled in phosphate buffer (pH 7.5) for 16 h. The conjugate was purified by gel filtration eluting with phosphate buffer (pH 7.1) and analyzed by LC-ESI-MS.

Conjugation to DNA. M13 phage DNA was thiolated using the commercial kit[6] then equilibrated with maleimido-acridinium-9-carboxamides (**1–5**) at pH 6.8.

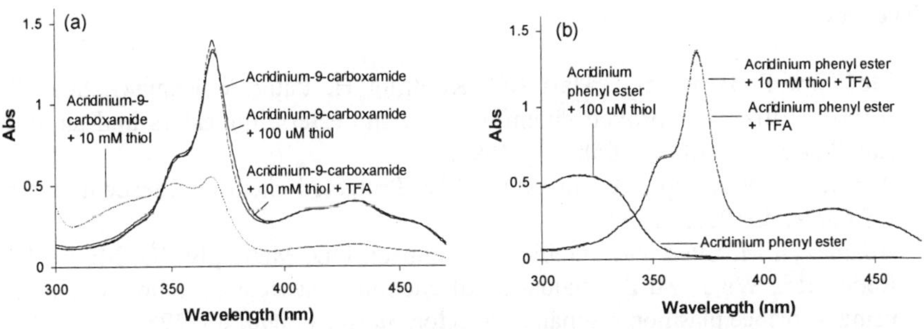

Fig. 2. Addition of *N*-acetyl-L-cysteine. (a) Acridinium-9-carboxamide (100 μM); (b) Acridinium-9-carboxylate aryl ester (100 μM).

3. RESULTS AND DISCUSSION

When the free carboxylic acid precursor corresponding to acridinium-9-carboxamide **1** was equilibrated with an equal concentration of *N*-acetyl-L-cysteine no addition to the 9-position could be detected by UV-visible spectroscopy (Figure 2a). However, with addition of a 100-fold excess of the thiol, addition was noted by the decrease in the 370/430 nm absorbances. In sharp contrast simple aryl N^{10}-methyl-acridinium-9-carboxylates[7] efficiently form the 9-position adduct in the presence of one equivalent of thiol (Figure 2b). When maleimido-acridinium-9-carboxamides (**1–5**) were equilibrated with one equivalent of *N*-acetyl-L-cysteine at pH 6.7, addition to the maleimide was exclusive as judged by the peak to peak conversion noted by RP-HPLC.

To test the efficacy and specificity of the new reagents in protein conjugation, bovine hemoglobin ($\alpha_2\beta_2$) was chosen as a model, since it contains only a single cysteine residue on the β-chain and is easily characterized by LC-ESI-MS.[8] Incubation of the protein with maleimide **1**, under the normal conditions for conjugation of protein thiols gave quantitative incorporation of the label as judged by LC-ESI-MS (*m/z* 16664 for the monoadduct vs 15954 for the native β-chain) No substitution was seen on the α-chain that contains no cysteine residues.

Thiols were introduced into M13 phage DNA (approx. 7000 bases) using the Vector Fasttag photolabeling kit.[6] Subsequent labeling of these thiols with maleimido-acridinium-9-carboxamides (**1–5**) resulted in the incorporation of 70–150 chemiluminescent labels (determined by UV).[9] The acridinium-9-carboxamide nucleus was independently evaluated for stability by HPLC and chemiluminescence intensity after equilibration under hybridization conditions (pH 7.8, 4 h, 65 °C). No degradation was noted.

References

1. McCapra F, Beheshti I, Hart RC, Koelling H, Patel A, Ramakrishnan K, Assays utilizing improved chemiluminescent esters, thioesters and amides. *Eur Patent Appl* EP0916658(A1), 1999.
2. Arnold LJ, Waldrop AA, Hammond PW, Protected chemiluminescent labels. *US Patent* 4950613, 1990.
3. Adamczyk M, Chen Y-Y, Gebler JC, Johnson DD, Mattingly PG, Moore JA, Reddy RE, Wu J, Yu Z. Evaluation of chemiluminescent estradiol conjugates using a surface plasmon resonance detector. *Steroids* 2000; **65**: 295–303.
4. Chorev M, Caulfield MP, Roubini E, McKee RL, Gibbons SW, Leu CT, Levy JJ, Rosenblatt M. A novel, mild, specific and indirect maleimido-based radioiodolabeling method. Radiolabeling of analogs derived from parathyroid hormone (PTH) and PTH-related protein (PTHrP). *Int J Pept Protein Res* 1992; **40**: 445–455.
5. Huber E, Klein C, Betz H-G, Preparation of aminoalkylmaleimides and their conjugates with haptens and antigens. *Ger Patent* 3919915A1, 1990.
6. Daniel SG, Westling ME, Moss MS, Kanagy BD. FastTag nucleic acid labeling system: a versatile method for incorporating haptens, fluorochromes and affinity ligands into DNA, RNA and oligonucleotides. *BioTechniques* 1998; **24**: 484-489.
7. Chang SCS, Klukas CK, Law SJ, Vitkauskas CA, Nucleophilic polysubstituted aryl acridinium esters, conjugates, production, and uses thereof in assays of biological molecules. *U S Patent* 5241070, 1993.
8. Adamczyk M, Gebler JC. Electrospray mass spectrometry of α and β chains of selected hemoglobins and their TNBA and TNB conjugates. *Bioconjugate Chem* 1997; **8**: 400–406.
9. Marshall RL, Farmer JL, personal communication, Molecular Diagnostics Business Unit, Abbott Laboratories, 1998.

QUANTITATION OF FREE CHEMILUMINESCENT ACRIDINIUM-9-CARBOXAMIDE SALTS IN BIOCONJUGATES

JOHN C. GEBLER, MACIEJ ADAMCZYK, KEVIN SHREDER, AND JIANG WU

Abbott Laboratories, Diagnostics Division, Department of Chemistry 9NM, Building AP 20, 100 Abbott Park Road, Abbott Park, Illinois, 60064-6016, USA

E-mail: maciej.adamczyk@abbott.com

1. INTRODUCTION

Chemical modification of proteins by small molecules to form bioconjugates is an important process in biotechnological, pharmaceutical, and diagnostic industry.[1] The characterization of such conjugates typically involves the measurement of average incorporation ratio of label to protein (hapten density),[2] and more recently, the distribution of label on the protein.[3] While the degree of protein modification is important, the presence of residual label might play a critical role in its assay performance.

The acridinium-9-carboxamide salts (Fig. 1) have been successfully employed in Abbott PRISM® and ARCHITECH® clinical analyzers as signal generators. Acridinylation of proteins is usually achieved by amidation of free amino groups on proteins (lysine, N-terminus). During the conjugation process, the acridinium active ester undergoes a competitive hydrolysis to form acridinium free acid (Fig. 1, **A**), which may noncovalently bind to the protein or any component of an assay system.

For the past several years, liquid chromatography-electrospray tandem mass spectrometry (LC/MS/MS) has been widely used for quantitative analysis of environmental toxins, pharmaceuticals, and metabolites at trace levels.[4] The technique of MS/MS provides molecular mass as well as specific structural information based on the collision-induced dissociation (CID) fragments characteristic to each analyte eluting from the HPLC column. After CID, only ions derived from the analyte of interest are selectively detected. This technique, called selected reaction monitoring (SRM), affords superior sensitivity and selectivity for the detection of components in complex mixtures.

Our continuing interest in developing acridinium compounds for use in diagnosis led us to develop a sensitive, accurate, and reproducible technique to evaluate the residual chemiluminescent label in protein conjugates. We report here on the use of a LC/MS/MS SRM technique to quantitatively determine the content of noncovalently bound acridinium-9-carboxamide in bioconjugates.

2. MATERIALS AND METHODS

Acridinium-9-carboxamide BSA, anti-T4 monoclonal antibody (aT4), and HIV envelope protein (HIVep) conjugates were prepared according to standard literature protocols.[5] Quantitation by mass spectrometry was carried out on a ThermoQuest LCQ ion trap equipped with an electrospray ion source. A microbore HPLC system with a 1 % 50 C_{18} column was interfaced directly to ion source. Samples were injected on the column and eluted with a two-step gradient. The mass analyzer was operated in the positive ion mode using standard parameters. For the generation of CID data, collision energy was fixed to 35% (of the maximum) for the fragmentation of the precursor ions of acridinium **A** (m/z 585.2) and internal standard **B** (m/z 601.2). In the SRM mode, the instrument was set to monitor the product ions at m/z 539.1 and 567.1 for acridinium free acid and at m/z 555.1 and 583.1 for internal standard, respectively. A calibration curve was produced with 7 different concentrations of **A** (50, 100, 200, 400, 600, 800, and 1000 nM), with each containing 200 nM **B**. A linear calibration curve (Fig. 2) was obtained by plotting the ratio of summed peak for the **A** m/z 585.2→539.1, 567.1 and internal standard **B** m/z 601.2→555.1, 583.1 ion pairs versus the concentrations of the calibration samples. The nanomolarity of all samples was calculated from the resulting peak area ratios and the regression equation of the calibration curve. Recovery was determined by spiking BSA, aT4, HIVep, BSA conjugate, and aT4 conjugate with **A** and 200 mM **B** (see Table 1).

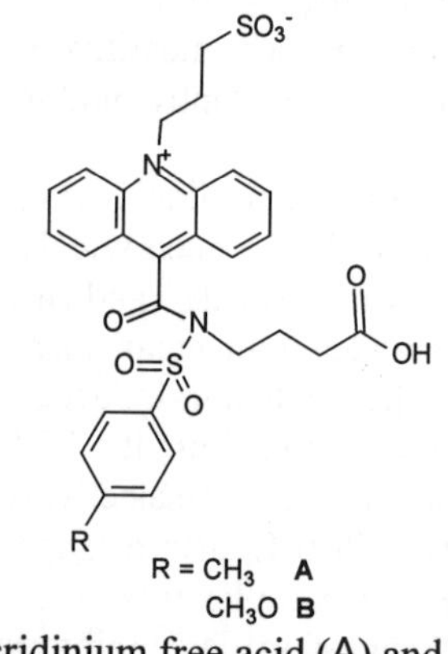

Fig. 1. Chemical structure for acridinium free acid (A) and internal standard, methoxylated acridinium free acid (B).

3. RESULTS AND DISCUSSION

The calibration curve was constructed using 7 different concentration standards and processed by weighted least-squares linear regression analysis (Fig. 2). The calibration was linear in the range of 50 to 1000 nM and afforded $R^2 = 0.9978$. The limit of detection (LOD) was 20 nM for **A**. Table 1 (1–14) summarizes the recovery, precision, and accuracy of the methodology for BSA, aT4, HIVep spiked with various concentrations of **A**. Acridinium conjugates prepared from BSA and aT4 did not contain measurable amounts **A** and gave similar results for recovery, precision, and, accuracy when spiked (samples 7–10). When spiking proteins or conjugates, nearly quantitative recovery and good precision were observed. Analysis of HIVep conjugates (11–14) showed significant amounts of **A**.

In conclusion, a simple, fast, and selective LC/MS/MS method has been developed for the quantitative determination of residual acridinium in protein conjugates. The method required no sample pretreatment, and affords the sensitivity, accuracy, and precision necessary for the quantitative measurement. The procedure described here is useful for advanced bioconjugate characterization.

Table 1. Recovery, precision, quantitative analysis of residual acridinium in BSA, aT4, HIVep, and their respective acridinium conjugates.

	Sample (spiked acridinium, nM)	**Mean ± S.D. (nM)**	**C.V.%**	**Recovery (%)**
1	BSA (45)	42.7 ± 2.8	6.6	94.8
2	BSA (300)	290 ± 9.2	3.2	96.6
3	BSA (480)	491 ± 18.8	3.8	102.3
4	aT4 (45)	41.8 ± 1.3	3.0	92.9
5	aT4 (300)	299 ± 21.8	7.3	99.9
6	aT4 (480)	489 ± 6.2	1.3	101.8
7	HIVep (30)	28.5 ± 7.0	24.8	95.0
8	HIVep (300)	278 ± 26.2	9.4	92.8
9	BSA conjugate (120)	114 ± 2.9	2.6	95.1
10	aT4 conjugate (45)	44.3 ± 3.1	7.0	98.5
11	HIVep conjugate 1	227 ± 9.0	4.0	—
12	HIVep conjugate 2	455.3 ± 20	4.4	—
13	HIVep conjugate 3	468 ± 21	4.5	—
14	HIVep conjugate 4	800 ± 16	2.0	—

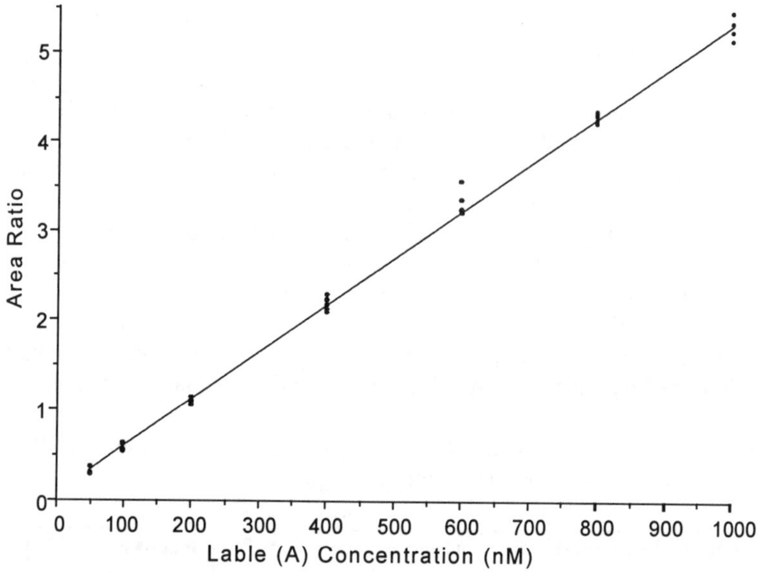

Fig. 2. Calibration curve for quantitation of acridinium free acid (A) in protein conjugates.

REFERENCES

1. Hermanson GT, *Bioconjugate Technique.*, Academic Press, San Diego, 1996, 419.
2. Adamczyk M, Buko A, Chen YY, Fishpaugh JR, Gebler JC, Johnson DD. Characterization of protein-hapten conjugates. 1. Matrix-assisted laser desorption ionization mass spectrometry of immuno BSA-hapten conjugates and comparison with other characterization methods. *Bioconjugate Chem* 1994; **5**: 631–5.
3. Adamczyk M, Gebler JC, Mattingly PG. Characterization of protein-hapten conjugates. 2. Electrospray mass spectrometry of bovine serum albumin-hapten conjugates. *Bioconjugate Chem* 1996; **7**: 475–81.
4. Purves RW, Gabryelski W, Li L. Investigation of the quantitative capabilities of an electrospray ionization ion trap/linear time-of-flight mass spectrometer. *Rapid Commun Mass Spectrom* 1998; **12**: 695-700.
5. Adamczyk M, Gebler JC, Shreder K, Wu J. Region-selective labeling of antibodies as determined by electrospray ionization-mass spectrometry (ESI-MS). *Bioconjugate Chem* 2000; **11**, 557-563.

CHARACTERIZATION OF ACRIDINIUM-9-CARBOXAMIDE—MONOCLONAL ANTIBODY BIOCONJUGATES BY ELECTROSPRAY IONIZATION MASS SPECTROMETRY

JOHN C. GEBLER, MACIEJ ADAMCZYK, KEVIN SHREDER, AND JIANG WU

Abbott Laboratories, Diagnostics Division, Department of Chemistry 9NM, Building AP 20, 100 Abbott Park Road, Abbott Park, Illinois, 60064-6016, USA
E-mail: maciej.adamczyk@abbott.com

1. INTRODUCTION

Antibody conjugates play a critical function in medical and diagnostics applications. When haptens are covalently attached to an antibody it is endowed with the ability to generate a signal or deliver a molecular payload. Recently chemiluminescent acridinium-9-carboxyamide monoclonal bioconjugates have been developed for ultra-sensitive immunoassays. The extent of acridiumylation has been observed to effect conjugate performance. Thus, accurate and robust characterization methods are needed. Traditionally, various spectroscopic techniques (UV-VIS/fluoroesence) have been employed to determine average number of attached labels. However, such methods are problematic and do not differentiate between covalently bound, non-specifically bound, and unbound acridinium molecules.

Over the past several years mass spectrometry has been successfully utilized for accurate mass analysis of proteins.[1, 2] Electrospray ionization mass spectrometry (ESI-MS) has been effective for determining the mass of bioconjugates.[3-6] We have developed and utilized an ESI-MS method to determine the extent, distribution, and regiospecificity of labeling monoclonal antibodies with acridinium-9-carboxamide salts.

2. MATERIALS AND METHODS

Antibody-acridinium-9-carboxamide conjugates were prepared by placing monoclonal anti-biotin antibody (3.0 mg in 1.4 mL of 100 mM sodium phosphate, 150 mM NaCl, pH 8.0) with different molar equivalents (4, 8, and 16 in 0.2 mL DMF) of acridinium-9-carboxamide active ester. The reaction mixture were briefly vortexed then allowed to stand for 18 hours at ambient temperature in the dark. The resulting reaction mixtures were purified by gel permeation chromatography (100 mM sodium phosphate, 150 mM NaCl, and 0.1% CHAPS, pH 6.3). Acridinium—antibody conjugates were digested with papain to obtain the Fab and Fc fragments.[7] Digest mixture were taken to 1N guanidine-HCl and then injected onto a reversed-phase HPLC column (PLRP-S, Polymer Laboratories) at a flow rate of 100 L/min. The fragments were eluted with a linear gradient of 10% to 90% acetonitrile in aqueous 0.1% formic acid. Column output was directed into the ESI-MS ion source. Mass analysis was carried-out with a single quadrapole instrument (PE-Sciex API 100 equipped with Turbo Ionspray ion source). All samples were analyzed from *m/z* 1000—3000 with a total acquisition time of 5.0 s. Molecular masses and their distributions of the fragments were determined by deconvolution of raw spectra. The resulting spectra were further processed to determine peak intensity (area) corresponding to the respective acridinium conjugates antibody fragments.

3. RESULTS AND DISCUSSION

Anti-biotin monoclonal antibody was covalently conjugated with acridinium active esters to achieve three different loading levels of label (Fig 1A). The resulting bioconjugates was purified and then digested with papain to its Fab and Fc fragments. The mixture of fragments was directly analyzed by HPLC ESI-MS. Two major total-ion-count peaks were observed originating from the Fc and Fab fragments (Fig. 1B). Extraction of raw spectra and their deconvolution gave *true mass* spectra for the antibody fragments before and after conjugation (Fig. 2). Fab spectra show evenly spaced peaks consistent with the addition of acridinium label. The spectra for Fc are more complicated due to glycosylation, however, repeating patterns consistent with molecular weight increase of the label (587 Da) are easily discriminated. Table 1 lists the results of labeling distribution. Coupling efficiency is ~50% with a slight labeling biases for Fab.

Fig. 1. **(A)** Chemical structure of acridinium label. **(B)** The total ion count chromatogram from LC/ESI-MS of the papain digested antibody conjugate.

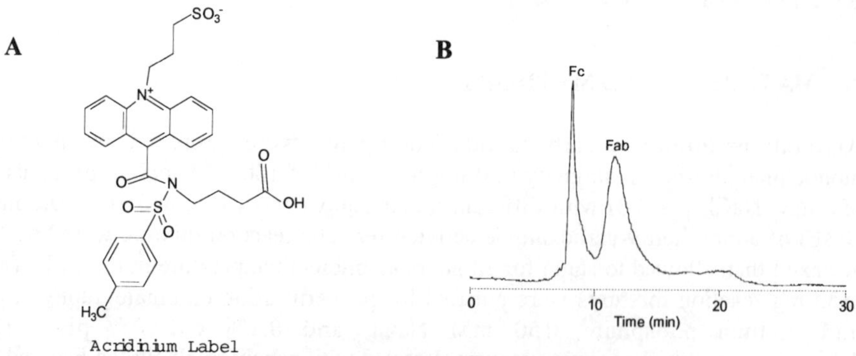

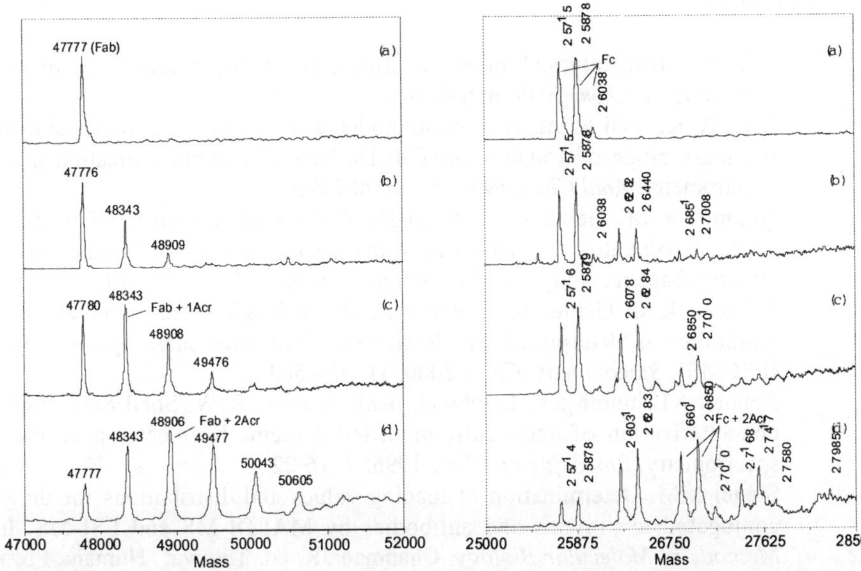

Fig. 2. Deconvoluted spectra of Fab region before and after conjugation (left panel). Deconvoluted spectra of Fc region before and after conjugation (right panel).

Table 1. Ratio of acridinium label to antibody region as determined by LC/ESI-MS.

Equiv of Acridinium	Fab Distribution	Fc Distribution	Labels per Fab region	Labels per Fc region	Labels per antibody region
4 (b)	0—3	0—2	1.06	0.82	1.9
8 (c)	0—3	0—3	2.10	1.75	3.9
16 (d)	0—6	0—4	4.40	3.60	8.0

Characterization of antibody conjugates can play an important role in their development as chemiluminescent immunoreagents. The ESI-MS approach demonstrated here will allow for the ready analysis of the average labeling number, distribution, and the region-modification of various labels conjugated to monoclonal antibodies. Such analysis can be applied to study, prepare, and characterize antibody conjugates to ensure the production of high quality immunoreagents.

REFERENCES

1. Loo JA. Bioanalytical mass spectrometry: Many flavors to choose. *Bioconjugate Chem* 1995; **6**: 644-665.
2. Feng R, Konishi Y. Analysis of antibodies and other large glycoproteins in the mass range of 150,000-200,000 Da by electrospray ionization mass spectrometry. *Anal Chem* 1992; **64**: 2090-2095.
3. Adamczyk M, Gebler JC, Mattingly PG. Characterization of protein-hapten conjugates. 2. Electrospray mass spectrometry of bovine serum albumin-hapten conjugates. *Bioconjugate Chem* 1996; **7**: 475-481.
4. Adamczyk M, Gebler JC, Shreder K, Wu J. Regio-selective labeling of antibodies as determined by electrospray ionization-mass spectrometry (ESI-MS). *Bioconjugate Chem* 2000, **11**, 557-563.
5. Bennett KL, Smith SV, Lambrecht RM, Truscott RJW, Sheil MM. Rapid characterization of chemically-modified proteins by electrospray mass spectrometry. *Bioconjugate Chem* 1996; **7**: 16-22.
6. Siegel MM. Determination of loading values and distributions for drugs conjugated to proteins and antibodies by MALDI-MS and ESI-MS. In *Methods in Molecular Biology.* Chapman JR, ed, Totowa: Humana Press Inc., 1996: 211-226.
7. Adamczyk M, Gebler JC, Wu J. Papain digestion of different mouse IgG subclasses as studied by electrospray mass spectrometry. *J Immunol Methods* 2000; **237**: 95-104.

CHEMILUMINESCENT IMMUNOASSAYS FOR THE DETECTION OF ORGANOCHLORINE PESTICIDES

S. GIROTTI AND F. FINI
Institute of Chemical Sciences, Via San Donato 15, I-40127 Bologna, Italy
E-mail: girotti@biocfarm.unibo.it

P. RAUCH, B. MICKOVA, L. KARASOVA AND L. FUKAL
Department of Biochemistry and Microbiology, Institute of Chemical Technology, Technika 5, 16628 Prague, Czech Republic

A. ABAD, J. J. MANCLUS, J. V. MERCADER, AND A. MONTOYA
Laboratorio Integrado de Bioingenieria, Politechnical University of Valencia, Camino de Vera s/n, 46022 Valencia, Spain

1. INTRODUCTION

One of the main problems of food safety is its contamination with biologically active compounds like pesticides, hormones and drugs. Traditional methods for monitoring of food samples are usually complicated and include several cleaning steps before analysis. Immunochemical techniques could represent alternative and/or complementary methods for the analysis of food samples, minimizing and simplifying the analysis. Among the immunoassay methods, the enzyme-linked immunosorbent assay (ELISA) met the requirements for large monitoring programs, i.e. to be technically simple, inexpensive and sensitive. Many ELISAs have been developed for pesticide determination using colorimetry as the detection system.[1] Chemiluminescence (CL), which already had a considerable impact on biomedical analysis,[2] has also been applied for the development of pesticide immunoassays in the agrochemical field.[3]

The aim of this work was the development and application of a chemiluminescent immunoassay for the determination of DDT and endosulfan, which are persistent organic pollutants (POPs), in food and environmental samples. POPs resist degradation and tend to bioaccumulate, this persistence in various media facilitates their transport over long distances to remote regions where they have never been used. In view of the risks posed by these compounds to human health and the environment, the international community has now, on several occasions called for urgent global actions to reduce and eliminate releases of these chemicals.[4]

Monoclonal antibodies (MAbs) were used in assay development. They offer greater selectivity and reproducibility than conventional antisera by being invariant reagents, facilitating improved quality control of the assay. The specificity of MAbs allows the selection of antibodies that either cross react with several metabolites or discriminate between them.

2. MATERIAL AND METHODS

Immunoreagents were obtained as reported by Abad *et al.*[5]: five haptens resembling as much as possible the general chemical structure of the DDT were synthesized following appropriate chemical strategies. Similarly, four haptens resembling endosulfan were obtained. All of the haptens were covalently attached to BSA using modified active ester method and to OVA (OVA-conjugates) using the mixed-anhydride method. BALB/c female mice were immunized with BSA-hapten conjugates using standard procedure. Mouse spleen lymphocytes were fused with murine myeloma cells (P3-X63-Ag 8.653) at a 5:1 ratio using PEG 1,500. According to cross-reactivity patterns, obtained anti-DDT Mabs may be classified as class-specific antibodies or DDT-specific antibodies. On the contrary all of the obtained anti-endosulfan Mabs were more or less non-specific for the endosulfan family.

Standard pesticides were from Riedel-de Haën (Seelze, Germany). For alkaline phosphatase activity detection, CDP Star (Amersham; CDP-A), CDP Star (Tropix; CDP-T), CSPD (Tropix) and Lumiphos plus (Lumigen; Lum+) were used. Sapphire II (Tropix) was employed as enhancer for both the Tropix reagents. For peroxidase activity detection, ECL and ECL-Plus from Amersham, and SuperSignal ULTRA (SSUltra) and SuperSignal™ ELISA Femto Maximum Sensitivity Substrate (SSELISA) from Pierce were employed. Peroxidase and alkaline phosphatase-labeled rabbit anti-mouse immunoglobulins were obtained from Dako (Glostrup, Denmark). High-binding microplates were from Costar-Corning (USA). Light emission was measured using a microplate luminometer (Victor, Wallac, Finland).

3. RESULTS AND DISCUSSION

Enzyme immunoassays to DDT and endosulfan with photometric and luminescent detection were developed. The optimal parameters and reagent concentrations were investigated and the competitive format of the assay was selected, because of the better results obtained. Two different enzymes were used as labels, horseradish peroxidase (HRP) and alkaline phosphatase (ALP), and several CL substrates were tested. Kinetic studies were performed for each of the substrates and several parameters were considered, i.e. the signal to noise ratio, the values of RLU max and RLU min, the incubation time, the stability, the preparation time. It was decided, mainly because of the delay time and of the simplicity of substrates preparation, to select HRP as labelling enzyme and SSElisa as substrate.

All these parameters fixed, several experiments were performed to study the dilution procedure and the solvent effects. These chlorinated pesticides are poorly soluble in water and organic solvents are necessary to extract them from samples, but at the same time this could damage immunoglobulins (IgG). The effect of several organic solvents (methanol, ethanol, acetone, DMF, DMSO) on IgG affinity was therefore investigated. Methanol (10-20 %) proved to be the best tolerated solvent and was therefore selected. This solvent decreased adsorption to the minimum and had also minimum effect on immunoglobulins nativity, i.e. no denaturation effects were observed.

The protocols for the ELISAs were as follows. A microplate was coated with ovalbumin-hapten conjugate. Then antibody and standard or sample were added and incubated for 2 or 1 h, for the DDT and endosulfan assay respectively. Next, the plate was incubated for 1 h with peroxidase-labelled rabbit anti-mouse immunoglobulin. Finally, peroxidase activity was determined either by the standard o-phenylendiamine procedure, or by a chemiluminescent assay. Results obtained with the two detection systems were compared, chemiluminescence displaying a lower consumption of reagents and a slightly better sensitivity. With the chemiluminescent detection, I_{50} was in the range 3-9 nM and 13-22 nM for DDT and endosulfan respectively. Repeatability and reproducibility were found to be in the range 10-30 %.

Several samples containing mixtures of the organochlorine pesticides under study were analyzed. The extent of matrix interference of methanol-extracts was determined in different samples, i.e. milk, soil, fish, vegetables, water. An acceptable recovery was usually obtained with the exception of some samples, as reported in Table 1.

Acknowledgements

This work was supported by the EC grant INCO Copernicus No. ERBIC15CT96100 and by grants from MURST (Ministero della Universit e della Ricerca Scientifica e Tecnologica, projects of National Interest) and from University of Bologna (ex-quota 60% and Funds for Selected Research Topics).

Table 1: Determination of DDT and Endosulfan in samples of different origin.

Sample	Expected value mg/kg	Obtained value mg/kg (Coefficient of variation) (%)	Recovery (%)
DDT			
CHARD	1.523	1.524 (23.2)	100
	2.531	2.683 (16.6)	106
SOIL	0.594	0.808 (14.2)	136
	0.879	1.002 (15.3)	114
STRAWBERRY	1.758	1.992 (14.2)	108
	2.339	2.244 (18.0)	96
SALAD	0.086	0.108 (26.2)	125
	0.109	0.157 (11.7)	144
ENDOSULFAN			
SOIL	0.074	0.088 (18.7)	119
	0.104	0.120 (28.7)	115
FISH	0.36	0.128 (29.3)	36
	0.50	0.39 (18.4)	78
CHARD	11.000	9.791	89
	8.708	7.924	91

References

1. Meulenberg EP, Mulder WH, Stoks PG. Immunoassays for pesticides. *Environ Sci Technol* 1995; **29**: 553-561.
2. Fini F, Gallinella G, Girotti S, Zerbini M, Musiani M. Development of a chemiluminescent competitive PCR for the detection and quantification of Parvovirus B19 DNA using a microplate luminometer. *Clin Chem* 1999; **45**: 1391-1396.
3. Navas-Diaz A, Garcia-Sanchez F, Lovillo J, Gonz lez-Garcia JA. Enhanced chemiluminescent kinetic ELISA of dichlorprop methyl ester. *Anal Chim Acta* 1996; **321**: 219-224.
4. United Nations Environment Programme. Decisions adopted by the Governing Council at its nineteenth session. February 7, 1997, Nairobi, Kenya.
5. Abad A, Manclus JJ, Mojarrad F, Mercader JV, Miranda MA, Primo J, et al. Hapten synthesis and production of monoclonal antibodies to DDT and related compounds. *J Agric Food Chem* 1997; **45**: 3694-3702.

DEVELOPMENT OF BIOLUMINESCENT ASSAY FOR PYRUVATE PHOSPHATE DIKINASE AND ITS APPLICATION TO BIOLUMINESCENT ENZYME IMMUNOASSAY

KATSUTOSHI ITO[1], KAZUMI NISHIMURA[1], SEIJI MURAKAMI[2], HIDETOSHI ARAKAWA[1] AND MASAKO MAEDA[1]

[1]*School of Pharmaceutical Sciences, Showa University, 1-5-8 Hatanodai, Shinagawa, Tokyo, 142-8555, Japan;* [2]*Research & Development Division, Kikkoman Corporation, 399 Noda, Chiba, 278-0037, Japan*

1 INTRODUCTION

Recently, we established a bioluminescent assay for acetate kinase (AK)[1] using a firefly luciferase-luciferin detection system. We applied it to bioluminescent enzyme immunoassay (BL-EIA),[1] and introduced the well-known streptavidin-biotin system to this assay and demonstrated a sensitive BL-EIA.[2] Furthermore, we detected 8.6×10^{-21} mol/assay of AK using thermostable mutant firefly luciferase and developing BL-EIA for various biological samples.[3-5] However, there is a possibility that the sensitivity of AK is decreased by bacterial contamination of the BL reagent for AK. Adenylate kinase (AdK) generates ATP from 2 mol of ADP and exists in various microorganisms. Since the BL assay of AK is required ADP as a substrate, AdK rises background luminescence in the BL assay of AK, and as a result, the sensitivity of AK is reduced. On the other hand, pyruvate phosphate dikinase (PPDK) catalyzes the formation of ATP from AMP, pyrophosphate (PPi) and phosphoenolpyruvate (PEP). Consequently, we speculated that the BL assay for PPDK under contamination of AdK would be highly sensitive. In this paper, we describe a novel bioluminescent assay for PPDK using firefly luciferase-luciferin reaction and BL-EIA using PPDK as a labeled enzyme of α-fetoprotein (AFP) and insulin for example of biological substances.

2 MATERIALS AND METHODS

Reagents

Pyruvate phosphate dikinase (PPDK; from *Microbispora rosea* subsp. *Aerata*, EC 2.7.9.1) and thermostable Thr 217 Ile mutant of *Luciola cruciata* firefly luciferase were from Kikkoman Co. (Chiba, Japan). D-Luciferin was purchased from Sigma Chemical Co. (St. Louis, MO), *N*-2-hydroxyethylpiperazine-*N'*-2-ethanesulfonic acid (HEPES) and fluorescein-4-isothiocyanate (FITC) were from Dojindo Laboratories (Kumamoto, Japan), PEP and AMP were from Boehringer Mannheim Co. (Tokyo, Japan). Rabbit anti-FITC IgG was obtained from DAKO Japan Co.

(Kyoto, Japan). An AFP, insulin, mouse monoclonal antibodies for AFP and insulin were obtained from Tosoh Co. (Tokyo, Japan).

BL assay for PPDK

An aliquot of 10 μL of standard PPDK solution was added to a microtiter plate. Then, a 100 μL of bioluminescent reagent (containing 5 μM AMP, 304 μM PPi, 560 μM PEP, 15 mM $MgSO_4$, 1 mM dithiothreitol, 7.5 mM ammonium sulfate, 1 mM EDTA, 10 % Block Ace, 0.35 mM D-luciferin and 1 mg/mL luciferase in 50 mM HEPES-KOH buffer, pH7.0) was added and incubated for 15 min at 37°C The bioluminescent intensity was integrated for 5 s by a MicroLumat LB96P luminescent reader (EG&G Berthold, Germany).

BL-EIA for AFP and insulin

BL-EIA for AFP was carried out following a one-step sandwich-type assay. The wells of microtiter plates were coated with 100 μL of a purified mouse anti-AFP or insulin monoclonal antibody (10 μg/mL) dissolved in a 0.05 M sodium bicarbonate buffer (pH 9.5). The plates were allowed to stand at room temperature for 16 h. After that, the antibody solution was removed, and the wells were then post-coated by adding 200 μL of 1 % water-soluble gelatin solution containing 0.05% NaN_3. The plates were stored at 4°C prior to use. After washing, the anti-AFP antibody coated plate, 50 μL of a standard or sample solution and 50 μL of a FITC labeled anti-AFP monoclonal antibody solution (1.6 μg/mL) were added to each well. The plates were incubated for 1 h at room temperature. After washing, 100 μL of PPDK labeled rabbit anti-FITC Fab' solution (98 ng/mL) was added and allowed to stand for 1h at room temperature. The microtiter plate was re-washed and measured by the bioluminometric detection method mentioned above. The insulin assay was likewise performed BL-EIA of AFP, except for the use of anti-insulin antibody coated plate and a FITC labeled anti-insulin monoclonal antibody solution (65.5 ng/mL).

3 RESULTS AND DISCUSSION

Bioluminescent assay of PPDK

To determine the optimum BL assay condition of PPDK using luciferase-luciferin reaction, various conditions were examined. The concentration of luciferase, PPi, AMP and PEP were explored. Finally, the optimal assay conditions of luciferase, PPi, AMP and PEP were 1 mg/mL, 304 μM, 5 μM and 560 μM, respectively, and other conditions as mentioned above.

As shown in Fig. 1 (a), the calibration linearity of the assay ranged from 3.5×10^{-20} to 4.5×10^{-16} mol/assay. The intra-assay coefficients of variation (CV) for 8 replicates with each standard point of PPDK were from 1.2 to 2.8%. The detection limit (at blank+3SD) was 1.36 x 10^{-20} mol/àssay. In addition, we observed the bioluminescence generated of the assay system. The BL intensity was stable for at least 120 min 37 °C, we have measured (Fig. 1 (b)). The values of the PPDK assay

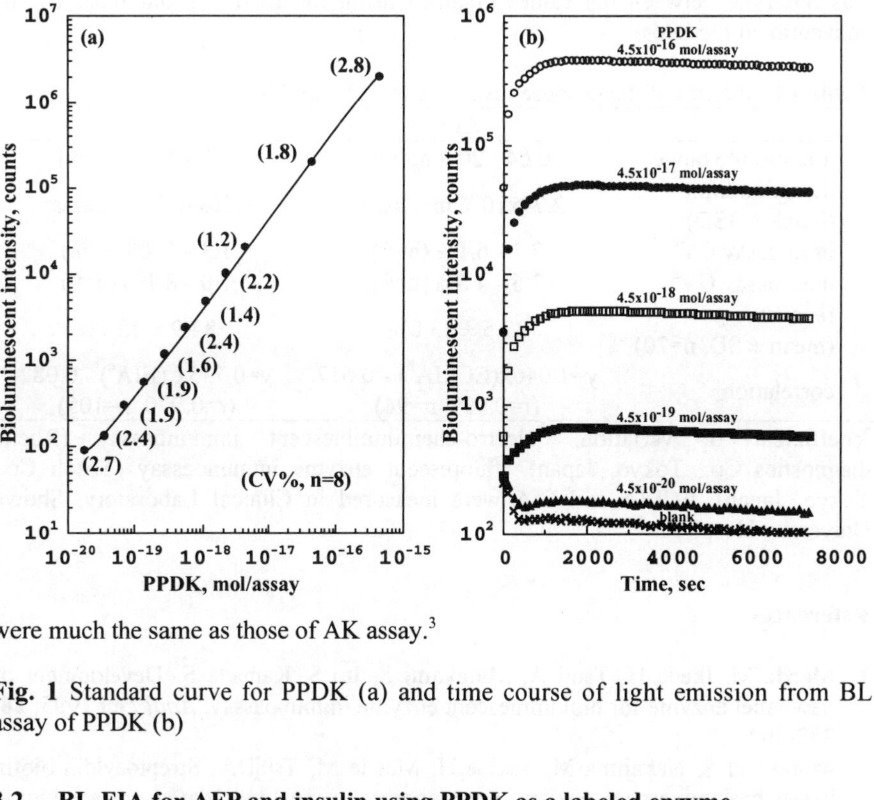

were much the same as those of AK assay.[3]

Fig. 1 Standard curve for PPDK (a) and time course of light emission from BL assay of PPDK (b)

3.2 BL-EIA for AFP and insulin using PPDK as a labeled enzyme

In the proposed BL-EIAs for AFP and insulin, we used PPDK labeled rabbit anti-FITC Fab as an enzyme labeled antibody. We speculate that PPDK can be conjugated to various antibodies or antigens without losing enzyme reactivity. FITC is easily attached to ε-amino groups of antibodies under moderate conditions. Therefore, it is convenient to construct BL-EIAs using the PPDK labeled rabbit anti-FITC Fab with FITC labeled antibody. Table 1 gives properties of the proposed BL-EIA for AFP and insulin. The measurable range of AFP and insulin were 0.04 - 20.8 ng/mL and 0.63 - 40.4 U/ mL, and the detection limit (blank+3SD) of AFP and insulin were 2.79 x 10^{-18} and 9.26 x 10^{-17} mol/assay, respectively. The intra-assay coefficients of variation (CV) of AFP and insulin with each standard point were 2.3 - 6.1% (n=6) and 1.3 - 5.7 (n=8), respectively. The sample were 8.1 and 7.0%, respectively. The mean recovery of AFP from plasma samples was 87.5% and insulin was 82.9%. In sample dilution test by the BL-EIA for AFP and insulin, good linearity was observed with each sample concerning the relation of extent of dilution and the assay result. The concentration of plasma samples could accurately be measured by the BL-EIAs. A high degree of correlation was observed between the values obtained using the BL-EIAs and those by the conventional methods.

Table 1 Properties of the proposed BL-EIA for AFP and insulin

	AFP	Insulin
measurable range	0.04 - 20.8 ng/mL	0.63 - 40.4 μU/mL
detection limit (blank + 3SD)	2.79×10^{-18} mol/assay	9.26×10^{-17} mol/assay
intra-assay CV[a]	2.3 - 6.1% (n=6)	1.3 - 5.7% (n=8)
inter-assay CV[a]	3.5 - 4.5% (n=9)	7.0 - 8.1% (n=9)
recovery (mean ± SD, n=20)	87.5 ± 13.8%	82.9 ± 12.7%
correlation	y=1.046x(ECLIA[b]) - 0.617 (r=0.999, n=94)	y=0.735x (EIA[c]) - 0.082 (r=0.970, n=105)

[a]coefficient of variation, [b]electro-chemiluminescent immunoassay (Roche diagnostics Co., Tokyo, Japan), [c]fluorescent enzyme immunoassay (Tosoh Co., Tokyo, Japan), ECLIA and EIA were measured in Clinical Laboratory, Showa University Hospital.

References

1. Maeda M, Ikeda H, Tsuji A, Murakami S, Ito S, Kamada S. Development of new label enzyme for bioluminescent enzyme immunoassay. *Anal Lett* 1995; **28**: 383-394.
2. Murakami S, Nakajima M, Sekine H, Maeda M, Tsuji A. Streptoavidin-biotin based bioluminescent enzyme immunoassay using biotinylated acetate kinase and recombinant firefly luciferase. *Anal Lett* 1996; **29**: 969-980.
3. Murakami S, Ito K, Goto T, Kamada S, Maeda M. Bioluminescent enzyme immunoassay using thermostable mutant luciferase and acetate kinase as a label enzyme. *Anal Chim Acta* 1998; **361**: 19-26.
4. Ito K, Goto T, Murakami S, Takaki A, Maeda M. Bioluminescent enzyme immunoassay for mouse interleukin-6 using acetate kinase. *Anal Sci* 1999; **15**: 91-93.
5. Ito K, Goto T, Murakami S, Maeda M, Shioda S, Arimura A. Bioluminescent enzyme immunoassay for pituitary adenylate cyclase activating polypeptaide 38 (PACAP38). *Anal Chim Acta* 1999; **382**: 245-251.

DEVELOPMENT AND APPLICATION OF A BIOLUMINESCENT IMMUNOASSAY USING AEQUORIN AS A LABEL

M. MIRASOLI, S. DEO, J.C. LEWIS AND S. DAUNERT
Departments of Chemistry and Pharmaceutical Sciences, University of Kentucky, Lexington, KY 40506-0055, USA
E-mail: mirasoli@libero.it

1. INTRODUCTION

Cortisol is a steroid hormone mainly involved in the glucose metabolism and released by the adrenal cortex. The determination of its serum concentration at time intervals during a 24-hour period is used to assess the adrenal cortex and the hypothalamic-pituitary-adrenal axis functions;[1] moreover cortisol serum levels are used as an indicator in stress evaluation studies. It has been demonstrated that cortisol concentration in saliva is proportional to that in serum and is independent from the rate of saliva production.[2] Salivary cortisol is about one order of magnitude lower than that in plasma, therefore its quantification requires a highly sensitive analytical method.

Aequorin, a calcium-dependent photoprotein isolated from the jellyfish *Aequorea victoria*, consists of apoaequorin, a single polypeptide chain of 189 amino acids, and the chromophore, a complex of coelenterazine and molecular oxygen. The photoprotein undergoes a conformational change upon binding to calcium ions. This leads to the bioluminescent reaction, that yields photon emission that lasts less than 10 seconds and has a maximum at $\lambda_{max}\approx 470$ nm.[3] The bioluminescent signal has virtually no background since no electromagnetic excitation is required and also because bioluminescence occurs very rarely in nature. For this reason aequorin is detectable down to attomole levels, thus representing a sensitive label for binding assays.[4]

In this project a bioluminescent immunoassay for cortisol was developed, where a recombinant mutant of aequorin, called mutant S, was chemically conjugated to the analyte and used as a label. The applicability of the immunoassay in the analysis of real samples, in particular saliva samples, was also investigated.

2. MATERIALS AND METHODS

An Optocomp I luminometer (GEM Biomedical, Carrboro, NC) equipped with a 100 μL fixed-volume injector was used for bioluminescent measurements.

Cortisol-3-(O-carboxymethyl)oxime-N-hydroxysuccinimidyl ester (cortisol-NHS) and mouse anti-cortisol-3-BSA Mab were purchased from United States Biological (Swampscott, MA). Pro-Active® Goat Anti-Mouse Coated Microspheres were obtained from Bangs Laboratories (Fishers, IN). Coelenterazine was bought from Biosynth International (Naperville, IL). All other reagents were of best grade and used as received. The buffers were as follows: bicarbonate buffer (100 mM $NaHCO_3$, 4 mM EDTA, pH=8.2); TE (30 mM Tris-HCl , 4 mM EDTA, 0.1% BSA, pH=7.5); TBS-EDTA (TE, 138 mM NaCl, 3 mM KCl, 0.05% Tween 20); $CaCl_2$ solution (200 mM $CaCl_2$, 100 mM Tris-HCl, pH=7.5).

The recombinant apoprotein of mutant S was expressed in *Bacillus subtilis*, purified, and incubated with coelenterazine to obtain the active photoprotein, as described elsewhere[5].

For the synthesis of the label, cortisol-NHS was dissolved in anhydrous DMSO, then dropwise added to a gently stirred solution of mutant S in carbonate buffer kept in an ice bath. The molar ratio of cortisol to photoprotein was 250. The solution was stirred overnight at 4°C, then the excess of conjugation agent was removed by gel-filtration with Sephadex G-25.

All the bioluminescent activity measures were performed as follows: 100 µL of the solution were dispensed in glass test tubes, then the calcium solution was added, and the resulting light emission was recorded for 6 s.

The stability of the label in different storage conditions was studied. Aliquots of a 1×10^{-6}M solution of the label in TE buffer were stored either at 4°C, at –20°C, and at –80°C. In addition, a 5×10^{-9}M solution was stored at 4°C. The bioluminescent activity was monitored over a period of 130 days.

After collection, the saliva samples were quickly frozen at –80°C, then thawed and centrifuged, in order to precipitate proteins and mucopolysaccharides. The supernatant was stored at –20°C until analysis.

For the development of the immunoassay, TBS-EDTA was used as assay buffer. All the incubation steps were performed at room temperature and mixing at 120 rpm. A volume of 100 µL of a 1 µg/mL solution of the anti-cortisol monoclonal antibody was incubated for 30 minutes with 100 µL of the Pro-Active® microspheres diluted to 1 mg/mL. After washing twice, the solid phase was incubated for 30 minutes with 100 µL of a cortisol standard solution in TBS or a saliva specimen. Then 100 µL of a 1×10^{-9}M label solution were added and the mixture was further incubated for 30 minutes. The solid phase was washed twice, then the bioluminescent activity of the bound fraction was measured. All measurements were performed in triplicate and were corrected for the blank.

3. RESULTS AND DISCUSSION

A heterogeneous immunoassay for cortisol was developed using magnetic microspheres, with goat anti-mouse Fc-specific antibodies immobilized on the surface, as a solid phase.

The mutant S of aequorin was expressed in *Bacillus subtilis*, purified, chemically conjugated to cortisol, and used as a label for the immunoassay. In this mutant the three cysteine residues of aequorin in positions 145, 152, and 180 were substituted by serine residues. The mutated photoprotein was shown to be slightly more active than the native one and also more stable, in the sense that no reducing agent, such as 2-mercaptoethanol, is required to prevent the formation of disulfide bonds[6]. The increased stability is an important advantage in the applicability of the photoprotein as a label in binding assays.

The synthesis of the label was performed using a 250 molar excess of cortisol over mutant S concentration. The conjugated photoprotein showed a flash-type bioluminescent emission similar to that of the unconjugated one and kept more than 55% of its activity. Moreover, it showed a good affinity for the anti-cortisol antibody, therefore it was suitable for the development of the immunoassay.

Stability studies in different storage conditions were performed. The label retained 100% of its activity when stored for 130 days at –20°C and at –80°C, while more than 75% and almost 60% of the activity was conserved after storage at 4°C respectively of a concentrated or a diluted solution.

Antibody and label concentrations and incubation time were optimized. The conditions that gave best performance in the shortest time were: 1 μg/mL of anti-cortisol antibody, 1×10^{-9}M for the label and 30 minutes for all the incubation steps.

Typical dose-response curves for cortisol obtained using different label concentrations are shown in figure 1. They have a working range of almost 5 orders of magnitude and a cortisol detection limit of 5×10^{-9}M was obtained using the label at the concentration 1×10^{-9}M.

The immunoassay was also applied to the analysis of saliva samples. In preliminar experiments it was observed that the standard deviation for the saliva samples was much higher than that of the cortisol standards prepared in buffer. This happened because the magnetic microspheres aggregated in the presence of the matrix from the sample. It was observed that increasing the detergency property of the buffer (by raising the Tween 20 concentration from 0.05% to 0.5%), prevented the microspheres from aggregating in the presence of saliva, thus significantly reducing the standard deviation in the determination of cortisol concentration.

In conclusion, we have developed a sensitive and rapid immunoassay for cortisol using a bioluminescent protein as a label, rather than an enzyme or a radioisotope. We have demonstrated that this method can be applied to the analysis of real samples, such as saliva specimens.

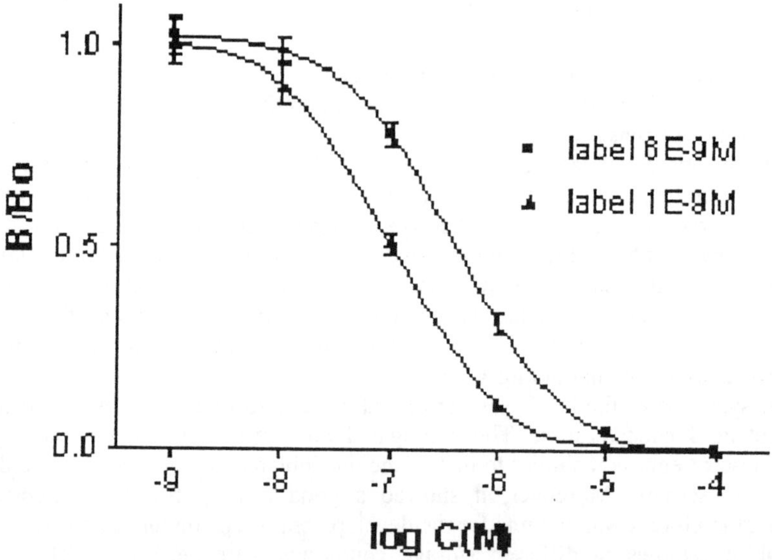

Fig. 1 Dose-response curves for cortisol obtained using 1 μg/mL of antibody and two different label concentrations.

Acknowledgements

We acknowledge the National Institute of Health and the University of Kentucky for funding this project.

References

1. Bonte HA, van den Hoven RJ, van der Sluijs Veer G, Vermes I. The use of free cortisol index for laboratory assessment of pituitary-adrenal function. *Clin Chem Lab Med* 1999; **37(2)**: 127-132.
2. Read GF, Walker RF, Wilson DW, Griffiths K. Steroid analysis in saliva for the assessment of endocrine function. *Ann N Y Acad Sci* 1990; **595**: 260-274.
3. Ohmiya Y, Hirano T. Shining the light: the mechanism of the bioluminescence reaction of calcium-binding photoproteins. *Chem Biol* 1998; **3**: 337-347.
4. Kendall JM, Badminton MN. *Aequorea victoria* bioluminescence moves into an exciting new era. *Trends Biotech* 1998; **16**: 216-224.
5. Lewis JC, López-Moya JJ, Daunert S. Bioluminescence and secondary structure properties of aequorin mutants produced for site-specific conjugation and immobilization. *Bioconjugate Chem* 2000; **11**: 65-70.
6. Kurose K, Inouye S, Sakaki Y, Tsuji F. Bioluminescence of the Ca^{2+}-binding photoprotein aequorin after cysteine modification. *Proc Natl Acad Sci USA* 1989; **86**: 80-84.

RAPID BACTERIAL DETECTION USING ADENYLATE KINASE (AK) IN A MAGNETIC BEAD IMMUNOASSAY

NV SELLICK-HARMON[1], MJ MURPHY[1], RL PRICE[1] AND DJ SQUIRRELL[1]
[1]*DERA Porton Down, Wiltshire SP4 0JQ, UK*

1. INTRODUCTION

Adenylate kinase (AK) is an essential enzyme which, inside the cell, catalyses the equilibrium reaction: Mg^{2+}.ATP + AMP ↔ Mg^{2+}.ADP + ADP. Supplied with purified ADP *in vitro*, AK will generate ATP which can then be measured by the bioluminescent reaction of firefly luciferase. In this way AK can be used as a generic cell marker giving detection limits for bacteria about 100 times better than can be achieved by bioluminescence based on simple ATP measurements.[1] Fewer than 10^2 bacterial cells may be detected in a 5 minute assay, with increased sensitivity being obtained from longer assay times. Because AK levels within cells fluctuate less than ATP levels do, correlation of the AK assays with cell numbers is more accurate than that given by direct ATP-based measurements. The work reported here covers the development of specific assays for bacteria based on bioluminescence using the AK approach. Specificity was imparted using immunocapture, bacteriophages or a combination of both. The immunoassays were initially developed in a microtitre plate format and compared with a conventional enzyme-linked immunosorption assay (ELISA). This was taken forward in a tube-based assay using antibody-coated magnetic beads. These can present a higher surface area for capturing the target cells than is provided by a microtitre plate well and also allow easier mixing. Detergents were used to release AK from captured cells so added ADP could be converted to ATP. Lysis using bacteriophages[2] rather than detergents was investigated for *Salmonella*, *E.coli* and *Listeria*. Experiments to optimise and understand the assays are described and results from assays using immunocapture coupled with phage lysis are given.

2. MATERIALS AND METHODS

AK assay reagents were purchased from Celsis-Lumac (Cambridge, UK).Dynabeads were obtained from Dynal (UK) Ltd. (Cheshire, UK). Bacterial strains were obtained from the National Collection of Industrial and Marine Bacteria, (Aberdeen, UK). The *Listeria* bacteriophage was an environmental isolate obtained from MAFF Central Science Laboratory (York, UK). **AK Assay**: 200μL bacterial dilution in 15mM magnesium acetate was added to 100μL Microbial Adenylate Kinase Assay solution, mixed and incubated for 5 minutes. 100μL bioluminescence reagent was then added and the light produced measured using a luminometer. **Microtitre Plate Immunoassays:** Microtitre plates were coated with antibody and blocked with BSA. 100μL bacterial dilution was added to each well and incubated at 37°C for 1 hour. After washing, magnesium acetate was added to each well and an AK assay carried out. Comparative colourimetric immunoassays were performed using HRP-labelled secondary antibodies and ABTS. **Magnetic Bead Immunoassay:** Bacterial suspensions were added to Dynabeads pre-coated with the appropriate antibody in a 1.5mL tube Eppendorf tube. After the required time, the beads were immobilised with a magnet and the supernatant removed. The beads were washed with PBS and standard AK assays were carried out. **Bacteriophage Assay**: Bacteriophage preparations were purified to remove ATP and AK contamination by centrifugation for 2 hours at 115,000g. Log phase bacterial cultures were split into two equal 10mL fractions: one for phage addition, the other as an uninfected control. Both were incubated at 37°C. At timed intervals, 100μL samples were removed. To determine total AK, these were assayed with extractant. When determining extracellular or released AK, no extractant was used.

3. RESULTS

In an AK assay, the detergents used to make added ADP available for conversion by the intracellular AK could act in two ways. They could either release the AK from the cells or could make the cell envelope permeable to both ADP and ATP whilst the AK was still retained. To distinguish these possibilities, a suspension of *E.coli* cells was treated with detergent and then centrifuged. Measurement of AK levels showed that 30 times more activity was associated with the supernatant than with the pelleted cells. The assays therefore work by releasing the enzyme into solution. A conventional sandwich ELISA for *E.coli* using horse radish peroxidase (HRP) was developed. This took 2½ hours and gave limit of detection of 1.8×10^6 cfu per mL (with a sample volume of 100μL). The endpoint of the assay was changed by omitting the HRP-labelled secondary antibody and performing an AK assay after the initial capture and washing steps. This assay approach gave a detection limit of 2.3×10^5 cfu per mL and took 1½ hours. Although better than the normal ELISA, the improvement afforded by using the AK endpoint was less than expected given that 3.5×10^2 cfu could be detected with a signal to noise ratio of 4:1 when 100μl of an *E.coli* suspension was added directly to a well on the microtitre plate. Tests with different conditions showed that the capture efficiency for whole cells was less than 1%. Antibody-coated magnetic beads were used to improve the AK-based immunoassay. Initial results for Salmonella using standard protocols for immuno-capture gave a limit of detection of 2.8×10^4 per mL (600μL sample volume) in 1½ hours. The assay time could be significantly reduced. Figs 1 and 2 show that incubating during capture, and multiple washes, provided little advantage.

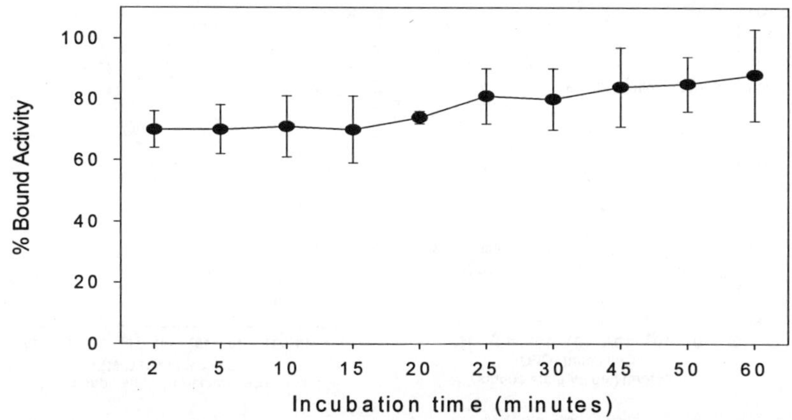

Fig. 1 Effect of Incubation time on the binding efficiency of magnetic bead immunoassays for *Salmonella typhimurium*. Mean (n=3) ± 1s.d.

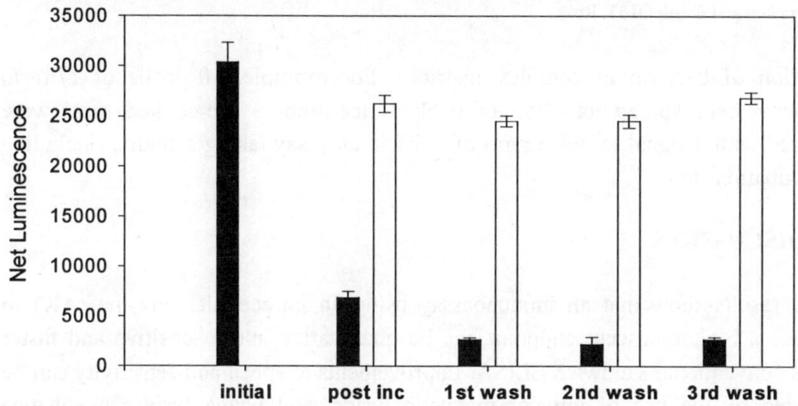

Fig. 2 Comparison of bound and free AK during the washing steps of a magnetic bead immunoassay for *Salmonella.* ■ Supernatant associated signal (unbound

The sensitivity of the assays was enhanced through improvements in reagent purity and optimisation of the concentration of beads. Current performance of the AK bead immunoassay is shown in Fig.3 for assays taking 10 minutes.

Instead of detergents, bacteriophages may be used to effect lysis after the target cells have been captured. The additional specificity provided can allow the

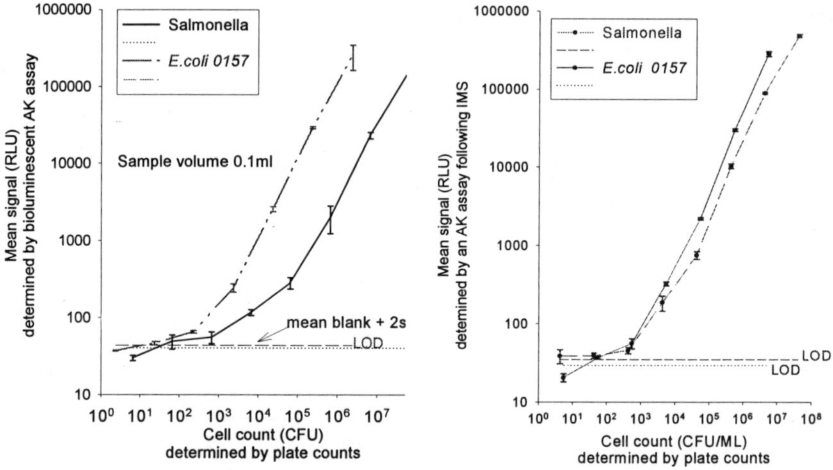

Fig. 3 Results of non-specific (left) and immuno-bead specific assays (right) for *Salmonella typhimurium* and *E.coli* O157. Row

detection of bacteria in complex matrices. For example, 10^4 cells of *Listeria monocytogenes* spiked into 25g of cooked rice from a pre-cooked meal were detected with a signal to noise ratio of 3.6:1 in an assay taking 5 hours, (including an incubation step).

4. DISCUSSION

These results show that an immunoassay using an intracellular enzyme (AK) to produce a bioluminescent endpoint can be quantitative, more sensitive and faster than a conventional sandwich ELISA. Improvements to speed and sensitivity can be provided by the use of immuno-magnetic beads, and phage lysis can enhance specificity. Assays with limits of detection below 10^2 cells mL^{-1} are possible. Applications in food microbiology and clinical testing are possible.

References

1. Murphy MJ and Squirrell DJ. *Encyclopedia of Food Microbiology*, Robinson RK, Batt CA & Patel PD eds., Academic Press, London, 1999; 16-24.
2. Blasco R, Murphy MJ, Sanders MF, Squirrell DJ. Specific assays for bacteria using phage mediated release of adenylate kinase. *Journal of Applied Microbiology* 1998; 84: 661-666.

COMBINED HIV ANTIGEN AND ANTIBODY ASSAY ON A FULLY AUTOMATED CHEMILUMINESCENCE BASED ANALYZER

DINESH SHAH, CHI-DEU CHANG, KEVIN CHENG, ANDREA WACHTER, AND JAMES STEWART

Abbott Laboratories, Abbott Park, IL 60064 USA
E-mail: Dinesh.shah@add.ssw.abbott.com

1. INTRODUCTION

Acridinium (Ac) derivatives have been utilized for highly sensitive immunoassays because of their stability and high chemiluminescence yield. Abbott PRISM® is a high throughput, fully automated serological analyzer to screen plasma or sera for Heptitis B surface antigen, antibodies to Heptitis B core, HCV, HIV, and HTLV.[1] The instrument uses acridinium label to tag analytes captured on microparticles, buffer to wash nonspecific binding, H_2O_2 to trigger chemiluminescence, and a photon multiplier to collect photon counts. Data are processed by a computer to align with sample barcode and to sort out the positive samples.

Tests for antibodies against the human immunodeficiency virus (HIV) have been used for the detection of HIV infection sine 1985. Recently, progress has been in improving the performance of HIV tests. Viral detectability has been broadened to cover HIV-1, HIV-2, HIV-O, and their variants, and the newer tests offer better detection of early seroconversion resulting in a reduction of the window between HIV infection and seroconversion. However, a window of 22 days[2] still persists before anti-HIV antibodies (HIVAb) become detectable with the third generation HIVAb tests. HIV core p24 antigen (HIVAg) appears earlier than HIV antibodies after HIV infection but is only temporarily detectable during the Window period in serological tests. A combined assay capable of detecting HIVAb and HIVAg simultaneously would shorten the HIV detection window when compared to assays that only detects HIV antibodies.

2. MATERIALS AND METHODS

2.1. Assay Reagents, Controls, and Test Panels

Solid-phase: It is a blend of two types of microparticles - 1) microparticles co-coated with recombinant antigens (rAg) of HIV-1 *env* p41, HIV-2 *env* p36, and HIV-O *env* p41 and 2) microparticles coated with monoclonal Anti-HIV p24 (Mab anti-p24).

Blend of Acridinium Conjugates: The blend consists of acridinium labeled HIV-1 *env* p41 rAg, HIV-2 *env* p36 rAg, HIV-O *env* p41 rAg, HIV-1 peptide, and Mab anti-p24. Each antigen or Mab was conjugated with 10-(3-sulfopropyl)-N-tosyl-N-(2-carboxyethyl)-9-acridinium carboxamide via N-hydroxysuccinimide ester in a phosphate or borate buffer. After conjugation, the conjugate was sized by HPLC to a selected Ac/protein molar ratio.

Controls: Negative Calibrator (NC) is re-calcified negative plasma. HIV-1 Positive Control (PC) and HIV-2 Positive Control (SPC) are prepared from patient plasma with anti-HIV-1 and anti-HIV-2, respectively, at selected dilutions. HIV-O Positive Control (OPC) is diluted from a Mab anti-O. Abbott's HIV-1 p24 Viral Antigen Quantitative Panel was used to assess the HIVAg sensitivity.

HIV Test Panels: HIV-1 Seroconversion Panels were purchased from Boston Biomedica, Inc. (BBI, West Bridgewater, MA), BioClinical Partners (Franklin, MA) and NABI (Boca Raton, FL). Specimens of HIV-1 subtypes from A to G, HIV-2 and HIV-O were acquired from Africa, Europe, and U.S.

2.2. Assay Format

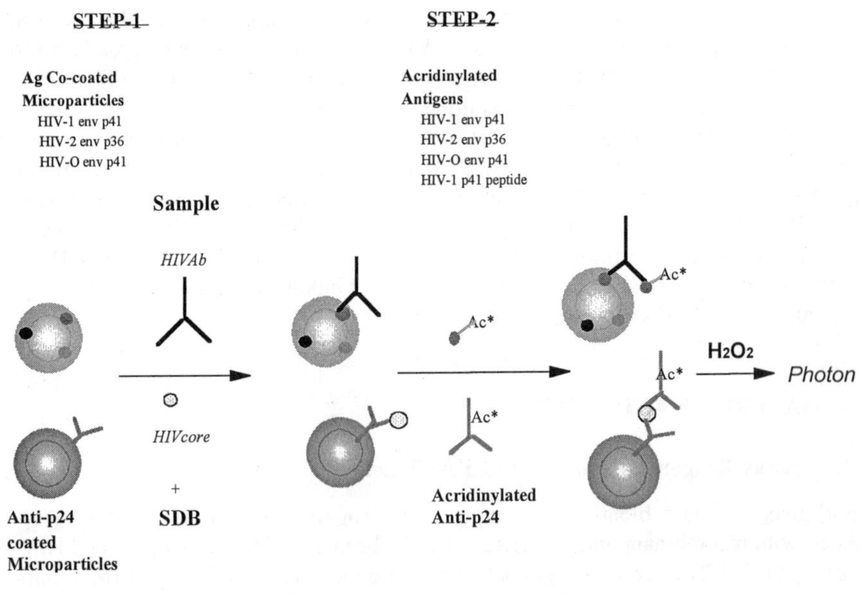

Fig. 1. ABBOTT PRISM® HIVAg/Ab Combo Assay Format

Assay steps and reactions are illustrated in Fig. 1. In Step-1, the assay starts with 100 μl of sample and incubates with specimen diluent buffer (SDB) and a blend of HIV rAg microparticles and Mab anti-p24 microparticles in the sample well on a reaction tray. After the first incubation, the tray is moved to the transfer station where the reaction mixture is flushed into the sandwich reaction well by transfer wash buffer and excessive fluid is absorbed by blotter underneath. In Step-2, the blend of conjugates is dispensed to the reaction well at the conjugate dispensing station. After the 2nd incubation, unbound conjugates are washed into the blotter. An alkaline hydrogen peroxide is then injected at the trigger/read station to trigger chemiluminescence from acridinium labels captured in the reaction well. The intensity of chemiluminescent signal is proportional to the amounts of HIVAb and HIVAg in the sample. The total assay time is under 1 h and 160 samples/h can be processed on each instrument channel.

3. RESULTS AND DISCUSSION

The cut-off calculation is based on 3.5x the average net NC counts. If sample response with net counts is equal to or greater than the cut-off or S/CO ratio $\geq$ 1.0, it is reactive; if less than the cut-off, it is negative. However, if sample S/CO ratio is between 0.9 and 1.0, it is considered "gray zone" and will be re-tested. In feasibility studies, the HIVAg/Ab combo assay showed a sensitivity of HIVAg about 5 pg/mL based on the Abbott HIV-1 p24 Antigen Quantitative Panel. Equivalent p24 sensitivities were shown across HIV-1 subtypes A-G. The assay detected all HIVAb positive samples including 28 HIV-1 A-G subtypes, 5 low titer HIV-2 samples and, 8 low titer HIV-O samples.

The combo assay was tested on 2,000 pre-screened negative serum and plasma specimens from Gulf Coast Regional Blood Center (Houston , TX). Four specimens were repeatedly reactive, yielding 0.2% false-positive or an assay specificity of 99.8%.

To determine the reduction of the HIV detection window, data of 28 HIV-1 seroconversion panels were generated and compared with a current third generation HIVAb assay and a HIVAg assay, the Coulter HIV-1 p24 EIA. The calculation was based on the average difference (days) with the 1st reactive member from the reference 3rd generation HIVAb assay. The results are presented in Table 1. The PRISM HIVAg/Ab combo assay showed a reduction in the seroconversion window of 5.8 days which is very close to the window reduction of 6 days by HIVAg testing estimated by Schreiber et al.[2] The combo assay showed excellent detection for HIV variants and enhanced seroconversion sensitivity. The Abbott PRISM® HIVAg/Ab Combo Assay is a single screening test which would significantly improve the safety of the blood supply by reducing the residual risk of HIV transfusion.

Table 1. Comparison of the 1st bleed detected on seroconversion panels

HIVAg/Ab Panel	Reference HIVAb 3rd Gen. Assay	HIVAg Coulter p24 Assay	*Diff. From 3rd Gen.*	Abbott PRISM® HIVAg/Ab Combo Assay	*Diff. From 3rd Gen.*
NABI	Day	Day	*Days*	Day	*Days*
SV241	**15**	1	***-14***	7	***-8***
SV251	**18**	22	***4***	15	***-3***
SV261	**13**	10	***-3***	10	***-3***
SV271	**15**	8	***-7***	8	***-7***
SV321	**12**	8	***-4***	8	***-4***
SV331	**20**	13	***-7***	13	***-7***
SV400	**24**	13	***-11***	13	***-11***
SV401	**12**	8	***-4***	12	***0***
SV404	**23**	19	***-4***	23	***0***
SV405	**17**	7	***-10***	10	***-7***
SV407	**17**	14	***-3***	17	***0***
BBI					
PRB941	**19**	10	***-9***	19	***0***
PRB943	**15**	8	***-7***	8	***-7***
PRB944	**15**	3	***-12***	8	***-7***
PRB949	**21**	19	***-2***	19	***-2***
PRB950	**29**	22	***-7***	22	***-7***
PRB951	**20**	9	***-11***	12	***-8***
PRB952	**15**	11	***-4***	11	***-4***
PRB953	**11**	8	***-3***	8	***-3***
PRB954	**22**	22	***0***	18	***-4***
PRB955	**13**	4	***-9***	8	***-5***
BCP					
ANT6243	**33**	26	***-7***	28	***-5***
ANT6245	**75**	68	***-7***	68	***-7***
ANT9017	**36**	29	***-7***	15	***-21***
ANT9020	**98**	88	***-10***	91	***-7***
ANT9021	**58**	48	***-10***	48	***-10***
ANT9022	**35**	28	***-7***	26	***-9***
ANT9025	**94**	88	***-6***	88	***-6***
Window Reduction Av Days:	**0**		***-6.5***		***-5.8***

Acknowledgements

We thank G. Schochetman, S. Devare, J. Hunt, X. Qiu, S. Lou, J. Byrant, K. Shreder, H.-P. Kapprell, D. Bussfeld, S. Marhoff, H. Schoebel, A. Goller, and A. Smith for their contributions and support.

References

1. Khalil OS, Zurek TF, Tryba J, et al. Abbott Prism: a multichannel heterogeneous chemiluminescent immunoassay analyzer. *Clin Chem* 1991; **37**: 1540-1547.
2. Schreiber GB, Busch MP, Kleinman SH, Korelitz JJ. The risk of transfusion-transmitted viral infections. The NIHLBI Retrovirus Epidmiology Blood Donor Study. *N Eng J Med* 1996; **334**: 1685-1690.

LUMINESCENT PCR-ELISA ASSAY TO DETECT TB

XIAOLIN YANG
People's Hospital of Beijing Medical University, Beijing, P. R. China, 100034
E-mail: yangzhng@public3.bta.net.cn

1. INTRODUCTION

Mycobacterium tuberculosis (TB) is a common disease, which is still a critical public health problem in some developing countries. Patients who receive appropriate treatments in the early stages of infection often recover better. Therefore, it is important to accurately detect and identify this disease as soon as possible. Culture assays are expensive, and slow and PCR offers a faster, low cost and high sensitivity alternative.[1] We describe an enhanced chemiluminescent (ECL) system,[2, 3] that combines PCR and ELISA into a new technique for identification and quantitative detection of TB and other Mycobacteria in clinical samples.

2. MATERIALS AND METHODS

Instrumentation Lumat LB9501, Micro Lumat LB96-P (EG&G Berthold, Wildbad, Germany); YH-2000 microplate luminometer (Beijing Pilotech Biotechniques, Beijing, P. R. China); BACTEC 460 (Becton Dickinson, Cockeysville, USA); PE-480 thermocycler (Perkin Elmer, Bedford, USA); Microplates (96-well) (NUNC, Denmark).

Reagents The ECL solution was prepared as described in our patent.[2, 3] Luminol, hydrogen peroxide, avidin, HRP (Sigma, St Louis, MO, USA); *Taq* DNA Polymerase (Sangon Ltd, Scarborough Ontario, Canada); uracil DNA glycosylase (GIBCO BRL, Gaithersburg, MD, USA); Mycobacteria and attenuated Tuberculosis standards (China National Institute for Drug and Biological Products Inspection, Beijing, P. R. China). Primers and TB specific probes,[4] upstream primer, CAC ATG CAA GTC GAA CGG AAA GG; downstream primer, Digoxin-GCC CGT ATC GCC CGC ACG CTC ACA; TB specific probe, GGT GGA AAG CGC TTT AGC GGT-Biotin. Probes for detecting Mycobacteria: GCG AAC GGG TGA GTA ACA CG-Biotin. All of DNA synthesis and labeling were undertaken by Cybersyn Co Ltd. (Lenni, USA).

Specimens TB positive clinical samples, e.g., phlegm, were collected from patients identified with active TB infection during clinical trials. TB negative samples were collected from normal individuals. Positive samples for TB or other Mycobacteria were prepared by adding the standard bacteria (1 to 1000 cells/sample) into TB negative samples; other Mycobacteria were added at >1000 cells/sample.

PCR Samples were disinfected and prepared by standard protocols.[1, 4] After mixing 20 μL buffer, 5 μL template, 2U Taq DNA Polymerase, 0.5 U uracil DNA glycosylase, 200 micromol/L d(A+C+G)TP, 750 micromol/L dUTP, 10 pMol/L of each primer and 20 μL mineral oil into each reaction tube. PCR amplification - 95°C for 240s, 35 cycles (anneal: 62°C for 30s; extension: 72°C for 60s; denature: 95°C 30s), finally, an extension at 72°C for 300s.

ELISA Biotin labeled probe (500 ng/mL) was added into an avidin-coated microplate (20 g/mL) ,[5, 6] incubated at RT for 1h, then washed 2-3 times. After heating the PCR products to 95°C for 60 s, hybridization buffer and 10 μL of the heated PCR products were mixed into microplates, 37°C for 1h, then washed 2-3 times, finally, HRP-anti-digoxin conjugate was added, incubated at 37°C for 1h, followed another wash. Light emission was detected for 1s/sample immediately after addition of the ECL solution. Luminescent signal (RLU/s) of the sample was divided by signal of negative control; S/N >2.1 = positive.

To exclude false negatives caused by PCR inhibiting contaminants, each clinical sample was divided into two, and as an amplification control, 10 TB cells were added into one of them. If this control was negative in the PCR assay, the sample was prepared again. As a routine precaution in PCR amplification, uracil DNA glycosylase and dUTP were added to prevent false positives induced by carry over.[7]

3. RESULTS AND DISCUSSION

The calibration curve with TB spiked phlegm samples indicated satisfactory correlation from 10 to 1000 cells/sample (r=0.9). For the spiked or clinical samples the luminescent results for the PCR-ELISA assay (cut-off value, S/N=2.1) compared favorably with culture. Coincidence rate (% correctly identified/number in total samples was - clinical samples 87% (negative: 102/102; positive: 77/104; total: 179/206); spiked samples, 84% (negative: 20/20; 1 TB cell per sample: 4/20; from 10 to 1000 TB cells per sample: 60/60; total: 84/100).

Tests for determine the specificity of the two probes with spiked samples showed that the Tuberculosis, Mycobacteria and other bacteria could be satisfactorily distinguished. At 1 *M Tuberculosis* cell/sample 4/20 samples tested positive with a TB specific probe and 6/20 were positive with a probe for Mycobacteria. At 10, 100, and 1000 *M Tuberculosis* cells/sample all samples tested positive with a TB specific probe and with a probe for Mycobacetria. At 1000 cells/sample the following were negative with a TB specific probe and positive with a probe for Mycobactria - *M. kansasii, M. cookii, M. marinum M. simiae, M. phlei, M. malmoense, M. intracellulae, M. gastri, M. flavescens, M. avium, Pneumococcus, Brucellosis, Colibacillus, Pertussis.* Table 1 shows that the luminescent PCR-ELISA had better sensitivity compared with the culture assay.

Table 1 Comparison between luminescent PCR-ELISA and culture assays

Assays*	Imitated Samples (TB Cells/sample)					Clinical Samples	
	0	1	10	100	1.000	+	-
Luminescent PCR-ELISA	20/20	4/20	20/20	20/20	20/20	77/104	102/102
Culture	20/20	0/20	12/20	19/20	20/20	59/104	102/102

* The data was showed correctly identified number/total samples.

In some active TB infection cases, we found a negative PCR-ELISA result but a positive Mycobacteria result, and similar results were obtained in other studies.[1, 4] This may be due to a gene mutation in the *M Tuberculosis* from these cases, or some of non-TB Mycobacteria may induce a TB-like infection.

Acknowledgements

Dr. Ding, Dr. Tu and other colleagues in Beijing Institute for Tuberculosis Control supply the clinical samples and all culture devices.

References

1. Wobeser WL WL. Krajden M. Conly J et al. Evaluation of Roche Amplicor PCR assay for Mycobacterium tuberculosis. *J Clin Microbiol* 1996; **34**: 134-139.
2. Yang XL, Wang SW. Ultra sensitive enhanced chemiluminescent solution. *Chinese Patent* 9110621.9, 1993.
3. Yang XL, Liu MZ. An improvement of enhanced chemiluminescence and its application to immunoassay. In: Roda A, Pazzagli M, Kricka LJ, Stanley PE eds. *Bioluminescence and Chemiluminescence-Perspectives for 21st Century*. Chichester: Wiley, 1998; 138-141.
4. Vincent J. et al. Detection of Mycobacterium tuberculosis by PCR amplification with pan-Mycobacterium primers and hybridization to an M. Tuberculosis—specific probe.*J Clin Microbiol* 1996; **34**: 918-923.
5. Yang XL, Wang H, Liu MZ. A chemiluminescent enzymimmunoassay with microplate to detect T3 and T4. *J Beijing Med Univ* 1997; **2**(Suppl C): 18-20.
6. Yang XL, Wang SW. The direct labelling chemiluminescence molecular hybridization system. *Chin J Biochem* 1995; **5**:483-486.
7. Kwok S, Hignhchi R. Avoiding false positives with PCR. *Nature* 1989; **339**: 237-239.

THE QUANTIFICATION OF REACTIVE OXYGEN SPECIES AND THEIR SCAVENGER FROM PHOTON EMISSION

Y. YOSHIKI, Y. AKIYAMA, K. ABE AND K. OKUBO
Department of Environmental Bioremediation, Graduate School of Agricultural Science, Tohoku University, 1-1 Tsutsumidori Amamiyamachi, Aoba-ku, Sendai 981-8555, Japan
E-mail: yoshiki@bios.tohoku.ac.jp

1. INTRODUCTION

Reactive oxygen species (ROS), represented by hydrogen peroxide (H_2O_2), superoxide (O_2^-) and hydroxyl radical ($HO^{\cdot}$), can induce serious cell damage including the degradation of DNA. Therefore, a number of studies on technique to assess the ROS and their scavengers contained in foodstuffs and biological substances has been reported. The chemiluminescence method using luminol and a *Cypridina luciferin* analogue is one of the emerging techniques for the detection of ROS.[1,2] Recently, we demonstrated that the ROS/hydrogen donor/mediator system (XYZ system) produced photon emission without a luminescent agent. The photon intensity of the XYZ system has a high concentration dependency to the reactive oxygen species (X), hydrogen donor (Y) and mediator (Z), which is indicated by $[P]=k[X][Y][Z]$ (k = photon constant).[3] This equation gives important clues in searching for the ROS, hydrogen donor and mediator. Based on this XYZ system, we developed a screening system in order to quantify the X, Y and Z species.

2. MATERIALS AND METHODS

2.1 Reagents and Materials

Hydrogen peroxide (30%) and MeCHO were purchased from Santoku Chem (Tokyo, Japan) and Merck (Darmstadt, Germany). Gallic acid, $KHCO_3$ and lipoxygenase (LOX) were obtained from Nakarai (Japan). One milligram of LOX used throughout this experiment had $9.7.10^3$ U activity. To analyze the ROS contents, gas phase components of the cigarette smoke (three pieces) was trapped with formic acid (1 mL). The condensed cigarette smoke solution (40 μL) was used in Photon-HPLC analysis. Fourteen kinds of Chinese and medical teas were extracted 10 fold in distilled water for 2h. For the hydrogen donor contents analysis, 30 μL of tea water extract were used. The mediator quantity was applied on the soybean seedlings. The soybean seeds sprouted in darkness at 25°C for 4 d, and divided into epicotyl, hypocotyl and cotyledon parts. Each freeze-dried segment was milled and added to water. Mediator contents were measured with supernatant (30 μL) after centrifugation (10,000 rpm, 40 s).

2.2 Photon-HPLC System

A chemiluminescence detector (CLD-110, Tohoku Electronic Ind., Sendai, Japan) was connected to a HPLC (model 515 pump and 2487 UV detector, Waters, Tokyo Japan) system (Photon-HPLC system). For the HPLC-fractionation, a gel column (Shodex Asahipak GS-520 7G, Shoko Co., Tokyo, Japan) and HPLC solvent (0.15 M Na_2SO_4 containing 0.03 M Na_2HPO_4, pH 7.0) were used. After the HPLC-fractionation, the elution was mixed with two out of three of the following solvents (X solvent; 196 mM H_2O_2 or 68 mM *tert*-BuOOH, Y solvent; 10 mM gallic acid was dissolved in 10 % MeOH, Z solvent; sat. $KHCO_3$ dissolved in 360 mM MeCHO) to produce the photon emission in a chemiluminescence detector.

3. RESULT AND DISCUSSION

The equation derived for the ROS/hydrogen donor/mediator system (XYZ system) indicates that a small amount of X species can be detected by the addition of high concentration of YZ reagent. An XZ reagent can be used to detect Y, and XY reagents for Z. We have already found some compounds which act as X, Y, and Z reagents.[4] For reasons of convenience, the H_2O_2 (X), *tert*-BuOOH (X), gallic acid (Y), $KHCO_3$ in MeCHO (Z) and lipoxygenase (Z) were used for standard XYZ reagents. The gel filtration column, used throughout this experiment, could fractionate H_2O_2, gallic acid and lipoxygenase (LOX) with retention times at 20 min, 30 min and 12 min. The photon intensity (P) shows a linear concentration dependency for H_2O_2 (P = 0.75X), gallic acid (P = 452.9Y) and LOX (P = 0.84Z for H_2O_2 and P = 4.1Z for *tert*-BuOOH), respectively, as shown in Fig. 1. The detection limits were found to be 78 nmol for H_2O_2 and 50 pmol for gallic acid. In addition, the results showed detection limits of LOX to be influenced by the X solvent, for example, 116.10^2 U/ml used 196 mM H_2O_2, and 12.10^2 U/ml used 68 mM *tert*-BuOOH.

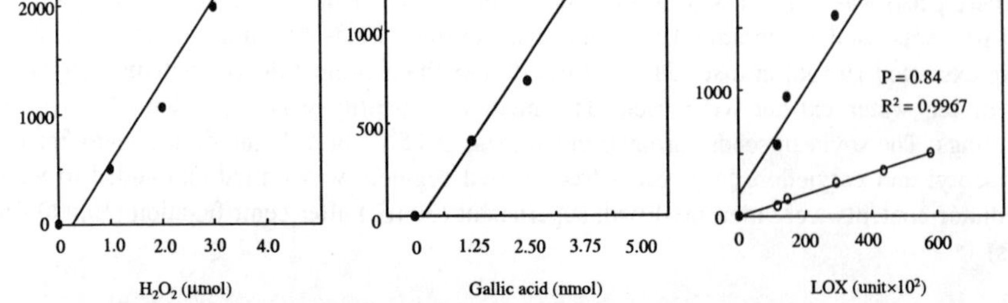

Fig. 1 Quantification of hydrogen peroxide (A), gallic acid (B), and lipoxygenase (C) from photon emission using Photon-HPLC system

Subsequently, we assessed quantification of the X, Y and Z species into some foodstuffs and other nonessential items. Since it has been reported that cigarette smoke contains ROS (X) such as H_2O_2 and O_2^-, we used cigarette smoke for the X quantification.[5] Comparable data for water and formic acid suggests that the X in the gas phase of cigarette smoke reacts more with formic acid than with water. Fig. 2 A shows HPLC and photon patterns of condensed cigarette smoke in formic acid. Despite a single peak in the HPLC pattern, two main peaks were observed in the photon pattern. Although we could not identify the first peak (retention time; 16 min), the second peak (retention time; 20 min) corresponded to H_2O_2. Investigation of the X effect in cigarette smoke from the peak high of the photon emission (*i.e.* P = 0.75X) suggested that ten kinds of Japanese cigarettes contained X in the range 60.9 nmol-9.9μmol H_2O_2/piece.

It has been suggested that X is an aggravation agent in cancer. Several epidemiological studies reported a diminishing cancer risk by the high consumption of tea, which is rich in hydrogen donors, including flavonoid and procyanidins.[6] In this study, hydrogen donor (Y) contents of Chinese and Japanese medical teas were quantified from photon intensity. The results of Photon-HPLC analysis showed that almost all the Chinese tea had the same photon peak of retention time at 18 min coinciding with HPLC peak (Fig. 2 B). In contrast, photon peaks from medical tea, either intensity or component, were influenced with a variety of medical stuffs. The Y effect, calculated with the photon standard curve of gallic acid (*i.e.* P = 452.9Y), was 2-7μmol gallic acid/g for Chinese tea and 0.3-2.0μmol gallic acid/g for medical tea.

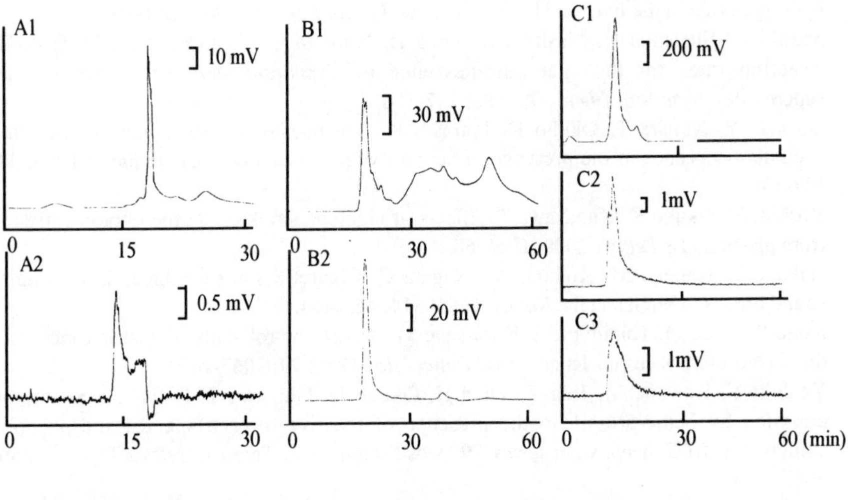

Fig. 2 Photon-HPLC chromatograms of condensed cigarette smoke (A), water extract of tea (B) and soybean seedling (C)

A1, B1 and C1 are HPLC patterns using gel filtration column. A2, B2, C2 and C3 are photon patterns in the gallic acid/$KHCO_3$ in MeCHO, H_2O_2/$KHCO_3$ in MeCHO, H_2O_2/gallic acid and *tert*-BuOOH/gallic acid systems.

Finally, we approached the detection and quantification of Z species using soybean seedlings which are a LOX rich source. In this Photon-HPLC system, LOX eluted at a retention time at 12 min. However, this gel-filtration column also eluted metalloproteins such as peroxidase and cytochrome c, which had a similar molecular weight to LOX and produced a photon emission in the presence of H_2O_2 and gallic acid. The quantification curve of LOX (Fig. 1 C) suggested that LOX was characterized by the high photon yield in the presence of *tert*-BuOOH (P = 4.1Z) rather than H_2O_2 (P = 0.84Z). In addition, metalloprotein expected LOX, did not produce photon emission in the presence of *tert*-BuOOH. For this reason, LOX contents of soybean seedlings were measured using 68 mM *tert*-BuOOH as X solvent and 10 mM gallic acid as Y solvent. The LOX contents were $16.3.10^2$ U/mg for epicotyl, $5.0.10^2$ U/mg for hypocotyl and $17.3.10^2$ U/mg for cotyledon.

We have demonstrated that some YZ combinations produce stronger effects (synergistic effect) on lipid antioxidation and O_2^- and $HO^·$ scavenging activities than that of Y only.[7,8] Photon-HPLC analysis based on XYZ system is effectual and provides a simple technique to search for ROS scavenger (Y and Z species), and ROS (X species). We anticipate that our new assay method will be able to contribute to the comparison of food stuffs and biomaterial directly.

References

1. Miyazawa T, Suzuki T, Fujimoto K, Yasuda K. Chemiluminescent simultaneous determination of phosphatidylcholine hydroperoxide and phosphatidylethanolamine hydroperoxide in the liver and brain of the rat. *J Lipid Res* 1992; **33**: 1051-1058.
2. Suzuki N, Suetsuna K, Mashiko S, Yoda B, Nomoto T, Toya Y, Inaba H, Goto T. Reaction rates for the chemiluminescence of *Cypridina luciferin* analogues with superoxide. *Agric Biol Chem* 1991; **55**: 157-160.
3. Yoshiki Y, Kahara T, Okubo K, Igarashi K, Yotsuhashi K. Mechanism of catechin chemiluminescence in the presence of active oxygen. *J Biolumin Chemilumin* 1996; **11**: 131-136.
4. Yoshiki Y, Okubo K, Kanazawa T. Effects of cigarette smoking on the photon emission from plasma. *ITE Letters* 2000; **1**: 65-68.
5. Nakano T, Kaneko M, Kodama M, Nagata C. Cigarette smoke induces DNA single-strand breaks in human cells. *Nature* 1985; **314**: 462-464.
6. Kono S, Ikeda M, Tokudome S, Kuratsune M. A case-control study of gastric cancer and diet in Northern Kyushu, Japan. *Jpn J Cancer Res* 1988; **79**: 1067-1074.
7. Yoshiki Y, Kawane M, Iida T, Yuan H, Okubo K, Ishizawa T, Kawabata S. Photon emission by heme-protein in the presence of reactive oxygen species and phenolic compounds. *ACS Symposium series 702*, eds Sibamoto T, Terao J, Osawa T, 1998; 256-265.
8. Yoshiki Y, Yuan H, Iida T, Kawane M, Okubo K, Ishizawa T, Kawabata S. Chemiluminescence of horseradish peroxidase and acetaldehyde related with gallic acid and hydrogen peroxide interaction. *Photochem Photobiol* 1998; **68**: 802-808.

DEVELOPMENT OF A CHEMILUMINESCENCE IMMUNOASSAY FOR CLARA CELL PROTEIN AT SUB-PM LEVELS

GIJSBERT ZOMER

National Institute of Public Health and the Environment,
PO Box 1, 3720 BA Bilthoven, The Netherlands
E-mail: Bert.Zomer@rivm.nl

1. INTRODUCTION

The symptoms of allergy, i.e. sneezing, sniffing and shortage of breath, are caused by a reaction to exposure to allergens. Medication is taken to diminish the symptoms, but its provision is based on how the patient feels, so overmedication can occur. An objective way to measure the severity of the allergic reaction is not available. With markers for allergy, patients can be monitored and their symptoms can be better controlled.

Clara cell protein (CC16), also known as Urinary Protein I (UP-1) is a 16-kD protein secreted at the surface of respiratory airways by nonciliated bronchial and bronchiolar cells, including Clara cells. Although its exact function is unknown there is increasing evidence that CC16 plays an important immunosuppressive and anti-inflammatory role in the lung.[1] Diagnostically, CC16 levels in serum have been used as biomarker[2] for assessing the integrity of the respiratory system. Decreased serum levels of CC16 have been found as a result of exposure to silica[3] while exposure to ozone[4] increased serum CC16 levels. Apart from lung fluids and serum, CC16 has also been detected in urine, semen, and nasal lavage fluids.[5] In lung fluids its concentration is very high (up to 5 mg/L in lung lavage fluid). Asthmatic patients have lower CC16 concentrations.[6] In serum and urine average CC16-concentrations drop to 5-50 μg/L and 10 μg/mmol of creatinine, respectively). Because of the great diagnostic interest for CC16, several immunoassays have been developed.[7,8]

In this paper the development of another immunoassay is reported with a greater sensitivity allowing the measurement of CC16 in saliva. The assay uses a sandwich immunoassay format with an HRP-labeled antibody for detection and quantification. The use of the recently developed HRP signal reagent based on acridan ester GZ-11[9-11] allowed the quantification of less than 0.01 μg/L of CC16. This detection limit is considerably less than using the other immunoassays reported thus far.

2. EXPERIMENTAL

Microtiterplates (solid white opaque high binding plates) were obtained from Corning Costar Corporation, Cambridge MA. Coating anti-CC16 antibodies and Horseradish Peroxidase labeled anti-CC16 antibodies were obtained from DAKO A/S, Glostrup, Denmark. CC16 standard was obtained from Prof. Bernard of the Université Catholique de Louvain, Brussels, Belgium. Chemiluminescence measurements were performed using a Lucy 1 plate luminometer (Anthos Labtec Instruments GmbH, Salzburg, Austria).

Microtiterplate coating using a 1:1,000 dilution of coating antibody in carbonate buffer (50mM, pH 9.4) was performed overnight at 4°C. The coated plates were washed with washing buffer (PBST, 10mM, pH 7.2 containing 0.1% Tween20). The plates were incubated with 0.1 mL of CC16-standards or samples in triplicate (saliva samples were diluted 10 times) prepared in assay buffer (wash buffer containing 0.5% bovine serum albumin). After incubating for 90 minutes at 37°C, the plates were washed with washing buffer (4x). A second incubation step was performed using a 1: 5,000 dilution of labeled antibodies (0.15 mL) in assay buffer. After 90 minutes at 37°C, the plates were washed with washing buffer (4x) and with PBS (1x). To each well 0.15 mL of signal reagent containing perfluoro-t-butyl 2-chloro-10-methylacridane-9-carboxylate (GZ-11) was added and the chemiluminescence signal measured in the plate luminometer after an incubation time of ca 10 minutes.

A calibration curve was constructed using the standards. The experimental data points were fitted using the function $y=a0*x/(a1+x)$. The developed assay was applied to samples of saliva collected from children.

Saliva samples were collected in January and at the start of the pollen season in May. Based on skin prick tests children were divided over four groups: house dust mite allergic children (group 1), grass pollen allergic children (group 2), children allergic to both house dust mite and grass pollen (group 3), and non-allergic children (group 4). In this preliminary study only saliva samples of children from groups 2 and four were analyzed. Saliva was collected with a cotton wool tampon (Salivette, Sarstedt, Sweden). Children were instructed to hold the tampon in the mouth for two minutes. After two minutes the saliva soaked tampon was put into the Salvette tube and cooled on ice. Within four hours the Salivette tube was centrifuged at 2,000g at 4 °C filled out in cryotubes and stored at –80 °C.

3. RESULTS AND DISCUSSION

A typical standard curve for CC16 is shown in Fig. 1. As can be seen from Fig. 1 CC16 can be quantified over a wide range from 0.01-10 ng/mL corresponding to 0.6-600 pM. This is considerably more sensitive than CC16 immunoassays reported earlier: Tomlinson et al —using the same sandwich assay with OPD as the substrate solution reported a detection limit of 0.45 ng/mL.[8] Because of the inherently more sensitive GZ-11 (Fig. 2) substrate a tenfold diluted labeled antibody could be used in the second incubation step. In the current assay the sensitivity is mainly

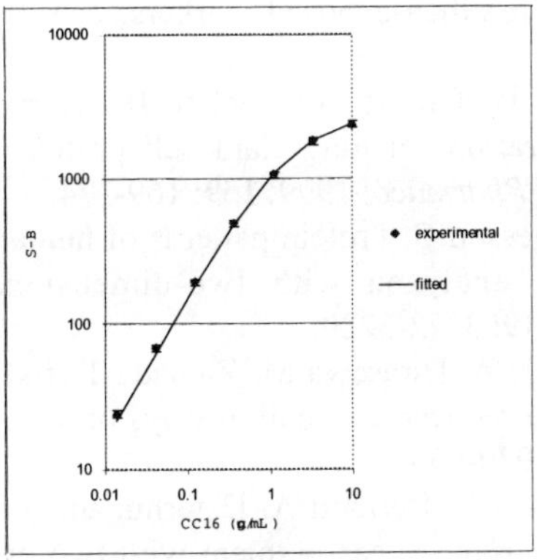

Figure 1 CC16 calibration curve

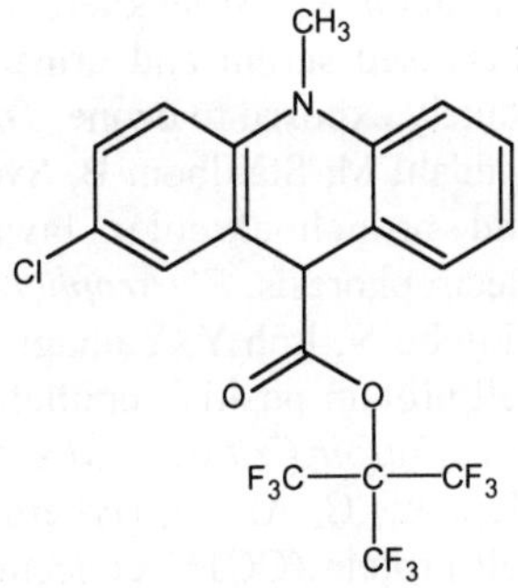

Figure 2 Structure of GZ-11

dependant on the aspecific binding.

CC16 concentrations in saliva collected in January in the control group averaged 1.7 ng/mL (range 0-6.7 ng/mL). In saliva of children allergic to pollen the average concentration of CC16 was 0.7 ng/mL (range 0-2.2 ng/mL). Statistically (Student-t test) this difference is significant ($p<0.05$). When the results of saliva samples collected in May were compared (controls mean 2.2 ng/mL, range 0.3-8 ng/mL, allergic to pollen mean 1.4 ng/mL, range 0-4.4 ng/mL) the difference is non significant. These results would indicate that salivary CC16 levels in children allergic to pollen rise from January to May.

Acknowledgements

Agnete Ingild of DAKO A/S, Glostrup, Denmark is gratefully acknowledged for a gift of the Peroxidase-conjugated anti CC16 antibody. Professor Alfred Bernard of the Université Catholique de Louvain, Brussels, Belgium is acknowledged for a gift of a standard of CC16.

References

1. Dierynck I, Bernard A, Roels H, De Ley M. The human Clara cell protein: biochemical and biological characterisation of a natural immunosuppressor. *Mult Scler* 1996; **1**: 385-387.
2. Broeckaert F, Bernard A. Clara cell secretory protein (CC16): characteristics and perspectives as lung peripheral biomarker. *Clin Exp Allergy* 2000; **30**: 469-475.
3. Bernard AM, Gonzalez-Lorenzo JM, Siles E, Trujillano G, Lauwerys R. Early decrease of serum Clara cell protein in silica-exposed workers. *Eur Respir J* 1994; 7: 1932-1937.
4. Arsalane K, Broeckaert F, Knoops B, Clippe A, Buchet JP, Bernard A. Increased serum and urinary concentrations of lung clara cell protein in rats acutely exposed to ozone. *Toxicol Appl Pharmacol* 1999; **159**: 169-174.
5. Lindahl M, Ståhlbom B, Svartz J, Tagesson C. Protein patterns of human nasal and bronchoalveolar lavage fluids analyzed with two-dimensional gel electrophoresis. *Electrophoresis* 1998; **19**: 3222-3229.
6. Shijubo N, Itoh Y, Yamaguchi T, Imada A, Hirasawa M, Yamada T et al. Clara cell protein-positive epithelial cells are reduced in small airways of asthmatics. *Am J Respir Crit Care Med* 1999; **160**: 930-933.
7. Hermans C, Aly O, Nyberg BI, Peterson C, Bernard A. Determinants of Clara cell protein (CC16) concentration in serum: a reassessment with two different immunoassays. *Clin Chim Acta* 1998; **272**: 101-110.
8. Tomlinson PA, Dalton RN, Turner C, Chantler C. Measurement of beta 2-microglobulin, retinol-binding protein, alpha 1-microglobulin and urine protein 1 in healthy children using enzyme-linked immunosorbent assay. *Clin Chim Acta* 1990; **192**: 99-106.
9. Zomer G, Jacquemijns M. On site screening: Application of novel acridan esters as HRP detection reagent in immunoassay for the detection of beta-agonists in bovine urine. In: Roda A, Pazzagli M, Kricka LJ, Stanley PE, ed. *Bioluminescence and Chemiluminescence. Perspectives for the 21st century*. Chichester: John Wiley & Sons, 1999: 87-90.
10. Osman AM, Zomer G, Laane C, Hilhorst R. Comparative studies of the chemiluminescent horseradish peroxidase-catalysed peroxidation of acridan (GZ-11) and luminol reactions: effect of pH and scavengers of reactive oxygen species on the light intensity of these systems. *Luminescence* 2000; **15**: 189-197.
11. Zomer G, Jacquemijns M. Application of novel acridan esters as chemiluminogenic signal reagent in immunoassay. In: García-Campaña AM, Baeyens WRG, ed. *Chemiluminescence in Analytical Chemistry*. New York: Marcel Dekker, in press.

PART 7

CELLULAR LUMINESCENCE

MOLECULAR OXYGEN, PHAGOCYTE MICROBICIDAL ACTION AND LUMINESCENCE

ROBERT C. ALLEN

Emory University School of Medicine, Department of Pathology & Laboratory Medicine, Grady Memorial Hospital Clinical Laboratory, 80 Butler St SE, Atlanta, GA, 30335 USA

E-mail: rallen3@emory.edu

1. INTRODUCTION

Phagocytic leukocytes are the first line of host cellular defense against infecting microbes. Phagocyte microbicidal action is dependent on NADPH oxidase- and myeloperoxidase-catalyzed electron transfer reactions that change the spin quantum number of molecular oxygen to realize its potential as a direct reactive oxygenating agent. Chemiluminescence (CL) is an energy product of microbicidal combustion, and is proportional to phagocyte generation of oxygenating agents and the CL quantum yield of the substrates oxygenated. Introducing a high yield substrate defines the reaction and allows CL-based measurement of the type and quantity of oxygenating agent generated. Lucigenin measures oxidase activity; luminol measures mostly, but not exclusively, myeloperoxidase activity. Sensitive real-time measurements of basal and stimulated phagocyte oxygenating activities require less than a drop of blood. Under properly controlled assay conditions, the functional state of phagocyte opsonin receptor expression is also determined. The composite CL data are analyzed by multiple discriminant functions and used to gauge the state of host immune activation, i.e., the inflammatory response.

2. MOLECULAR OXYGEN AND COMBUSTION

Considering the relative electronegativity of oxygen and the favorable thermodynamic potential for its reaction with organic molecules, the fact that we survive and prosper in an atmosphere that is one-fifth O_2 presents a puzzle. The electron spin symmetry of O_2 is the explanation. The O_2 we breathe is a diradical paramagnetic molecule with a total spin quantum number (***S***) equal to +1 or —1. The multiplicity of an atom, molecule or ion is equal to $|2S| + 1$, and as such, the multiplicity of O_2 is triplet.[1] The typical molecule encountered in biochemistry is a nonradical with ***S*** equal to 0. This difference in spin symmetry imposes a reactive barrier that protects biologic molecules from direct reaction with O_2, i.e., spontaneous combustion.[2]

Radical character does not in itself impart reactivity. Chlorine (Cl_2), a nonradical (***S*** = 0) singlet multiplicity molecule, is highly reactive with organic substrates. Reaction of Cl_2 with ethylene yields a net free energy (ΔG) of -42 kcal/mole. Although the reaction of O_2 with ethylene is almost twice as exergonic

(Δ**G** + -93 kcal/mole), reaction with O_2 is not spontaneous.[2] For any reaction the overall spin symmetry of both reactants and products must be conserved. The Wigner spin conservation rules define the tendency of a reacting system to conserve spin angular momentum.[3]

Exergonicity provides information with regard to the direction and magnitude of reactive change, but additional factors govern reaction rate. In the context of the absolute reaction rate theory, Rate = $\kappa(\boldsymbol{k}T/h)K^{*}\boldsymbol{e}^{-Ea/RT}$, where κ is the transmission coefficient, $\boldsymbol{k}$ is the Boltzman entropy constant, T is the absolute temperature, h is Planck s constant, K* is the activation equilibrium constant (K* = F /FaFb, where Fa and Fb are the initial states of the reactants and F is the activation complex), Ea is the activation energy per mole at absolute zero (approximately equivalent to the experimental energy of activation of the classical Arrhenius relationship) and R is the gas constant. Not all species with the necessary free energy to cross the potential energy barrier do so. The transmission coefficient κ defines the ratio of the number of complexes that yield products relative to the number having reached the activated state. According to the Wigner spin conservation rule, a reaction system will resist any change in spin angular momentum. As such, κ will approximate unity when the reactants, the reaction complex and the products have the same symmetry properties and the same quantum numbers for $\boldsymbol{S}$ and $\boldsymbol{L}$ (the total orbital angular momentum). For reactions that involve a change in multiplicity the κ values less than 10^{-4}.[3]

3. PHAGOCYTE MICROBE KLLING AND LUMINESCENCE

Blood phagocytes serve as the cellular effectors of microbe killing in the acute inflammatory response against infection. Neutrophil leukocytes, monocytes, and eosinophil leukocytes respond to molecular information generated by the humoral, endothelial and cellular immune systems. Information in the form of cytokines and other molecular mediators directs phagocyte-endothelial contact and adhesion, diapedesis, and migration to the inflammatory site. Phagocytes recognize opsonin-labeled microbes by opsonin-specific membrane receptors. Ligation triggers phagocytosis, degranulation and phagolysosome formation.[4]

Phagocytosis is associated with a greater than tenfold increase in glucose oxidation via the dehydrogenases of the hexose monophosphate (HMP) shunt and a proportional increase in non-mitochondrial O_2 consumption.[4] These metabolic changes drive electron transfer reactions that change the $\boldsymbol{S}$ value of O_2. Univalent reduction of O_2 by NADPH oxidase yields the doublet multiplicity products HO_2 and O_2^- ($\boldsymbol{S}$ = +_ or -_). Acid disproportionation of HO_2 and O_2^- yields the $\boldsymbol{S} = 0$ products singlet multiplicity H_2O_2 and possibly singlet molecular oxygen ($^1O_2^*$). Myeloperoxidase (MPO) catalyzes the oxidation of Cl^- ($\boldsymbol{S} = 0$) by H_2O_2 ($\boldsymbol{S} = 0$) yielding the singlet multiplicity reactant HOCl ($\boldsymbol{S} = 0$). Additional spin allowed reaction of HOCl with H_2O_2 yields the reactive dioxygenating agent $^1O_2^*$.[5]

By changing the value of $\boldsymbol{S}$ phagocytes realize the reactive potential of O_2 and catalyze combustive microbicidal action. Native phagocyte luminescence, *i.e.*, photon emission in the visible spectral range, is an energy product of phagocyte oxygenation reactions.[6]

4. CHEMILUMINIGENIC SUBSTRATES

Native phagocyte luminescence demonstrates that mammalian biochemistry can liberate energies far in excess of those considered in classical treatments of bioenergetics. Phagocyte luminescence provides a window to view the temporal kinetics of phagocyte microbicidal action. Measurement is non-destructive and can be made in real time. When the opsonized microbe is held constant, phagocyte luminescence is directly proportional to HMP shunt metabolism and is oxygen-dependent. As a diagnostic technique, native luminescence measurement has limitations. The yield of excited product per native substrate oxygenation is low in comparison to bioluminescence reactions, and the substrates vary with the nature of the target microbe.

Use of chemiluminigenic substrates (CLS) overcomes these limitations. A CLS is an exogenous substrate susceptible to oxygenation by one or more of the phagocyte-derived agents, and generating electronically excited product in high yield. Luminescence is dependent on phagocyte generation of oxygenating agents and on the nature and quantum yield of the substrate oxygenated. As such, the sensitivity for detecting oxygenating agents is proportional to the luminescence quantum yield of the substrate oxygenated.[5]

4.1 Acridinium Compounds (Lucigenin)

The acridinium compound lucigenin can serve as substrate for luminescence measurement of the products of NADPH oxidase activity, O_2^- and H_2O_2. Lucigenin must be univalently reduced to a doublet multiplicity radical before it can react with doublet multiplicity O_2^-.[7] The reaction is the spin-allowed doublet-doublet, annihilation reaction yielding a moloxide intermediate that disintegrates yielding one N-methylacridones (NMA) in ground state and another NMA in the singlet excited state that relaxes by light emission. Under alkaline conditions lucigenin can directly react with the HO_2^- to generate the same products and luminescence. Reactants and products are all singlet multiplicity.

4.2 Cyclic Hydrazide Compounds (Luminol)

The cyclic hydrazide luminol was the first CLS employed in the study of phagocyte oxygenation activity. Addition of luminol increases neutrophil leukocyte luminescence by several orders of magnitude.[8] Although the luminol luminescence of neutrophils is highly dependent on MPO, it is not exclusively MPO-dependent.

5. PHAGOCYTE LUMINESCENCE AND INFLAMMATION

Phagocyte opsonin receptor expression is proportional to the degree of exposure to inflammatory agents. The state of immune activation is reflected by the opsonin receptor expression per circulating phagocyte. Under kinetically defined conditions phagocyte opsonin receptor expression can be assayed as the luminescence associated with the initial phase of phagocyte oxygenation of luminol following exposure to a non-limiting concentration of opsonized particles. As such, blood phagocyte luminescence can be successfully applied as a gauge of the systemic inflammatory state.[9]

References

1. Matthews PSC. *Quantum Chemistry of Atoms and Molecules*. Cambridge: Cambridge University Press, 1986.
2. Allen RC. Role of oxygen in phagocyte microbicidal action. *Environ Health Perspect* 1994; **102** (Suppl 10): 201-208.
3. Laidler KL. *The Chemical Kinetics of Excited States*. Oxford: University Press, 1955.
4. Klebanoff SJ, Clark RC. *The Neutrophil: Function and Clinical Disorders*. Amsterdam: North-Holland, 1978.
5. Allen RC. Phagocytic leukocyte oxygenation activity and chemi-luminescence: A kinetic approach to analysis. *Methods Enzymol* 1986; **133**: 449-493 (1986).
6. Allen RC, Stjernholm RL, Steele RH. Evidence of the Generation of (an) Electronic excitation state(s) in human polymorphonuclear leukocytes and its participation in bacterial activity. *Biochem Biophys Res Commun* 1972; **47**: 679-684.
7. Allen RC, Loose LD. Phagocytic activation of a luminol-dependent chemi-luminescence in rabbit alveolar and peritoneal macrophages. *Biochem Biophys Res Commun* 1976; **69**: 245-252.
8. Allen RC. Lucigenin chemiluminescence: A new approach to the study of polymorphonuclear leukocyte redox activity. In: DeLuca MA, McElroy WD, eds. *Bioluminescence and Chemiluminescence: Basic Chemistry and Analytical Applications*. New York: Academic, 1981; 63-73.
9. Allen RC, Dale DC, Taylor FB. Blood phagocyte luminescence: Gauging systemic immune activation. *Methods Enzymol* 2000; **305**:591-629.

THE CHEMIEXCITATION OF THE PHOTOSENSITIZERS IN THE CELLULAR CULTURES

MARTINA BANCIROVA,[1] HANA KOLAROVA,[2] RENE LENOBEL,[2] JAN LASOVSKY[1]
[1] *Faculty of Science, CMBM UP, Tr. Svobody 26, Olomouc, 771 46, Czech Republic*
[2] *CMBM UP Olomouc, Slechtitelu 11, Olomouc 783 71, Czech Republic*
E-mail: bancir@risc.upol.cz

1. INTRODUCTION

Weak chemiluminescence accompanies the catalytical decomposition of hydrogen peroxide. It is possible to enhance this light emission and therefore also population of the excited states by means of oxidizable substrates with the high quantum yield of excitation (energizers e.g. phthalhydrazide (PH) and luminol (LU).[1]

It was found that the chemiluminescent oxidation of luciferin is able to generate sufficiently intense and long-live emission to induce antiviral activity of hypericin,[2] well-known photosensitizer.

Photodynamic therapy (PDT) is a treatment that is used for destruction of certain tumors. The absence of light in many regions of the body may preclude the use of the photosensitizers as the therapeutic compounds for PDT.

Photodynamic therapy involves the combination of visible light and a organic dye - photosensitizer. Each factor is harmless by itself, but when combined with oxygen, can produce lethal cytotoxic agents that can inactivate tumor cells. This enables greater selectivity towards diseased tissue as only those cells that are simultaneously exposed to the cytotoxic effect. Upon illumination, the photosensitizer is excited from the ground state to the first excited single state, followed by conversion to the triplet state via intersystem crossing. The excited triplet state can react in two ways. A Type I mechanism involves hydrogen-atom abstraction or electron-transfer reactions between the excited state of the photosensitizer and a substrate to yield free radicals and radicals ions. A Type II results from an energy transfer between the excited triplet state of the photosensitizer and the ground state molecular oxygen, generating the first excited state of oxygen, singlet oxygen.[3]

2. EXPERIMENTAL

Instrumentation: Luminometer BioOrbit 1250 (Finland), Spectrofluorimeter F 4500, (Hitachi, Japan), Luminoskan and Fluoroskan *Ascent* (Labsystems Finland) and inversion fluorescent microscope Olympus IX 70.
Reagents: phthalocyanines ($CuPcS_4$, $NiPcS_4$), phthalhydrazide, protoporphyrine IX, methylene blue, eosin B, $CuSO_4$, H_2O_2 (Sigma-Aldrich, Germany), hypericin, hypocrellin A, luminol, calcein AM (Molecular Probes, USA).

Methods. Chemiluminescent reaction of phthalhydrazide or luminol was initiated with the injection of hydrogen peroxide solution at presence of $CuSO_4$. The total chemiluminescence was measured with a luminometr BioOrbit 1250 and Luminoskan Ascent (measurements were done in volume 2 mL respective 200 μL).

NIH3T3 (mouse fibroblast), MCF7 (human breast adenocarcinoma) (Masaryk Memorial Cancer Institute, Brno-Czech Republic) were used as a standard testing system. Twice washed trypsinized cells were divided in amount 10^4 to each well and filled in DMEM with 10% FCS in a total volume 80 μL. After 24 hours of cultivation at 37°C in 5% CO_2 the photosensitizer with chemiluminescent system (20 μL) was added. The total volume 100 μL (cells with photosensitizer) were cultivated for 24 hours. For identification of live and dead cells the fluorescent probes calcein AM was used. For the elimination of the influence of own fluorescence of the photosensitizer the testing wells were aspirated, washed with 100 μL PBS and filled by 100 μL PBS and 100 μL calcein AM. The results were proved by fluorometric measurement with Fluoroskan Ascent. Morphological changes of the cells during cultivation have been observed in an Olympus IX 70 inversion fluorescent microscope.

3. RESULTS AND DISCUSSION

The chemiexcitation of the xanthene dyes and photosensitizers (e.g. hypericin, phthalocyanines) induced by the hydrogen peroxide decomposition and autooxidation by dissolved oxygen in the micellar solutions of the cetyltrimethylammonium cation was studied.[4] We assume, similar to the CIEEL theory, a formation of the micellar complex of electron transfer between hydrogen peroxide and dyes and formation of their excited states in the chemiluminescent system. Observed facts can be phenomenologically explained using the idea of the formation of micellar complexes of dyes according to Poisson Distribution Model (PDM). Both, chemiluminescent intensity and chemiexcitation rate depend on the rate of hydrogen peroxide decomposition. The chemiexcitation does not depend on the structural type of dye. The final step is peroxide decomposition, back electron transfer to dyes and formation of their excited states. The energy transfer by energy pooling is only the minority way. The goal of this work is to find out the significance of the chemiexcitation for the photodynamic therapy.

We have found that photosensitizers chemiexcited by phthalhydrazide could produce the photodynamic damage on the rabbit s spleen (test of vitality of the migration activity of splenocytes derived from rabbit spleen fragments)[5] and on the Gram-negative and Gram-positive bacteria (determination of the MIC - minimal inhibitory concentration).[6] In the present work we focus on the chemiexcited photodynamic damage in the different cellular cultures. Viability of cells was determined by means of the molecular probes for fluorescence microscopy (calcein AM).

The chemiexcited photodynamic damage was studied in the follow cell cultures: NIH3T3 (mouse fibroblast) and MCF7 (human breast adenocarcinoma). The photosensitizers were chemiexcited by phthalhydrazide or by luminol. We used phthalhydrazide 5.10^{-6} M, luminol 5.10^{-6} M, $CuSO_4$ 5.10^{-6} M, H_2O_2 1.10^{-6} M.

The Table 1 shows the cytotoxicity (IC 50 for NIH3T3) of the photosensitizer only (without excitation) and of the photosensitizer chemiexcited by luminol or phthalhydrazide.

Table 1. Cytotoxicity (IC 50) for NIH3T3

	Photosensitizer only	Photosensitizer chemiexcited by phthalhydrazide	Photosensitizer chemiexcited by luminol
Protoporphyrin IX	0,3870 mg/mL	0,3500 mg/mL	0,3500 mg/mL
Hypericin	0,0123 mg/mL	0,0093 mg/mL	0,0093 mg/mL
Hypocrellin A	0,0063 mg/mL	0,0016 mg/mL	0,0016 mg/mL
$CuPcS_4$	0,7000 mg/mL	0,3500 mg/mL	0,3500 mg/mL
$NiPcS_4$	0,2970 mg/mL	0,0110 mg/mL	0,0110 mg/mL
Methylene Blue	0,0220 mg/mL	0,0110 mg/mL	0,0110 mg/mL
Eosine B	0,2640 mg/mL	0,3500 mg/mL	0,3500 mg/mL

The Table 2 shows the cytotoxicity (IC 50 for MCF7) of the photosensitizer only (without excitation) and of the photosensitizer chemiexcited by luminol or phthalhydrazide.

Table 2 Cytotoxicity (IC 50) for MCF7

	Photosensitizer only	Photosensitizer chemiexcited by phthalhydrazide	Photosensitizer chemiexcited by luminol
Protoporphyrin IX	0,7000 mg/mL	0,0420 mg/mL	0,0310 mg/mL
Hypericin	0,0224 mg/mL	0,0300 mg/mL	0,0190 mg/mL
Hypocrellin A	0,0063 mg/mL	0,0016 mg/mL	0,0016 mg/mL
$CuPcS_4$	0,7000 mg/mL	0,3500 mg/mL	0,3500 mg/mL
$NiPcS_4$	0,3200 mg/mL	0,0360 mg/mL	0,0660 mg/mL
Methylene Blue	0,0220 mg/mL	0,0110 mg/mL	0,0110 mg/mL
Eosine B	0,0910 mg/mL	0,3500 mg/mL	0,3500 mg/mL

The results confirm the possibility of the chemiexcitation of the photosensitizer in the cellular cultures. Of course, there is a different efficiency of the chemiexcitation. It depends on the kind of the photosensitizer and on the kind of the cellular culture. It means there is different ability of the photosensitizers to remain in the cells and also the different ability of the photosensitizers to produce the reactive oxygen species. The influence of the kind of the energizer (phthalhydrazide or luminol) seems to be negligible.

Acknowledgments

This research was supported by the grant from Ministry of Education No. VS 96 154, post-doc grant No. 203/98/P257 of the Grant Agency of the Czech Republic and by the "CONTACT" No. 17/26, programme based on the agreement between the governments of Italy and the Czech Republic.

References

1. Lasovsky J, Rypka M, Slouka J. Chemiluminescence enhanced by intramicellar energy transfer. *J Lumin* 1995; **65**: 25-32.
2. Carpenter S, Fehr MJ, Kraus GA, Petrich JW. Chemiluminescent activation of the antiviral activity of hypericin: A molecular Flashlight. *Proc Natl Acd Sci USA* 1994; **91**:12273-12277.
3. Sharman WS, Allen CM, van Lier JE. Photodynamic therapeutics: basic principles and clinical applications, *Drug Discov Today* 1999; **4** (11): 507-517.
4. Lasovsky J, Bancirova M, Hrbac J, Otyepka M. Chemiexcitation of the xanthene dyes by phthalhydrazide and hydrogen peroxide in the micellar environment of cetyltrimethylammonium bromide. In: Roda A, Pazzagli M, Kricka LJ, Stanley PE eds. *Bioluminescence and Chemiluminescence: Perspectives for the 21st Century*. Chichester: John Wiley & Sons. Ltd. 1998: 555-558.
5. Bancirova M. PhD. Thesis. *Micellar Catalysis*. Palacky University Olomouc, Faculty of Science 1997.
6. Bancirova M, Lasovsky J, Kolarova H, Frebortova J, Medkova J. Chemiexcitation of phthalocyanines for the photodynamic therapy by phthalhydrazide on the gram-positive and gram-negative bacteria. In: Roda A, Pazzagli M, Kricka LJ, Stanley PE editors. *Bioluminescence and Chemiluminescence: Perspectives for the 21st Century*. Chichester: John Wiley & Sons. Ltd. 1998: 567-557.

EFFECT OF METHOTREXATE ON LEUCOCYTE RESPIRATORY BURST IN RHEUMATOID ARTHRITIS PATIENTS

P. DE SOLE, G. NOCCA, R. FRESU, A.POMPA, AND A. ZOLI
Department of Clinical Chemistry and Rheumatology Division
Catholic University of Sacred Heart
L.go A. Gemelli, 8, - 00168 Rome, Italy
Email: p.desole@uniserv.ccr.rm.cnr.it

1. INTRODUCTION

Accumulation of polymorphonuclear leukocytes (PMNs) in synovial fluid and of monocytes in synovium is a characteristic feature of rheumatoid arthritis (RA). These cells release proteolytic enzymes and reactive oxygen species (ROS) which are strongly involved in inflammation of joints. ROS seem to play an important role in pathogenesis of articular damage through their action on collagen and hyaluronic acid. Methotrexate (MTX) is actually the drug of choice in the treatment of RA. Its clinical efficacy is probably related to a anti-inflammatory action mediated by a decrease of proinflammatory factors such as ROS. The aim of our study was to clarify the mechanism of action of MTX, evaluating its effect on PMN and monocyte production of ROS, measured by means of chemiluminescence.

2. PATIENTS AND METHODS

Ten patients (6 women and 4 men, aged 44.9 ± 13.4 years), affected by RA, were enrolled in our study. Disease duration was 1.6 ± 1.2 years. All the patients had an active disease. Ten healthy subjects (5 men and 5 women), whose mean age was 37.2 ± 11.6 years, represented the control group. Patients were treated with parenteral MTX (10 mg once a week). Our follow up lasted two months and included the assessment of laboratory and clinical parameters of disease activity (ESR, C-reactive protein, number of swollen joints, Ritchie index). Venous blood (10 mL) from patients was withdrawn 24-48 hours after MTX intramuscolar injection and collected in EDTA. It was diluted with saline (1:2; v: v) and dextrane 6% in saline was added to diluted blood (1:5; v: v). After 30 minutes, the white blood cells suspension was centrifuged on Lymphoprep (Pharmacia, Sweden), according to the manufacture's instructions. Residual erythocytes in PMN precipitate were lysed (30 seconds NaCl 0.23%) and then granulocytes were suspended to a cell concentration of 2 x 10^5/mL in a modified Krebs Ringer Phosphate (KRP).[1] Lympho-monocytes were suspended to a cell concentration of

2 x 10^6/mL in modified KRP. Chemiluminescence was assayed with a Autolumat LB953 luminometer (Berthold, Germany) at 37¡ C for 60 minutes, by adding 100 nanomoles of luminol or lucigenin as amplifier substances to 0.1 mL of cellular suspensions in 1.0 mL final volume. Chemiluminescence was assessed in basal condition and after phagocyte stimulation, adding 0.5 mg of zymosan or 5 nmoles of phorbol myristate acetate (PMA) (Sigma Chemical CO, St Louis, MO) in presence of luminol or 50 picomoles of PMA in presence of lucigenin. This evaluation was repeated before (T0), one (T1) and two months (T2) after the beginning of MTX treatment. Chemiluminescence cell activity was expressed as total counts in 60 minutes. Statistical analysis was performed with a commercially available package (Statistica, Statsoft, Microsoft Co.). Data are expressed as medians and quartile ranges. Significance analysis was obtained by KrusKal-Wallis Anova (median test). Pearson product-moment was used for correlation analysis.

3. RESULTS

Table 1 shows the monocyte luminescence values of patients and control subjects, measured before, one and two months after MTX treatment. It clearly indicates that monocyte luminescence of patients at T0 is substantially higher than that of controls, while it decreases during the MTX treatment. On the contrary, Table 2 shows that neutrophil luminescence activity is not significantly different from that of controls and it is not modified by MTX. Correlation analysis between the monocyte luminescence parameters and clinical and laboratory disease activity indexes (ESR, CRP, Ritchie index and number of swollen joints) is given in Table 3. A statistically significant positive correlation is only obtained between the monocyte luminescence parameters and Ritchie index.

4. DISCUSSION

Reactive oxygen species derived from PMNs and monocytes have been demonstrated to be involved in inflammation and joint destruction in RA patients. Previous studies do not give unequivocal information about MTX capacity of decreasing phagocytic ROS production in RA patients.[2-6] Our results clearly point out that monocyte oxidative burst is significantly higher in patients with active RA than controls. Moreover, MTX treatment (10 mg/ week) decreases monocyte ROS production. Nesher et al.[2] have described the inhibitory action of MTX on monocyte oxidative burst in vitro. This effect is obtained at concentration higher than micromolar, while MTX serum concentration is in the range 10^{-6}-10^{8} mol/L.[7] Therefore, the in vivo inhibitory effect on monocyte is observed also at very low concentration. This result can be explained by different kinetics of MTX passage through cell membrane and /or its intracellular accumulation in vivo and in vitro or by the high concentration of folic acid present in the fetal serum usually used for monocyte cultures. Increased folate may interfere with MTX effect owing to a competition on the enzyme dihydrofolate reductase. Besides, our study points out that MTX does not interfere significantly with PMN oxidative burst, in accordance with the results of the in vivo study performed by Storgaard et al.[5] On the contrary, other studies observed a significant inhibition of ex vivo neutrophil oxidative responses from RA patients treated with low dose of MTX.[3-6] The different methods used for the evaluation of the oxidative burst and particularly the different clinical features of patients could explain these discrepancies. A different behaviour of the two types of phagocytic cells emerges from our study: monocytes have a higher respiratory burst in active patients than in controls, while PMNs show a small and not significant increase of it. Moreover, MTX inhibits chemiluminescence response of monocytes, but has a less significant effect on PMN oxidative response. Serological and immunological conditions of RA patients may explain this different behaviour of monocytes and PMNs. Among all cytokines involved in the persistence of joint inflammation, IL-1 and TNF-α seem to play a fundamental role. They stimulate much more monocyte-macrophage system than neutrophil leukocytes. Therefore, the high serum concentration of these cytokines may explain the elevated luminescence activity of monocytes observed in RA patients. MTX seems to exert its anti-inflammatory action with different mechanisms that include the inhibition of dihydrofolate reductase, cyclo-oxygenase, lipo-oxygenase and the production of various cytokines. While some inhibitory effects should involve monocytes as well as PMNs, the action on IL-1 production should result more selective for monocytes than PMNs. Since in our study MTX exhibits a strong inhibitory effect on monocyte luminescence activity but not on PMNs, its main mechanism of action in RA could be linked to its effect on IL-1. However, the lack of correlation between monocyte CL and serum disease indexes (ESR, CRP), together with its significant correlation with the less specific Ritchie index, points out a multifactorial MTX mechanism of action.

Table 1: Monocyte luminescence values of patients and control subjects (counts/ 60 minutes x 10^{-6}); Median (lower-upper quartile range)

	Patients T0	Patients T1	Patients T2	Controls
Luminol	64 (20-103)^a	11 (5-40)^b	17 (7-28)	4.8 (1-24)
Luminol + Zym	333 (130-343)^a	95 (54-200)^b	108 (74-143)	47 (17-193)
Luminol + PMA	617 (182-720)^a	139 (31-202)^b	106 (46-120)	107 (17-117)
Lucigenin	16 (7-21)^a	6 (5-7)^b	14.8 (11-18)^a	4.8 (2-6)
Lucigenin+PMA	32 (14-35)^a	18 (9-32)	23.6 (20-37)^a	6.6 (4-20)

a = p vs controls < 0.05
b = p vs T0 < 0.05

Table 2: PMN luminescence values of patients and control subjects (counts/60 minutes x 10 $^{-6}$); Median (lower-upper quartile range). $p > 0.05$ in all the cases

	Patients T0	Patients T1	Patients T2	Controls
Luminol	49 (18-200)	23 (17-120)	68 (15-223)	29 (13-75)
Luminol + Zym	249 (92-296)	107 (73-366)	154 (100-305)	99 (45-218)
Luminol + PMA	259 (118-332)	159 (29-454)	205 (107-347)	97 (40-158)
Lucigenin	14 (12-24)	18 (12-24)	20 (11-35)	14 (8-19)
Lucigenin+PMA	22 (13-28)	19 (17-34)	27 (24-54)	19 (10-32)

Table 3: Correlation between different parameters of monocyte CL and disease activity indexes. Results are expressed as Pearson product-moment correlation. Marked values are significant at $p < 0.05$

	ESR	CRP	Swollen joints	Ritchie Index
Luminol	0.12	0.03	0.25	0.49*
Luminol + Zym	0.18	0.02	0.18	0.51*
Luminol + PMA	0.10	0.06	0.04	0.40*
Lucigenin	0.12	021	0.18	0.41*
Lucigenin + PMA	0.36*	0.19	0.15	0.44*

References

1. De Sole P, Lippa S, Littaru GP: Whole blood chemiluminescence: a new technical approach to assess oxygen-dependent microbicidal activity of granulocytes *J Clin Lab Automation* 1983; **3**: 391-400.
2. Nesher G, Moore TL, Dorner RW. In vitro effects of methotrexate on peripheral blood monocytes: modulation by folinic acid and S-adenosylmethionine. *Ann Rheum Dis* 1991; **50**: 637-641.
3. Laurindo IM, Mello SB, Cossermelli W. Influence of low doses of methotrexate on superoxide anion production by polimorphonuclear leukocytes from patients with rheumatoid arthritis. *J Rheum* 1995; **22**: 633-638.
4. Balla SA, Johnston C, Davis P. The in vivo effect of nonsteroidal anti-inflammatory drugs, gold sodium thiomalate and methotrexate on neutrophil superoxide radical generation. *Clin Exp Rheum* 1990; **8**: 41-45.
5. Storgaard M, Jensen MP, Stengaard-Pedersen K et al. Effects of Methotrexate, Sulphasalazine and Aurothiomalate on polymorphonuclear leucocytes in Rheumatoid Arthritis. *Scand J Rheumatol* 1996; **25**: 168-173.
6. Okuda A, Kubota M, Watanabe K et al. Inhibition of superoxide production and chemotaxis by methotrexate in neutrophils primed by TNF-α or LPS. *Eur J Haematol* 1997; **59**: 142-147.
7. Edelman J, Biggs DF, Jamali F, Russel AS. Low dose methotrexate kinetics in arthritis. *Clin Pharmacol Ther* 1984; **35**: 382-386.

CRITICAL LEG ISCHAEMIA: *IN VITRO* MODULATION OF PERIPHERAL LEUCOCYTE ACTIVITY BY (+)-CATECHIN

P DE SOLE

Clinical Chemistry, Catholic University, Largo A. Gemelli 8, 00168 Roma, Italy
E-mail: p.desole@uniserv.ccr.rm.cnr.it

ML VUOTTO, R MIRANDA, C NAPPO, MTL IELPO, V MOSCATIELLO, E LAGHI, G ASSO AND L RICCIARDI

Institute of General Pathology and Oncology, II University Napoli, Italy
E-mail: vuotto@unina2.it

A BASILE

Department of Plant Biology, University Federico II , Napoli, Italy

1. INTRODUCTION

Critical ischaemia of the lower limbs (CIL) is a severe ischaemia which endangers a leg in whole or in part. The criteria to define CIL are listed in Dormandy and Stock.[1]

Ischaemia-reperfusion is the most widely used experimental model to study CIL. Several investigations have shown that leukocytes play a key role in experimental conditions matching the model[1,2] and that drugs modulating leukocyte activation protect tissues from reperfusion damage.[1]

Catechins are among the most studied flavonoids and their use in different fields of medicine is expected.[3-5] Flavonoids modulate leukocyte activity[6] as antioxidants (scavengers or chain breakers)[7] or as biological response modifiers (BRMs) since they interfere with enzymes or cell receptors.[4,7] Many authors have studied flavonoids in prevention and repair of ischaemic damage and have demonstrated their utility.[3,8]

The aim of this study was to determine the effects of (+)-catechin on peripheral leukocyte functions in patients with CIL. The investigation was carried out by measuring chemiluminescence (CL) emission, a value that well correlates with the paramount phagocyte function.[9]

2. PATIENTS AND METHODS

<u>Patients</u>: 6 patients, 4 men and 2 women, (age 66 – 4.56 years; mean – sd) and 6 control subjects (mean age 67 – 5.11 years) were studied. Two patients had well

controlled non insulin-dependent diabetes mellitus. The pain onset was from 3.83 ± 1.47 weeks. Two of the subjects were in stage IV Fontaine, and 4 were in stage IIIb. The trophic lesions in the two stage IV subjects appeared without clinical signs of infection and none developed after recruitment. Each patient underwent physical examination to exclude any condition that could compromise the validity of the CL test.

Blood: heparin anticoagulated blood was collected from patients and controls between 08.00 and 09.00 a.m. and diluted 1:100 in KRP* for whole blood (WB) CL assay.

Modified Krebs-Ringer phosphate medium (KRP*): modified Krebs-Ringer phosphate medium was composed of 119 mM NaCl, 4.8 mM KCl, 0.5 mM $CaCl_2$, 1.2 mM $MgSO_4$, 5.6 mM glucose and 16.6 mM sodium phosphate buffer pH7.4.

Luminol: 5-amino-2,3-dihydro-1,4-phthalazinedione (luminol, Sigma, Chemical Co., St. Louis, MO, USA) was stored at room temperature as $5x10^{-2}$ M stock solution in dimethyl sulfoxide (DMSO). Immediately before use, it was diluted with KRP* to give a 10^{-3} M working solution.

(+)-Catechin: the (+)-catechin (Sigma) was diluted in KRP* at 50μM concentration. Such a concentration has been shown to be effective by preliminary studies.

Phorbol myristate acetate (PMA): the stock solution was prepared by diluting PMA (Sigma Chemical Co.) at 5×10^{-3} M in DMSO and stored at -20° C. The working solution was 1.5 μM in KRP*.

Luminol-dependent chemiluminescence assay : chemiluminescence assays were performed following a modification of the protocol described by De Sole et al.[9] using an automatic luminometer (Autolumat LB 953, Berthold, Wildbad, Germany). Whole blood CL emission was evaluated within three hours after venipuncture. Each vial contained 100 μL of 10^{-3} M luminol, 100 μL of the 50μM (+)-catechin, 100 μL of 1.5 μM PMA, 100 μL of diluted WB suspension and sufficient KRP* to yield a final volume of 1.0 mL. Samples without stimuli were also included as controls. The emission of CL by circulating phagocytes was measured in absence or in presence of (+)-catechin. All measurements were performed in triplicate. CL activity was determined as total photon emission per neutrophil, per 90 minutes.

Statistical analysis: The results from the three experiments were expressed as total counts (mean ± SD). Levels of significance were determined using the ANOVA test. Values of $p<0.05$ were regarded as significant.

3. RESULTS

Results of this study are reported in Tables 1 and 2. Chemiluminescence recorded without stimulants (basic CL) was higher in patients than in controls. Addition of (+)-catechin to unstimulated cells caused a statistically significant reduction of basic emission in both groups (Table 1).

The stimulation of the leukocytes with PMA caused a remarkable increase in photon emission in both groups that was higher in the patients than in the controls. When the leukocytes were activated with PMA in presence of (+)-catechin, the level of CL recorded was lower in both groups, but again significantly higher in patients than controls (Table 1).

Table 2 shows the increase of CL emission over the basal emission, that is the ratio between stimulated chemiluminescence and basal chemiluminescence activity (stimulation index, SI). The SI with PMA and without (+)-catechin was significantly lower in the group of patients than in the healthy controls. The SI with PMA, in presence of (+)-catechin, did not significantly differ from that in absence of (+)-catechin with both groups.

Table 1 Effects of (+)-catechin on whole blood CL (counts/neutrophil/90 minutes x 10^{-3}; mean ± SD)

	Basic CL	Basic CL	PMA-stimulated CL	PMA-stimulated CL
	without catechin	with catechin	without catechin	with catechin
Controls	0.32 ± 0.05	0.20 ± 0.07^	9.29 ± 0.96	3.36 ± 1.16^
Patients	1.66 ± 0.88*	0.86 ± 0.37 *^	21.06 ± 7.95*	11.14 ± 7.31*^

* p (vs controls) < 0.05; ^ p (with catechin vs. without catechin) < 0.05

Table 2 Leukocyte stimulation index with PMA (mean ± SD)

	without catechin	with catechin
Controls	29.5 ± 2.3	17.1 ± 4.1
Patients	14.3 ± 6.2*	13.1 ± 4.6

* p (vs controls) < 0.05

4. DISCUSSION

There is a great deal of evidence that leukocytes play an important role in arterial disease because they are implicated in atherogenesis, thrombosis and tissue damage from ischaemia. Our findings show that leukocytes from patients with vascular disease emit more chemiluminescence in basal conditions than healthy controls. Such a finding may be taken as a sign of leukocyte activation in peripheral blood according to other Author observations.[10,11]

PMA produces PKC activation, which follows NADPH-oxidase activation in the leukocyte membrane and therefore respiratory burst. Added PMA produced a greater increase of photon emission expressed as absolute values in our patients than in controls, but lower when expressed as stimulation index, clearly showing a functional failure of phagocytic cells of patients with vascular disease. It is interesting to note that (+)-catechin exerts the same effect on resting and stimulated cells of controls and patients, leaving unchanged the values of the stimulation indexes.

In conclusion, catechins are flavonoids that can modulate the leukocyte function and could play a role in the treatment of occlusive ischaemia of the limbs.

References

1. Dormandy JA, Stock G, Critical leg ischaemia. Its pathophysiology and management. New York: Springer-Verlag, 1990.
2. Lefer AM, Endothelial dysfunction and neutrophil adherence as critical events in the development of reperfusion injury. *Agents Actions* 1993; Suppl. **41**: 127-135.
3. Rice-Evans CA, Packer L, Flavonoids in health and disease. New York: Marcel Dekker, 1998.
4. Rice-Evans C, Miller, Paganca G, Structure-antioxidant activity relationships of flavonoids and phenolic acids. *Free Radic Biol Med* 1996; **20**: 933-956.
5. Hollman PC, Katan MB, Absorption, metabolism and health effects of dietary flavonoids in man. *Biomed Pharmacother* 1997; **51**: 305-310.
6. Limasset B, Le Doucen C, Dore JC, Ojasoo T, Damon M, De Paulet AC, Effects of flavonoids on the release of reactive oxygen species by stimulated human neutrophils. *Biochem Pharm* 1993; **66**: 1257-1271.
7. Rice-Evans C, Plant polyphenols: free radical scavengers or chain breaking antioxidants? *Biochem Soc Symp* 1995; **61**: 103-116.
8. Inanami O, Oral administration of (-)catechin protects against ischemia-reperfusion-induced neuronal death in the gerbil. *Free Radic Res* 1998; **29**: 359-365.
9. De Sole P, Lippa S, Littaru, GP. Whole blood cheluminescence: a new tecnical approach to assess oxygen-dependent microbial activity of granulocytes. *J Clin Lab Automat* 1983; **3**: 391-400.
10. Nash GB, Thomas PRS, Dormandy JR, Abnormal flow properties of white cells in patients with severe ischaemia of the leg. *Br Med J* 1988; **296**: 1699-1701.
11. Ciuffetti G, Lombardini R, Pasqualini L, Vaudo G, Leukocyte-endothelial interactions in chronic critical limb ischaemia. *Ann Ital Med Int* 1996; **11**: 170-174.

THE FUNCTION OF NEUTROPHIL GRANULOCYTES AFTER RADIOTHERAPY

P. HENGSTER, G. BRANDACHER and S. SCHNEEBERGER
Innsbruck University Hospital, Department of Surgery, Daniel Swarovski Research Laboratory,
E-mail: Paul.Hengster@uibk.ac.at

A. DE VRIES
Innsbruck University Hospital, Department of Radiotherapy, Anichstrasse 35, 6020 Innsbruck, Austria

1. INTRODUCTION

Neoadjuvant radiochemotherapy is the treatment modality for stage III and IV rectal cancer. The risk of local infection due to this treatment and the disturbance of granulocyte function followed by surgery is under debate.[1] Generally two different attempts are in consideration. One is to perform surgery immediately after a radiochemotherapy, and another is to wait approximately two weeks in order to have a restored immune system.

Granulocytes bear two main functions, first, migration into the place of inflammation and second, the release of reactive oxygen species. Migration can be detected by the progression of cells through a membrane filter against the chemo-attractant formyl-methionyl-leucine-phenylalanine. Furthermore, granulocytes release oxygen-free radicals during phagocytosis. The release of oxygen-free radicals can be quantitatively detected by chemiluminescence (CL). The function of these cells has been considered as a prognostic factor in breast cancer patients.[2] Defective migration and inhibition by tumor cells have been reported.[3-5]

2. MATERIALS AND METHODS

Fresh blood from volunteer donors was used as leucocyte source after separation of plasma and red blood cells. Buffy coat (35 mL) was layered on 15 mL Lymphoprep[R] (Nycomed Pharma, Oslo, Norway) in a conical 50-mL tube (Falcon, Becton Dickinson Labware, NJ, USA) and centrifuged at 1000 g for 20 min. The band containing mononuclear cells was removed, washed with phosphate-buffered saline

(PBS) and centrifuged once for 10 min at 199 g and twice for 10 min at 155 g to remove platelets. Cell count was adjusted to $1x10^6$/mL in RPMI without FCS and, when necessary, diluted according to the protocol. A total of 50 μL of cell suspension was then seeded into each well.

Granulocytes were incubated with 50 L of the colon carcinoma cell line HRT 18 adjusted to 5 x 10^4 cells/mL in RPMI for 6, 4, 2, 1, 0.5, 0 hours at 37°C in a humidified incubator. Measurement was performed on the Luminometer Lucy III (Anthos, Salzburg, Austria) at a constant temperature of 24°C in a white 96-well plates (Isoplate 1450-581, Wallac, Turku, Finland) were measured at a constant temperature of 24°C every 5 minutes for a one-hour period. After a background measurement 50 μL of luminol (Sigma, Deisenhofen, Germany) 1:1000 dilution of 1.77 mg dissolved in 1 mL dimethylsulfoxide (final concentration 0,0445mM) was added and measurement performed over an one hour period. To determine the remaining capacity to release reactive oxygen species the cells were finally stimulated by PMA (final concentration of 0,00081mM) and measurements taken again.

In parallel cells were incubated with tumor cells for or 6, 3, 1.5, 0.5 and 0 hours and layered on 0.3μm porous membrane filters for migration (Leukotec, Klagenfurt, Austria) with 3 sites each containing formyl-methionine-leucine-phenylalanine or controls without chemoattractant. Microscopy was performed on an automated microscope (Olympus, Tokyo, Japan) and a specially developed software (Immotech, Sauerlach, Germany) which recognizes the three dimensional structure of cells and measures the depth of infiltration into the membrane.

The Excel^R program was used for analysis and graphics. All values were obtained in duplicate. Tests were performed at least three times; mean values of consecutive measurements are listed in all results.

3. RESULTS

Radiation of granulocytes up to 32 Gy did only slightly diminish their phagocytic function. The addition of colon carcinoma caused activation of these cells, which lasts more than 6 hours. Granulocytes were not stimulated exhaustively and kept an ability to be further activated by the addition of PMA which chemically stimulates these cells irreversibly to release their reactive oxygen species.

The colon carcinoma cell line HRT 18 activates granulocytes to release oxygen free radicals. This effect occurs rapidly within few minutes and lasts during the time of measurement of 6 hours. This activation, however, is not exhaustive, because the addition of PMA after the 6 hour period is still able to further activate the cells and causes radical release.

Radiation of granulocytes up to 32 Gy did diminish their phagocyte function in a dose depended manner with a maximal decrease of 35% after radiation with 32 Gy. The addition of colon carcinoma caused an immediate activation of these cells regardless of their radiation dose and this effect was even detected after 6 hours.

For migration counts of migrating cells was reduced by radiation and tumour cells. The depth of invasion was found increased with cells radiated with high dosage (32Gy) and incubation time with tumour cells up to 1.5 hours.

The results of all parameters tested in cancer patients did not show any difference between pre treatment, during treatment and after treatment samples.

4. DISCUSSION

In order to find out whether radio-chemotherapy may alter the function of tumor adjacent phagocytes and to find a possible influence on the local control of infection we chose this in vitro model. Granulocyte function in peripheral blood was shown to be impaired in patients with solid tumors.[3,6] Furthermore radiation of these cells decreases free radical release.[7,8] These data may be conclusive, in so far, that phagocytic neutrophilic granulocytes are replaced every three days in vivo.

Although radiation causes a 35% decrease in the luminescence signalafter radiation with 32 Gy no change has been demonstrated with 2 Gy and only little effect with 4 Gy. If one assumes a life time of granulocytes in the tissue of two days a maximal radiation dose of 4 Gy can be expected for a cell population until they are replaced by new cells. Nevertheless, the ability to be activated by tumour cells remains in all test situations, without reduction during incubation time. All the granulocytes were not stimulated exhaustively and kept an ability to be further activated by PMA which chemically stimulates these cells irreversibly to release their reactive oxygen species.

Migration of all cells is reduced after radiation or tumour cell treatment. There are some cells which are radiated with a very high dose and incubated with tumour cells less than 1.5 hours which are able to migrate deeper into the filter than untreated cells.

The investigations of granulocytes in the blood stream did not show any differences during the course of treatment.

It could be demonstrated that neither systemically (in vivo) nor locally (in vitro) radiation or tumour cell contact causes a severe disturbance of granulocyte function. From these data it must be assumed that local radiotherapy does not change the function of circulating granulocytes. Even with a partial reduction in the granulocyte function after radiation these cells can be stimulated and are capable to phagocyte. Cells in tissue are replaced within a few days and their life time does not allow a high dose radiation as used in our experiment. This allows to chose the time point for surgery regardless of granulocyte function.

References

1. Kaban AP, Lisetskii VA, Novokhatskii KI. Changes in the cytochemical indices and in the phagocytic activity of the leukocytes during the combined therapy of rectal cancer. *Lik Sprava* 1992;**3**:41-44.
2. Lukac J, Lechpammer S, Kusic Z, et. al. Prognostic significance of phagocytic functions in breast cancer patients. *Eur J Cancer* 1994; **30A:** 2185-2186.
3. Frumento G, Bonvini E, Minervini F, et. al. Defective neutrophil mobilization to skin chambers in cancer patients. *J Cancer Res Clin Oncol* 1984; **107:**53-56.
4. Hengster P, Eberl T, Hechenleitner P, et. al. Influence of tumour cells on luminol enhanced granulocyte chemiluminescence. In: Roda A, Pazzagli M, Kricka LJ, Stanley PE, ed. *Bioluminescence and Chemiluminescence; Perspectives for the 21st Century.* Chichester;Wiley, 1999: 267-269.
5. Sulowicz W. Random migration of leukocytes and phagocytic activity of neutrophils in patients with cancer of gastrointestinal tract at different clinical stages. *Rev Esp Oncol* 1983; **30:**317-321.
6. Jablonska E, Pietruska Z, Furman et al. Polymorphonuclear leukocytes (PMN) activity in peripheral blood in patients with esophageal carcinoma. *Arch Immunol Ther Exp Warsz* 1994; **42:**393-397.
7. Lejeune M, Sariban E, Cantinieaux B, et. al. Granulocyte functions in children with cancer are differentially sensitive to the toxic effect of chemotherapy. Pediatr Res 1996; **39:**835-842.
8. Hancock BW, Bruce L, Heath J, et. al. The effects of radiotherapy on immunity in patients with localized carcinoma of the cervix uteri. *Cancer* 1979; **43:**118-123.

IMPAIRED GRANULOCYTE FUNCTION PRECEDES INFECTION IN IMMUNO-COMPROMISED ALLOGRAFT RECIPIENTS

PAUL HENGSTER, WALTER MARK AND RAIMUND MARGREITER
Department for Transplantsurgery, Innsbruck University Hospital, Anichstrasse 35, A-6020 Innsbruck, Austria
E-mail: Paul.Hengster@uibk.ac.at

GERD EGGER
Department for Experimental Pathology, Graz University Hospital, Graz, Austria

1. INTRODUCTION

Immunosuppressive agents exert their effects on the specific immune system, but there are influences also on unspecific mechanisms. Therefore, infections are frequent complications after organ transplantations. This study investigated various parameters to characterize the function of polymorphonuclear neutrophilic granulocytes (PMN) and to show the predictive value.

2. MATERIAL AND METHODS

Thirty one patients (mean 48.9 (25-70) years) were enrolled into this study. Twenty two received a liver, 9 kidney pancreas simultaneously, 5 kidneys, 3 hearts, 1 liver-kidney and 1 a lung. All had a perioperative antibiotic therapy for 3 days. Immunosuppression was carried out with a triple drug regime with either cyclosporine or tacrolimus and steroids and either azathioprine or mycophenolat mofetil. Rejections were treated either with steroid pulses or anti-thymocyte globulins. Patients with infections were divided into three groups. Group1, with local infections, group 2, with systemic infection, and group 3, with sepsis.

White blood cell count was performed on a Coulter Counter^{R} T-540. Elastase, soluble ICAM-1and soluble VCAM-1 were measured by ELISA. Reactive oxygen species (ROS) production by luminol enhanced chemiluminescence. Molondialdehyde as a marker for membrane damage was measured by a colorimetric assay. The ability to migrate was determined with a newly developed membrane filter assay. In brief, cellulose nitrate membrane filters, 3μm pore size, 140 μm thick and the chemoattractant fMLP were used to investigate number of cells, depth of cells entering the filter with an automated microscope which measure the three dimensional appearance of cells in 8 m layers.

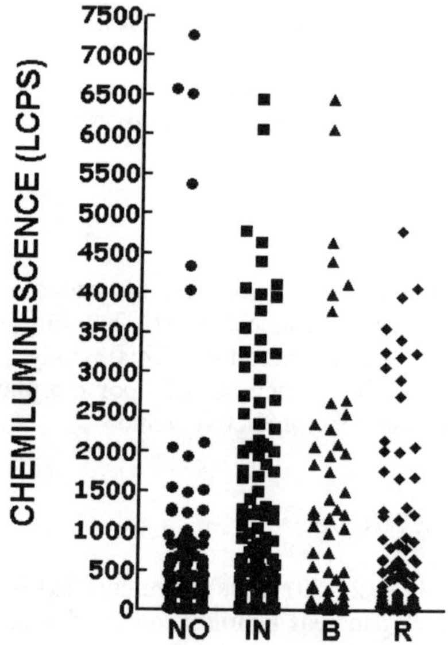

Fig. 1 Chemiluminescence - NO —No infection; IN —Infection; B —Infection and Rejection; R Rejection

3. RESULTS

The white blood cell count showed no difference between all groups. Reactive oxygen species (Figure 1) as well as s-ICAM-1, s-VCAM-1, neopterine and malondialdehyd were elevated with the grade of severity in some infections. Nevertheless, these changes were reactive and could not clearly discriminate healthy persons from patients with infections.

A predictive value was found for the first elastase determination. If this early after operation examined value was found to be normal there was a lower risk for infectious complications as compared to high postoperative values. The total migratory index dropped below the normal value in group 1 in 13%, in group 2 in 45% and in group 3 in 85.7%. The decrease in value preceded infection 5-11 days (Figure 2).

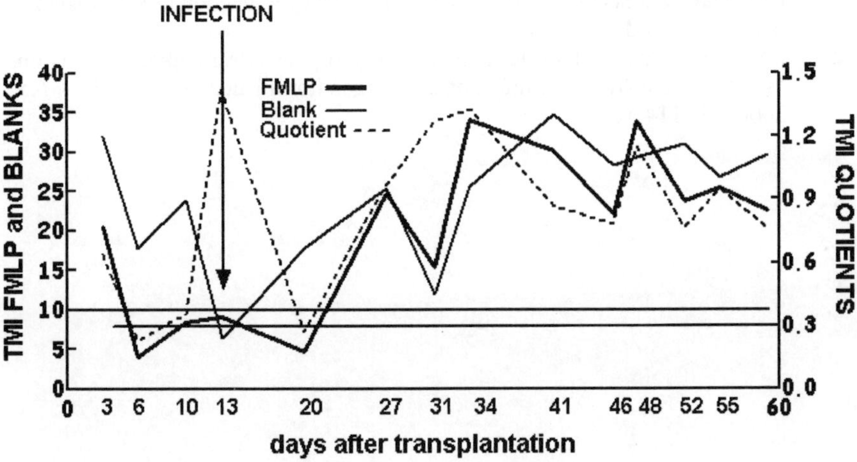

Fig. 2 Migration of PMN: Low TMI precedes infection

4. DISCUSSION

Immunosuppressive therapy is intensive after the first three month after transplantation. The appearance of symptoms of the rejection process, as well as, the infection are often very similar, the treatment, however, is totally different. During a rejection episode immunosuppression will be even more intensified One would like to discriminate and even more to establish a measurable risk for infection either to indent to treat or to reduce immunosuppression. Most parameters tested were reactive after severe infection. Elastase after surgery was partially predictive. It might be an indicator for traumatizing surgery which is associated with infections.

Migration of granulocytes can be influenced rapidly. The loss of migratory capacity is found prior to severe infection with a specificity and sensitivity of about 80%.

References

1. Egger G, Burda A, Hengster P, et. al. Definition of the sepsis risk in transplant patients by polymorphonuclear leucocyte function test. *Virchows Archiv* 1999; **435:** 321-312.
2. Egger G, Burda A, Hengster P, et. al. Polymorphonuclear leukocyte functions as predictive markers for infection after organ transplantation. *Transplant Intl* 2000; **13:** 114-121.

INHIBITORY EFFECT OF TERBINAFINE ON REACTIVE OXYGEN SPECIES (ROS) GENERATION BY *CANDIDA ALBICANS*

U.-CH. HIPLER, CH. SCHROETER, B. HIPLER[1], U. WOLLINA
Department of Dermatology and Allergology, Friedrich Schiller University Jena, D-07743 Jena Germany
[1]BMG Lab Technologies GmbH, D-77656 Offenburg, Germany

1. INTRODUCTION

Candida albicans is the most important opportunistic fungal pathogen. The yeast is able to cause candidiasis, a disease that can vary from superficial mucosal to life-threatening systemic disorders. Several virulence factors of *C.albicans* are known including the dimorphism, which is the ability to grow in either a yeast-like (blastoconidial) or a mycelial (hyphal) form in response to different environmental factors.[1] The argument that hyphae represent an advantage in tissue invasion seems obvious and widely accepted, but definitive experimental evidence is lacking.[2]

Reactive oxygen species (ROS) receive increasing attention as short acting compounds involved in inflammation and tissue damage. ROS can react with polyunsaturated fatty acids in cellular membranes, nucleotides, and sulfhydryl bonds in proteins[3], and have been related to the tissue injury in yeast infections.[4]

In *C. albicans* mitochondria are capable of generating extracellulary released ROS.[5] In our previous work we showed that ROS generation by *C. albicans* is dependent on morphogenesis and highest in hyphal form cells.[6] Thus, ROS generation may play a major role in tissue invasion and infection.

Terbinafine, a synthetic allylamine antifungal agent, inhibits ergosterol biosynthesis at the level of squalene epoxidase. When exposed to terbinafine fungi accumulate squalene while becoming deficient in ergosterol, an essential component of fungal cell membranes. The inhibition of squalene epoxidase leads to fungistatic and fungicidal action.[7]

Furthermore, there is evidence that terbinafine at therapeutic concentrations, can be considered a free radical interceptor *in vitro* and could exert an anti-inflammatory activity *in vivo.*[8] The substance is capable of interacting with cell membranes and inhibiting the generation of reactive oxygen species.[9,10]

In this study, the effect of terbinafine on ROS generation by *C. albicans* has been analysed. We investigated whether terbinafine influences the oxidant properties of this pathogen.

2. MATERIAL AND METHODS

C. albicans strain 3153A was pre-cultured in yeast extract peptone glucose (YEPG) medium 21 h at 25°C to obtain cells in the yeast phase (blastoconidiae). Samples at the concentration of $1x10^6$ cells/mL to $1x10^9$ cells/mL were prepared in saline and

YEPG medium. *Candida* cells at a concentration of 1x10^8 cells/mL were incubated in saline with terbinafine at concentrations of 1 µg/mL, 10 µg/mL and 100 µg/mL dissolved in DMSO. Incubation was performed for 10 min and 60 min at 25°C in a shaking incubator. The duration was restricted to exclude effects of terbinafine on cell number and growth.

ROS measurements were carried out by means of lucigenin amplified chemiluminescence (Cl) as described before.[11] Lucigenin solution was prepared in PBS at a concentration of 0.1 mmol/L. 200 L of fungal cell suspension were measured in the tube luminometer LB 953 (Berthold, Germany). Chemiluminescence signals were recorded using the slow kinetic method for yeast cells at different cell numbers and for 1x10^8 cells/mL incubated with terbinafine. The calculation was performed using the average counts per minute (cpm) over measurement periods of 20 min. All data were expressed as means ± standard deviation. Each test was replicated. Data were analyzed using Student's t-test for paired data. $P < 0.05$ was considered statistically significant.

3. RESULTS AND DISCUSSION

Generation of reactive oxygen species was detectable in *C. albicans* blastoconidiae. The results were dependent on yeast cell number and media used for measurement. Chemiluminescence signals could be found at concentrations ≥ 1x10^6 cells/mL of *C. albicans* blastoconidiae. The maximum was measured at concentrations >1x10^8 cells/mL (Fig. 1). At higher concentrations ROS generation was inhibited in both media, which is very likely to be a protective mechanism.

The media used for measurements had strong effects on total amounts of reactive oxygen species . Measurements in saline revealed the highest values, which could be explained by the lack of nutrients in this solution and resulting stress within the cells. Furthermore, YEPG medium presumably possesses light quenching effects in chemiluminescence measurements.[6,12]

Due to these results, further experiments were carried out with blastoconidiae at a cell number of 1x10^8 cells/mL in saline.

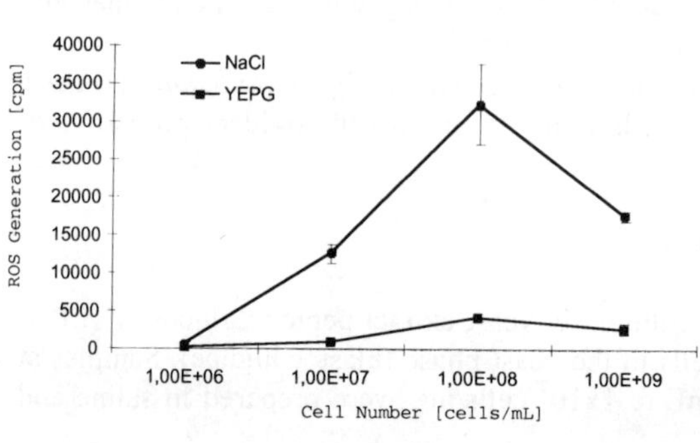

Fig. 1: Effect of increasing numbers of *Candida albicans* blastoconidiae on generation of ROS, measured by lucigenin amplified chemiluminescence in NaCl solution and YEPG medium.

After 10 min of incubation with terbinafine already a dose-dependent inhibition of ROS generation was measured. DMSO itself as the soluble reagent revealed antioxidant properties. However, the inhibition of ROS generation was considerably dependent on the used terbinafine concentration. At a terbinafine concentration of 1 μg/mL (T1) the inhibition corresponded to 9.4 % and at 100 μg/mL (T 100) to 24 % (Fig. 2).

The inhibition was significant for the comparison of the control incubated with DMSO and the highest chosen terbinafine concentration of 100 μg/mL (T 100) ($p < 0,05$).

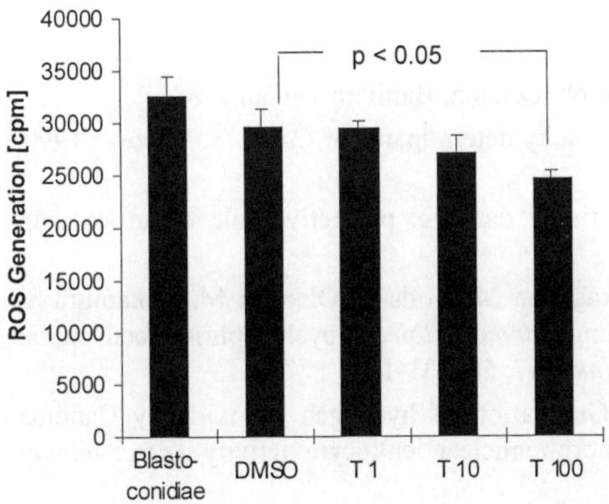

Fig. 2:
Dose-dependent inhibition of ROS generation by *Candida albicans* after 10 min incubation with different concentrations of terbinafine.
Blastoconidiae: cell number 1×10^8 cells/mL in saline.
All other samples contain yeast cells and additionally DMSO or terbinafine.
DMSO: + 1 % DMSO,
T1: + 1 μg/mL terbinafine,
T10: + 10 μg/mL terbinafine,
T100: + 100 μg/mL terbinafine.

After 60 min of incubation with terbinafine the inhibition was amplified. The final inhibitory effect was 18 % for 1 μg/mL terbinafine (T 1) and 41.1 % for 100 μg/mL terbinafine (T 100) (Fig. 3). The effect was significant for terbinafine concentrations of 10μg/mL (T10) and 100 μg/mL (T 100) ($p < 0.05$). The dose of 10 μg/mL (1 %) correspondes to the dose applied for topical preparations.[7]

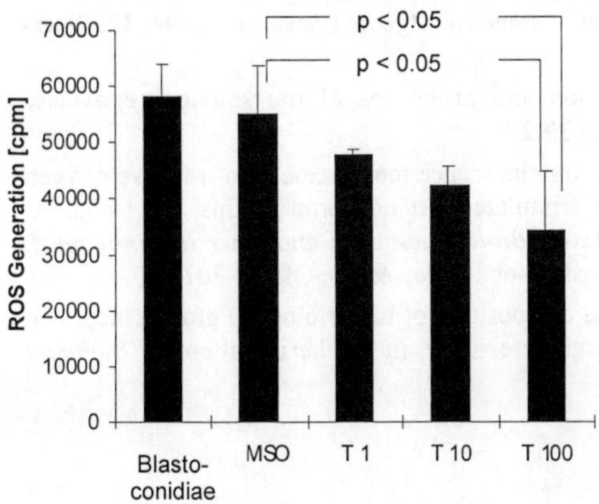

Fig. 3:
Dose-dependent inhibition of ROS generation by *Candida albicans* after 60 min incubation with different concentrations of terbinafine.
Blastoconidiae: cell number 1×10^8 cells/mL in saline.
All other samples contain yeast cells and additionally DMSO or terbinafine.
DMSO: + 1 % DMSO,
T1: + 1 μg/mL terbinafine,
T10: + 10 μg/mL terbinafine,
T100: + 100 μg/mL terbinafine.

These results reveal a time- and dose-dependent inhibition of ROS generation by *C. albicans*. Due to the time-depency it is likely that not only radical scavenging properties but also an interaction with the generation of reactive oxygen species is responsible for the terbinafine-induced inhibition.

In summary, terbinafine reduces the ability of *C. albicans* to generate reactive oxygen species. Besides the known influence on ergosterol biosynthesis, this mechanism may contribute to antifungal action and improvement of the inflammatory situation within the host. Further investigations have to prove whether terbinafine inhibits generation of reactive oxygen species as well in the hyphal phase, which is associated with markedly increased ROS generation.[6]

References

1. Odds FC. *Candida* and Candidosis. 2. Edition, London; Bailliere Tindall 1988.
2. Ghannoum MA, Abu-Elteen KH. Pathogenicity determinants of *Candida*. *Mycoses* 1990; **33**: 265-282.
3. Machlin LJ, Bendich A. Free radical tissue damage: protective role of antioxidant nutrients. *FASEB J* 1987; **1**: 441-445.
4. Nishikawa T, Tokunaga S, Fuse F, Takashima M, Noda T, Ohkawa M, Nakamura S, Namiki M. Experimental study of ascending *Candida albicans* pyelonephritis focusing on the hyphal form and oxidant injury. *Urol Int* 1997; **58**: 131-136.
5. Danley DL, Hilger AE, Winkel CA. Generation of hydrogen peroxide by Candida albicans and influence on murine polymorphonuclear leukocyte activity. *Infect Immun* 1983; **40**: 97-102.
6. Schroeter C, Hipler UC, Wilmer A, Kuenkel W, Wollina U. Generation of reactive oxygen species by *Candida albicans* in relation to morphogenesis. *Arch Dermatol Res* 2000; **292**: 260-264.
7. Ryder NS, Favre B. Antifungal activity and mechanism of action of terbinafine. *Rev Contemp Pharmacother* 1997; **8**: 275-386.
8. Camera E, Cannistraci C, Briganti S, Colombo D, Picardo M. Scavenging effects of terbinafine on free radicals in vitro. *Br J Dermatol* 1999: **140**: 640-644.
9. Vago T, Baldi G, Colombo D, Barbareschi M, Norbiato G, Dallegri F, Bevilacqua M. Effects of naftifine and terbinafine, two allylamine antifungal drugs, on selected functions of human polymorphonuclear leukocytes. *Antimicrob Agents Chemother* 1994; **38**: 2605-2611.
10. Ryder NS. Terbinafine: mode of action and properties of the squalene epoxidase inhibition. *Br J Dermatol* 1992;126 Suppl **39**:2-7.
11. Hipler UC, Wollina U, Mayser P. Chemiluminescence measurements of reactive oxygen species (ROS) generated by different stimulated Trichosporon strains. In: Roda, A, Pazzagli, M, Kricka, L J, Stanley, PE (eds). *Bioluminescence and Chemiluminescence. Perspectives for the 21st Century*. Chichester: John Wiley & Sons, 1999: 307-310.
12. Van Poucke SO, Nelis HJ. Effects of the composition of bacteriological growth media on a chemiluminometric assay of beta-galactosidase in Escherichia coli. *J Biolumin Chemilumin*

FLUORESCENCE ANALYSIS OF REACTIVE OXYGEN SPECIES (ROS) GENERATED BY SIX ISOLATES OF *ASPERGILLUS FUMIGATUS*

U-CH. HIPLER, U. WOLLINA, D. DENNING[1], B. HIPLER[2]
Department of Dermatology, Friedrich Schiller University Jena, D-07743 Jena Germany
[1]Department of Microbiology, University of Manchester , M8 8RB Manchester, United Kingdom
[2]BMG LabTechnologies GmbH, D-77656 Offenburg, Germany

1. INTRODUCTION

Reactive oxygen species (ROS) are essential intermediates in oxidative metabolism. Nonetheless, when generated in excess, ROS can damage cells by peroxidizing lipids and disrupting structural proteins, enzymes and nucleic acids.

Excess ROS are generated during a variety of cell stresses, including ischemia / reperfusion, exposure to ionizing and ultraviolet radiation and / or inflammation. Reactive oxygen species (ROS) may contribute to inflammation and tissue damage.

The generation processes of reactive oxygen species can be monitored using the luminescence analysis or also fluorescence methods. The intracellular ROS generation of cells can be investigated using the 2',7'-dichlorfluorescein-diacetate (DCFH-DA) as a well-established compound to detect and quantify intracellular produced H_2O_2.[1] The conversion of the nonfluoresecent 2',7'-dichlorfluorescein-diacetate (DCFH-DA) to the highly fluoresecent compound 2',7'-dichlorfluorescein (DCF) happens in several steps. First, DCFH-DA is transported across the cell membrane and deacetylated by esterases to form the non-fluorescent 2',7'-dichlorfluorescein (DCFH) . This compound is trapped inside of the cells. Next, DCFH is converted to DCF trough the action of peroxide (H_2O_2). The conversion to DCF is greatly accelerated by the presence of peroxidase.[2]

Aspergillus species are of further interest in the pathogenesis of several dermatological diseases. It is uncertain whether *Aspergillus* itself may generate ROS and therefore actively induce tissue damage. The present study investigates whether *Aspergillus* species are capable to produce ROS by themselves and if there are differences between the several strains.

2. MATERIALS AND METHODS

Six isolates of *Aspergillus fumigatus* (AF 65, AF 71, AF 72, AF 91, AF 210, AF 294) cultured 5 weeks on Sabouraud-Glucose-Agar (BAG, Lich, Germany) were investigated. After addition of isotonic NaCl solution and centrifugation (100 rpm / 10 min) the blastospore concentration could be estimated by counting in CASY 1 (Schärfesystem GmbH, Germany). These cell suspensions with concentrations of 10^5 to 10^7 cells /mL were measured in the fluorometer FLUOstar (BMG

Labtechnologies, Germany) using for each single test 100 μL of fungal cell suspension after incubation with 100 μL DCFH-DA (0.4 nM/ mL). To eliminate LBS induced effects polymyxin B (3 mg / mL) was added to all experimental suspensions. Each measurement was done at least sixteen times in duplicate for calculation of the mean and the standard error of the mean.

3. RESULTS AND DISCUSSION

The ability of various *Aspergillus* species to generate reactive oxygen species (ROS) was investigated. For all fungal cells a linear increasing fluorescence activity could be observed depending on the incubation time with DCFA-DA (Fig. 1). Because of the small sized *Aspergillus fumigatus* (2-3 μm) detectable fluorescence was observed only at concentrations > 10^5 cells /mL. The ROS generation showed a linear and direct proportional dependence on cell numbers (Fig. 2). The highest value could be found at concentrations of 10^7 cells /mL. The isolates of *A. fumigatus* AF 91 (Fig. 3) and AF 72 (Fig. 4) are restistent against itraconazole and AF 65 is amphotericin B resistent.[3,4] We investigated whether there are connections between the restistance and the ROS generation. Interisolate differences could be found (Fig. 5).

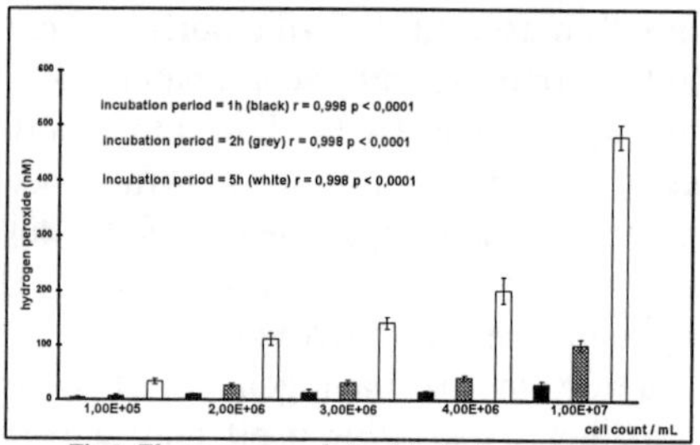

Fig.1 Fluorescence analysis of A . fumigatus AF 71

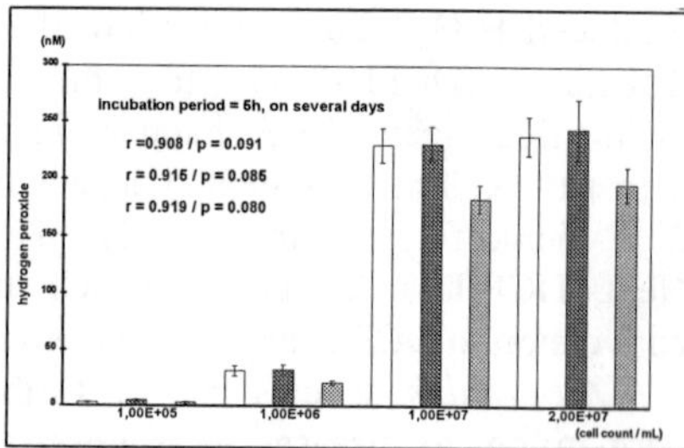

Fig.2 *Fluorescence analysis of AF 71 (Incubation period 2 h, 37°C), Correlation between cell counts and the amount of H_2O_2*

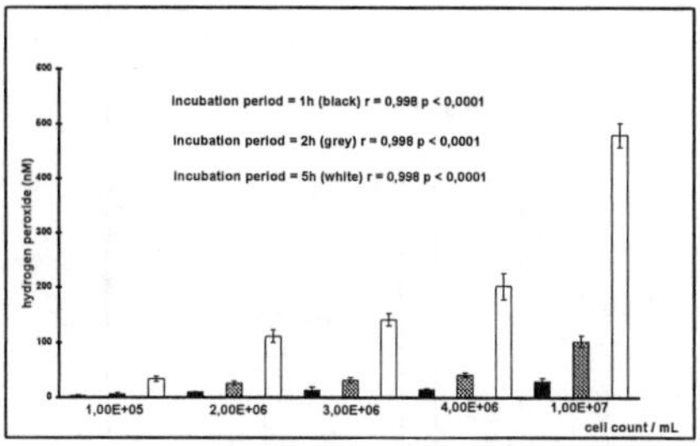

Fig.3 Fluorescence analysis of AF 91 (Incubation period 2 h, 37°C), Correlation between cell counts and the amount of H_2O_2

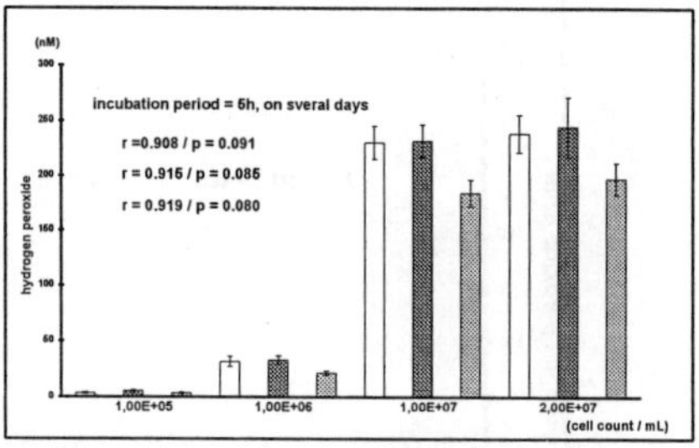

Fig.4 Fluorescence analysis of AF 72 (Incubation period 2 h, 37°C), Correlation between cell counts and the amount of H_2O_2

4. CONCLUSION

In conclusion, the morphological event of fungi, usually acknowledged as a major factor of virulence, is associated with increased intracellular ROS formation, which are most likely to be secreted. This process my be contribute to invasivness of *A. fumigatus* and inflammatory response of the host.

Clearly, the pathogenicity of *Aspergillus* species is a function of a multitude of parameters working together in a sequential and cooperative manner to establish infection. Together with phospholipases ROS are capable of destroying the host cell membranes.

It has been shown by several authors, that water as well as superoxide, hydrogen peroxide and OH-radicals can be generated in the course of mitochondrial electron transport process.[5] Fungi also possesses the normal and the alternative pathway of electron transport. The interesting connection between the monovalent oxygen reduction and the energy conservation in isolated chloroplastes was described. The hydrogen peroxide formation at the phosphorylation points I and II is disposed by a cytochrom c-oxidation process

In this study the method of ROS fluorescence measurements was utilized on different unstimulated *Aspergillus* species for the first time.There were found linear correlations between ROS- levels and blastospore concentrations. A pathophysiological meaning of the released oxygen metabolites in the complicated system of inflammatory reactions as additional factor of virulence is not to be excluded which was also estimated in *Saccharomyces cervevisae.*[6]

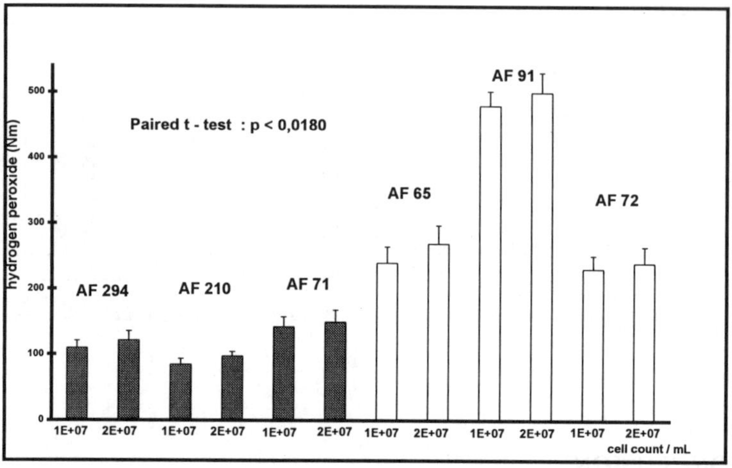

***Fig.5** Fluorescence analysis of several strains of Aspergillus fumigatus Comparison of AF 294, AF 210, AF 71, AF 65, AF 91, AF 72*

References

1. Cathcart R, Schwiers E, Ames BN. Detection of picomole levels of hydrogenperoxide using a fluorescent dichlorofluorescein assay. *Anal Biochem* 1983; **134**:111-116.
2. LeBel CP, Ischiropoulos H, Bondy SC. Evaluation of the probe 2',7'-dichlorofluorescein as an indicator of reactive oxygen species formation and oxidative stress. *Chem Res Toxicol* 1992; **5**: 227-231.
3. Denning DW, Radford SA, Oakley KL, Hall L, Johnson EM, Warnock DW.Correlation between in-vitro susceptibility testing to itraconazole an in vivo outcome of *Aspergillus fumigatus* infection. *J Antimicrob Chemother* 1997; **40**: 401-414.
4. Oakley KL, Moore CB, Denning DW. In vitro activity of the echiocandin antifungal agent LY 303,366 in comparison with itraconazole and amphotericin B against *Aspergillus* spp. *Antimicrob Agents Chemother* 1998; **42**: 2726-2730.
5, Ruotsalainen M, Hyvärinen A, Nevalainen A, Savolainen KM. Productions of reactive oxygen metabolites by opsonized fungi and bacteria isolated from indoor air and their interactions with soluble stimuli, f MLP or PMA. *Environ Res* 1995; **69**: 122-31.
6. Jakubowsky W, Bartosz G. Estimation of Oxidative stress in *Saccharomyces cerevisae* with fluorescent probes. *Int J Biochem Cell Biol* 1997; **29**: 1297-1301.

ANALYSIS OF NEUTROPHIL-DERIVED REACTIVE OXYGEN SPECIES USING A HIGHLY SENSITIVE CHEMILUMINESCENCE PROBE L012

I. IMADA, E. F. SATO, Y. KIRA, Y. MINAMIYAMA, R. KONAKA, Y. ICHIMORI,*
S. HONDA* AND M. INOUE

Osaka City Univ. Medical School, 1-4-3 Asahimachi, Abenoku, Osaka 545-8585, and
**Takeda Chem. Ind. Ltd., 2-17-85, Juso-honmachi, Yodogawaku, Osaka 532-8686, Japan*
E-mail: imada@med.osaka-cu.ac.jp

1. INTRODUCTION

Reactive oxygen species (ROS) produced by various biological systems play important roles in the regulation of physiological processes and in the pathogenesis of various diseases. ROS have been studied by various methods including chemiluminescence (CHL). Although CHL analysis is a potent and convenient method for the determination of ROS, the sensitivy of commercially available CHL probes in complex biological systems is fairly low. We previously reported that a highly sensitive CHL probe 8-amino-5-chloro-7-phenylpyrido[3,4-d]pyridazine-1,4-(2H,3H)dione (L012) reacted with various types of ROS generated by activated neutrophils in blood samples.[1] The present work describes analysis of ROS generation by neutrophils from human blood and the oral cavity using L012. Biochemical properties of L012 and its reactivity with nitric oxide (NO) are also studied.

2. ROS-DEPENDENT L012-CHL

Properties of L012-CHL: We have synthesized a luminol like cyclic hydrazide-derivative having 2,3-dihydro-1,4-pyridopyridazine dione in its structure. L012 showed the strongest CHL among the luminol-related compounds tested. The presence of superoxide dismutase (SOD) decreased the CHL intensity of L012, suggesting that L012 reacts with the superoxide radical and/or its metabolite(s).[2]

Assay for the generation of ROS by cell free systems: To clarify the chemical properties of ROS responsible for L012-dependent CHL, ROS were generated by cell-free systems. Xanthine oxidase-generated ROS increased L012-CHL by some mechanism which was inhibited by either SOD or deferoxamine (Fig. 1A). It is well known that H_2O_2 is converted to hypochlorite by neutrophils. As shown in Fig. 1B, L012 also developed CHL by reacting with hypochlorite. The presence of Fe^{2+} and H_2O_2 strongly increased the L012-dependent CHL by a mechanism which was inhibited by deferoxamine (Def), dimethyl sulfoxide (DMSO) or mannitol (Fig. 1C).

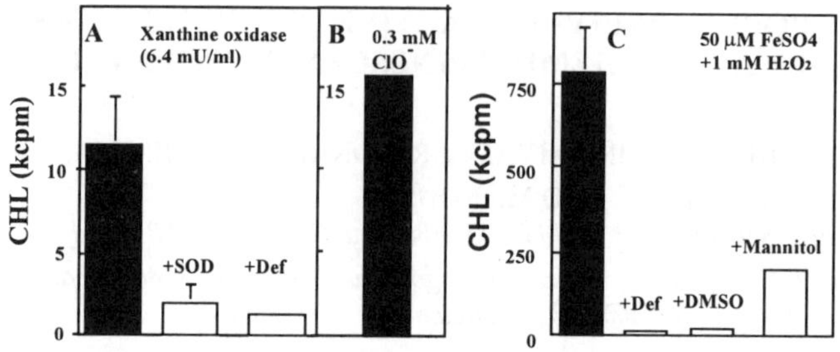

Fig. 1. Effects of various compounds on ROS dependent L012-CHL. ■ Control, □: +SOD (32 mU /ml), +Def (0.1mM deferoxamine), +DMSO (8 mM) or +Mannitol (8 mM).

Generation of ROS by human neutrophils: We obtained human blood samples and oral neutrophils from healthy volunteers. After incubation of cells at 37°C for 3 min in the presence of 400 μM L012, the reaction was started by adding various stimuli. When stimulated by opsonized zymosan (OZ), CHL intensity from L012 strongly increased; under identical conditions, the CHL intensity of luminol and MCLA was about 1% and 5% of L012, respectively (Fig. 2). Phorbol-12-myristate-13-acetate (PMA) also increased L012-CHL in human blood. Effects of various scavengers and inhibitors on OZ-stimulated CHL were also studied. L012-dependent CHL of blood neutrophils was decreased significantly either by SOD, catalase, uric acid, deferoxamine or azide (Fig. 3). Because neutrophils-dependent CHL was strongly inhibited by azide, deferoxamine and uric acid, hypochlorite and hydroxyl radical might be involved in L012-dependent CHL. Because azide and uric acid also scavenge singlet oxygen, the possible involvement of this species should be studied further.

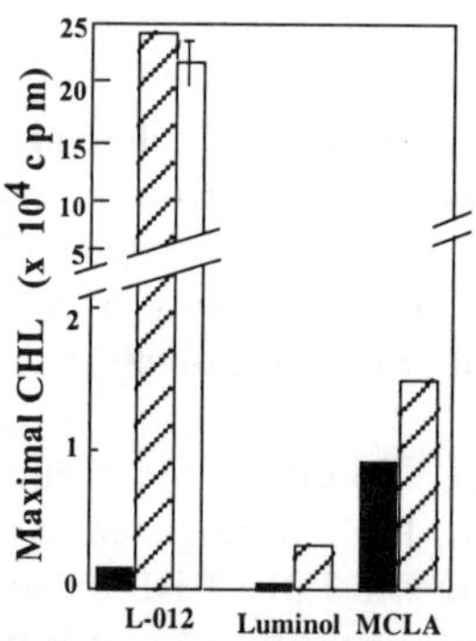

Fig. 2. CHL of three probes in ligand-stimulated blood samples. ■: Control, ▨:+Zymosan (2 μg/mL), □ : +3 nM PMA.

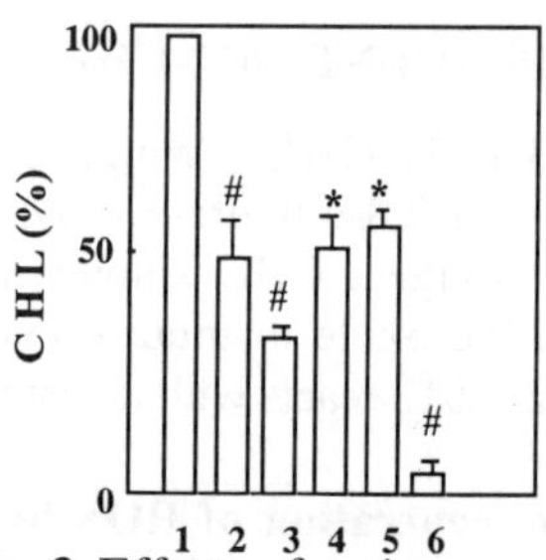

Fig. 3. Effects of various agents on OZ- enhanced L012-CHL in blood samples. 1. OZ-stimulated, 2.+150U SOD, 3.+22U catalase, 4. + 5 μM urate, 5. +0.1 mM Def, 6. +20 μM azide. *P<0.05, #P<0.01 compared with 1.

Because human neutrophils are strongly primed during and/or after infiltration to the oral cavity, they spontaneously generate ROS; CHL intensity from L012 was about 50 and 3 times greater than those of luminol and MCLA, respectively (Fig. 4A). When stimulated by PMA, ROS generation also increased. PMA enhanced L012-dependent CHL by about 12-fold. The enhanced CHL of L012 was about 100 and 10-times stronger than those of luminol and MCLA, respectively. As shown in Fig. 4B, L012-dependent CHL (1) of oral neutrophils was decreased significantly either by SOD (2), catalase (3), uric acid (4), deferoxamine (5) or azide (6).

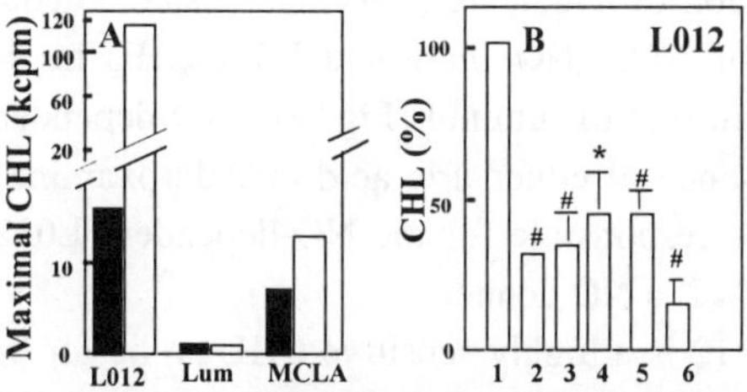

Fig. 4. (A), CHL of PMA-stimulated oral PMN. ■: Control (spontaneous), □: +PMA. (B) Effects of various agents on L012-CHL. Experimental conditions for 1 to 6 were as shown in Fig. 3. $*P<0.05$, $\#P<0.01$ compared with 1.

Age-dependent changes in ROS generation by neutrophils: We also studied the relationship between aging and ROS production by neutrophils. Rats were fed standard diet for two years. During this time, the number of PMN in blood remained unchanged. However, the maximum intensity of OZ-stimulated L012-CHL increased with age of the animals (Table 1). The maximum intensity of L012-CHL of PMA-stimulated oral PMN was higher with neutrophils from old subjects than from younger subjects (Table 2). In contrast, the time required for the maximal intensity of CHL was significantly delayed in older people as compared to younger subjects. The slower response of PMN from older subjects may reflect a decrease in resistance to pathogens. Because ROS generated by activated neutrophils differ from one ligand to another[3], the metabolic fate of ROS and their effects on microorganisms may change depending on the stimulants.

Table 1. Age-dependent changes in PMN numbers and their ROS generation

Age of male Wistar rats	PMN numbers ($x10^5$/mL)	CHL (cpm/cell)
12 weeks	40.51±7.97	0.015± 0.006
15 weeks	59.67± 20.01	0.436± 0.2.3
30 weeks	49.86± 14.23.	2.331± 0.847
100 weeks	47.35± 12.36	4.194± 2.89

Table 2. Age-dependent changes in ROS generation by oral PMN

Age of male subjects	CHL (cpm/cell)		Time for maximal CHL (sec)
	Spontaneous	+PMA	
22-29 (n=5)	25.4–17.2	111.8–58.6	327.2–68.1
37-68 (n=4)	37.25–16	357–153	602.5–64.5

3. NO-DEPENDENT L012-CHL IN THE PRESENCE OF H_2O_2

NO also plays a role in bacteriocidal action of leukocytes. We also tested whether L012-CHL is useful for the analysis of NO. NO increased L012-CHL in the presence of H_2O_2; CHL was stronger than that of luminol (Fig. 5). NO-dependent CHL was inhibited strongly by the presence of either uric acid or deferoxamine (Fig.6). Thus, hydroxyl radical might be responsible for the NO-dependent L012-CHL. This was confirmed by using NOC-12, a NO donnor.

The present work demonstrates that L012 is a highly sensitive CHL probe for the analysis of ROS generated in complex biological systems.

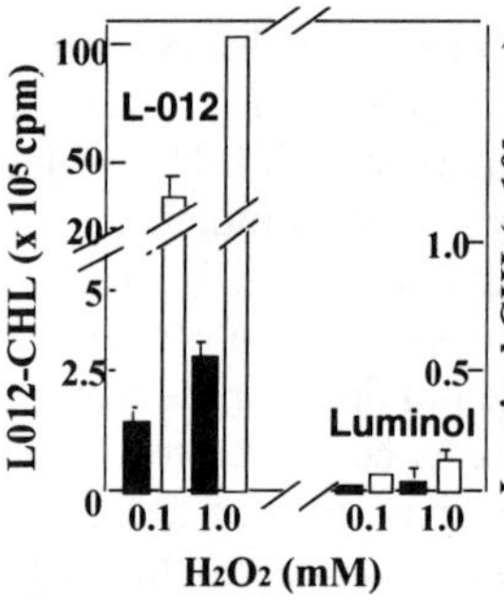

Fig. 5. Effects of NO and H_2O_2 on L012 and luminol-CHLs. ■: Control, □: + 20 μM NO

Fig. 6. Effects of uric acid and Def on L012-CHL elicited by NO and H_2O_2.

References

1. Imada I, Sato EF, Miyamoto M, Ichimori Y, Minamiyama Y, Konaka R, Inoue M. Analysis of ROS generated by neutrophils using a CHL probe L-012. *Anal Biochem* 1999; **271**: 53-58.
2. Nishinaka Y, Aramaki Y, Yoshida H, Masuya H, Sugawara T, Ichimori Y. A new sensitive CHL probe, L-012, for measuring the production of superoxide anion by cells. *Biochem Biophys Res Commun* 1993; **193**: 554-559.
3. Takahashi R, Edashige K, Sato EF, Inoue M, Matsuno T, Utsumi K. Luminol chemiluminescence and active oxygen generation by activated neutrophils. *Arch Biochem Biophys* 1991; **285**: 325-330.

NADPH OXIDASE ACTIVITY CHANGES BY ATHEROGENIC AND ATHEROPROTECTIVE LIPOPROTEINS

S. KOPPRASCH, J. GRAESSLER, H.E. SCHRÖDER
Department of Internal Medicine 3, Path. Biochemistry, C.G.Carus Medical School, Fetscherstr. 74, D-01307 Dresden, Germany
E-mail: Steffi.Kopprasch@mailbox.tu-dresden.de

J. PIETZSCH
Institute of Metabolic Research C.G.Carus Medical School, Fetscherstr. 74, D-01307 Dresden, Germany
E-mail:julius@rcs.tu-dresden.de

1. INTRODUCTION

One of the most important cellular sources of reactive oxygen species (ROS) in the organism is NADPH oxidase. Several types of cells in the circulation and in the vessel wall are equipped with this enzyme complex such as phagocytes, endothelial cells, and vascular smooth muscle cells. As such, they are able to promote oxidative stress which is a causal factor in the pathogenesis of atherosclerosis. Native and modified lipoproteins are thought to play an important role in atherogenesis either by enhancing cell-mediated inflammatory processes or by suppressing the progression of atherosclerosis.[1,2]

In the present study we tested the effect of native and oxidized low-density lipoproteins (LDL) on the NADPH oxidase activity of polymorphonuclear leukocytes (PMNL). High-density lipoprotein (HDL) has been shown to be inversely correlated with the risk of coronary artery disease,[3] but the mechanisms of its protective effect are not entirely understood. Therefore, we evaluated if native HDL (natHDL) is able to reverse the proinflammatory effects of oxidized LDL (oxLDL) on PMNL.

2. METHODS

Native, albumin-free LDL (density 1.006-1.063 g/mL) and HDL (1.063-1.210 g/mL) were isolated from human plasma by very fast ultracentrifugation as previously described.[4] Apolipoprotein AI (ApoAI), the main protein component of HDL, was isolated by delipidation of freshly prepared human HDL as described by Scanu and Edelstein.[5] Oxidation of native LDL (0.2 g/L protein) with sodium hypochlorite (1 mmol/L NaOCl final concentration) was carried out at 37°C for 40 min. This resulted in a molar lipoprotein to oxidant ratio of 1: 2900. Using these experimental oxidation conditions the NaOCl content declined exponentially and was no more detectable after 30 min. Native HDL suspensions (0.2 g/L protein) were treated with trypsin/EDTA (Sigma; 278 μg trypsin per mg protein) for 1 h at 37°C.

PMNL of healthy volunteers were isolated from heparinized blood by density gradient centrifugation with equal volumes of Histopaque-1077® and Histopaque-1119®. Cells were washed twice and in an additional step erythrocytes were lysed with ice-cold water.

NADPH oxidase activity changes of PMNL were measured by means of chemiluminescence (CL) with an Autolumat LB 953 (Berthold Co., Germany) using polypropylene tubes with a 1.0 mL reaction mixture. CL response of PMNL was obtained by incubating 10^6 cells in Krebs-Ringer phosphate buffer (pH=7.3) with luminol (10^{-4} mol/L final concentration) for 5 min at 37°C and subsequent addition of 0.04 g/L oxLDL, opsonized zymosan (0.5 mg/mL final concentration), phorbol myristate acetate (PMA, 5 x 10^{-8} mol/L final concentration, or formyl-methionyl-leucyl-phenylalanine (FMLP, 10^{-7} mol/L final concentration).

To investigate the influence of native or modified (oxidized or trypsinized) HDL or ApoAI on the oxLDL-induced ROS generating activity PMNL were preincubated with the corresponding lipoprotein or protein suspensions at a final protein concentration of 0.04 g/L for 5 min and after completion of the reaction mixture light emission was recorded.

PMNL CL activity was expressed as maximum counts/min per 10^6 cells.

3. RESULTS AND DISCUSSION

After oxidation with 1 mmol/L hypochlorite native LDL showed considerable modifications of their lipid and protein components as indicated by elevated diene and malondialdehyde formation and increased relative electrophoretic mobility.[6]

Oxidized LDL were able to substantially activate the respiratory burst of resting PMNL and to modulate the CL response of cells to various stimulants.

As shown in Table 1 oxLDL considerably amplified NADPH oxidase activity stimulated with FMLP. In contrast, the oxidized lipoprotein inhibited zymosan-induced CL response and showed no effect on PMA-induced chemiluminescence. Native LDL and native HDL did not activate resting PMNL. Native HDL, however, diminished the effects of oxLDL on PMNL respiratory burst activities.

Several mechanisms underlying the neutralizing effects of HDL have been proposed: (1) uptake of cellular cholesterol ester;[7] (2) incorporation of proinflammatory phospholipids (lysophosphatidylcholine, platelet-activating factor, PAF) from oxidized LDL;[8] (3) breakdown of oxidized phospholipids from oxLDL by enzymes closely associated with HDL (paraoxonase, PAF acetylhydrolase, lecithin cholesterol acyltransferase);[9] (4) enhancement of the cellular resistence against the toxicity of ox LDL by induction of the synthesis of yet unknown proteins in the cells.[10] In our study the suppressing effect of native HDL did not require prior incubation with PMNL or with oxLDL suggesting a fast-acting protective mechanism in the minute range.

Table 1 Effect of native and oxidized lipoproteins on the respiratory burst of resting and activated PMNL

Incubation conditions	CL response (max. counts/min x 10^6)
PMNL + natLDL	1.3 ± 0.2
PMNL + natHDL	1.3 ± 0.2
PMNL + oxLDL	24.1 ± 1.9
PMNL + oxLDL + natHDL	7.1 ± 1.3
	CL response (% control without lipoproteins)
PMNL + FMLP + oxLDL	1400 ± 236
PMNL + FMLP + oxLDL + natHDL	345 ± 55
PMNL + zymosan + oxLDL	56 ± 1
PMNL + zymosan + oxLDL + natHDL	63 ± 2
PMNL + PMA + oxLDL	101 ± 9
PMNL + PMA + oxLDL + natHDL	104 ± 7

To assess the contribution of the lipid and protein components of the HDL particle in preventing activation of PMNL we tested the effects of ApoAI and trypsinized HDL on our cell system. While native HDL inhibited the oxLDL-induced CL response by 64 ± 4 %, trypsinized HDL (57 ± 5 %) and ApoAI (43± 4 %) showed a slightly lower inhibitory capacity. From these results we conclude that not only the protein but also the lipid fraction of HDL, possibly by removing oxidized phospholipids, could play a significant role in the protection of HDL against oxidized LDL.

Recently, Cahilly et al. provided data supporting an association between a mutation in the $p22^{phox}$ subunit of NADPH oxidase and increased severity and progression of coronary artery disease.[11] The results of our study show that estimation of PMNL NADPH oxidase activity changes by luminol-amplified chemiluminescence is a sensitive tool to follow modulations of this enzyme complex by active biomolecules. Thus, polymorphonuclear leukocytes may serve as a cell model to investigate stimulus-induced changes in the generation of oxidative stress by cells equipped with this enzyme complex.

4. REFERENCES

1. Francis GA. High density lipoprotein oxidation: in vitro susceptibility and potential in vivo consequences. *Biochim Biophys Acta* 2000; **1483:** 217-235.
2. Steinberg D. Low density lipoprotein oxidation and its pathobiological significance. *J Biol Chem* 1997; **272:** 20963-20966.
3. Barter PJ, Rye KA. High density lipoprotein and coronary heart disease. *J Cardiovasc Risk* 1994; **3:** 217-221.
4. Pietzsch J, Subat S, Nitzsche S, Leonhardt W, Schentke KU, Hanefeld M. Very fast ultracentrifugation of serum lipoproteins: influence on lipoprotein separation and composition. *Biochim Biophys Acta* 1995; **1254:** 77-88.
5. Scanu AM, Edelstein C. Solubility in aqueous solutions of ethanol of the small molecular weight peptides of the serum very low density and high density lipoproteins: relevance to the recovery problem during delipidation of serum lipoproteins. *Anal Biochem* 1971; **44:** 576-588.
6. Kopprasch S, Leonhardt W , Pietzsch J, Kühne H. Hypochlorite-modified low-density lipoprotein stimulates human polymorphonuclear leukocytes for enhanced production of reactive oxygen metabolites, enzyme secretion, and adhesion to endothelial cells. *Atherosclerosis* 1998; **136:** 315-324.
7. Uittenbogaard A, Shaul PW, Yuhanna IS, Blair A, Smart EJ. High density lipoprotein prevents oxidized low density lipoprotein-induced inhibition of endothelial nitric-oxide synthase localization and activation in caveolae. *J Biol Chem* 2000; **275:** 11278-11283.
8. Matsuda YK, Hirata K, Inoue N, Suematsu M, Kawashima S, Akita H, Yokoyama M. High density lipoprotein reverses inhibitory effect of oxidized low density lipoprotein on endothelium-dependent arterial relaxation. *Circ Res* 1993; **72:** 1103-1109.
9. Watson AD, Berliner JA, Hama SY, La Du BN, Faull KF, Fogelman AM, Navab M. Protective effect of high density lipoprotein associated paraoxonase. Inhibition of the biological activity of minimally oxidized low density lipoprotein. *J Clin Invest* 1995; **96:** 2882-2891.
10. Suc I, Escargueil-Blanc I, Troly M, Salvayre R, Negre-Salvayre A.. HDL and ApoA prevent cell death of endothelial cells induced by oxidized LDL. *Arterioscler Thromb Vasc Biol* 1997; **17:** 2158-2166.
11. Cahilly C, Ballantyne CM, Lim DS, Gotto A, Marian AJ. A variant of $p22^{phox}$, involved in generation of reactive oxygen species in the vessel wall, is associated with progression of coronary atherosclerosis. *Circ Res* 2000; **86:** 391-395.

TRAINING EFFECTS ON CHEMILUMINESCENT RESPONSE OF ALVEOLAR MACROPHAGE IN MATURED RATS

T. KUMAE AND H. ARAKAWA

Department of Industrial Health, National Institute of Public Health, Minato-ku, Tokyo 108-8638, Japan

E-Mail: kumae@iph.go.jp

1. INTRODUCTION

Recent research trends indicate that moderate exercise may increase immune responsiveness, but high-level competition sport, especially if it involves extensive endurance training, may lead to a degree of immunosuppression.[1] Although jogging/running is a convenient and popular exercise, several epidemiological reports have suggested that athletes engaging in marathon-type events are at increased risk of upper respiratory tract infection.[2] The principal objective of this study is prevention or decreasing the risk of injury/disease in healthy aged persons engaging in endurance training for the purpose of increasing their health levels. Phagocytic cells, polymorphonuclear leucocytes (PMN) and macrophage, play an important role in a non-specific cellular immunity which is first defense line for infectious disease.[3] In this study, to evaluate effects of training started after maturation on non-specific immunity, we used the chemiluminescent technique to measure reactive oxygen species (ROS) generation from rat alveolar macrophage (AM) as an indicator of AM activity.

2. MATERIALS AND METHODS

2.1. Experimental animal and training program

Male Wistar JcI strain (SPF, five weeks old) rats were obtained from Nippon Clea Breeding Laboratories. The rats were maintained in SPF condition using an isolated room in barriered area of the breeding laboratory in National Institute of Public Health. We divided the rats into following two groups; one group of rats started training at 11 weeks old (Group A) and another group started at 17 weeks old (Group B). Each group was divided into 4 sub-groups: 1) forced training group

(Forced); exercised on a treadmill (35 m/min, 15 min/day, 5 days/week), 2) voluntary training group (Volun.); housed separately with running wheel, 3) control of voluntary training group (V-Cont.); housed separately without running wheel, and 4) control group (Control); housed under sedentary condition, same as usual breeding condition for rats. All animals were allowed rodent food (F-2; Funabashi Farms) and water ad libitum, and kept on a 12-h light-dark cycle. Experimental period was 12 weeks and about half of rats in the A and B groups were sacrificed after 6 weeks of training. After 6- or 12-week training, the rats were anesthetized with pentobarbital sodium (40 mg/kg, ip), and AMs were collected by the broncho-alveolar lavage method.[4]

2.2. Chemiluminescence measurement

The chemiluminescence was measured by a parallel luminometer (Lumibox H-1000, Microtec Niti-On Co.) using a cooled (-40°C), integrating charged coupled device (CCD) camera with an improved software.[5] AMs (150 μL, 9.0 x 10^4 cells) were placed in each well of a 96-well black microtiter plate (Greiner Japan Co.), which prevented scattering chemiluminescence from other wells.[5] Zymosan (Sigma) was opsonized using a pooled rat serum and suspended in physiological saline (10 mg/mL) as described elsewhere.[4] Lucigenin (120 μL, 1 mmol/L) or luminol (120 μL, 1 mmol/L) was mixed with the opsonized zymosan (60 μL) in each well of a 96-well V bottom microtiter plate (BioTec Co.). The microtiter plates, black and V bottom, were pre-incubated for 3 min at 37°C. The mixture (150 μL) of enhancers and opsonized zymosan was added into corresponding wells of the black microtiter plate using the automatic dispenser in the Lumibox H-1000. Luminol- and lucigenin-dependent chemiluminescence (LmCL and LgCL, respectively) were then automatically recorded for 60 min at 37°C with shaking. The chemiluminescent response was indicated by peak height (PH; photon/sec), which corresponds to the maximum releasing activity of ROS for AMs, and peak time (PT; min), which corresponds to the mean time from the recognition of foreign matters to the maximum release of ROS.[4]

3. RESULTS AND DISCUSSION

The Forced and Volun. groups were significantly smaller than the Control group in body weight after 6- and 12-week training. However, no significant differences of cell numbers in diameter 10.0 to 16.0 μm, suggesting AMs,[4] in BALF were observed among all groups. Furthermore, after 6-week training in the Forced group in the Groups A, the AM and monocytes populations investigated by the micrographs were increased and decreased significantly, respectively. These results suggest that tightly controlled host defence systems exist in the lungs, and the forced exercise affect the cell populations in younger animals.

Table 1. Changes of chemiluminescent response

	Group A (11 weeks old)				Group B (17 weeks old)			
	Forced	Volun.	V-Cont.	Control	Forced	Volun.	V-Cont.	Control
After 6-weeks	(13)	(20)	(11)	(20)	(8)	(10)	(8)	(15)
LgCL; PT	14.4	13.1	14.1	14.3	12.0*	11.8#	11.4#	14.1
(min)	±0.5	±0.5	±0.6	±0.8	±0.4	±0.5	±0.5	±0.7
LgCL; PH	801	834	882	892	770*	762*	787*	970
(photon/sec)	±48	±27	±49	±25	±60	±46	±109	±40
After 12-weeks	(10)	(10)	(8)	(15)	(9)	(10)	(8)	(8)
LgCL; PT	14.2	14.4	13.4	14.1	19.5	20.7	18.5	18.2
(min)	±0.9	±1.3	±0.6	±0.7	±2.5	±1.8	±1.4	±2.1
LgCL; PH	874	949	957	970	1127	1078	1100	1014
(photon/sec)	±38	±44	±62	±40	±72	±39	±85	±32

Numbers in parentheses represent sacrificed rat numbers.
Numbers represent mean and + standard error.
Statistical significances between the Control and other groups are represented as follows: *; $p< 0.05$, #; $p< 0.01$

Changes of chemiluminescent response of AM are summarized in Table 1. The mean levels of PT of LgCL and LmCL were almost same levels, however, those of PH of LgCL were 6 to 12 times higher than those of LmCL. We reported that mean levels of PH of LmCL were almost 20 times higher than those of LgCL when PMNs were used.[5] These results may agree with previous studies that have suggested that LgCL is specific to superoxide and LmCL is mainly dependent on myeloperoxidase (MPO) activity, and AM has lower MPO activity compared to PMN. We thought that the lower MPO activity caused no practical difference on the LmCL. In Group A, no significant difference in the chemiluminescent response of AM among the sub-groups was observed after 6-week training. To the contrary, in Group B, significant changes in the mean levels of PT and PH of LgCL between the Control and other sub-groups were observed after 6 weeks (Table 1). The mean levels of PT and PH of the Forced, Volun., and V-Cont. groups were increased and suppressed significantly, respectively. The same trend in the mean levels of PH of LgCL was observed in Group A but it was not significant. The chemiluminescent responses suggest that AMs in those sub-groups are primed, and the maximum releasing activities of ROS are reduced. These changes in AM activity may be caused by the stressors, that is the forced training and separation. After 12-week training, significant delay ($p < 0.05$) in the mean levels of PT of LmCL between the Control and the Volun. groups was observed in Groups A. Except for this significant difference, no practical difference among the sub-groups in Groups A and B was observed after 12 weeks. These results suggest that the rats may adapt to the stressors after 12 weeks. The chemiluminescent technique is thought to be useful to evaluate the changes of AM activity.

Acknowledgments

This work was supported in part by a research grant from the Ministry of Education, Science, Sports and Culture of Japan (No. 10470101).

References

1. Sharp NCC, Koutedakis Y. Sport and the overtraining syndrome: Immunological aspects. *Br Med Bull* 1992; **48**: 518-533.
2. Nieman DC. Exercise, infection, and immunity. *Int J Sports Med* 1994; **15**: S131-S141.
3. Pyne DB. Regulation of neutrophil function during exercise. *Sports Med* 1994; **17**: 245-258.
4. Kumae T, Arakawa H, Uchiyama I. The combined effects of stressful exercise and sleep disturbance on rat. Part 1. Combined effects started from childhood on cells in bronchoalveolar lavage fluids. *Jpn J Phys Fitness Sports Med* 1996; **45**: 301-310.
5. Kumae T. A study for reactive oxygen species production from neutrophils in the mouth. In: Roda A, Pazzagli M, Kricka LJ, Stanley PE, eds. *Bioluminescence and Chemiluminescence. Perspectives for the 21st Century*, Chichester: Wiley, 1999: 273-276.

INHIBITION ANALYSIS OF CHICKEN HETEROPHIL LUMINOL AND LUCIGENIN LUMINESCENCE

GERALD A. MERRILL AND ROBERT C. ALLEN
Brooke Army Medical Center, Ft. Sam Houston, TX
and Emory University, Atlanta, GA

1. INTRODUCTION

Chicken polymorphonuclear(heterophil)leukocytes (PMNLs) have NADPH oxidase but lack myeloperoxidase (MPO) activity.[1] Chemiluminescence (CL) is an energy product of the oxygenation activity following chicken PMNL activation by chemical or phagocytic stimuli[1-4]. The sensitivity for detecting these oxygenation activities is increased several magnitudes by introduction of high quantum yield chemiluminigenic substrates[5] (CLSs). CLSs, such as luminol (5-amino-2,3-dihydro-1,4-phthalazinedione) and lucigenin (dimethyl biacridinium (DBA^{++}) dinitrate), have different reactivities and measure different pathways of PMNL metabolism[5]. Both CLSs have been used to measure respiratory burst activities of both human and chicken PMNLs.[1-8]

DBA^{++} CL measures functional superoxide-producing oxidase activity of stimulated human and chicken PMNLs.[1,5,8] Luminol is a very efficient CLS for measuring peroxidase activity, although luminol dioxygenation and luminescence can also occur by peroxidase-independent pathways.[5,8] Stimulated chicken PMNLs produce a luminol-dependent CL that is small relative to human PMNLs.[1] Although chicken PMNLs are devoid of myeloperoxidase,[1,9,10] chicken eosinophils contain an eosinophil peroxidase that is capable of producing luminol CL. Luminol CL was previously found not well correlated with the presence of eosinophils.[1]

2. METHOD

An inhibitor-based approach to characterizing the oxidase and peroxidase activities of chicken PMNLs was investigated. PMNLs were obtained from Peterson roosters by isopycnic centrifugation following dextran sedimentation. Inhibition assays were performed using $5x10^4$ rooster PMNL in veronal (barbital) buffer containing specified CLS at one of two concentrations (60 and 120 μM DBA, 2.5 and 10 μM luminol) and inhibitor at one of eight concentrations. CL was determined in a single photon counter operated in a cyclic mode (approximate 5 min intervals) for 120 min following stimulation with either phorbol 12-myristate 13-acetate (PMA) or chicken opsonified zymosan (cOZ). Where the inhibition data was consistent with Dixon inhibition analysis, the inhibitor constant (K_i) was graphically determined,[11] otherwise, the concentration of inhibitor producing 50% inhibition (I_{50}) was estimated.

3. RESULTS

Azide and cyanide selectively inhibit luminol-dependent CL in peroxidase positive human PMNLs.[12] Neither sodium azide or cyanide had an inhibitory effect on either the luminol or DBA^{++} CL responses of stimulated rooster PMNLs. Eosinophil peroxidase is susceptible to both azide and cyanide inhibition, thus, failure of azide or cyanide to inhibit the luminol CL of stimulated rooster PMNLs is consistent with lack of both MPO and eosinophilic peroxidase activities. The DBA^{++} CL of rooster PMNLs is not inhibited by either azide or cyanide. These inhibitors are not active against NADPH oxidase and are not expected to inhibit superoxide formation and subsequent superoxide-dependent DBA^{++} CL.

A general metabolic inhibitor, sodium fluoride (NaF), was investigated. NaF (0.3-2.4 mM) inhibits both DBA^{++}-dependent and luminol-dependent rooster PMNL CL. The curves are non-linear and the data can not be interpreted with classical Dixon inhibition kinetics. NaF inhibition involves pre-oxidase and pre-MPO metabolism. As such, NaF is expected to inhibit both the DBA^{++} and luminol CL activities of rooster PMNLs by blocking the generation of reducing equivalents and depriving the oxidase of its substrate, NADPH, thus inhibiting the subsequent production of superoxide.

A scavenger enzyme, catalase, was also employed as an inhibitor. Catalase, which catalyzes the breakdown of H_2O_2 to water and O_2, had no inhibitory effect on the stimulated rooster PMNL CL with either CLS. Hydrogen peroxide is the secondary product of NADPH oxidase generated by disproportionation of superoxide. Under alkaline conditions, H_2O_2 can directly react with DBA^{++} to yield luminescence. However, the lack of catalase inhibition suggests that a H_2O_2 reaction pathway does not contribute significantly to either the DBA^{++} or luminol luminescence in rooster PMNLs.

4. CONCLUSION

Although not specific for superoxide, DBA^{++} is selective in measuring PMNL oxidase activity. Peroxidases, including MPO, do not catalyze DBA^{++} luminescence[8]. Univalently reduced DBA^{++} can react directly with superoxide yielding chemiluminescence.[5,8] Superoxide is a reductant and can not directly react with luminol to yield light. However, univalent oxidation of luminol yields a radical which can react directly with superoxide via radical-radical annihilation to yield CL. The relative efficiency of non-peroxidase luminol dioxygenation yielding luminescence is relatively low as reflected by the minimal luminol CL of the peroxidase deficient rooster PMNLs.

References

1. Merrill GA, Bretthauer R, Wright-Hicks J, Allen RC. Oxygenation activities of chicken polymorphonuclear leukocytes investigated by selective chemiluminogenic probes. *Lab Anim Sci* 1996;**46**:530-538.
2. Desmidt M, Van Nerom A, Haesebrouck F, Ducatelle R, Ysebaert MT.

Oxygenation activity of chicken blood phagocytes as measured by luminol- and lucigenin-dependent chemiluminescence. *Vet Immunol Immunopathol* 1996;**53**:303-311.

3. Kogut MH, Genovise KJ, Stanker LH. Effect of induced molting on heterophil function in white leghorn hens. *Avian Diseases* 1999;**43**:538-548.
4. Kogut MH, Holtzapple C, Lowry MS, Genovise KJ, Stanker LH. Functional responses of neonatal chicken and turkey heterophils following stimulation by inflammatory agonists. *Am J Vet Res* 1998;**59**:1404-1408.
5. Allen RC. Biochemiexcitation: Chemiluminescence and the study of biological oxygenation reactions. In *Chemical And Biological Generation of Excited States* (Adam W and Cilento G, eds.) New York:Academic Press, 309-344, 1982.
6. Allen RC. Lucigenin chemiluminescence: A new approach to the study of polymorphonuclear leukocyte redox activity. In *Bioluminescence and Chemiluminescence*(DeLuca MA and McElroy WD, eds.) New York: Academic Press, 63-73, 1981.
7. Muller-Peddinghaus R. Comparison of luminol- and lucigenin-amplified chemiluminescence (CL) of phagocytes. In *Oxygen Radicals In Chemistry and Biology* (Bors W, Saran M, and Tait D, eds.) Berlin: Walter deGruyter and Co., 881-882, 1984.
8. Allen RC, Dale DC, Taylor FB, Jr. Blood phagocyte luminescence: Gauging systemic immune activation. *Methods Enzymol* 2000; **305**: 591-629.
9. Brune,K., Leffell,M.S., and Spitznagel,J.K.; Microbicidal activity of peroxidaseless chicken heterophile leukocytes. *Infect Immun* 1972; **5**:283-287.
10. Penniall R, Spitznagel JK. Chicken neutrophils: oxidative metabolism in phagocytic cells devoid of myeloperoxidase. *Proc Nat Acad Sci* 1975;**2**:5012-5015.
11. Dixon M. The determination of enzyme inhibitor constants. *Biochem J 1953;***55**:170-171.
12. Allen R.C. 1986. Phagocytic leukocyte oxygenation activities and chemiluminescence: A kinetic approach to analysis. *Methods Enzymol* 1986;**133**; 449-493.

NEUTROPHIL CL ACTIVITY IN A HYPER-IgE ASSOCIATED LEUKOCYTOCLASTIC VASCULITIS CASE

ML VUOTTO, R MIRANDA, MTL IELPO, C NAPPO, D MONTANARO, E LAGHI AND L RICCIARDI

Istituto di Patologia Generale e Oncologia, II Universit degli Studi, Napoli, Italy
E-mail: vuotto@unina2.it

A GATTONI, C ROMANO AND A CECERE

Dipartimento di Medicina Sperimentale F. Magrassi , II Universit degli Studi, Napoli, Italy

1. INTRODUCTION

Studies show that PMNs play a key role in the pathogenesis of leukocytoclastic vasculitis.[1,2] Using CL emission as functional parameter, the aim of our study was to determine leukocyte activity in peripheral blood from a patient with this disease and find signs of systemic PMN activation. Most CL activity in circulating leukocytes occurs in PMNs. Other cell types produce less. This patient was a 31-year-old man with long-standing undiagnosed and untreated petechiae on the lower limbs. At the age of 25, plantar and external malleolar blisters appeared and ulcered. While hospitalized in a dermatology ward, plantar ulcers in a hyper- IgE patient [1] were diagnosed. Platelet count was 96,000/μL. He underwent cortisonic therapy for suspected collagenopathy. At the age of 28, he was splenectomized and the histologic exam of the spleen revealed consolidation of the red pulp and mild hyperplasia of the spleen follicules. After surgery, the platelets increased remarkably to 580,000/μL, however, the petechiae and necrotic lesions on the lower limbs worsened. At age 30 the patient was readmitted to the hospital because of ulcers on his feet and left hip. The histologic exam of the leg skin revealed a regular epidermic line and extended areas of fibrinoid necrosis that destroyed arterial walls and caused inflammatory infiltrates mostly of neutrophils. Later, signs of regression and apoptosis of PMNs were seen that clinched the diagnosis of leukocytoclastic vasculitis. The patient was placed on long-term cortisonic and immunosuppressive treatment (azatioprin) with no improvement. Afterward, he was started on alpha-interferon (IFN-α; 6,000,000 U/week) for another 2 months. His skin lesions ameliorated. The patient came to our attention at recovery and end of IFN treatment. Interestingly, the patient s son, in apparent good health, presented elevated IgE. The same laboratory work was repeated on the son because of the deranged IgE without other signs of disease.

2. MATERIALS AND METODS

Blood Peripheral blood (WB) was collected from subject and controls between 08.00 and 09.00 a.m., to minimize day-time variability of phagocytic blood cell respiratory burst.[3]
Modified Krebs-Ringer phosphate medium (KRP*) KRP* was composed of 119 mM NaCl, 4.8 mM KCl, 0.5 mM $CaCl_2$, 1.2 mM $MgSO_4$, 5.6 mM glucose and 16.6 mM sodium phosphate buffer pH 7.4.
Luminol 5-amino-2,3-dihydro-1,4-phthalazinedione (luminol, Sigma, Chemical Co., St. Louis, MO, USA) was as $5x10^{-2}$ M stock solution in dimethyl sulfoxide (DMSO); immediately before use, it was diluted with KRP* to give a 10^{-3} M working solution.
Stimuli phorbol myristate acetate (PMA) was used as soluble stimulus. The stock solution was prepared by diluting PMA (Sigma Chemicals Co.) at 5 x 10^{-3} M in DMSO and stored at -20° C. The working solution was 1.5 μM in KRP*. Zymosan (ZYM, Sigma) and zymosan opsonized with autologous (OZA) or pooled homologous sera (OZH) were used as particulate stimuli at 5 mg/mL final concentration.
Luminol-dependent chemiluminescence assay Chemiluminescence assays were performed following the protocol described by De Sole et al.[4], using an automatic luminometer (Autolumat LB 953, Berthold, Wildbad, Germany). The reaction mixtures were prepared in 4-mL polypropylene vials. Each vial contained 100 μL of 10^{-4} M luminol, 100 μL of the 50μM catechin, 100 μL of the different stimuli, 100 μL of diluted WB suspension and sufficient KRP* to yield a final volume of 1.0 mL. Samples without stimuli were also included as controls. The resulting light emission was recorded for 0.5 sec on each vial over a 90 min period. All measurements were performed in triplicate. The CL was determined as total photon emission for each neutrophil. This parameter is the integration of the light signal recorded during the reaction.
Statistical analysis Levels of significance were determined using the ANOVA test. Values of $p<0.05$ were regarded as significant.

3. RESULTS

The Table reports the CL values recorded in the patient, in his son and in the control. The total and relative CL counts were higher than normal at baseline. The basal CL emission per PMN was higher in the patient than in the healthy control. PMA stimulation further increased the response. If the CL emission value after stimulation is expressed as the emission versus the basal value (stimulation index, SI), it can be shown that this was lower in the patient than in the control. A greater response instead was observed using zymosan opsonized with autologus or homologous serum, with even a lower SI compared to the control. The patient s son had a baseline activity and a response to stimuli like his father.

4. DISCUSSION

PMNs are the principle cells in histologic sections of areas with leukocytoclastic vasculitis.[1,5] The functional role of PMNs in experimental vasculitis has been demonstrated by monoclonal antibodies.[1] The high basal activity is the expression of systemic leukocyte activation *in vivo*. The milder increase in leukocyte activity compared to basal value in cells treated with PMA may be interpreted as a sign of functional failure in already primed cells.[6] The OZA and OZH response could depend on the increased expression of Fc-receptors after neutrophil activation.[7] In our patient, vasculitis was coupled with elevated IgE which persisted after treatment and remission with IFN-α.[8] The healthy son also had elevated IgE. Like his father, the high levels of basal activity may represent systemic leukocyte activation *in vivo*. No explanation is available to interpret this finding, and the possible correlation with elevated IgE remains unclear.

Table Chemiluminescence recorded in the patient (*father*), in his son (*son*) and in a control subject (Counts x 10^{-3} x 90 minutes; mean ± SD).

	father	*son*	*control*
WB	3155,36 ± 960	3007,5 ± 390	1008,44 ± 350
WB+PMA	8976,19 ± 1210	12150 ± 2530	6066,67 ± 795
WB+ZYM	5952,38 ± 1025	9572 ± 940	3666,67 ± 575
WB+OZA	11863,1 ± 1320	12050 ± 2050	5488,89 ± 950
WB+OZH	12767,9 ± 1100	13140 ± 1215	8788,89 ± 1020

WB = whole blood, PMA = phorbol myristate acetate, ZYM = zymosan, OZA = zymosan opsonized with autologous serum, OZH = zymosan opsonized with pooled homologous sera

References

1. Qasim FJ, Mathieson PW, Sendo F, Thiru S, Oliveira DB. Role of neutrophils in the pathogenesis of experimental vasculitis. *Am J Pathol.* 1996; **149**: 81-89.
2. Sams WM Jr, Thorne EG, Small P, Mass MF, McIntosh RM, Stanford RE. Leukocytoclastic vasculitis. *Arch Dermatol.* 1976; 112: 219-226.

3. Heberer M, Ernst M, During M, Algower M, Fisher H. Measurement of chemiluminescence in freshly drawn human blood. *Klin Wochenschr* 1982; **60**:1443-1448.
4. De Sole P, Fresu R, Frigeri L, Pagliari G, De Simone C, Guerriero C. Effect of adherence to plastic on pheripheral blood monocyte and alveolar macrophage chemiluminescence. *J Biolumin Chemilumin* 1993; **8**:153-158.
5. Zax RH, Hodge SJ, Callen JP. Cutaneous leukocytoclastic vasculitis. Serial histopathologic evaluation demonstrates the dynamic nature of the infiltrate. *Arch Dermatol* 1990; **126**: 69-72.
6. Grunwald MH, Shriker O, Halevy S, Alkan M, Levy R. Impaired neutrophil functions in patients with leukocytoclastic vasculitis. *Int J Dermatol* 1997; **36**: 509-513.
7. Hundt M, Zielinska-Skowronek M, Schmidt RE. Fc gamma receptor activation of neutrophils in cryoglobulin-induced leukocytoclastic vasculitis. *Arthritis Rheum* 1993; **36**: 974-982.
8. Zimmermann R, Konig V, Bauditz J, Hopf U. Interferon alfa in leukocytoclastic vasculitis, mixed cryoglobulinaemia, and chronic hepatitis C. *Lancet* 1993; **341**: 561-562.

PHAGOCYTE ACTIVITY IN OLD AND YOUNG MICE UNDERGOING SURGICAL STRESS

ML VUOTTO, MTL IELPO, R MIRANDA, C NAPPO, D SPATUZZI AND L RICCIARDI
Istituto di Patologia Generale e Oncologia, II Universit degli Studi, Napoli, Italy
E-mail: vuotto@unina2.it

P DE SOLE
Istituto di Chimica e Chimica Clinica, Universit Cattolica, Roma, Italy

C ASTARITA, D GARGANO AND F MANGUSO
Dip. Internistica Clinica e Sperimentale Flaviano Magrassi , II Universit , Napoli, Italy

1. INTRODUCTION

Surgical stress induces systemic endocrine-metabolic modifications and influences leukocyte function, leading to various systemic responses that appear to depend on severity of surgical stress and on the type of surgical preocedure.[1] Less attention has been paid to the function of polymorphonuclear leukocytes after surgical stress, despite the fact that phagocytes are known to play a major role in innate host defense against infections. These cells use well known mechanisms such as chemotaxis, phagocytosis, a final respiratory burst and killing. The same cells also play an important role in the specific immune response, contributing to its modulation with inhibiting or stimulating effects through proinflammatory cytokine secretion.[2,3] The few studies concentrating on luminol-dependent chemiluminescence (CL) response of neutrophils in patients undergoing surgery have had conflicting results. Decreased, normal, or increased CL levels have all been described. [1,4,5] Recently, some studies have examined the applicability of neutrophil luminol-dependent CL response to assess the degree of stress in operated patients.[6,7] Conflicting results have chiefly come from the variability of parameters under study. Together with the non surgical aspects, such as immunological profile, postoperative monitoring times, and different study approaches, parameters strictly linked to surgery such as duration, anesthesia, antibiotic prophylaxis, transfusions, and, above all, the degree and entity of operation and stress should be taken into account. The use of a murine model offers great advantages such as inbreeding, homogeneity and availability of strains with short and long lifespan. The purpose of this study was to investigate the in vitro reactive oxygen species (ROS) production by whole blood phagocytes, mainly PMNs, in mice undergoing elective mild surgical stress and the effects of age on the oxidative burst of circulating phagocytes. ROS generation was evaluated by measuring CL emission, in a luminol-enhanced system, in young and old mice during the perioperative period.

2. MATERIALS AND METHODS

Mice. 2-month-old (young) or 18-month-old (old) female C57BL/6 mice (specific pathogen-free) were purchased from Charles River Laboratories, under contract with the National Institute of Aging.
Surgery and peripheral blood collection. Ten mice were included in the study: 5 young and 5 old. All of them underwent mild surgical stress with dorsal skin incision, in general anesthesia, on the same day. Each mouse had peripheral blood drawn at baseline (PRE-O) and after 24, 48, 72, 96 and 168 hours from surgical stress (PO24, PO48, PO72, PO96, and PO168).
Whole blood CL. CL assays of whole blood phagocytes was performed using a modification of the technique described by De Chatelet and Shirley [8,9] and an automatic luminometer (Autolumat LB953, Berthold, Wildbad, Germany). Leukocytes were resting or activated with phorbol myristate acetate (PMA) or with zymosan opsonized (OZ) or with autologous serum (OZA).
Data Analysis. The results of the tests were expressed as means ± SD. General Linear Model (GLM) Repeated Measures with Bonferroni test for multiple comparisons and mice age (young/old) as between-subject factor, and Student's t test were used for statistical analysis. For all analyses, probability values of 5% or less were taken to be statistically significant. The statistical analysis was performed using the SPSS software package for Windows (release 10.0.7, 1 June 2000; SPSS Inc., Chicago, Ill., USA)..

3. RESULTS

The effect of surgical stress on whole blood leukocyte CL activity. The results are summarized in Figure 1. In all mice a significant statistical difference was found for CL activity of unstimulated leukocytes among all observation times ($p<0.0001$). In particular, CL activity significantly decreased at PO48 compared to PO24 ($p=0.009$), and revered to PRE-O values at PO72 with a significant increase in respect to PO48 ($p=0.01$). Even in the case of CL emission in response to PMA a significant difference was found among all observation times ($p<0.0001$). In this case a significant decrease was observed at PO72 in respect to PO24 and PO48 ($p=0.007$ and $p=0.02$, respectively) with values at PO168 higher than baseline ($p=0.04$). CL emission of cells opsonized with autologous serum (OZA) did not differ before and after surgery.
The effects of age on leukocytes CL activity. The GLM results also indicated the influence of age only in the case of PMA-stimulated leukocytes ($p<0.0001$). An extremely significant increase of CL activity in old mice compared to young ones was found throughout the perioperative period. In fact, the mean CL values were higher in old than in young mice at any time except at PO72 (Figure 2). No significant differences were found in CL emission of OZA-stimulated cells in both groups of animals, before and after surgery (data not shown).

4. DISCUSSION

PMNs play an important role in innate and acquired immunity, mainly in primary host defense against infectious diseases. [3] The present study explored the effects of surgical stress on respiratory burst and the different behavior of young and old mice in response to stress. The data demonstrate that surgical stress influences the phagocyte function inducing a significant baseline and PMA-stimulated depression of PMN activation. With regard to the effects of age on CL emission in operated on mice, when we used PMA-stimulation the CL activity of aged mice was significantly increased compared to the young controls before and after stress. This suggests that CL production through the PMA-pathway is in some way modified with an excessive ROS production. On the contrary, the CL responses of unstimulated as well as of the OZA-stimulated cells are statistically non significant, confirming that the PMN defect is stimulus- and function-specific. The alteration in PKC isoenzymes during aging may contribute to aberrancies in intracellular signal transduction via PMA activation.[10]

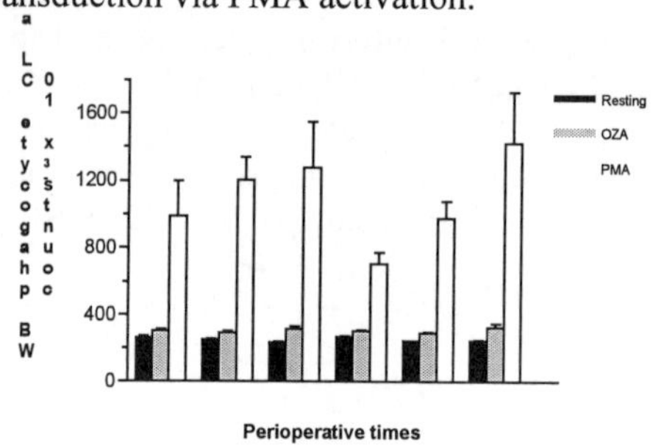

Fig 1. Effects of surgery on CL activity of resting ($p < 0.0001$) and PMA-activated ($p < 0.0001$) cells of all operated mice ($n = 10$) at the various times of the perioperative period. CL emission of OZA-stimulated cells did not differ before and after surgery. Data are expressed as mean ± standard deviations.

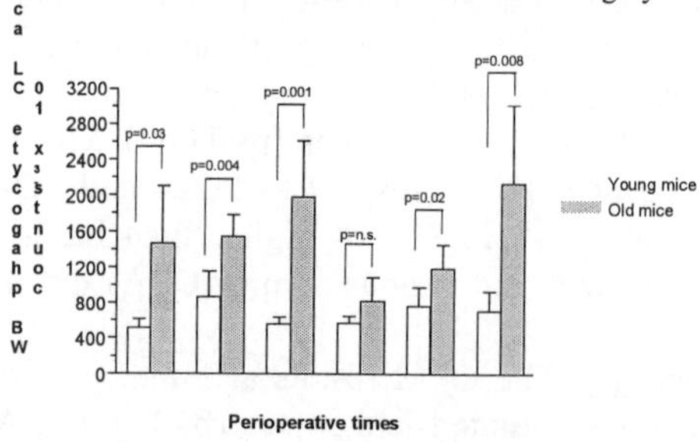

Fig 2. Effects of age on phagocyte function in mice undergoing elective surgical stress. CL responses to PMA in young ($n = 5$) and old ($n = 5$) mice. The mean CL counts were higher in old than in young animals at each time except PO72 (p=not significant). Data are expressed as mean ± standard deviations.

References

1. Aosasa S, Ono S, Mochizuchi H, Tsujimoto H, Osada S, Takayama E, SekiS, Hiraide H. Activation of monocytes and endothelial cells depends on the severity of surgical stress. *World J Surg* 2000; **24**:10-16.
2. Cassatella MA. Neutrophil-derived proteins: selling cytokines by the pound. *Adv Immunol* 1999; **73**:369-509.
3. Rosenberg WE, Gallin JI. Inflammation. In: Paul W, ed. *Fundamental Immunology*, 4th ed. Philadelphia: Lippincott-Raven, 1999:1051-1066.
4. Kowatari K, Suzuki K, Kudo S, Yamada M, Liu Q, Danjo K, Umeda T, Nakaji S, Sugawara K. Applicability of chemiluminescence to assess the degree of operative stress in patients undergoing spinal surgery. *Luminescence* 1999; **14**:331-334.
5. Yokota K, Nishihira T, Shineha R, Ueda H, Mori S. Perioperative alterations in polymorphonuclear leukocyte function mediated by protein kinase C. *Nippon Geka Gakkai Zasshi* 1994; **95**:149-153.
6. Barry MC, Condron CM, watson RW, Redmond HP, El Jack M, Watson RG, Bouchier Hayes D. Pre-operative neutrophil and monocyte activation state predicts post-operative neutrophil and monocyte function. *Eur J Surg* 1997; **163**: 739-745.
7. Isozaki H, Nishino H, Sako S, Naomoto Y, Tanaka N. The relationship between levels of neutrophil-related factors and the degree of surgical stress. *Int Surg* 1999; **84**: 105-110.
8. De Chatelet LR and Shirley PS. Evaluation of chronic granulomatous disease by chemiluminescence assay of microliter quantities of whole blood. *Clin Chem* 1981; **27**:1739-1741.
9. Vuotto ML, Ielpo MT, Liguori G, Benedetto R, Moscatiello V, Maselli R. Activation of peripheral phagocytes in BCG-vaccinated subjects. *Luminescence* 2000; **15**:153-157.
10. Pascale A, Govoni S, Battaini F. age-related alteration of PKC, a key enzyme in memory process: physiological and pathological example. *Mol Neurobiol* 1998; **16**:49-62.

PART 8

REPORTER GENES IN CELL BIOLOGY AND ANALYSIS

SECRETED *RENILLA* LUCIFERASE AS A MARKER OF GENE EXPRESSION IN MAMMALIAN SYSTEMS

M. FILIPPOVA[1], J. LIU[1,2], AND A. ESCHER[1,2]

[1]*Center for Molecular Biology and Gene Therapy and* [2]*Department of Microbiology and Molecular Genetics, 11085 Campus Street, Loma Linda University, Loma Linda* CA 92350, USA
E-mail: aEscher@som.llu.edu

1. INTRODUCTION

Secreted marker proteins permit the monitoring of gene expression without destruction of cells containing their encoding genes. This property greatly simplifies measurements of gene expression in mammalian cells grown *in vitro*, and has potentially useful applications for following gene expression in living animals over time. We have previously compared a secreted form of the sea pansy *Renilla reniformis* luciferase[1] (SRUC3) with human placental secreted alkaline phosphatase (SEAP) to identify advantages and disadvantages of these proteins as markers of gene expression in mammalian cells.[2] In this work, retrovirus-packaging cells grown *in vitro* were transduced with retroviral particles carrying modified cDNAs encoding SRUC3 and SEAP to investigate the use of these marker proteins as indicators of high virus titer. In addition, plasmid DNA carrying cDNAs encoding various light-catalyzing enzymes was injected intramuscularly into mice, and light emission from muscle tissue and sera was measured over time to determine the usefulness of those marker proteins in the context of plasmid DNA injection. Results indicated that, in contrast with SEAP, SRUC3 could not be used to screen for packaging cell lines producing high titers of retroviral particles. Results from intramuscular injection of plasmid DNA revealed that while catalyzed light emission could be detected from muscle tissue, no activity from reporter proteins could be detected in animal sera.

2. MATERIALS AND METHODS

Plasmid construction and isolation: For retroviral packaging, *sruc3* and *seap* cDNAs were cloned into plasmid pLNCX. For intramuscular injection, cDNAs encoding firefly luciferase (*luc*), *Renilla* luciferase (*ruc*), secreted *Renilla* luciferase (*sruc3*), and human placental secreted alkaline phosphatase (*seap*, from Clontech Laboratories, Palo Alto, CA) were cloned into plasmid pND2 (a gift from R. Malone, University of Maryland). Plasmid DNA used for intramuscular injection was isolated using two successive cesium chloride gradient. All cDNAs were placed under transcriptional control of the cytomegalovirus promoter.

Packaging of retroviral particles: Ecotropic packaging cell line GP+E86 was transfected with plasmid pLNCX carrying *sruc3* or *seap* cDNAs using the calcium phosphate transfection method. Viral supernatants that were obtained were then used to transduce amphotropic packaging cell line PA317. Titering of supernatants was done on murine cell line NIH 3T3 after selection with G418. All cell lines were grown in DMEM medium + 10% fetal bovine serum.

Intramuscular DNA injection: Plasmid DNA was injected into quadriceps muscles (200 microg/leg) of 7 week old BALB/c mice (Jackson Laboratories) with a 27-gauge needle under general anesthesia (Ketamine, 66 mg/Kg body weight, Oxylazine, 7.5 mg/Kg body weight, and Acepromazine Maleate, 1.5 mg/Kg body weight). Each plasmid DNA carrying a cDNA was injected into 32 animals, and plasmid vector pND2 was injected into 16 mice as negative control. Four mice from each group (two from control group) were killed on day 1, 3, 5, 7, 10, 12, 15 and 20 after injection for analysis of reporter protein in muscle tissue and sera. Quadriceps muscles (~ 0.12 mg) were stored at -70^0 C. Blood was collected through cardiovascular puncture, stored at 4^0 C overnight, and centrifuged at 12,000 x g for 10 minutes to collect sera, which were stored at -70^{0}C.

Analysis of reporter enzyme activity: Muscle tissue samples were homogenized in 1 mL Cell Culture Lysis Reagent (25 mM Tris-phosphate, pH7.8, 2 mM DTT, 2 mM 1,2-diaminocyclohexane-N,N,N',N'-tetraacetic acid, 10% glycerol, 1% Triton X-100) using an OMNI 2000 homogenizer (Omni International Inc., VA, USA). Firefly luciferase and SEAP assays were performed according to manufacturers protocols (Promega, Madison, WI, and Clontech, Palo Alto, CA, respectively). The *Renilla* luciferase assay was performed as previously described.[1] Light emission from 50 microL of homogenized tissue and serum samples and from various chosen volumes of cell culture media were measured in a Turner TD-20e luminometer.

3. RESULTS AND DISCUSSION

3.1. Retroviral transduction

Amphotropic packaging cell line PA317 was transduced with viral supernatants from ecotropic cell line GP+E86 that had been transfected with plasmids pLNCX-SRUC3 or pLNCX—SEAP. Media from 61 and 20transduced PA317 stable lines were then screened for SRUC3 and SEAP activity, respectively. Among cell lines transduced with retroviruses carrying the *sruc3* cDNA, 20 % did not secrete detectable levels of *Renilla* luciferase activity. In contrast, culture media from all clones of PA317 cells transduced with retroviruses carrying the *seap* cDNA contained SEAP activity well above background (data not shown). Supernatants from PA317 cell lines secreting the highest levels of SRUC3 and SEAP activity were used to transduce murine NIH 3T3 cells for titering, and titers were compared with SRUC3 and SEAP activities in culture media of the same packaging cell lines (Fig.1). Data indicated that increasing levels of SEAP activity were associated with increasing titers of viruses secreted by

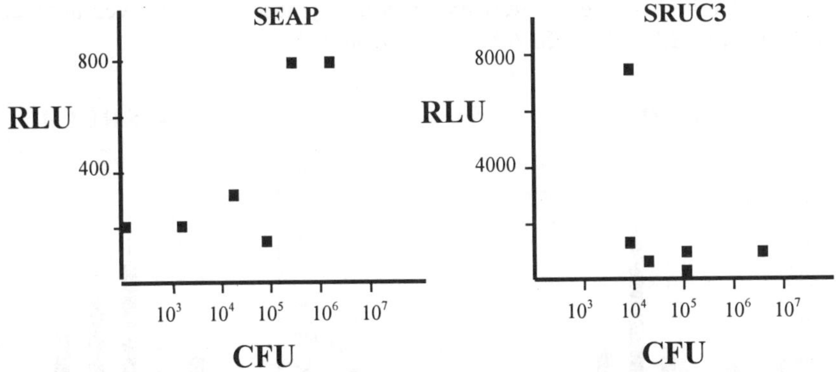

Fig. 1. Comparison between viral supernatant titer (CFU, or colony forming units) and activity of marker protein (RLU, or relative light units) secreted by clones of packaging cell line PA317. Each square represents an individual clone.

packaging cells lines expressing the *seap* cDNA. In contrast, decreasing levels of SRUC3 activity were associated with increasing titers of retroviruses secreted by packaging cells lines expressing the *sruc3* cDNA. These results suggest that SEAP activity in cell culture media of retrovirus packaging cell lines can be used to screen for producers of high titer viral supernatants, but that SRUC3 activity cannot. Loss of luciferase activity may have been due to inactivation of *sruc3* expression due to toxicity of SRUC3 protein. This phenomenon was reported previously with cell lines packaging retroviruses carrying the *gfp* cDNA.[3] However, in the case of GFP, green fluorescence was observed only in cells producing low titers of retroviral particles, and presence of *gfp* cDNA could not be detected in viruses from high titer supernatants. The finding that SRUC3 activity could be measured with cells producing high titers of viruses suggests that a mechanism other than gene inactivation may be responsible for the observed loss of luciferase activity.

3.2. Intramuscular injection of plasmid DNA

Light emission from all four marker proteins could be detected in muscle tissue after injection of their encoding plasmid DNA. Peak emission for both intracellular and secreted *Renilla* luciferase was 7 days post injection (Fig. 2 A and B, respectively), 12 days for firefly luciferase (Fig. 2 C) and 10 days for alkaline phosphatase (Fig. 2 D). Interestingly, *Renilla* luciferase activity appeared to be higher than firefly luciferase activity, indicating that it could be used as an alternative to firefly luciferase for studying gene expression in the context of plasmid DNA injection. SEAP activity in muscle tissue varied greatly, and its assay showed highest levels of background activity. No marker protein activity was detected in mice sera, suggesting that higher levels of gene expression may need to be achieved in order to detect activity of SRUC3 and SEAP in blood of animals.

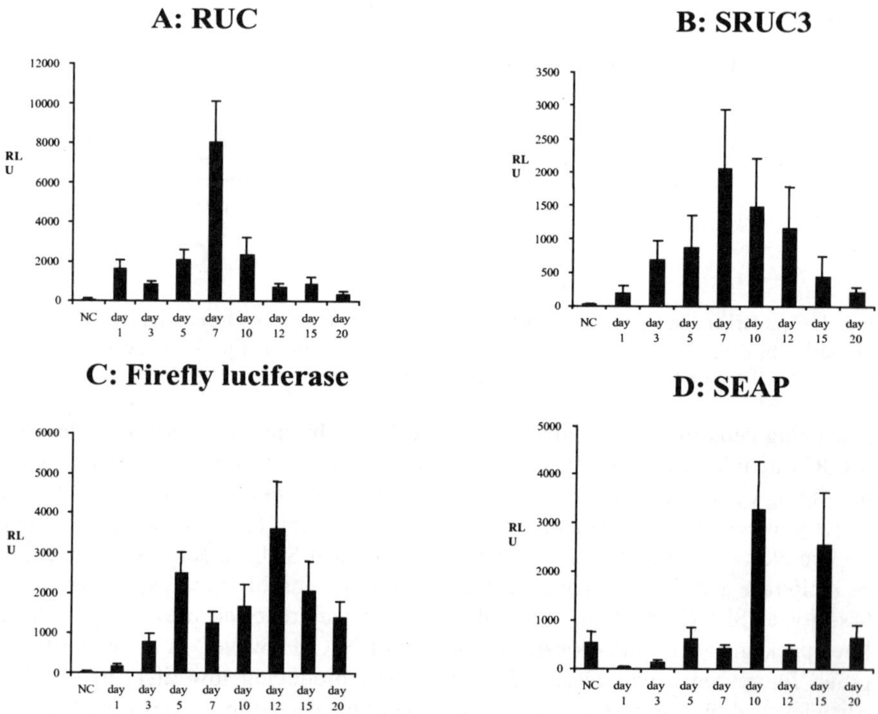

Fig. 2. Time-course of reporter protein activity in muscle tissue after injection of plasmid DNA encoding A, intracellular *Renilla* luciferase, B, secreted *Renilla* luciferase, C, firefly luciferase, and D, secreted alkaline phosphatase. RLU, relative light units, NC, negative control.

References

1. Liu J, Escher A. Improved assay sensitivity of an engineered secreted *Renilla* luciferase. *Gene* 1999; **237**: 153-159.
2. Liu J, Filippova M, O Kane DJ, Escher A. Secreted *Renilla* luciferase : A new marker of gene expression. In: Roda A, Pazzagli M, Kricka LJ, Stanley PE, eds. *Bioluminescence and Chemiluminescence: Perspectives for the 21st Century*, Chichester: Wiley, 1999: 471-474.
3. Hanazono Y, Yu JM, Dunbar CE, Emmons RVB. Green fluorescent protein retroviral vectors: low titer and high recombination frequency suggest a selective disadvantage. *Hum Gene Ther* 1997; **8**:1313-1319.

THE GLIAL CELL-LINE DERIVED NEUROTROPHIC FACTOR-LUCIFERASE FUSION IS FULLY FUNCTIONAL UPON SECRETION FROM MAMMALIAN CELLS

H. LIU[1,2] AND A. A. SZALAY[1]

Dept. of Biochemistry[1], *Dept. of Physiology and Pharmacology*[2]
School of Medicine, Loma Linda University, Loma Linda, CA 92350
USA
E-mail: aszalay@som.llu.edu

1. INTRODUCTION

Glial cell-line derived neurotrophic factor (GDNF) is a recently identified neural growth factor purified from conditioned media from the rat glioma cell line, B49, and later C6 glioma cell line.[1] Because of its relative selective survival effect on dopaminergic (DA) neurons,[2] GDNF has been tested in numerous studies with the hope to establish effective drug therapy and gene therapy approaches for Parkinson's Disease (PD). Overwhelming evidence has demonstrated the effectiveness of GDNF in promoting dopaminergic neuron fiber outgrowth and its resistance to neurotoxins in animal PD models.[3]

Previous studies using transformed cells to deliver GDNF failed to monitor the viability and functionality of the implanted cells *in vivo* since the monitoring must be done by post-mortem biopsy. The conventional methods to quantify GDNF in given specimens is either Western blotting, or ELISA. The detection limit of ELISA using an available commercial kit (GDNF E_{max} ImmunoAssay System, Promega) is 32 pg/mL, while the concentration of GDNF in culture to support survival of DA neurons is approximately 10 to 10000 pg/mL.[3] Since dilution by at least 10^3 folds occurs once a protein diffuses from tissue matrix into cerebrospinal fluid which is the only specimen that can be used to monitor implanted cells for GDNF production, it is obvious that ELISA is not sensitive enough to monitor GDNF production by implanted cells in the cerebrospinal fluid.

Using the *Renilla* luciferase (RUC) as a reporter system to monitor the expression levels and secretion of a transgene encoded protein *in vivo* is thus highly desired. It is sensitive, has no or minimal background in mammalian cells, has no toxicity to mammalian cells, has no interference with GDNF activity, and is relatively inexpensive. *Renilla* luciferase has been used as a marker of gene expression in bacteria, plant, and mammalian cells.[4] The *Renilla* luciferase cDNA can be fused with the cDNA of GDNF so that GDNF and luciferase can be co-expressed as fusion protein, thus enabling the measurement of GDNF expression based on light emission.

In this paper, we describe the engineering and testing of a mammalian cell line that overexpresses RUC-GDNF fusion protein. The fusion protein retains the function of both RUC and GDNF. The expression of GDNF at the protein level can be directly determined by RUC bioluminescence assay in the culture medium.

2. MATERIAL AND METHODS

Constructs: The vector DNA pLNCX-ruc-gdnf (Fig. 1) was engineered by placing RUC cDNA, GDNF cDNA fusion cassette into the plasmid vector pLCNX (Clontech). To facilitate the secretion of RUC-GDNF fusion protein, the human interleukin-2 signal peptide was fused to the N-terminus of Renilla luciferase.[5] The rat cDNA encoding GDNF was excised from the vector pGDNF3a, a gift from Dr. Bohn of Northwestern University. The RUC-GDNF-fusion gene contains a 30 bp linker linking 3' and 5' ends of RUC and GDNF respectively. To ensure in frame translation and secretion, the stop codon of RUC and signal peptide codons of GDNF were removed.

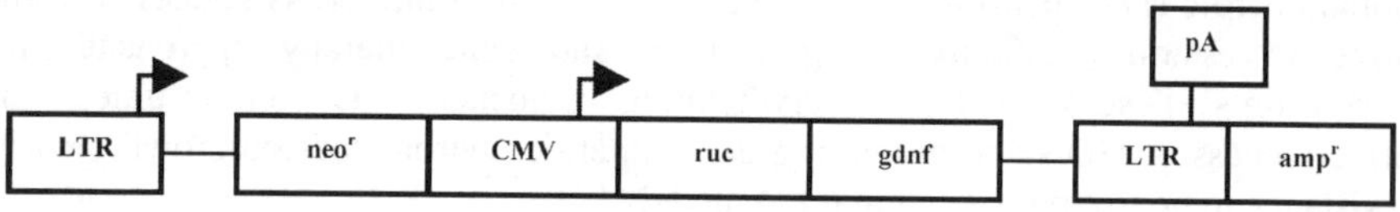

Fig. 1 The structure of pLNCX-ruc-gdnf construct

Transfection of PO-D17 cells with pLNCX-ruc-gdnf DNA: Lipofection approach (Fugene6, Roche) was used to transfect PO-D17 cells kindly supplied by Dr. Frisa of Case Western Reserve University. Cells were cultured in DMEM supplemented with 10% fetal bovine serum (Cellgro) and 1x antibiotic solution (Sigma). Puromycin (Sigma) 1 to 5 μg/mL was applied to culture medium to select stably transfected cells (RG-1).

Monitoring of secretion of RUC-GDNF by RUC bioassay: RG-1 cells were grown to monolayer in DMEM supplemented with fetal bovine serum, followed by wash with ice-cold PBS and changing to fresh medium. An aliquot of 100 μL (1% total volume) of culture medium, or 10 μL (1% total volume) of cell lysate from the same petri dish was then mixed with 1 μL of 1 μM coelenterazine. The light emission was measured using Hamamatsu Argus-100 with 3-minute exposure.

GDNF detection, quantification, and bioassay: Western blotting was performed to determine the molecular weight of the fusion proteins. Chicken anti-human GDNF antibody was used as the primary antibody. ELISA was performed using GDNF E_{max} ImmunoAssay System to quantify GDNF concentration in culture medium.

GDNF bioassay was performed using near-term sheep dorsal root ganglion cell culture.

3. RESULTS

Upon transfection of mammalian cells with the fusion construct, luciferase activity was detected in the cells as well as in the culture medium (Fig. 2).

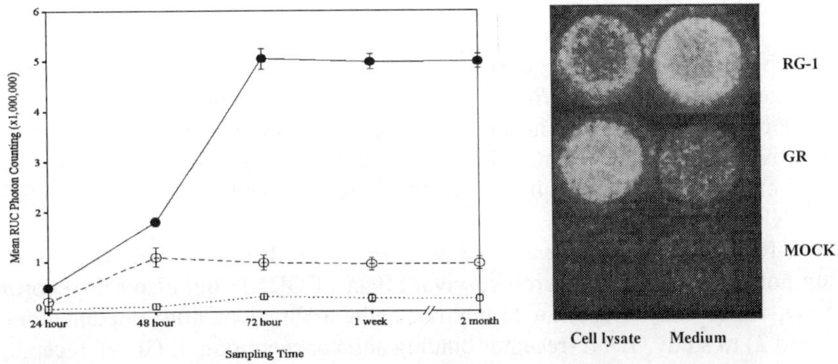

Fig. 2 Graph and images show secretion reported by RUC bioluminescence

The graph shows time course of RUC activity in culture medium of RG-1 (•), GR (o, a cell line transfected with construct that contains GDNF-RUC fusion), and MOCK transfectant ( ). The image shows RUC assay in medium and cell lysate after 72 hours of culture. Data clearly show RUC-GDNF fusion is secreted, and GDNF-RUC is non-secretive.

The molecular weight of the fusion protein is between 70 – 75 kD as determined by Western Blotting using polycolonal antibodies against GDNF (Fig. 3), suggesting the correct size of the fusion protein.

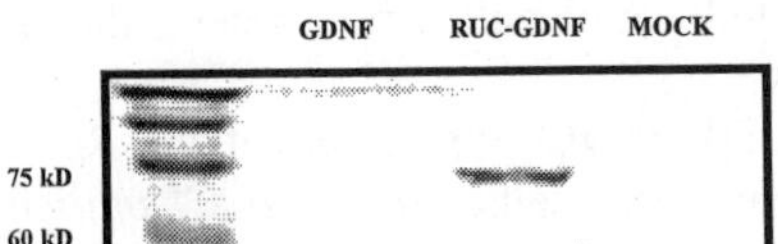

Fig. 3 Western blot analysis shows the position of RUC-GDNF fusion protein

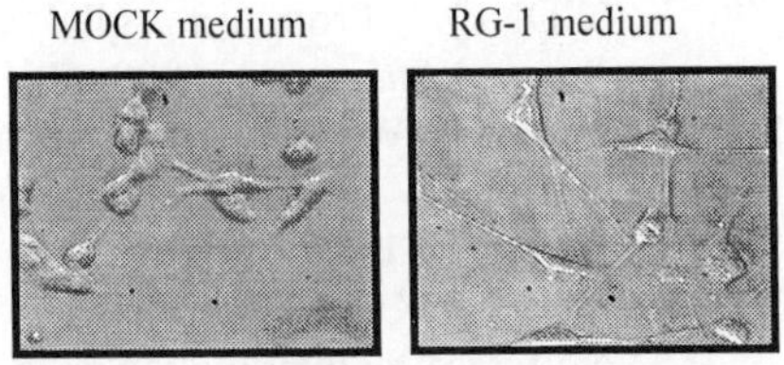

Fig. 4 Stimulation of neurite outgrowth of cultured near-term sheep dorsal root ganglion cells 72 hours after treatment with culture medium of RG-1 or MOCK transformed cells

The bioactivity of GDNF of the fusion protein was confirmed by the fetal sheep dorsal root ganglion assay which resulted in distinct stimulation of neurite outgrowth *in vitro* (Fig. 4).

The amount of GDNF was determined by ELISA assay which correlates (Spearman's rho = 0.95, $p < 0.01$) with RUC photon counting, suggesting that luciferase activity in the culture medium accurately reflects the concentration of GDNF in the same culture.

4. CONCLUSION

In this report, we describe the generation of a cell line, RG-1, which is permanently transformed and carries the *Renilla* luciferase-GDNF fusion construct encoding RUC-GDNF fusion protein. The culture medium of transformed cells has displayed 1) bioactivities of both RUC and GDNF and shows no decline in RUC activity after 60 days of cell culture; 2) high correlations were observed between RUC activity and GDNF concentration.

The RG1 cell line may be utilized in *ex vivo* gene therapy cellular experiments having not only dopamine neuron survival effect of GDNF, but also self-reporting property, which can be used to 1) monitor cellular functions after implantation in vivo; and 2) to study GDNF-receptor binding and concentration of GDNF receptors.

Reference:

1. Lin LF, Doherty DH, Lile JD, *et al.*, GDNF: a glial cell line-derived neurotrophic factor for midbrain dopaminergic neurons [see comments]. *Science* 1993; **260**:1130-1132.
2. Hudson J, Granholm AC, Gerhardt GA, *et al.*, Glial cell line-derived neurotrophic factor augments midbrain dopaminergic circuits in vivo. *Brain ResBull* 1995; **36**:425-432.
3. Bohn MC, A commentary on glial cell line-derived neurotrophic factor (GDNF). From a glial secreted molecule to gene therapy. *BiochemPharmacol* 1999; **57**:135-142.
4. Lorenz WW, Cormier MJ, O'Kane DJ, *et al.*, Expression of the Renilla reniformis luciferase gene in mammalian cells. *JBioluminChemilumin* 1996; **11**:31-37.
5. Liu J, O'Kane DJ,Escher A, Secretion of functional Renilla reniformis luciferase by mammalian cells. *Gene* 1997; **203**:141-148.

VISUALIZATION OF BACTERIA IN LIVE ANIMALS USING LUCIFERASE LABELING

SHAHROKH SHABAHANG[1,2] AND ALADAR A. SZALAY[1]

[1]*Department of Biochemistry, School of Medicine, Loma Linda University, Loma Linda, California 92350, USA*
[2]*Department of Microbiology and Molecular Genetics, School of Medicine, Loma Linda University, Loma Linda, California 92350, USA*
E-mail: sshabahang@sd.llu.edu
E-mail: aszalay@som.llu.edu

1. INTRODUCTION

Many systemic illnesses have been attributed to local infections in the body. The oral cavity has one of the highest number and variety of bacterial species in the human body.

Infections of distant sites, such as deep and superficial tissues of the head and neck, orbital space, intra-cranial locations, the respiratory tract and the cardiovascular system have been reported for susceptible patients.[1] More recently, research in the field of periodontics has linked oral infections to systemic diseases such as atherosclerosis,[2,3] diabetes mellitus, and low-weight childbirth.[4] Infective endocarditis (IE) is the infection of the myocardium or heart valves, which may result from invasive medical or dental procedures in susceptible patients. Physicians and dentists commonly encounter the dilemma of whether to prescribe prophylactic antibiotic coverage for these patients to prevent the occurrence of IE.

The goal of this investigation is to develop a model, which can be used to observe the fate of microorganisms in a living mammalian host. Using a five-genic *luxCDABE* cassette from *Xenorhabdus luminescens*, this model would allow *in vivo* assessment of human diseases such as IE.

2. MATERIALS AND METHODS

All animals used in this project were anesthetized with an intra-peritoneal injection of Nembutal (60 mg/kg body weight). The incision sites were cleaned with Betadine.

2.1 Intramuscular injections of luminescent bacteria

A C57 mouse was injected with four attenuated bacterial strains containing the lux cassette. A total of 10^6 bacterial cells were intramuscularly injected in each limb. The animals were monitored under a low light imager (Night Owl) immediately after the injections and over a one-week period to determine the stability of the plasmid (Figure 1).

2.2 Intra-venous injections of luminescent bacteria

Nude mice were intravenously injected with an attenuated strain of *Vibrio* (10^6 cells) via the tail vein. The animals were then visualized under the low light imager (Night Owl) sixty and ninety minutes after injection.

2.3 Visualization of bacteria in rat tibia

In order to determine whether luminescent bacteria can be visualized through hard tissues, rat tibias were injected with 10^6 *E. coli* cells through a 1 mm hole drilled on one end and subsequently sealed. The tibia was then visualized under the low light imager (Night Owl) immediately following injection.

2.4 *In vivo* visualization of luminescent bacteria in hard tissues

C57 mice were injected with 10^6 *Vibrio* in the skull cavity through the foramen magnum. The animals were then visualized immediately under the low light imager (Night Owl).

2.5 Infective endocarditis experiments

Sprague-Dawley rats were catheterized through the right carotid artery to the level of the mitral valve as previously described.[5] Control animals did not receive carotid catheters. All animals were injected with 10^7 *E. coli* via the femoral vein. Three days following bacterial inoculation, the heart tissues were excised and grown in selection media overnight. The specimens were then visualized under the low light imager (Hamamatsu).

3. RESULTS

3.1 Intramuscular injections of luminescent bacteria

This experiment demonstrated that visualization of luminescent bacteria in muscle tissues of live mice is possible. The four strains showed varying degrees of light emission as visualized in the four limb muscles (Figure 1).

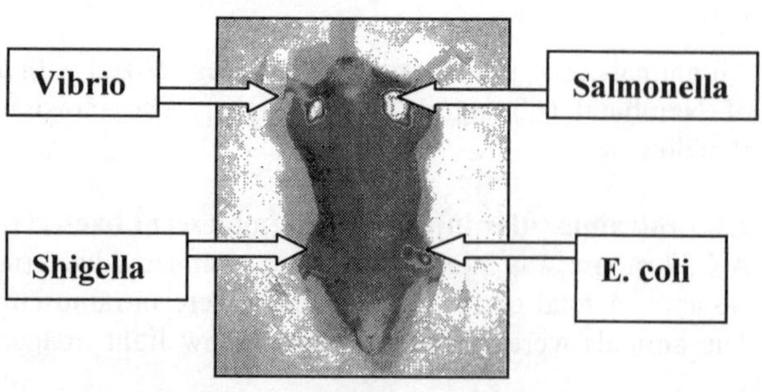

Figure 1. Muscle injection

3.2 Intra-venous injections of luminescent bacteria

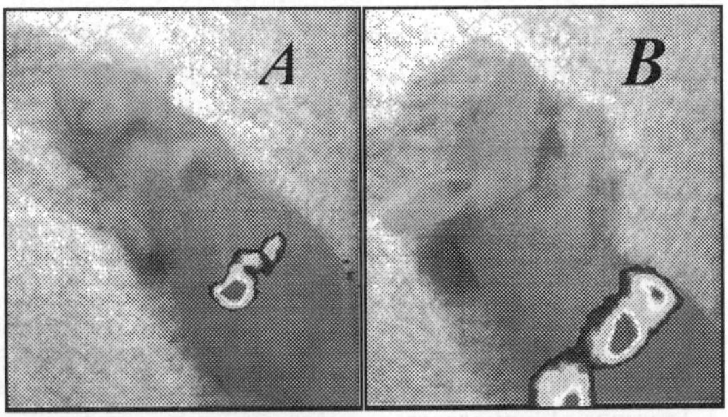

Fig. 2. Luminescent bacteria were visualized under the low light imager (Night Owl) in the liver and kidney after sixty (*A*) and ninety (*B*) minutes. This image represents one minute of photon collection.

3.3 Visualization of bacteria in rat tibia

Rat tibias were injected with the liquid media alone as control. In a second experiment, bacteria were injected into the tibia through a hole drilled on one end. The hole was filled with wax and the light emission determined in real time using the low light imager. Luminescent activity through the bone was observed only in the experimental group (Figure 3).

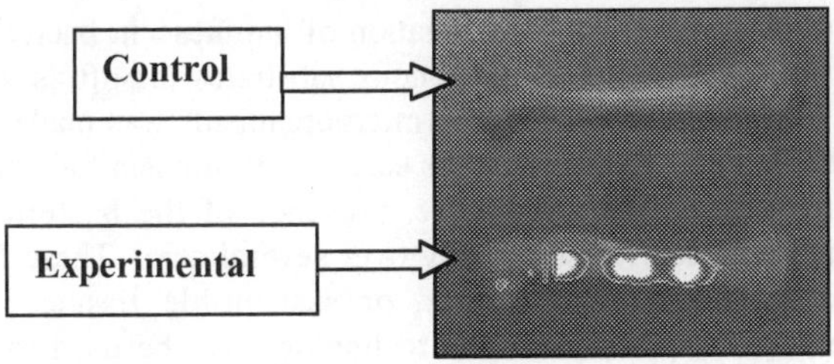

Fig. 3. Luminescent bacteria in rat tibia.

3.4 *In vivo* visualization of luminescent bacteria in hard tissues

Luminescent bacteria were visualized through the skull of the animal within thirty minutes of inoculation (Figure 4).

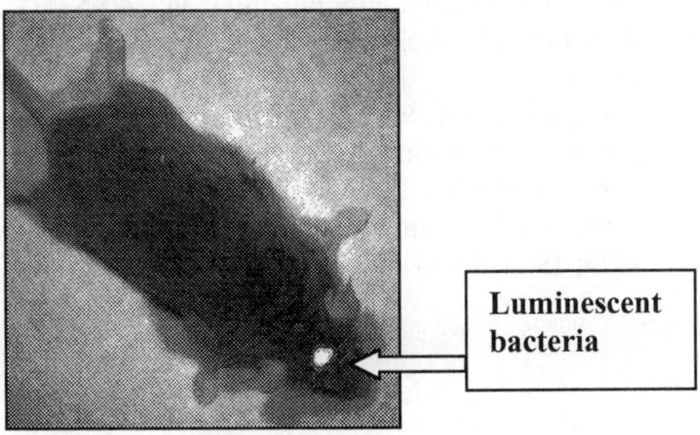

Fig. 4. Visualization of luminescent bacteria through the mouse skull, *in vivo*. This figure represents one minute of photon collection.

3.5 Infective endocarditis experiments

Based on previous reports of infective endocarditis models, animals are observed for the presence or absence of infection of the heart three to seven days following bacterial inoculation.[5] In our study, the hearts were excised three days following inoculation and grown in liquid media overnight to amplify the luminescent cells. The cells were then spun down and visualized under the low light imager (Hamamatsu) for one minute. Luminescent bacteria were only visible in the experimental sample.

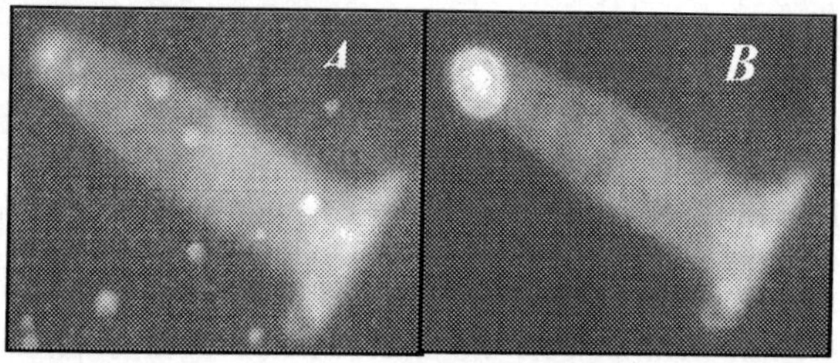

Fig. 5. Visualization of luminescent bacteria isolated from excised hearts. *A,* Presence of luminescent bacteria in experimental group. *B*, No luminescent bacteria in control Group.

4. CONCLUSION

Based on these results, visualization of luminescent bacteria is possible without the need for the addition of exogenous substrates in soft as well as hard tissues. Real time *in vivo* observation of the microorganisms was made under a low light imager (Night Owl and Hamamatsu) as early as 10 minutes following intra-venous delivery of bacterial cells. Furthermore, presence of the bacteria in various organs was verified with luciferase activity over several days. These findings open the way to study colonization of injured or susceptible tissues in animals following a bacteremia. Furthermore, this technology can be used to monitor the survival of cellular implants in bone by non-invasive methods.

References

1. Newman HN. Focal infection. *J Dent Res* 1996; **75**: 1912-1919.
2. Genco RJ. *J Periodontal Res* 1999; **34**: 340-5.
3. Valtonen VV. Role of infections in atherosclerosis. *Am Heart J* 1999; **138**: S431-3.
4. Scannapieco FA. Position paper of The American Academy of Periodontology: periodontal disease as a potential risk factor for systemic diseases. *J Periodontol* 1998; **69**: 841-50.
5. Santoro J, Levison ME. Rat model of experimental endocarditis *Infect Immun* 1978; **19**: 915-918.

GFP IN PLANT BIOTECHNOLOGY AND AGRICULTURE

C. NEAL STEWART, JR., MATTHEW D. HALFHILL, AND HAROLD A RICHARDS
University of North Carolina-Greensboro, Dept of Biology, Greensboro, NC 27402-6174, USA
Email: nstewart@uncg.edu

1. OVERVIEW

Plant biotechnology is here to stay despite the controversies surrounding its current and future roles in agriculture[1]. In fact, private companies and the US government are allocating record-high funding for both the improvement and biosafety research in plant biotechnology. In the USA, the National Academy of Sciences has given its approval of agricultural biotechnology in general.[2] There are a number of concerns surrounding plant biotechnology. Among these are the monitoring of transgene flow from crops to weeds, the early detection of plant diseases and other stresses, and food biosafety. These seemingly disparate topics converge when considering the utility and application of the green fluorescent protein (GFP). Therefore, we will briefly outline these and associated applications of GFP in agriculture with an eye toward a green future.

2. GFP IS USEFUL TO MONITOR GENE FLOW AND EXPRESSION IN TRANSGENIC PLANTS

One of the looming biosafety issues about the use of transgenic plants is the unintended transgene flow from crop to weed. Indeed, GFP was first proposed as a useful monitor for such an application in 1996.[3] Our group has explored the monitoring the expression and movement of an insecticidal *Bacillus thuringiensis* (Bt) transgene in both tobacco and canola. In these cases a synthetic *Bt cry1Ac* transgene was linked to the mGFP5er variant of GFP, with all transgenes under the control of the constitutive CaMV 35S promoter.[4] Plants were transformed using *Agrobacterium*-mediated transformation. GFP transgenic plants fluoresce green when excited by the proper wavelength (Fig. 1).

2.1 Monitoring transgene flow and expression intraspecifically

mGFP5er is excitable with UV or blue light. Therefore, screening progeny of mutative transgenic plants with a handheld UV lamp (UVP 100AP, Upland ,Cal) is a convenient way to distinguish transgenic and nontransgenic plants. The progeny of both tobacco and canola transformants segregated at 3:1 Mendelian ratios, and the presence of GFP corresponded with PCR analysis.[4] More surprisingly, the expression of the Bt transgene corresponded with the expression of GFP, even though the transgenes were merely linked.[4] Threrefore, GFP could be utilized as a visual assay to distinguish transgenic and nontrangenic plants from the same parent, and the approximate expression level of a linked, economically and ecologically important transgene. More recently, we have fused GFP to a pollen-specific promoter and GFP-transgenic tobacco pollen grains could be seen under an epifluorescent microscope segregating at a 1:1 ratio (Laura C. Hudson et al. unpublished data).

Fig. 1 GFP transgenic canola leaf (left) and wildtype leaf (right) excited at 470 nm and viewed with an emission longpass filter at 515 nm..

2.2 Monitoring interspecific transgene movement

A handcross hybridization event of canola (*Brassica napus*) and a wild relative (*Brassica rapa*), containing Bt/GFP could be detected by GFP visualization with a handheld UV light.[4] In the interest of documenting transgene flow from crop to weed, we are currently performing field experiments in Georgia and NC (USA) and collaboratively in Saskatchewan and Ontario (Canada) using the same Bt/GFP construct. In this way, GFP will facilitate large, field-level experiments that will integrate insect herbivore pressure, geography, transgenic line, and weed (*B. rapa* and *Raphanus raphanistrum*), in real time. There is no other *in vivo* universal transgene marker that would allow such experiments to be performed at necessarily large scales and using remore sensing. Such an approach would not only facilitate transgene monitoring, but also identity preservation as well.

2.3 GFP is a neutral ecological marker

Early reports suggested that GFP may be cytotoxic to plant cells.[4] In order for it to be useful as a transgene marker or in other ecological or physiological applications GFP must be neutral. Therefore, we tested if plants expressing GFP have any decreased fitness. We used tobacco plants transformed for either mGFP5er, mGFP4, or S65T (codon optimized for humans).[4] Under field conditions in Reidsville, NC there was no decreased biomass or seed yield associated with GFP synthesis. However, on the ER targeted and S65T variant (non-targeted) were expressed to the degree that GFP could be detected in the field.[4] We have now performed similar experiments with GFP canola in the field and find there is no cost for the plant expressing GFP. Expression levels approach 0.25% of total soluble protein.

3. GFP AS AN INDUCIBLE INDICATOR IN PLANTS

We believe GFP will be increasingly used in precision agriculture and as an enabling technology in agricultural biotechnology.[5] Since it is a neutral marker and can be easily detected in plant organs, it is a good candidate for exploitation. For example, we should be able to take advantage of a plant's ability to sense pathogens through the hypersensitive response to produce an early-detection system to mitigate crop damage from plant disease. Research is underway to fuse pathogenesis related protein gene promoters to GFP, and characterize the response among diseases and crops. Such a system has been shown to work with PR1-GUS fusions, but GUS detection requires destructive sampling and cannot be used in real time.[6] We envision interspersing sentinel GFP-induced crop plants among standard crop plants. Such a system could be monitored at a standoff using laser-induced fluorescence imaging systems.[7] Plants have the unique characteristics as being long-lived, sessile, and root systems that act as concentrators, and have great promise in bioremediation. Various inducible promoters fused with GFP might be used in transgenic plants for the detection of landmines, and the continuous monitoring of environmental contaminants such as heavy metals, perchlorides, and radionuclides. Such uses of biotechnology in bioremediation could indeed make the world a greener place.

4. GFP IN FOOD

From the previous section, it should be evident that, because its great utility, GFP may someday be in the human food supply. In fact, it could be possible that some peoples have routinely consumed GFP in aboriginal diets (fungi, marine organisms, worms). Certainly other animals consume GFP. Since it is a novel protein in human diet, it must be tested for allergenicity and toxicity.[1] Feeding studies in rats will be used as a first approximation of acute oral toxicity of GFP—both synthesized in transgenic plants and as purified protein.

Allergenicity is not as straightforward to test. One accepted method is to compare the allergen candidate to known allergens. We have compared the GFP amino acid sequence to all known allergens in the Genprot, Swiss prot and Protein ID Reserve databases (200 allergens) and have not found potential epitope matches. The closest match has been peanut agglutinin with two-thirds similarity among nine consecutive peptides. Low identity to a known allergen does not guarantee that a protein will not cause an allergic reaction, but it is suggestive that GFP is not a major allergen. Allergens are generally stable to digestion in the vertebrate stomach. We subjected GFP transgenic plants and purified GFP to in vitro digestion using acidic gastric solutions. The results show that GFP in transgenic plants would be completely digested in the stomach (Fig. 2). GFP should not be allergenic in food.

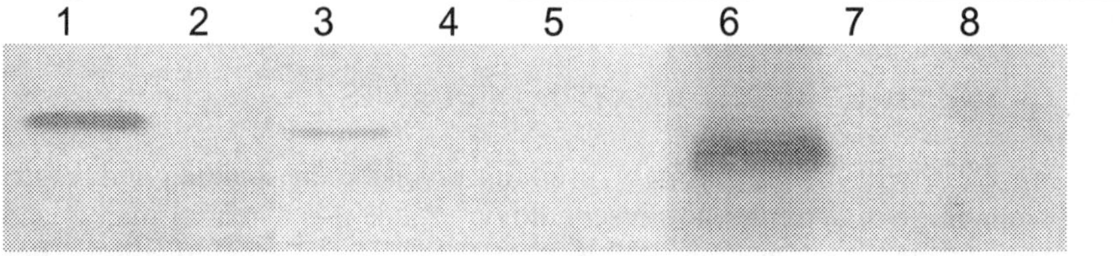

Fig. 2 Protein blot analysis of *in vitro* digestions. Lane 1 contains GFP standard and lane 2 contains control canola digesta. Lanes 3 and 6 are homogenate fractions. Lanes 4 and 7 are gastric phase digesta. Lanes 5 and 8 are intestinal phase digesta. The blot indicates that GFP is broken down during the gastric phase.

Acknowledgements

We wish to acknowledge Reggie Millwood and Steve Mabon for technical assistance in several of the studies that have led to this paper. We thank Staci Leffel and Brian Harper for much of the field work that has contributed to this paper We acknowledge the support of NSF grant 9604528, USDA Biotechnology Assessment grant 98-335226797, and funding from Dow AgroSciences.

References

1. Stewart CN Jr, Richards HA, Halfhill MD. Transgenic plants and biosafety: science, misconceptions and public perceptions. *BioTechniques,*2000; in press.
2. National Research Council. *Genetically Modified Pest Protected Plants Science and Regulation.* National Academies Press, Washington, DC, 2000: 315 pp.
3. Stewart CN Jr. Monitoring transgenic plants with *in vivo* markers. *Nature Biotech* 1996; **14:**682.
4. Harper BK, Mabon SA, Leffel SM, Halfhill MD, Richards HA, Moyer KA, Stewart CN Jr. Green fluorescent protein in transgenic plants indicates the presence and expression of a second gene. *Nature Biotech* 1999; **17:**1125-1129.
5. Leffel SM, Mabon SA, Stewart CN Jr. Applications of green fluorescent protein in plants. *BioTechniques* 1997; **23:** 912-918.
6. Mitter N, Kazan K, Way HM, Broekaert WF, Manners JM. Systemic induction of an Arabidopsis plant defensin gene promoter by tobacco mosaic virus and jasmonic acid in transgenic tobacco. *Plant Sci* 1998; **136:**169-180.
7. DiBenedetto J. Laser-induced fluorescence remote sensing. *Second International Symposium on GFP Program and Abstracts* 1999.

VISUALIZATION OF VACCINIA VIRUS INFECTION USING THE *RENILLA*-LUCIFERASE-GFP FUSION PROTEIN

TATYANA M. TIMIRYASOVA[1], YONG A.YU[2], SHAHROKH SHABAHANG[2,3], ISTVAN FODOR[1,3], ALADAR A. SZALAY[2]

Center for Molecular Biology & Gene Therapy[1], Department of Biochemistry[2], Department of Microbiology & Molecular Genetetics[3], Loma Linda University & School of Medicine, Loma Linda, CA 92350, USA

E-mails: ifodor@som.llu.edu; aszalay@som.llu.edu

1. INTRODUCTION

The application of reporter genes in virology has become particularly important for the construction of mutant viruses, study of gene expression, identification of infected cells, examination of viral pathogenesis in animals and performing gene therapy. Several reporter genes have been engineered into the virus genomes for monitoring gene expression: the *Escherichia coli* β-galactosidase gene[1,2], the firefly (*Photinus pyralis*) luciferase gene[2-4], the green fluorescent protein (GFP) of jellyfish *Aequorea victoria*[5,6]. The GFP is applicable for *in vitro* and *in vivo* analysis and in addition, it allows *in situ* detection of gene expression. In contrast to other marker proteins, GFP is not an enzyme, and thus, does not require any cofactors or substrates[7] for the light emission.

The *Renilla* luciferase-GFP (Ruc-GFP) fusion protein was shown previously to be as a dual reporter in cell culture and in live animals detected by low light image analyses[4] and fluorescence microscopy[8]. In this study, we used a fused protein composed of GFP and Renilla luciferase as a reporter of vaccinia virus replication *in vivo*. We have constructed a recombinant vaccinia virus rVV-RG effectively over-expressing Ruc-GFP fusion protein in infected cells. A high-resolution reporter gene allows us to monitor replicating viral particles in real time in infected CV-1 cells by imaging and fluorescence microscopy.

2. MATERIALS AND METHODS

Cell culture. Cell lines were obtained from the American Type Culture Collection (Rockville, MD). The rat C6 nitrosourea-induced glioma cell line, originally established and described by[9], was cultured in RPMI - 1640 medium (Cellgro/Mediatech, Inc., Herndon, Virginia) supplemented with 10% fetal bovine serum and antibiotic-antimycotic solution (Sigma Chemical Co., St. Louis, MO). CV-1 African green monkey kidney fibroblasts were maintained in Dulbecco s modified Eagle s medium containing 5% fetal bovine serum and antibiotic-antimycotic solution (Sigma).

Viruses. Lister (LIVP) strain of vaccinia virus[2] was used in these experiments as a wild-type virus. Recombinant vaccinia virus rVV-RG (Fig. 1) was constructed by homologous recombination which will be described elsewhere (Timiryasova et al, manuscript in preparation). rVV-RG contains fusion of *Renilla* luciferase (*ruc*) and *gfpc DNAs* in nonessential region of wild-type vaccinia virus under the control of vaccinia strong synthetic early/late (PE/L) promoter. Infection of the cells was carried out at a multiplicity of infection (MOI) of 0.1 plaque-forming units (PFU) per cell. Virus suspension was added to cell monolayers and incubated at 37^0C for 1 h with brief agitation every 10 min. Then, the virus was removed, cells were washed once with serum free medium, complete growth medium was added, and the cells were incubated at 37^0C. VV was propagated in CV-1 cells and purified through sucrose gradient. Standard plaque assay has been used to determine the titer of virus. At 72 h post-infection (p.i.), the cells were stained with 50% of crystal violet solution in ethanol. The viral titer was expressed as $\log_{10}$ PFU/mL.

Fluorescence microscopy of GFP. Analysis of GFP expression by CV-1 cells grown in 60 mm plates (Corning) infected with recombinant rVV-RG virus at MOI of 1 was performed by inverted Axiovert 100TV microscope (Zeiss, Oberkkochen, Germany) using fluorescein isothiocyanate (FITC) filter. Mock-infected cells were used to determine the background fluorescence due to medium, cells and plastic.

Animals. Athymic nude mice (outbred background, *nu/nu*; Charles River Breeding Laboratories, Inc., Wilmington, MA) were purchased at 4-5 weeks of age and acclimatized for approximately 3 weeks before beginning the experiments. Biosafety level 2 precautions were implemented during housing and handling of the animals, as described in Guidelines for Research involving Recombinant DNA Molecules (NIH Guidelines, 1994). Euthanasia by rapid CO_2 asphyxiation was performed in compliance with the NIH Guide for the Care and Use of Laboratory Animals. All animal care was performed under the direction of a licensed veterinarian. The experimental protocol was approved by the University Animal Research Committee.

3. RESULTS AND DISCUSSION

We have constructed recombinant vaccinia virus strain Lister rVV-RG carrying a fusion of *Renilla* luciferase and GFP under the control of strong vaccinia synthetic PE/L promoter by homologous recombination (Fig. 1)

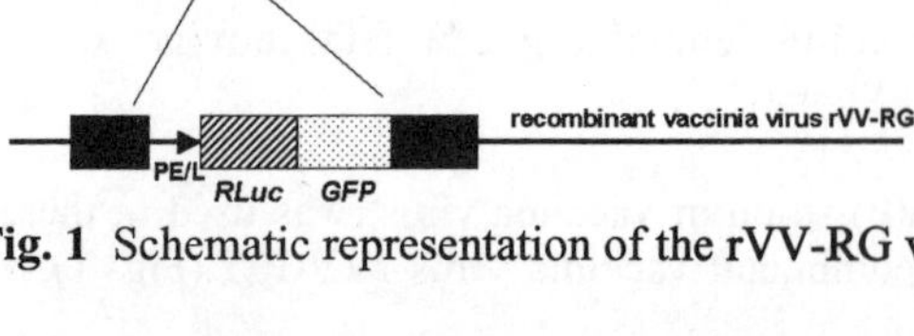

Fig. 1 Schematic representation of the rVV-RG virus.

High levels of GFP expression in cells infected with recombinant rVV-RG virus at MOI of 0.1 were monitored by a fluorescence microscopy 48 h post-infection (Fig. 2). Expression of *Renilla* luciferase by rVV-RG virus in *in vitro* was measured by luminometer and monitored using a Hamamatsu Low Light Video Camera or a Nightowl Low Light Imager (EG&G). CV-1 cells grown in 24-wells plates (2 x 10^5cells/well) were infected with 16 single rVV-RG virus plaques. At 48 h post-infection the plate was frozen and thawed three times and the RUC activity was measured in 5 μL from 1 mL of the growth medium. The range of luciferase activity was between 9 x 10^5 to 2 x 10^6 RLU/mL.

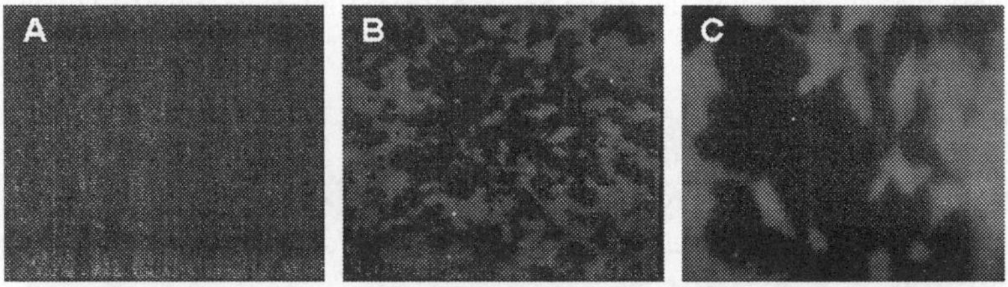

Fig. 2 Vaccinia virus-mediated GFP expression *in vitro*. A, CV-1 cells infected with wild-type vaccinia virus; B and C, CV-1 cells infected with recombinant vaccinia virus rVV-RG, 100x and 400x magnification, respectively.

In order to visualize the efficiency of vaccinia virus infection in the animals, 1x10^8 PFU of recombinant virus rVV-RG or control wild-type virus was injected subcutaneously to nude mice in 0.1 mL of PBS.

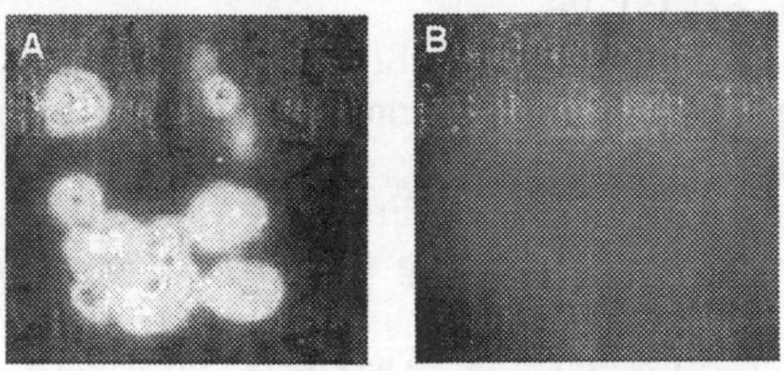

Fig. 3 Detection of recombinant rVV-RG virus infection in animals. A, luciferase expression; B, GFP expression.

Virus infection was monitored at 48 h after virus injection. The mice were anesthesized, and 1 μg of coelenterazine in 50 μL of luciferase assay buffer was injected into the site where the viral infection occurred. Activity of recombinant luciferase and GFP was clearly detectable in live animals (Fig. 3) using a low-light video camera and fluorescence microscopy, respectively.

4. CONCLUSIONS

We have constructed a recombinant vaccinia virus strain rVV-RG which over-expresses RUC-GFP fusion protein. Vaccinia virus replication was easily monitored in cell culture by fluorescence microscopy. Vaccinia virus infection and dissemination can be monitored in live animals by low light imaging and fluorescence microscopy.

Acknowledgements

This work was supported by the Basic Science Research Grants from LLU School of Medicine to I.F. and A.A.S.

References

1. Boldogkoi Z, Medveczky I, Gl vits R, Braun A, Fodor I. *In vivo* studies on Aujeszky's disease virus mutants. *Acta Microbiol. Immun. Hung.* 1996; **43:** 307-318.
2. Timiryjasova TM, Kopylova-Sviridova TN, Fodor I. Analysis of expression of the reporter genes in the different regions of vaccinia virus genome. *Molec Biol (in Russian)* 1993; **27:** 392-401.
3. Kovacs F, Mettenleiter.TC.. Firefly luciferase as a marker for herpesvirus (pseudorabies virus) replication *in vitro* and *in vivo*. *J. Gen. Virol.* 1991; **72:** 2999-3008.
4. Langridge WHR, Krausova VI, Szalay AA, Fodor I.. Detection of baculovirus gene expression in insect cells and larvae by low light video image analysis. *J Virol Meth* 1996, **61:** 151-156.
5. Eriksson S, Raivio E, Kukkonen JP, Eriksson K, Lindquist K. Green fluorescent protein as a tool for screening recombinant baculoviruses. *J Virol Methods* 1996; **59:** 127-133.
6. Boldogkoi Z, Erdelyi F, Sik A, Freund TF, Fodor I. Construction of a recombinant herpesvirus expressing the jellyfish green fluorescent protein. *Luminescence* 1999; **14:** 69-74.
7. Chalfie M, Tu Y, Euskirchen G, Ward WW, Prasher DC. Green fluorescent protein as a marker for gene expression. *Science* 1994; **263:** 802-805.
8. Wang Y, WangG, O Kane DJ, Szalay AA. The *Renilla* luciferase-modified GFP fusion protein is functional in transformed cells. In: Hastings JW, Kricka LJ, Stanley PE, editors. Bioluminescence and chemiluminescence: molecular reporting with photons. Chichester: Wiley 1997; 419-422.
9. Benda P, Lightfoot J, Sato G, Levine L, Sweet W. Differentiated rat glial cell strain in tissue culture. *Science* 1968;**161:** 370-371.

GENOME-WIDE EXPRESSION PROFILING WITH *luxCDABE* GENE FUSIONS

TINA K. VAN DYK[1], GREGORY E. GONYE[1,3], MARY JANE G. REEVE[1], ELLEN J. DEROSE[1], MAUREEN DOLAN[2], MICHAEL K. HANAFEY[2], YAN WEI[1,4], J. ANTONI RAFALSKI[2], AND ROBERT A. LAROSSA[1]

[1]DuPont Company, Central Research and Development Department, Wilmington, DE 19880-0173 USA

E-mail: Tina.K.Van-Dyk@usa.dupont.com

[2]DuPont Agricultural Enterprise, Genome Science, Newark, DE 19714 USA

[3]Current address: Thomas Jefferson University, Department of Pathology, Anatomy and Cell Biology, Philadelphia, PA 19107 USA

[4]Current address: Roche Vitamins Inc., Genomics and Screening, Nutley, NJ 07110 USA

1. INTRODUCTION

Organisms respond to stressful conditions by various mechanisms including changes in gene expression. Monitoring of these transcriptional alterations has found utility in several areas including molecular toxicology and environmental detection. Most applications to date, however, have monitored one or a small number of selected stress responses within an organism.[1] With the availability of complete genomic sequences for many microorganisms, comprehensive transcriptional analyses with greatly expanded information content are now possible. Here we describe a genomic approach to gene fusion technology using bioluminescence specified by *luxCDABE,* a sensitive and easily assayed reporter for gene expression. DNA sequence determination for each member of a collection of random fusions of *Escherichia coli* genomic DNA to this reporter allowed precise identification of each fusion. Use of such genome-registered collections allows detailed analyses of gene expression changes induced by stress conditions. The resultant patterns will be useful to characterize the biological effects of chemicals and toxicants and will lead to discoveries of novel stress responses useful for environmental monitoring.

2. SEQUENCING AND MAPPING THE LUX-A COLLECTION

The Lux-A Collection consists of 8066 individual *E. coli* transformants, each containing an *E. coli* genomic fragment in plasmid pDEW201, which carries promoterless *Photorhabdus luminescens luxCDABE* genes.[2,3] Plasmid DNA isolated from each transformant was used for DNA sequencing reactions performed with previously described primers.[3] To map each member of the

Lux-A collection with respect to the *E. coli* genome, a homology search for the sequence from the beginning and end of each Lux-A clone was performed against the complete *E. coli* sequence.[4] The location of each Lux-A clone on the *E. coli* genome was then based on the computed homologies for both the beginning and end of the insert. The results were used to generate a web-based, tabular interface to the data and a Java-based graphical interface that allowed visualization of positional and directional relationships (Figure 1).

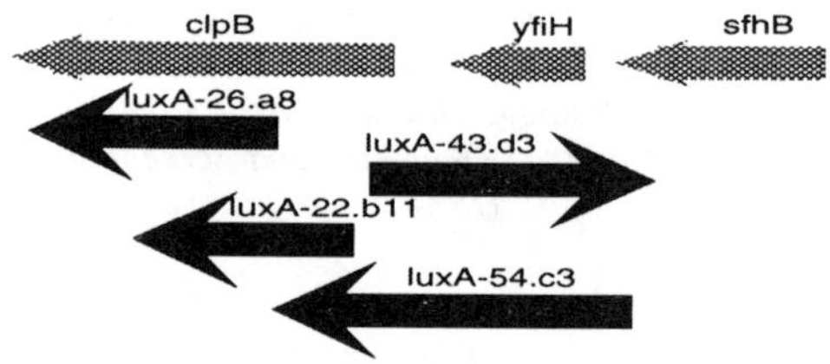

Fig. 1. An example of the graphical view for the *E. coli* chromosomal region near the heat shock gene *clpB*. The hatched arrows indicate the chromosomal genes. The solid arrows indicate the location and orientation of the chromosomal segments in LuxA library plasmids mapped to this region. The direction of the arrowhead indicates if the DNA of the cloned chromosomal segment is oriented such that a promoter, if present, would be driving expression of genes on the direct (arrowhead to the right) or complementary (arrowhead to the left) strand. In this example, the gene fusion luxA-54.c3 contains a functional fusion of the *clpB* promoter region to the *luxCDABE* reporter.

3. VERIFICATION OF GLOBAL REGULATORY RESPONSES

In order to validate utility of this genome-wide registered collection of gene fusions, the biological responses of selected strains containing *luxCDABE* fusions to members of several well-characterized regulatory circuits were tested. For example, the Lux-A Collection contains fusions to four of twelve σ^{32}-controlled heat shock promoters. Each of these were tested by stressing actively growing cultures in LB medium with ethanol or captan, potent chemical inducers of the heat shock response in *E. coli.*[1] Figure 2 shows typical results for the *E. coli* strain carrying the *clpB-lux* gene fusion. Similar results were obtained for the four other members of the heat shock regulon from the LuxA collection. Likewise, gene fusions to the *lac* operon, and to members of the SOS, SoxRS, and OxyR regulons were found in the Lux-A Collection and responded as anticipated to known inducers of these global regulatory circuits. Thus, while not yet comprehensive, this collection mirrors the transcriptional wiring diagram of *E. coli* providing functional gene fusions to major regulatory circuits. In the same way, representation of other, perhaps as of yet undiscovered, regulatory circuits should be found the LuxA collection.

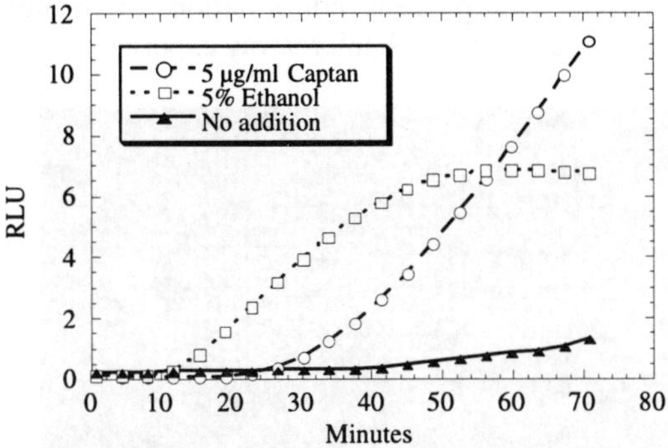

Fig.2. Response of the *E. coli* strain with the gene fusion luxA-54.c3, *clpB-luxCDABE,* to chemical inducers of the heat shock response. Actively growing cultures in LB medium were mixed with the specified chemical at time zero. Light was measured in a ML3000 luminometer (Dynex) at 37 °C. No addition, closed triangles; 5 μg/mL captan, open circles; 5% ethanol, open squares. The bioluminescence of the untreated culture increases with time due to growth. Following a lag time in which the stress response is initiated and the *lux* reporter is transcribed and translated, the bioluminescence of the chemically treated cultures increases indicating increased transcription initiated at the *clpB* promoter in response to the stress.

4. USES OF MAPPED COLLECTIONS OF GENE FUSIONS

The LuxA collection was used to develop a highly parallel cell-based platform for gene expression analysis. A maximal nonredundant subset containing functional reporter constructs was selected using a computational filter in conjunction with a list of definitions of 2584 documented and predicted operons in *E. coli.*[5] The set of strains containing these 714 identified plasmids, termed LuxArray1.0, consists of 42% fusions to operons of known genes and 58% fusions to operons of uncharacterized open reading frames. Cellular arrays generated using high density printing to membranes on agar plates (Figure 3) can be used and analyzed in a fashion analogous to hybridization assays to monitor transcriptional changes.

The genome-wide collection of gene fusions has several further applications. Previously unknown and unpredicted promoter activities can be discovered and characterized by measuring bioluminescence from the large random collection of gene fusions. Conversely, promoters that are predicted from genomic sequence data may be found non-functional in specified conditions. Thus, bioluminescent reporter activity in conjunction with the map position of the reporter gene fusion can be used to suggest operon structure. In addition, the availability of a large collection of gene fusions allows selection of appropriate gene fusions to provide independent confirmation of results from other expression analyses, such as from DNA microarray experiments.

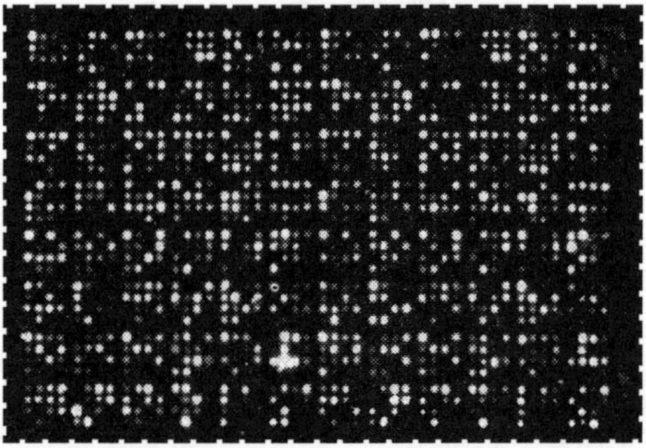

Fig. 3. LuxArray 1.0. LuxArray1.0 cultures grown overnight at 37°°C in microplates were used as the cell source for printing. A BioMek 2000 (Beckman Coulter) equipped with a high density replication tool was used to generate a 4 X 4 density on sterilized 8x12°cm membranes placed in contact with LB media in a culture dish. Thus, 1536 spots were placed on the membrane allowing each clone to be represented in duplicate along with controls. This image was taken after 14°h of growth using a cooled CCD camera (FluorChem 8000: f0.85 lens, 2min exposure. AlphaInnotech). The intensity of each spot correlates with promoter activity of the *lux* gene fusion at that location.

References

1. LaRossa, RA, Van Dyk, TK. Applications of stress responses for environmental monitoring and molecular toxicology. In: Storz, G. and Hengge-Aronis, R., eds. *Bacterial Stress Responses,* Washington, DC: ASM Press, 2000: 455-468.
2. Van Dyk, TK, Rosson, RA. *Photorhabdus luminescens luxCDABE* promoter probe vectors. In: LaRossa, R. A., ed. *Methods in Molecular Biology: Bioluminescence Methods and Protocols,* Towowa, NJ: Humana Press Inc., 1998: 85-95.
3. Van Dyk, TK, Ayers, BL, Morgan, RW, LaRossa, RA. Constricted flux through the branched-chain amino acid biosynthetic enzyme acetolactate synthase triggers elevated expression of genes regulated by *rpoS* and internal acidification. *J Bacteriol* 1998; **180**:785-792.
4. Blattner, FR , Plunkett, G III, Bloch, CA *et al.* The complete genome sequence of *Escherichia coli* K-12. *Science* 1997; **277**:1453-1462.
5. Thieffry, D, Salgado, H, Huerta, AM, Collado-Vides, J. Prediction of transcriptional regulatory sites in the complete genome sequence of *Escherichia coli* K-12. *Bioinformatics* 1998; **14**:391-400.

INDUCIBLE GENE EXPRESSION *IN VIVO* USING A *RENILLA* LUCIFERASE – GFP FUSION CONSTRUCT

YONG A. YU[1,2], SHELLEY CALTHARP[1], ALADAR A. SZALAY[1]
Department of Biochemistry[1,2], Division of Human Anatomy and Pathology[2],
Loma Linda University, Loma Linda, CA 92350
USA
E-mail: aszalay@som.llu.edu

1. INTRODUCTION

In vivo studies of numerous eukaryotic genes have been carried out in animal models through over-expression of the gene-of-interest. However, constitutive expression of non-physiological genes may not be desirable in situations where accumulation of the over-expressed proteins starts to interfere with normal physiological processes. Furthermore, many of the studies have aimed to study protein functions during specific stages of development of the organism in a spatially and temporally controlled manner. Such experiments cannot be properly performed using transgenic animals that carry the constitutively expressed genes. To address this problem, inducible gene expression systems have been developed.[1,2] Among them, "Gene Switch" is a two-plasmid system, which uses a progesterone antagonist (RU486) as the exogenously added inducer drug to induce gene expression.[1] In this system, one of the two plasmids is the reporter plasmid encoding the inducible reporter gene, and the other plasmid is the inducer plasmid encoding the inducer chimeric protein. Binding of RU486 to the inducer protein activates the inducer to trigger transcription of the reporter gene.

We use Gene Switch system to study the regulated expression of a *Renilla* luciferase-GFP fusion (RG) construct in cell cultures and in animal models.[3] We showed that significant levels of RG marker gene expression can be induced in tissue cultures through the addition of RU486 inducer drug. We also showed that similar inducible RG expression could be observed from transformed cellular implants in live animals.

The establishment of this rapid and noninvasive method to monitor inducible gene expression in living organisms will help investigators to develop imaging procedures to continuously follow the location of therapeutic proteins during gene therapy.

2. MATERIALS AND METHODS

Construction of plasmid vectors. A fragment of pCEP-4-RG, containing the RG expression cassette under the control of CMV promoter, was inserted into the reporter plasmid p17×4-TATA of the Gene Switch system (from Yaolin Wang,

Department of Cell Biology, Baylor College of Medicine, Houston, TX) to create p17×4-TATA-RG.

Cell cultures and transient transfection assays. 5×10^5 COS cells were seeded in 6-well plates, grown overnight in DMEM supplemented with 10% fetal bovine serum. Cells were transfected with designated amounts of DNA using calcium phosphate precipitation method (Promega). Twenty-four hours after transfection, RU486 (in 100% ethanol) was added to the media to initiate gene expression inductions. Following RU486 induction, cells were tested for luciferase activity (using coelenterazine as substrate) and GFP signal (under a Zeiss UV microscope).

Luciferase assays. The culture medium were removed by aspiration, and replaced with 1 ml of luciferase assay buffer (0.5 M NaCl; 1 mM EDTA; and 0.1 M potassium phosphate, pH7.4) along with 1 μg of coelenterazine (0.5 μg/μL methanol solution). The cells were then imaged using a Hamamatsu Low Light Video Camera or a Nightowl Low Light Imager (EG&G). The light emissions from luciferase assays can also be monitored using a Turner TD 20e luminometer.

Transplantation of *ex vivo*-modified cells into muscles of nude mice. COS cells (~ 5×10^6) were grown in two 10-cm plates. The cells were transfected using the same calcium phosphate method as used for 6-well plates. Twenty-four hours after transfection, the cells are trypsinized, and harvested by centrifugation. Cell pellet was resuspended in serum-free DMEM medium to yield final concentration of 1×10^8 cells/mL. One hundred microliter of the cell suspension were injected into the limb muscles of 7-8 week old male nude mice anesthetized with 60 mg/kg of Nembutal. Meanwhile, 250 nM (50 μL in volume) of RU486 was injected into the intraperitoneal cavity of the mice. Thirty-six hours after injections, the mice were re-anesthetized, and 1 μg of coelenterazine in 50 μL of luciferase assay buffer was injected into the muscles where the cells were transplanted. The mice were then imaged for five minutes under a Hamamatsu Low Light Video Camera.

3. RESULTS

Inducible expression of *Renilla* luciferase – GFP fusion (RG) construct in transiently transfected cells. Inducible RG gene expression was first tested in cell cultures. Approximately 18% of the cells exhibited green fluorescence after treatment with 250 nM of RU486. Virtually no GFP signals were observed in control cells lacking RU486 induction. The highest inducible level of luciferase expression we observed was ~ 30-fold (data not shown).

RU486-dependent target gene induction by inducer plasmid. Various ratios of inducer and reporter plasmids were co-transformed into the cells to test gene expression (Fig. 1). Under the same amount of RU486 (250 μM) treatment, it became evident that by limiting the amount of inducer plasmid, the leakiness of gene expression could be kept at a relatively low level. A reporter to inducer ratio of 20-100 : 1 allowed more optimal inductions while maintaining a basal level of leakiness than application using a 1 : 1 ratio.

The RU486 dose dependent manner of the regulatory system. We treated the transiently transformed cells with a wide range of RU486 concentrations (Fig. 2). We observed that treatment with 250 nM RU486 yielded the highest levels of reporter gene expression.

Kinetics of the inducible system. Induction was evident just three hours after exposure to RU486 (Fig. 3). The induced gene expression occurred over a linear range following treatment with RU486 during the 17-hour exposure time. From then on, the activities enter the plateau phase although signals were still increasing throughout the 38 hour-durations of our experiments. In the control group, even with the presence of both plasmid components, gene expression remained at basal levels in the absence of the inducer drug RU486.

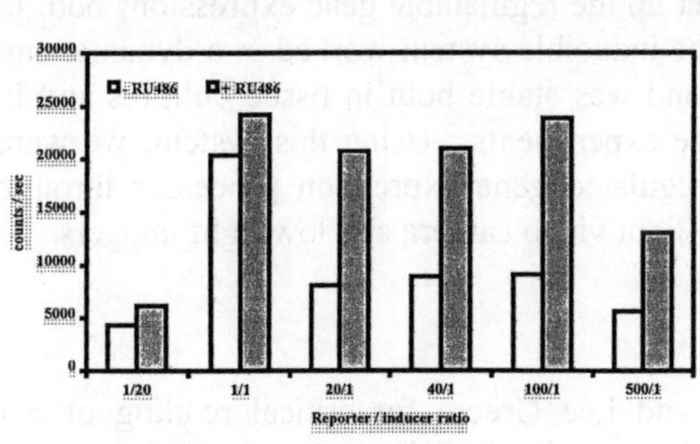

Fig. 1 Various ratio of inducer and reporter plasmid were used for gene expression.

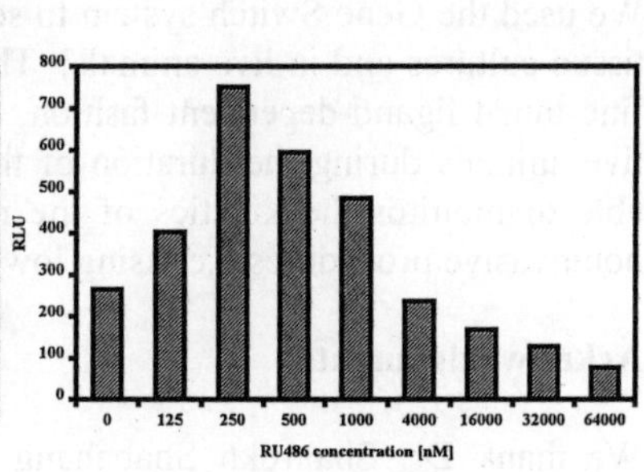

Fig. 2 RU486 dose dependent curve.

***In vivo* studies of the induction system.** We co-injected naked DNA of both reporter and inducer plasmids (20 : 1) into the limb muscles of 7-8 week old nude mice (Fig. 4A). RU486 (50 μg/kg) was delivered by intraperitoneal injections. When both plasmids of the inducible system were present, RG marker gene expression was clearly detectable, with an estimated >12 fold induction. In contrast, without the inducer plasmid (Fig.4A-b), the reporter plasmid alone remained quiescent and no marker gene expression was observed. In addition to direct DNA injections experiments, a cellular implantation approach using *ex vivo*-modified COS cells was performed in the limb muscles of nude mice (Fig. 4B). COS cells were co-transformed with reporter and inducer plasmids (20 : 1) before they were injected into the limb muscles. We showed that RG expression from these implanted cells could also be induced by intraperitoneal injection of RU486.

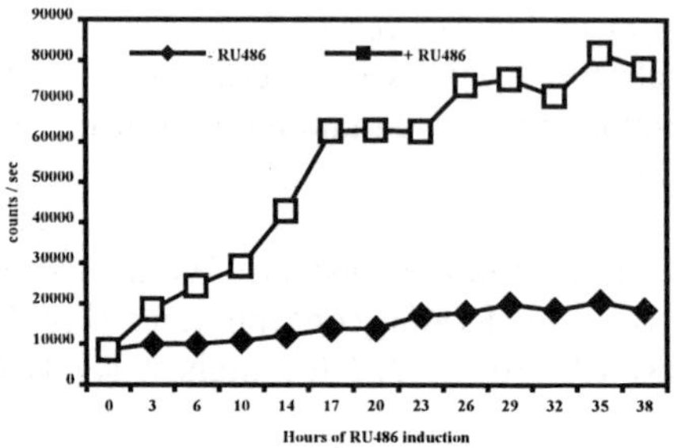

Fig. 3 Kinetics of the inducible system.

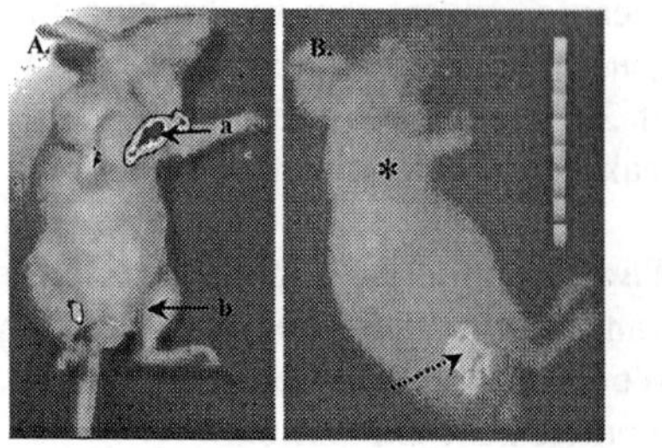

Fig. 4A Gene expression from injected DNA. **4B** Gene expression from cellular implants.

4. CONCLUSION

We used the Gene Switch system to set up the regulatable gene expressions both in tissue cultures and in live animals. The inducible system worked in a dynamic and fine tuned ligand-dependent fashion, and was stable both in tissue cultures and in live animals during the duration of the experiments. Using this system, we were able to monitor the kinetics of the regulated gene expression processes through noninvasive procedures, i.e. using low-light video camera and low light imagers.

Acknowledgements

We thank Dr. Shahrokh Shabahang and Lee Greere for critical reading of our manuscript. This work was supported by the Basic Science Research Grant from LLU School of Medicine.

References

1. Wang Y, O'Malley BWJr., Tsai SY, O'Malley BW. A regulatory system for use in gene transfer. *Proc Natl Acad Sci USA* 1994; **91**: 8180-8184.
2. Delort JP, Capecchi MR. TAXI/UAS: A molecular switch to control expression of genes *in vivo*. *Hum Gene Ther* 1996; 7, 809-820.
3. Wang Y, Wang G, O'Kane DJ, Szalay AA. The *Renilla* luciferase-modified GFP fusion protein is functional in transformed cells. In: Hastings, JW, Kricka, LJ, Stanley, PE, editors. *Bioluminescence and chemiluminescence: molecular reporting with photons.* Chichester: Wiley 1997; 419-422.

SYNTHETIC LUCIFERASE GENES AS BETTER REPORTER MOLECULES

YAO ZHUANG, AILEEN P. PAGUIO, MONIKA G. GRUBER, KEITH V. WOOD
Promega Corporation, 2800 Woods Hollow Rd., Madison, WI 53719, USA

1. INTRODUCTION

Reporter gene technology has been widely used for the study of transcriptional regulation, and it can also be used to monitor other cellular events such as signal transduction pathways or in cell-based assays in drug discovery.[1-3] The true value and the proper utility of this system are highly dependent on its sensitivity and reduced anomalous expression. A systematic approach has been used to design the synthetic luciferase genes aimed to improve such reporter properties, which involves using optimized mammalian codons for the synthetic gene and remove transcription regulatory elements. We demonstrate here that the resulting synthetic genes exhibit better characteristics that are desirable for better reporter functions.

2. MATERIALS AND METHODS

Synthetic genes for *Renilla* luciferase and two click beetle luciferases were designed by a systematic approach. Briefly, mammalian preferred codons were used first to assemble the gene sequences for luciferase proteins; then the undesirable sequences, in particular, transcriptional regulatory elements such as transcription factor binding sites were identified through available database and were subsequently removed by using alternative codons for the same amino acid. The resulting gene sequences are largely devoid of these undesirable sites while the overall codon usage has been improved for mammalian expression. Three synthetic luciferase genes were designed and constructed coding respectively for *Renilla* luciferase gene, red light emitting and green light emitting click beetle luciferases (RD and GR). These synthetic genes were cloned into mammalian expression vectors by standard cloning methods and were transfected into mammalian cells using lipid-mediated transfection methods.[4,5] Cells were harvested 24 hours after transfection and luciferase assays were carried out using DLR kit from Promega Corporation.[6]

3. RESULTS AND DISCUSSION

To test whether the synthetic luciferase genes have improved expression in mammalian cells, we cloned synthetic *Renilla* luciferase gene as well as native *Renilla* luciferase gene into a pGL3-control vector backbone under strong viral SV40 promoter and enhancer, the synthetic gene always exhibits increased expression when compared to its wild-type counterpart (Table 1). Large increases in expression levels were also observed for the synthetic click beetle luciferase.

Table 1. Enhanced expression of synthetic *Renilla* luciferase gene in mammalian cells

	Amount Vector	Fold Expression Increase
CHO	0.2 μg	142
	2.8 μg	145
NIH3T3	0.2 μg	326
	2.0 μg	593
Hela	0.2 μg	185
	1.0 μg	103
CV-1	0.2 μg	68
	2.0 μg	72

With reduced spurious expression, the synthetic gene should exhibit less basal level transcription in a promoterless vector; it should also exhibit less transcriptional activity in the presence of an enhancer. So we cloned synthetic and native *Renilla* luciferase genes into the pGL3-basic and pGL3-enhancer vectors to compare the expression from synthetic gene and the native gene. Because the synthetic gene itself has increased expression efficiency, we cannot compare the activity of the two directly, rather, we take this into consideration by comparing their relative activity in reference to the control vector (expression from the basic or enhancer vector divided by the expression from the fully functional expression vector with both promoter and enhancer elements). The data shows that the synthetic genes have reduced basal level transcription and also eliminated the anomalous expression in the presence of an enhancer (Table 2). Similar results were obtained for the synthetic click beetle luciferases (data not shown).

Table 2. Reduced risk of anomalous expression for the synthetic *Renilla* luciferase gene

	Percent activity of control vector		
Vector	CHO cells	NIH3T3 cells	Hela cells
control-native	100	100	100
control-synthetic	100	100	100
basic-native	4.1	5.6	0.2
basic-synthetic	0.4	0.1	0.0
promoter-native	5.9	7.8	0.6
promoter-synthetic	15.0	9.9	1.1
enhancer-native	42.1	123.9	52.7
enhancer-synthetic	2.6	1.5	5.4

Transient transfections are often carried out in the so-called co-transfection experiments, in which along with the primary reporter construct containing the testing promoter, a second control reporter under a constitutive promoter is transfected into cells as an internal control to normalize experimental variations including transfection efficiencies between the samples. Control reporter signal, potential promoter cross talk between the control reporter and primary reporter, as well as potential regulation of the control reporter by experimental conditions, are important factors that affect the reliability of the results, and the synthetic gene can provide great advantage.

A. Increased reporter signal. With better expression efficiency, the synthetic luciferase genes give out higher signal when transfected into mammalian cells (table 1). The increase in light output will increase the sensitivity of the reporter assay, and making it possible to study weak promoters or rare events; study in cells with low transfection efficiency; reduce the amount of sample required or the time needed for sample reading to improve the assay throughput.

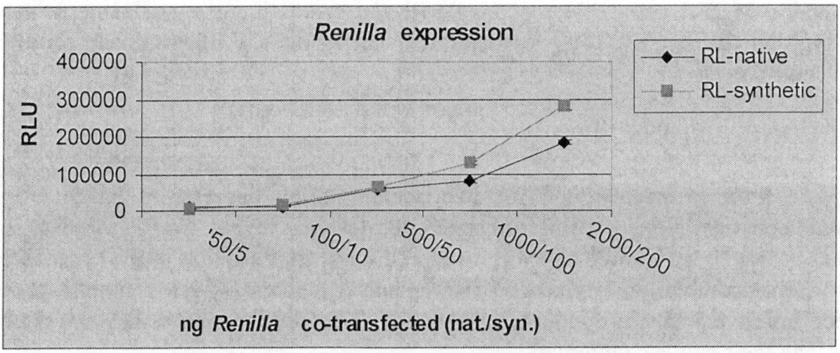

Figure 1a. Expression of *Renilla* luciferase from native and synthetic gene

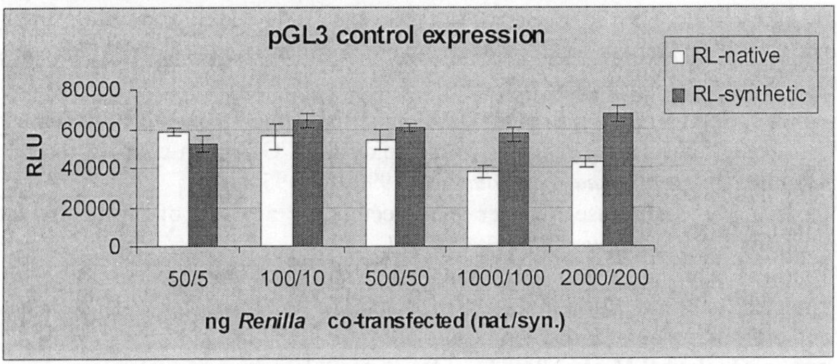

Figure 1b. Reduced promoter interference with synthetic *Renilla* luciferase gene

Table 3. Induction of control reporter by TPA

Vector	Induction
RL-TK	4.4
RG-TK	4.4
hRG-TK	0.9
RG-B	8.0
hRG-B	1.5

B. Reduced risk of promoter cross talk. Because higher signal is obtained, less promoter activity is required to achieve the same reporter signal, this reduces risk of promoter interference.[7] Shown in Figure 1 is an experiment demonstrating that 10 fold less synthetic *Renilla* luciferase is required to give similar or more signal as the native gene, without inhibiting expression from the primary reporter pGL3-control.
C. Reduced risk of physiological interference. Experimental treatment sometimes may activate cryptic sites within the gene and cause induction or suppression of the co-reporter expression, which would compromise its function as co-reporter for normalization of transfection efficiencies. One example is that TPA induced expression of control reporter vectors harboring the wild-type gene when transfecting MCF-7 cells. By replacing with the synthetic gene, such interference was avoided (table 3). This demonstrates that using a synthetic gene reduces anomalous expression.

4. CONCLUSIONS

Synthetic luciferases designed for better mammalian expression and reduced anomalous expression, exhibit enhanced expression levels from a variety of expression backbones, and also exhibit reduced basal transcription in the absence of an apparent promoter and presence of an enhancer element. Such synthetic genes provide improved sensitivity and reliability and therefore function as better reporter genes than their native counterpart.

References

1. Naylor LH. Reporter gene technology: the future looks bright. *Biochem Pharmacol* 1999; **58**: 749-757.
2. Stables J, Scott S, Brown S, et al. Development of a dual glow-signal firefly and *Renilla* luciferase assay reagent for the analysis of G-protein coupled receptor signaling. *J Recept Signal Transduct Res* 1999; **19**: 395-410.
3. Nordeen SK. Luciferase reporter gene vectors for analysis of promoters and enhancers. *Biotechniques* 1988; **6**: 454-457.
4. Promega Technical Bulletin **260**. TransFast™ Transfection Reagent.
5. Promega Technical Bulletin **216**. Tfx™-10, Tfx™-20 and Tfx™-50 Reagents for the Transfection of Eukaryotic Cells.
6. Promega Technical Manual **046**. Dual Luciferase¤ Reporter Assay System.
7. Farr A, Roman A. A pitfall of using a second plasmid to determine transfection efficiency. *Nucleic Acids Res* 1991; **20**: 920.

PART 9

LUMINESCENCE IN SCIENCE AND EDUCATION

PUBLIC UNDERSTANDING OF SCIENCE AND HEALTH: THE FLORENCE PROJECT

P. PINZANI, C. ORLANDO, B. SALVADORI AND M. PAZZAGLI
Dip. Fisiopatologia Clinica, Viale Pieraccini 6, 50139 Florence, Italy.
E-mail: m.pazzagli@dfc.unifi.it

F. TORRICELLI
U.O. of Cytogenetics and Genetics A.O. of Careggi, Florence, Italy

L. MERCATI
Provveditorato agli Studi di Firenze, Italy

1. INTRODUCTION

Science is the engine that drives our economy and, for over four centuries, has been one of the dominating influences on the evolution of Western culture. Yet the public remain confused, bemused, and, in some European countries, even hostile to science. Public Understanding of Science and Health (PUSH) is thus one of the key issues for Europe to lead in the new Millennium. Over the past ten years, PUSH initiatives have become very popular in several countries within the European Union and outside. Italy, UK and Sweden have initiated an international PUSH programme through two societies, the International Federation of Clinical Chemistry and the International Society for Bioluminescence and Chemiluminescence. PUSH initiatives want to educate both school pupils and adults about the successes of modern medicine, how these affect the lives of everyday people, and how medical inventions are a vital part of our economy.

1.1 . The PUSH initiative in Florence

Several PUSH events have been organized in Florence, including lecture-demonstrations to teachers and students, DNA training for teachers, student projects, and firefly hunts. In 1997, during the Meeting "Bioluminescence and Chemiluminescence 1997" an open session with students and teachers represented the basis for the project PUSH 1999. This initiative was included in the continuing educational programmes of the teachers in the Florence area and some basic experiments of biotechnology (two models were selected: 1. transformation experiments of E.coli with green fluorescent protein; 2. effect of pesticides on DNA) were chosen as models for an integrated approach to the scientific education. 16 teachers of secondary and high school of the area of Florence voluntarily accepted to participate at the project. They were first trained under the supervision of researchers of the University of Florence or of the Hospital of Careggi (fall 1998) and then started to work with their students (beginning of 1999). The products of the project, which has involved about 350 students, 16 teachers and 12 researchers of different disciplines, have been presented at the IFCC-Worldlab 99 Congress as a special event. A Scientific session, a poster session and awards for best results and presentations were included in the programme of the Congress. This session was carried out under the auspices of the IFCC Education and Management Division. Parents also were an important part of the event. The students also met representatives from industry (Abbott, Bio-Merieux, Dade-Behring, Menarini diagnostics, Roche) and were shown several major instruments and technologies used in clinical diagnosis throughout the world. A close activity was also performed during this year and the results presented by the students and teachers at the Italian National Congress on Biotechnologies TEBIO (Genoa, May 2000) during a special event session dedicated to the Teaching Biotechnologies in Schools.

2. STRUCTURE OF THE PUSH INITIATIVE IN FLORENCE

The project was carried out in order to address the following main points.

1. To have students of the secondary school directly performing scientific experiments under the supervision of their teachers and with the collaboration of scientists from academy and research institutions .
2. To indicate the importance of an integrated approach to the explanation of the scientific phenomena and experiments. This approach was considered suitable to help students to be aware of the role of the different disciplines in understanding a scientific phenomenon and thus to facilitate the choice of the university faculty.
3. To induce a more positive approach in the students and in the general population (their parents) on the role of science in human life.

3. THE VIDEO: DNA VISION

A video (8 min in a VHS format) was produced in order to furnish the students of a cultural vision of the potentials and limitations of biotechnologies. It introduces four basic themes: EXPLORE, PRESERVE, BUILD, PARTICIPATE.

EXPLORE!! Know the DNA and how genes work.
This is the basis for understanding the use of biotechnologies. A practical class has been performed with students using the Explorer kits from Bio-Rad based on the use of pGLO plasmid and expression of the GFP in E.coli[1]. A number of lessons both at theoretical and practical level were addressed to the teachers and then presented in schools. These activities showed the role of gene transfection and expression.

PRESERVE!! Protect DNA.
The functionality of living organisms is related to DNA integrity. Protecting DNA integrity from pesticides, UV radiations or radioactivity is essential to preserve life. Students evaluated the effect of pesticides in cultured human lymphocytes using the sister-chromatid exchange test[2].

BUILD!!! Potential of biotechnologies.

Biotechnologies are presented to students as a turning point in modern sciences. The comprehension of the role of genes and of the methods for gene transfer are the basis for the development of new living organisms possessing specific characteristics. Biotechnologies open new perspectives in health care. Laboratory Medicine is presented as one of the fields in which Biotechnologies are producing important developments. This represents the link between the experiences performed by the students at school and their active participation at the Worldlab 99 Meeting).

PARTICIPATE!!! Limitations of biotechnologies.
On the other hand we do not know yet the effects of gene manipulation on living organisms. The OGM in agriculture in particular are considered as a controversial application of biotechnologies. Students are invited to consider the precautionary principle as a necessary rule which in any case should not be considered as an immobilizing principle.

4. CONCLUSIONS

If Europe is to benefit from the discoveries and inventions arising from our very strong science base, it is vital that the public understand why we do what we do, and that ethical and social decisions can only be made when science is fully understood. Furthermore, the health of Europe now depends critically on people understanding their own physiology and chemistry. PUSH initiatives are addressed to the education of both school pupils and adults about the successes of modern medicine, how these affect the lives of everyday people, and how medical inventions are a vital part of our economy. The implementation of an educational module requires a considerable effort and repeated interactions between teachers and scientists from academic institutions. Yet, the experience gained is rarely spread, or even known, by other teachers, educational institutions, or universities. Moreover bioethical and cultural issues, other than diffusion of scientific knowledge, require further implementation in these educational models. These activities need to be organized not only at local or regional level but require a network to link students and teachers with academical institutions and scientific societies interested in the diffusion of science.

Acknowledgments
This project has been supported by the MURST, Region of Toscana, Provincia and Comune di Firenze. BIO-RAD supported the project with the reagents for the experiments in schools. A member of the Italian National Bioethical Committe (Dr. E. Mancini) has been appointed by the President of this Committe to participate as an external member to the PUSH Project. We acknowledge the following schools and teachers for their effort in the project: Scuola media 'Manzoni' (Prof. D. Ciatti, Prof. U. Dejana); Scuola media 'Pieraccini-Rosselli-Verdi' (Prof. C. Colonna, Prof. F. Fossi); Scuola media 'Rosai-Calamandrei': (Prof. D. Frilli); Istituto "Sassetti-Peruzzi" (Prof. K.

Schieder); I.T.C. "Calamandrei" (Prof. M. Borrani); Liceo Sperimentale SS. Annunziata (Prof. M. Olivo).

References

1. BIO-RAD Biotechnology Explorer Kit: Secrets of the Rain Forest. Cat. N. 166-0006-EDU. www.bio-rad.com
2. Dolara P, Salvadori M, Capobianco T, Torricelli F. Sister-chromatid exchange in human lymphocytes induced by dimethoate, omethoate, deltamethrin, benomyl and their mixture. *Mutation Res* 1992; **283:** 113-118.

GFP AS AN EDUCATIONAL TOOL IN BIOTECHNOLOGY DISTANCE LEARNING

W. W. WARD, G.C. SWIATEK AND C. M. THOMSON
Dept. of Biochemistry and Microbiology, Rutgers University, Cook College, New Brunswick, NJ 08901, USA
E-mail: crebb@rci.rutgers.edu

1. INTRODUCTION

In the ever-advancing pharmaceutical and biotechnology industries, it is imperative to maintain a highly skilled workforce by providing quality instruction about new and evolving techniques. Companies that provide opportunities for regular continuing education are at a distinct competitive advantage. Such instruction is best accomplished via short courses taken in a university setting where modern research techniques are transferred in a timely, non-commercial fashion from the research laboratory into the classroom. To this end, we at CREBB have developed a series of biotechnology short courses for industrial scientists that utilize green-fluorescent protein (GFP) as a novel and highly aesthetic teaching tool[1,2,3]. New research results, techniques, and inventions are regularly transferred, with minimal delay, into the instructional program.

Regardless of our feelings about college-level distance learning initiatives, many of us have come to believe that distance learning is here to stay. Thus, it makes sense for university faculty to begin developing continuing education curricula for industrial scientists and undergraduate students, alike, that can be accommodated in a "virtual university" format. Developing such curricula in the biotechnology lab sciences is a demanding challenge as the industrial/student audience will not usually have access to equipment and materials to conduct their own experiments. The instructor will have to conduct experiments and demonstrations in a remote setting, generally inaccessible to the student. For such a program to be effective, it would seem that the exercises would have to arouse great curiosity, generate visual excitement, and provide many opportunities for discovery learning and interactive student feedback. These goals we believe have been reached in our current series of hands-on, industrially oriented short courses.

In distance learning versions of such courses, more emphasis will have to be placed on teaching students to think, reason, and deduce in a simulated laboratory setting. Consequently, less emphasis will be placed on traditional expository teaching and lecturing. The distinctions between university undergraduate instruction and continuing education are likely to blur in the future or, perhaps, to disappear entirely. To satisfy both industrially oriented continuing education needs and undergraduate needs, it may be useful to develop a series of single-concept distance learning modules. In the field of biotechnology, for example, such topics as, "The Generation of Affinity-Tagged Recombinant Proteins" or "Methods Development in HPLC Separations of Biomolecules", could be produced. For such modules to be most useful, they must be visually and intellectually exciting and they must be able to stand alone or to be mixed and matched according to the needs of the student population. In GFP, we have the necessary tool to accomplish these objectives, yet developing such a series of distance learning courses is a daunting challenge

2. PROTEIN PURIFICATION

In 1989, our center, CREBB, at Rutgers University launched a 5_ day, hands-on short course called "Protein Purification: Isolation, Analysis, and Characterization of GFP", a continuing education course geared to the industrial scientist. In its 11-year history, "Protein Purification" has been taken by nearly 600 industrial scientists and 50 science educators and their students. The aim of the course is to teach protein purification from start to finish using whenever possible a problem-solving, interactive approach. Each group of registrants begins the course with a pellet of transformed *E. coli* cells expressing green-fluorescent protein (GFP) and purifies the GFP to homogeneity with a combination of techniques including: ammonium sulfate precipitation, gel filtration, ion exchange chromatography, hydrophobic interaction chromatography, size exclusion HPLC, and immobilized metal ion affinity chromatography (see Fig. 1). Results of purification are analyzed by absorption spectrophotometry, native and SDS gel electrophoresis, and isoelectric focusing. A full description of the course is available via the internet:

http://www.rci.rutgers.edu/~meton/protein.html

3. GREEN-FLUORESCENT PROTEIN (GFP) AS A VISUAL AID

The unique feature of this course and the main reason for its success is that GFP is used, throughout, as a way to generate excitement and as a tool to follow every step of protein purification and characterization in real time. Having GFP as the centerpiece facilitates instant feed-back for each procedure and provides a means for easy and rapid trouble-shooting of instruments and methods. Now that multiple colors are available (blue, green, yellow, and red - - Clontech 'Living Colors'), each is now being incorporated into the course, providing a "technicolor" experience unsurpassed by any other biotechnology teaching tool.

4. GFP IN DISTANCE LEARNING

It is our current plan to create an electronic version of "Protein Purification" so as to reach a wider (international) audience at a lower cost to each registrant. This may be of particular value for those who must travel long distances to take the "live" version of the course; in the current year alone, for example, people have traveled from as far away as Brazil, Japan, England, and India to take "Protein Purification".

Because "Protein Purification" unfolds as a series of sequentially related problem-solving exercises, beginning with cell transformation and concluding with protein sequencing by mass spectrometry, it is our aim to create, from the same general outline, as many as 10 or 15 related and integrated electronically accessible short courses that, together, cover virtually the entire field of biotechnology.

We are currently making the transition to an electronic version of "Protein Purification" first by incorporating into our website "still shots" of GFP-based laboratory exercises and by developing a CD that can be used to advertise the initiative and to solicit corporate support for the venture. Others who have pioneered in the use of GFP in education include Dr. James Slock of Kings College, Wilkes Barre, Pennsylvania (who uses both GFP and bacterial luciferase in laboratory instruction) and Dr. Tim Herman of the Milwaukee School of Engineering (who has developed GFP-based molecular models and computerized molecular graphics exercises for the classroom).

Acknowledgements

NSF grant # DUE-9602356 to WWW and DFD. Modular approach to biotechnology laboratory instruction based on a novel green fluorescent protein.

References

1. Cutler MW, Davis DF, and Ward WW. GFP as an instructional tool for biochemistry laboratory courses. In: Campbell AK, Kricka LJ, and Stanley, PE, editors. *Bioluminescence and Chemiluminescence: Fundamentals and Applied Aspects*. Chichester: John Wiley & Sons, 1994: 383-386.
2. Yernool DA, Reddy PV, Davis DF, and Ward WW. GFP gene as a model gene expression marker for rDNA laboratory exercises. In: Hastings JW, Kricka LJ, and Stanley, PE, editors. *Bioluminescence and Chemiluminescence: Molecular Reporting with Photons*. Chichester: John Wiley & Sons, 1997: 192-195.
3. Ward WW, Swiatek GC, and Gonzalez DG. Green fluorescent protein in biotechnology education. *Methods Enzymol* 2000; **305**: 672-680.

PART 10

INSTRUMENTATION AND IMAGING OF LUMINESCENCE

THE NORTHSTAR™ HTS WORKSTATION LUMINESCENCE DETECTION SYSTEM

CORINNE E. M. OLESEN, DEBORAH M. BOLDT-HOULE, SUSAN YAN, BRIAN J. D'EON, BETTY LIU, BONNIE TILLOTSON, JOHN C. VOYTA AND IRENA BRONSTEIN

Tropix, Inc./Applied Biosystems, 47 Wiggins Avenue, Bedford MA 01730, USA
E-mail: jvoyta@appliedbiosystems.com

1. INTRODUCTION

The NorthStar™ HTS workstation is a fully integrated high throughput screening (HTS) luminescence detection system developed to expedite lead discovery with whole-plate imaging for both glow and flash luminescence chemistries. Features include a cooled charge-coupled device (CCD) camera, rapid 96- and 384-well reagent injection capability, plate handling 96-, 384-, and 1536-well microplate imaging formats, and data analysis. A microplate-matched collimator with an integrated Fresnel lens functions as a telecentric lens and is removed to permit imaging of flat membranes. A motor driven filter wheel contains multiple interchangeable band pass filter positions, which enables concurrent measurement of multiple wavelength signals. The NorthStar HTS workstation enables highly sensitive imaging of a range of luminescence detection applications.

2. MATERIALS AND METHODS

<u>cAMP Quantitation</u>: β2 Adrenergic receptor-expressing cells (10,000 cells/well) were cultured for 18 h in a 96-well microplate. Cells were stimulated with isoproterenol for 10 min and then lysates assayed according to the cAMP-Screen™ protocol (Tropix). The plate was imaged for 1 min on the NorthStar™ HTS workstation.

<u>Aequorin-based Ca^{++} quantitation</u>: CHO-Aequorin-Gα16-5HT2B cells (AequoScreen™ cell line) were kindly provided by Euroscreen (Brussels, Belgium). Cells were cultured for 18 h in clear-bottom, white-side 96-well tissue culture plates (25,000 cells/well) and were then loaded with coelenterazine h (Molecular Probes, Eugene, OR) in the presence/absence of 0.5 μM BAPTA-AM (Molecular Probes) in culture media for 4 h. Cells were incubated for 30 min with HBSS, and then α-Me-5HT (Sigma, St. Louis, MO) was injected into wells in HBSS containing 15 mM Ca^{++}. Light emission was imaged immediately for 20 s on the NorthStar.

<u>Cytotoxicity assay</u>: NIH/3T3 cells were cultured for 24 h (10,000 cells/well in 100 μL media) and then treated with staurosporine for 26 h. A low pH reaction buffer containing Glucon™ substrate was added directly to cells in the presence of media,

and incubated for 30 min. A high pH accelerator solution containing luminescence enhancer was then added and light emission measured with the NorthStar. For the alamarBlue™ assay, alamarBlue reagent (BioSource International, Camarillo, CA) was added to cells (10 μL/well) and incubated for 4 h, and then fluorescence was measured with the FLUOstar (BMG LabTechnologies, Durham, NC).
Dual wavelength detection: Dilutions of purified calf intestinal alkaline phosphatase (BioZyme, San Diego, CA) and bacterial β-galactosidase (Sigma) were assayed in a single reaction buffer containing both Galacton-*Star*® and BZPD substrates and Sapphire-II™ luminescence enhancer (Tropix).

3. RESULTS AND DISCUSSION

Luminescent assays systems, including homogeneous cell-based assays and heterogeneous immunoassay formats, utilizing several different luminescence chemistries, have been developed and performed on the NorthStar™ workstation.[1] Reporter gene assays include Gal-Screen™ for β-galactosidase quantitation in both mammalian and yeast cells, and Luc-Screen™ for firefly luciferase quantitation in mammalian cells. Both homogeneous assay systems provide glow light emission, are performed directly in the presence of cell culture medium, and have been demonstrated in 96-, 384- and 1536-well microplate formats on the NorthStar.

The cAMP-Screen™ competitive immunoassay system provides ultrasensitive quantitation of cAMP in cell lysates.[2] The system incorporates a cAMP-AP conjugate, CSPD® substrate and Sapphire-II™ enhancer. Glow light emission kinetics are obtained, which reach maximum intensity 30 min after substrate addition. The cAMP-Screen assay has been performed on the NorthStar workstation in both 96-well and 384-well microplate formats. With cAMP-Screen, a sensitivity of 0.06 pmol/well of purified cAMP is achieved with a dynamic range of detection of over four orders of magnitude (not shown). Quantitation of cAMP levels induced in response to isoproterenol stimulation of β2 adrenergic receptor-expressing C2 cells with the cAMP-Screen assay is demonstrated in Fig. 1. cAMP production is induced by a G-protein-coupled pathway in response to agonist.

A cell-based aequorin Ca^{++} quantitation assay has been developed, incorporating the Ca^{++} chelator BAPTA-AM, to provide an extended flash reaction capable of being imaged on the NorthStar. The CHO-Aequorin-5HT2B cell line, stably expressing both mitochondrial-targeted aequorin and the 5HT2B GPCR were used to demonstrate the effect of BAPTA-AM on the aequorin light emission reaction. In kinetic plots, performed with a PMT-based luminometer (not shown), BAPTA-AM provides a delay of approx. 8 s in the peak of light emission; for other cell lines the delay achieved depends both on the BAPTA-AM concentration and the specific agonist/GPCR response. The dose response and EC50 obtained with α-methyl-5HT are unchanged in the presence of BAPTA-AM and the EC50 is similar to that reported (1.4 nM[3]) (Fig. 2). For this agonist response, the NorthStar is capable of quantitatively imaging the luminescent flash, even without the delay afforded by

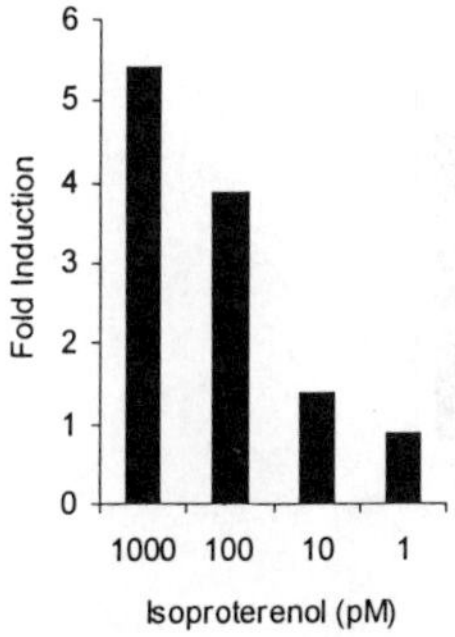

Fig. 1. cAMP induction in β2 adrenergic receptor-expressing cells.

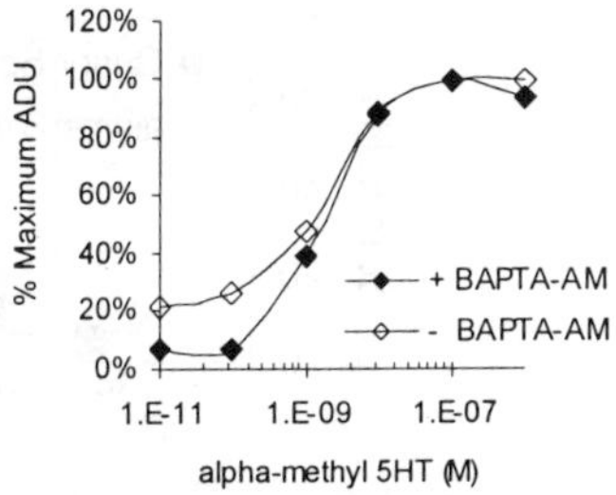

Fig 2. Ca^{++} quantitation with BAPTA-AM-modified aequorin assay. Agonist stimulation of CHO-Aeq-5HT2B cells.

BAPTA-AM. However, with other agonist/GPCR systems, the use of BAPTA to delay the light emission, together with the very rapid injection capability of the NorthStar, will enable Ca^{++} quantitation with this imaging system. The total injection time per plate (96-well) is 6 s, with imaging commencing 8 s following the start of injection.

Cell quantitation/cytotoxicity assays are performed by quantitating endogenous β-glucosidase in mammalian cells and alkaline phosphatase in yeast (not shown), using Glucon™ and CDP-*Star*® 1,2-dioxetane substrates, respectively. NIH/3T3 cells were incubated in the presence of the cytotoxic agent staurosporine for 26 h, and then assayed in the presence of culture medium with Quant-Screen™ reagents (Fig. 3). Cytotoxicity is measurable over a wide range of drug concentration, and quantitation of endogenous β-glucosidase levels with Glucon substrate shows good correlation to the alamarBlue™ assay, which measures cellular respiration.

Simultaneous dual-wavelength detection has been demonstrated using the blue-emitting Galacton-*Star*® 1,2-dioxetane substrate for quantitation of β-galactosidase and the green-emitting BZPD 1,2-dioxetane for quantitation of alkaline phosphatase (Fig. 4). A range of detection of each enzyme of four orders of magnitude is achieved with a simultaneous reaction, imaged using two band pass filters, 435BP70 and 600BP120. Potential dual wavelength applications include simultaneous quantitation of multiple enzyme read-outs from a single cell sample or multiplex detection of biomolecules using multiple enzyme-labeled probes.

The NorthStar HTS workstation is a versatile integrated system, capable of imaging flash and glow luminescence assays, and is compatible with multiple microplate formats and interfacing with robotic handling. The NorthStar workstation, together with these luminescence assays, provides the capability to monitor biological signaling pathways, quantitate biomolecules and perform compound screening with highly sensitive, high-throughput, facile assay systems.

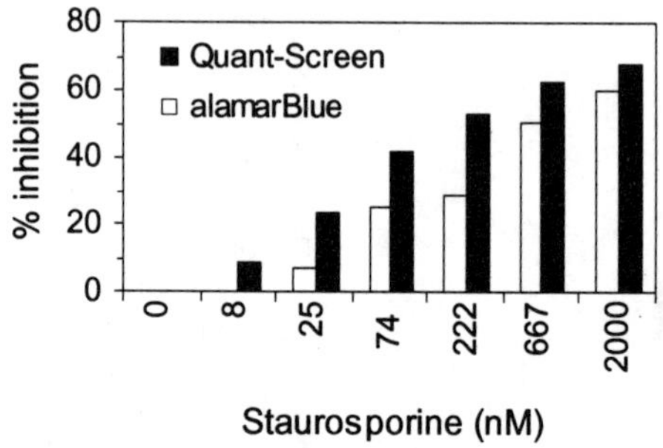

Fig. 3. Quant-Screen™ cytotoxicity assay with staurosporine-treated NIH/3T3 cells.

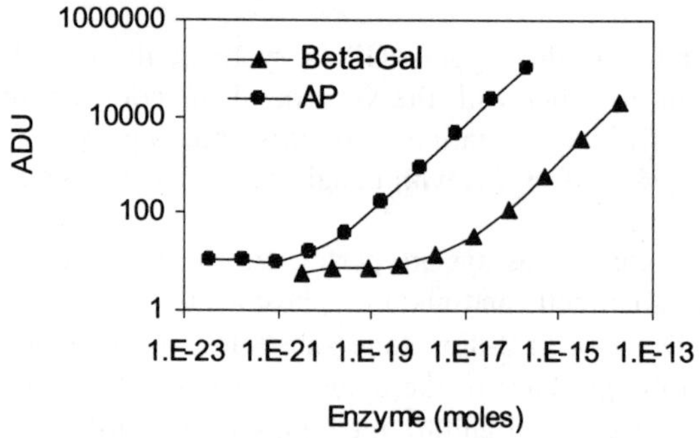

Fig. 4. Dual simultaneous wavelength enzyme quantitation.

References

1. Boldt-Houle D, Yan S, Olesen C, D'Eon B, Lee J, Liu B, Bodzuich U, Chiulli A, Atwood J, Gambini M, Voyta J, Bronstein I. A CCD-based integrated platform for accelerated high-throughput screening. *Am Lab* 2000; **32**: 60—70.
2. Chiulli A, Trompeter K, Palmer M. A novel high-throughput chemiluminescent assay for the measurement of cellular cAMP levels. *J Biomol Screening* 2000; **5**:239-247.
3. AequoScreen™ product literature, Euroscreen (Brussels, Belgium).

INSTRUMENTATION FOR *IN VIVO* IMAGING OF BIOLUMINESCENT REPORTERS

B.W. RICE, M. CABLE, B. NELSON, AND D. STEARNS
Xenogen Corporation, 860 Atlantic Ave, Alameda, CA 945011, USA
E-mail: brice@xenogen.com

1. INTRODUCTION

The use of light-emitting enzymes, such as luciferase, as a reporter of gene expression in living cells is a well established technique in molecular biology studies. Bioluminescent reporters have been most widely used *in vitro* where light detection is easily accomplished using standard inexpensive CCD arrays. Detection of bioluminescent reporters *in vivo* is also possible due to the semi-transparent nature of mammalian tissue[1,2], but requires high-sensitivity low-noise detectors and a more sophisticated imaging system. At Xenogen, we have developed the IVIS™ imaging system, which is specifically optimized for in vivo imaging. With this system, bioluminescent cells can be observed up to a few cm deep in small laboratory animals such as mice. Applications include the study of infectious disease, tumor growth, gene expression in transgenic animals, viral infection, and toxicology. In this paper, we examine the issues associated with in vivo imaging and describe the design of Xenogen's IVIS™ system.

2. OPTICAL PROPERTIES OF TISSUE

Mammalian tissue is a turbid medium that both scatters and absorbs photons. Photons are scattered due to changes in refractive index at cell membranes and internal organelles. Absorption varies greatly with cell type, but melinan and hemoglobin are two dominant absorbers. Oxyhemoglobin, in particular, absorbs strongly in the blue-green region of the visible spectrum, but is fairly transparent in the red at wavelengths > 600 nm. Typically, for λ>600 nm, the 1/e absorption length is 1-2 cm, while the length for scattering is much shorter at ~0.5 mm. In this regime, where scattering is dominant, a significant amount of bioluminescent emission can escape the tissue, but the emission is highly diffuse as illustrated schematically in Fig. 1a.

The propagation of light under these conditions is well described by a diffusion model[3]. In this model, the photon fluence, ϕ, decays exponentially away from the source according to

$$\phi(r) = \frac{P}{4\pi Dr}\exp(-\mu_{eff}r) \tag{1}$$

where r is the distance away from the source, P = source power, $D = 1/[3(\mu_s' + \mu_a)]$, and $\mu_{eff} = \sqrt{3\mu_a(\mu_s' + \mu_a)}$ with μ_s' = effective scattering coefficient and μ_a = absorption coefficient for the tissue.

The transmission intensity observed at the tissue surface can be calculated from Eq. 1 above with appropriate boundary conditions[4]. For the conditions given in Fig. 1a, the intensity profile at the surface is shown in Fig. 1b.

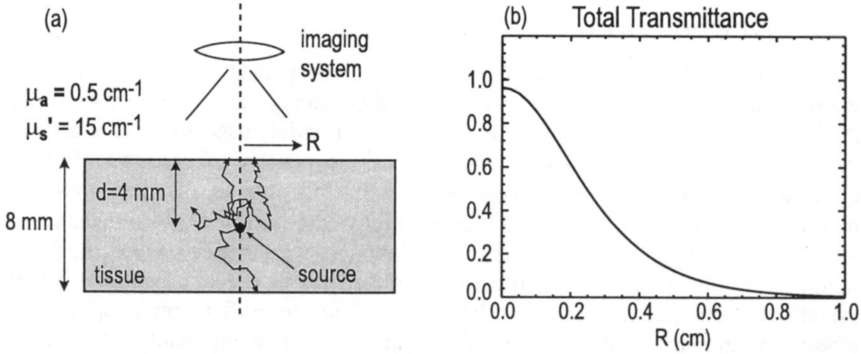

Fig. 1: a) Illustration of light propagation through a turbid medium (tissue); b) diffusion model calculation of the surface intensity profile that would be measured by an external imaging system using the parameters shown in (a).

3. Spectral Emission of Luciferase

Design of an *in vivo* imaging system requires knowledge of the spectral properties of the bioluminescent source. It is important to realize that the emission spectrum of luciferase is significantly different *in vivo* compared with *in vitro*. At Xenogen, the majority of our studies to date have been performed with three types of luciferase whose *in vitro* spectral properties are shown in Fig. 2a. Bacterial luciferase, *Photorhabdus luminescens*, peaks at 490 nm, while the yellow-green firefly luciferase, *Photinus pyralis*, peaks at 560 nm. The dashed curve in Fig. 2a shows the *in vivo* spectrum of the yellow-green firefly luciferase for a subcutaneous tumor that is ~1 mm under the skin of a nude mouse. The strong attenuation of the signal at wavelengths < 600 nm is apparent and is a characteristic property of oxyhemoglobin. In Fig. 2a the tumor is only ~1 mm deep so some of the emission below 600 nm is still present; however, for sources deeper into the tissue, for example internal organs such as the liver, kidney, and lungs, the emission below 600 nm would be completely extinguished leaving only the red-shifted light above 600 nm.

The maximum depth that bioluminescent cells can be observed depends on the number of cells and the wavelength of emission (type of luciferase). The variation of attenuation with depth can be predicted from diffusion models (Eq. 1) as shown in

Fig. 2b for three cases corresponding roughly to blue (μ_{eff} = 11 cm^{-1}), green (μ_{eff} = 3.5 cm^{-1}), and red (μ_{eff} = 0.87 cm^{-1}) light. For short wavelengths ~450 nm attenuation is 10^{-4} for depths of only 4 mm, while long wavelengths ~650 nm can pass through 2cm of tissue with the same attenuation.

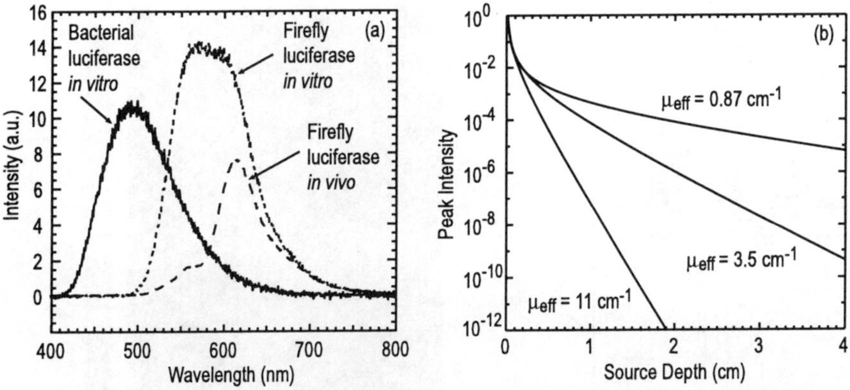

Fig. 2: a) *In vitro* spectral measurements (solid lines) for bacterial luciferase psuedamonis aeruginosa and firefly luciferase in PC3M protstate tumor cells. The bacteria is measured on agar while the PC3M cells are measured in PBS solution. The dashed line represents the in vivo spectrum of PC3M tumor cells injected subcutaneously in a nude mouse. (b) Peak intensity at the tissue surface versus depth of source in tissue for 3 different tissue parameters.

4. IVIS™ Imaging System

The IVIS™ imaging system has been optimized for *in vivo* imaging of small animals as illustrated in Fig. 3a. As discussed earlier in Fig. 2, the CCD for in vivo imaging must have good sensitivity in the red and near-infrared part of the spectrum (>600 nm) in order to detect sources of light buried deep in tissue. The CCD must also have the low noise and dark current to maximize detection of low-light signals. For this purpose we utilize a 1300×1340 pixel CCD in a 1 inch square format. The CCD is back-thinned for maximum quantum efficiency of ~90% and is cooled to -120 deg C to minimize dark current and noise. Light is coupled to the CCD using a fast f/1 lens. The sample stage moves up and down to vary the field-of-view from 10 to 25 cm. A 7 position filter wheel is also available so the user can image discreet portions of the spectrum. The sample stage, lens focus, f/stop, and filter wheel are all stepper motor controlled through Xenogen's LivingImage™ software. Other convenient features include a gas anesthesia system and a heated sample shelf to help maintain the animals temperature during anesthesia.

A typical *in vivo* image of a nude mouse is shown in Fig. 3b where bacterial luciferase was injected into each thigh. The underlying black and white image is a photographic image of the animal taken with illumination lamps on. After the photographic image is acquired, a 30 sec luminescent image is acquired, shown in false color representation in Fig. 3b, where red represents the highest intensity and purple represents the lowest intensity. With longer exposure times (~5 min) subcutaneous cell counts of < 1000 cells are still observable *in vivo*.

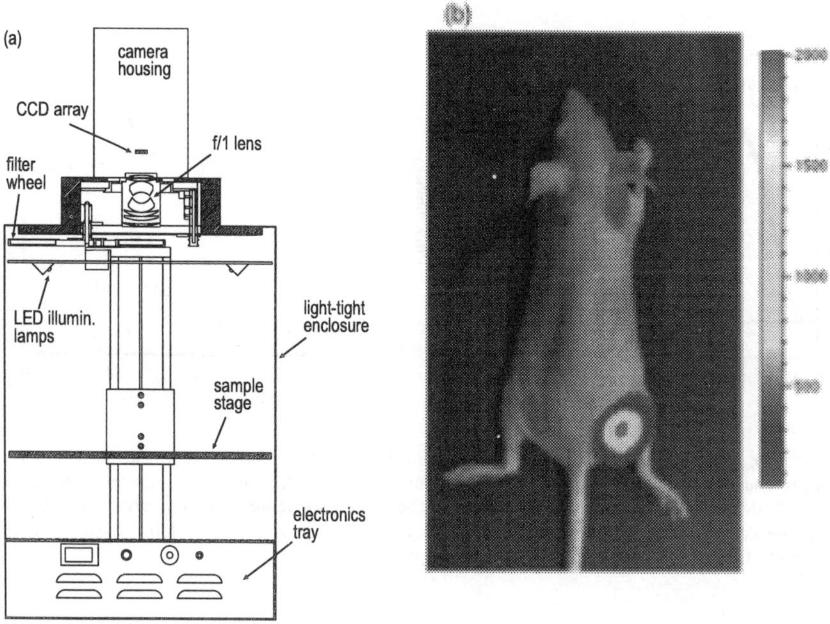

Fig. 3: a) Overview of the IVIS™ imaging system; (b) 30 second exposure of nude mouse with 10^7 *Pseudomonas aeruginosa* (*lux CDABE*) cells injected into the right thigh and 10^7 *Staphyloocccus aureus* (*lux ABCDE*) cells (less bright) injected into the left thigh.

References

1. Contag CH, Contag PR, Mullins JI, Spilman SD, Stevenson DK, Benaron DA. Photonic detection of bacterial pathogens in living hosts. *Mol Microbiol* 1995; **8**: 593-603.
2. Contag CH, Jenkins D, Contag PR, Negrin RS. Use of reporter genes for optical measurements of neoplastic disease *in vivo*. *Neoplasia* 2000; **2**: 41-52.
3. Ishimaru, A. *Wave Propagation and Scattering in Random Media*. New York: Academic Press, 1978.
4. Haskell RC, Svaasand LO, Tsay T, Feng T, McAdams MS, Tromberg BJ. Boundary conditions for the diffusion equation in radiative transfer. *J Opt Soc Am A* 1994; **11**: 2727-2741.

DEVELOPMENT OF A TIME-RESOLVED FLUORESCENCE MICROSCOPE FOR IMAGING ANALYSIS

A. RODA, M. GUARDIGLI, P. PASINI AND M. BARALDINI

Dept. of Pharmaceutical Sciences, University of Bologna, Via Belmeloro 6, Bologna 40126 Italy

E-mail: mguardig@alma.unibo.it

1. INTRODUCTION

Imaging techniques relying on fluorescent labels coupled with optical microscopy are commonly used to localize specific molecules or gene sequences in tissue sections or single cells.[1] However, sample autofluorescence, excitation light scattering and emission of optical components often determine a background signal that decreases the detectability of the labeled bioprobe.

This drawback can be overcome by using luminescent labels with long-lived emission and time-resolved (TR) fluorescence measurement techniques.[2] In such techniques, sample excitation is performed using a pulsed light source and the measurement of the emission is started after a delay upon excitation, so allowing all the short-lived sample autofluorescence (ns-μs range) and light scattering to dissipate. Luminescent lanthanide compounds are probably the most interesting labels for such applications because they are characterized by long luminescence lifetimes, in the order of milliseconds.[3] Several methods for exploiting these compounds based on TR fluorescence measurement techniques have been commercialized and applied in immunoassays and nucleic acid hybridization assays.[4,5] In recent years luminescent lanthanide labels have been successfully proposed for TR fluorescence imaging microscopy,[6-8] but instruments designed for such an application are not yet commercially available. In this paper we describe the development of a TR fluorescence microscope and the results obtained on model systems and in the immunohistochemical localization of antigens in tissue samples.

2. MATERIALS AND METHODS

Oxyrane acrylic macroporous beads (diameter about 250 μm) were purchased from Sigma Chemical Co. (St. Louis, MO, USA). Nylon net (100 ± 3 threads/cm) was supplied by Assoprint s.r.l (Modena, Italy). Immobilization of luminescent species

(Eu^{3+} chelates and fluorescein) was obtained by immersion of the solid supports in a solution containing these species, followed by air drying. All the other chemicals used were of analytical grade.

Localization of von Willebrand factor was performed in human tonsil tissue sections by means of a rabbit anti-human von Willebrand factor primary antibody (Dako A/S, Glostrup, Denmark). Primary antibody was detected using the LSAB®2 System, Peroxidase (also supplied by Dako), which is a detection system for mouse and rabbit primary antibodies based on biotin-streptavidin interaction and horseradish peroxidase (HRP) as a label. Colorimetric and chemiluminescent detection were perfomed using colorimetric (3,3'-diaminobenzidine, from Sigma) or chemiluminescent (ECL®, from Amersham Pharmacia Biotech, Buckingamshire, England, UK) HRP substrates. A Eu^{3+}-BCPDA-labeled goat anti-rabbit secondary antibody (BCPDA: 4,7-bis(chlorosulphonylphenyl)-1,10-phenanthroline-2,9-dicarboxylic acid) was used in the TR fluorescence detection.

In order to perform TR fluorescence microscope imaging, we have modified a conventional epifluorescence microscope (Model BX 60, Olympus Optical, Tokyo, Japan) by adding an auxiliary Xenon lamp and two mechanically synchronized choppers in the light excitation and emission pathways. The relative phase of the choppers has been adjusted in order to obtain non-overlapping excitation and measurement time windows, with a delay of about 1 ms. The microscope is equipped with an ultrasensitive, cryogenically cooled CCD camera (LN/CCD Princeton Instruments, Roper Scientific, Trenton, NJ, USA) for the acquisition of both transmitted light and TR fluorescence (or chemiluminescence) images.

3. RESULTS AND DISCUSSION

The instrument has been tested using a model system consisting of a luminescent Eu^{3+} chelate immobilized on oxirane acrylic beads and a fluorescein-containing nylon net. The comparison of the conventional and TR fluorescence images (Figure 1) proved that the microscope allowed to discriminate between the long-lived lanthanide chelate emission and the short-lived fluorescein one. On the basis of the luminescence intensity values, it has been calculated that the TR fluorescence acquisition mode allowed a 1000-fold increase in the Eu^{3+}/fluorescein emission ratio in comparison to the conventional fluorescence measurement.

In order to compare the analytical performance of TR fluorescent, colorimetric and chemiluminescent detection we have performed the immunohistochemical localization of von Willebrand factor in human tonsil tissue sections using the three different detection procedures. As shown in Figure 2, the different detection procedures gave consistent results. In comparison with conventional fluorescence measurement (image not shown), the TR fluorescence detection allowed to efficiently suppress sample autofluorescence. It should be emphasized that TR fluorescent detection does not require addition of reagents in the detection step; this simplifies the experimental procedure and avoids problems related to the presence of a liquid film over the sample. In conclusion, the developed TR fluorescence microscope allowed to suppress sample autofluorescence, and therefore can be successfully used in TR luminescence imaging to localize molecules down to the attomole level.

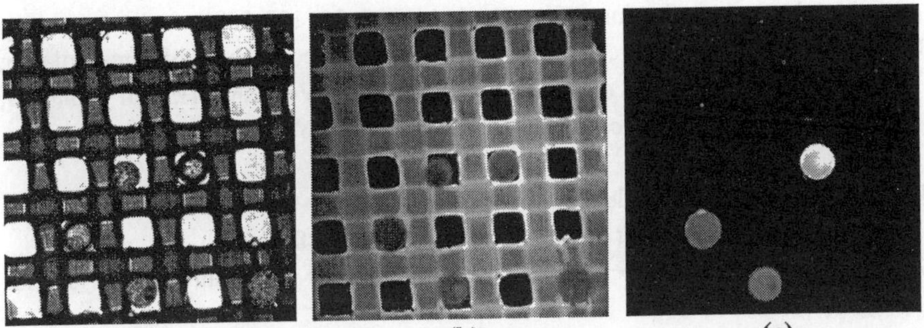

(a) (b) (c)

Fig. 1 Imaging of a model system consisting in a luminescent Eu^{3+} chelate immobilized on acrylic beads and a fluorescein-containing nylon net. (a) Transmitted light image (b)Conventional fluorescence image (c) TR fluorescence image.

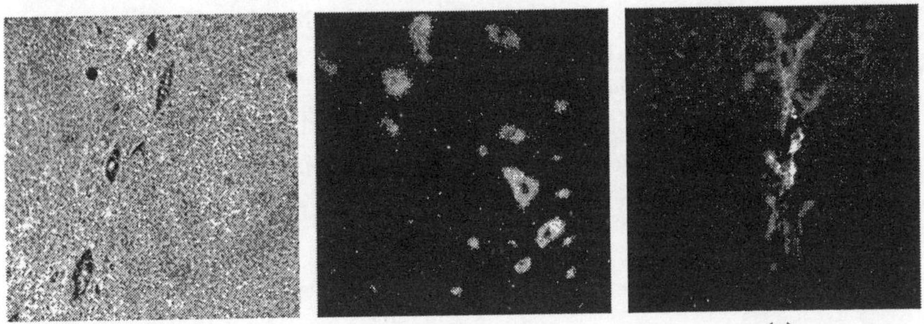

(a) (b) (c)

Fig. 2 Localization of von Willebrand factor in human tonsil tissue sections using different detection principles. (a) Colorimetric detection. (b) Chemiluminescent detection. (c) TR fluorescent detection.

Acknowledgements

We thank Dr. A. Ius for providing the BCPDA-labeled goat anti-rabbit secondary antibody and Prof. P. Chieco for the tonsil tissue samples. This work was partially supported by MURST.

References

1. Rost FWD. *Fluorescence microscopy.* Cambridge: Cambridge University Press, 1995.
2. Diamandis EP. Immunoassays with time-resolved fluorescence spectroscopy: principles and applications. *Clin Biochem* 1988; **21**: 139-150.
3. Sabbatini N, Guardigli M, Manet I. Lanthanide complexes of encapsulating ligands as luminescent devices. In: Neckers DC, Volman DH, von Bünau G, eds. *Advances in Photochemistry*, Vol. 23, New York: John Wiley & Sons, 1997, 213-278.
4. Sammes PG, Yahioglu G. Modern bioassays using metal chelates as luminescent probes. *Nat Prod Reports* 1996; **13**: 1-28.
5. Mathis G. Rare earth cryptates and homogeneous fluoroimmunoassays with human sera. *Clin Chem* 1995; **41**: 1391-1959.
6. Seveus L, Väisälä M, Syrjänen S, Sandberg M, Kuusisto A, Harju R, Salo J, Hemmilä I, Kojola H, Soini E. Time-resolved fluorescence imaging of europium chelate label in immunohistochemistry and in situ hybridization. *Cytometry* 1992; **13**: 329-338.
7. Marriott G, Heidecker M, Diamandis EP, Yan-Marriott Y. Time-resolved delayed luminescence image microscopy using an europium ion chelate complex. *Biophys J* 1994; **67**: 957-965.
8. Rulli M, Kuusisto A, Salo J, Kojola H, Simell O. Time-resolved fluorescence imaging in islet cell autoantibody quantitation. *J Immunol Methods* 1997; **208**: 169.

THE STUDY OF LUMINESCENCE RESONANCE ENERGY TRANSFER IN CELL EXTRACTS OF TRANSFORMED CELLS USING A LOW LIGHT IMAGING SYSTEM

YUBAO WANG, ALADAR A. SZALAY
Department of Biochemistry, School of Medicine, Loma Linda University, Loma Linda, CA 92350, USA
E-mail: aszalay@som.llu.edu

1 INTRODUCTION

Fluorescence resonance energy transfer (FRET) results from the transfer of excited state energy from one fluorophore (the donor) to a second fluorophore (the acceptor).[1, 2] FRET requires an overlap between the emission spectrum of the donor and the excitation spectrum of the acceptor. The efficiency of FRET is dependent on the distance separating the participating partners. Significant energy transfer occurs over distances not exceeding 80 Angstroms [2], thus provide a method which can be used for the study of interactions between macromolecules. Recently, Burmeister and associates reported the use of luminescent reagents as donors for FRET study to determine the distance between two domains of myosin and used the name luminescence resonance energy transfer (LRET) .[3] Li and colleagues have discussed the advantage of LRET over FRET extensively[4]. LRET can be used for the study of protein-DNA, and protein-protein interaction. The use of a luminescent protein and green fluorescent protein as the donor and the acceptor for LRET study to probe the interaction between IGFBP6 and IGF-II was first reported in 1998.[5] Xu and his colleagues reported the use of this system to study the interactions of the circadian clock proteins using LRET (or BRET).[6] The dimerization of β_2-adrenergic receptors was also shown by using luciferase and GFP energy transfer.[7] All those studies showed that luminescent protein could be used as the donor in LRET. The detection of protein-protein interactions in living cells was documented based on the shift in wavelengths spectroscopically. We suggest that in the future and energy transfer study based on image analysis will be more sensitive in detecting LRET at a single cell level. To date, there is no single cell luminescence image method available. Therefore, we employed a low-light image analysis system to capture the images of luminescent light emission from cell extracts. The light emission was quantitatively measured pixel by pixel based on the image acquired. The interaction between IGF-II and IGFBP-6 is shown based on the energy transfer from *Renilla* luciferase to *Aequorea* GFP. This approach may serve as the first step for the future development of a quantitative single cell luminescent image analysis system and the determination of energy transfer in individual mammalian cells.

2 MATERIALS AND METHODS

2.1 Plasmids and Cell Lines

Three plasmids were used in the experiments. The construction of the plasmids was implemented as described by Wang et al. [5] The different fusion constructs were named as follows: pC-IGF-II-GFP, pC-INS-GFP and pC-IGFBP6-RUC. They encode the fusion proteins of IGF-II with GFP, insulin with GFP, and mature IGFBP-6 with *Renilla* luciferase respectively. CHO cells were transiently transfected using the calcium precipitation method. The expression of the above three fusion proteins was identified by Western blot analyses using antibodies against *Renilla* luciferase or GFP. CHO cells were grown in DMEM supplemented with glutamine, 10% fetal bovine serum with penicillin and streptomycin.

2.2 Preparation of cell extracts

Twenty-four hours after transient transfection, CHO cells were harvested by using a cell scraper. Cells were washed in PBS once and lyzed in a buffer consisting of 10 mM Tris-HCl, pH 7.6, 100 mM NaCl, 1 mM $MgSO_4$, 1 mM EDTA and 100 μg/mL PMSF by sonication (3 times for 10 seconds with 10-second intervals). The cell extracts were kept on an ice during the experiments. The soluble proteins were recovered from the supernatant of cell extracts after centrifugation of the homogenates at 12,000 rpm for 10 minutes.

2.3 Luminescence image acquisition and analysis

The light emission from cell extracts was quantitatively measured using a low light imaging system (Hamamatsu Corporation, Bridgewater, NJ, USA), which includes a specimen chamber and Hamamatsu CCD intensified camera C2400. An absorbent glass filter (Chroma Technology Corp., Brattleboro, VT, USA), absorbs light with wavelength greater than 480 nm, was mounted on the camera to block the emission of light with wavelengths greater than 480 nm. Thus the light emission from GFP was excluded because GFP emits at a wavelength greater than 480 nm. The MetaMorph Imaging System (Universal Imaging Corporation, West Chester, PA, USA) was used for image processing and data acquisition. First, extracts from two individual mammalian cell lines were mixed and incubated at room temperature for 10 to 20 minutes to allow protein-protein interactions to occur. After the pre-incubation one μL of 1 mM coelenterazine was added to the extracts followed by the imaging acquisition. Figure 1 shows the details of the quantifying of the light emission. The images from the luciferase reaction were captured using the intensified CCD camera and monitored in real-time.

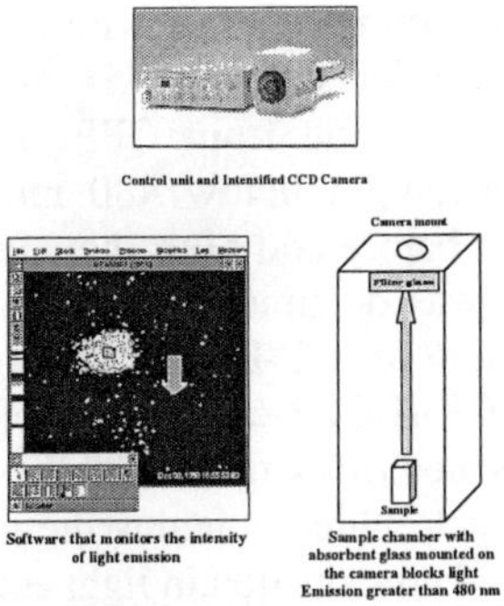

Fig. 1. Quantifying the levels of energy transfer using a low light imaging system. An intensified CCD camera is mounted to the specimen chamber, an absorbency glass is fitted to the camera to block light emission wavelength greater than 480 nm. The image was acquired using the MetaMorph Imaging System. The shaded arrow shows the area to be quantified.

The light that transmitted through the glass filter formed the images. As described before, the light is from luciferase catalyzed reaction, thus the image represents the emission of *Renilla* luciferase. A defined area from the image was quantitatively measured at real time according the intensity of the light emission using the MetaMorph Imaging System as shown in fig. 2 A.

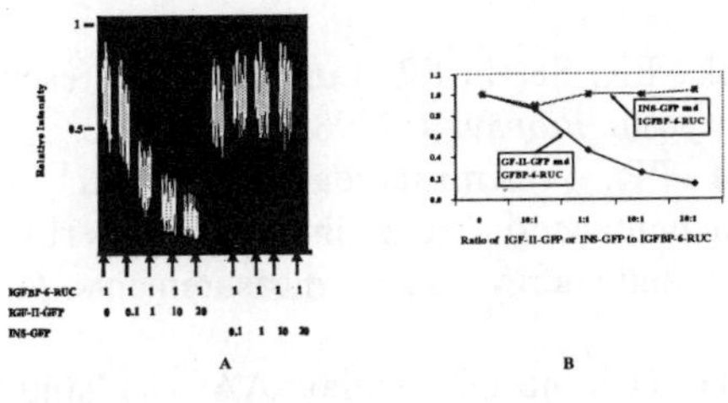

Fig. 2. A. Measurement of the intensity of the shaded area in real-time after the addition of IGF-II-GFP or INS-GFP extract to IGFBP-6-RUC containing extract. B. The middle point of light intensity in A was taken and the relative light intensity was calculated by dividing the light intensity of individual reactions by basal light emission intensity, which is the intensity of IGFBP-6-RUC-containing cell extract.

3 RESULTS AND DISSCUSSION

We used the low light imaging system to measure the energy transfer from IGFBP6-RUC fusion to IGF-II-GFP fusion quantitatively upon the addition of coelenterazine. The entire emission is captured with an intensified camera. Only the emission produced by *Renilla* luciferase reaction is captured by the camera, because the glass filter blocked the emission from GFP. The light emission from the luciferase reaction with wavelength below 480 nm generated the images. By quantifying the light emission before and after the addition of the binding partner, IGF-II-GFP, energy transfer was documented by the reduction in light emission from luciferase after the addition of IGF-II-GFP. Figure 2 show the energy transfer between IGFBP6-RUC and IGF-II-GFP-containing cell extracts in the presence of coelenterazine but in the absence of exogenous light. In contrast, the addition of INS-GFP-containing cell extract, which contains the non-binding partner (Insulin) GFP fusion as a control, showed no reduction in light emission below 480 nm.

In summary, based on these data, luminescence energy transfer can be quantitatively measured using a low light image system. As the development of a tunable filter and high sensitivity camera system in progress, it may be possible to image the luminescence of a single cell and the intensity of light emission can be quantified pixel by pixel. This progress will be very significant for monitoring protein-protein interactions in living cells.

References

1. Stryer, L, Haugland RP. Energy transfer: a spectroscopic ruler. *Proc Natl Acad Sci U S A*. 1976; **58**: 719-26.
2. Selvin P.R. Fluorescence resonance energy transfer. *Methods Enzymol.* 1995; **246**: 300-34.
3. Burmeister G, Cooke ER, Selvin PR. Luminescence resonance energy transfer measurements in myosin. *Biophys J*. 1998: **74**: 2451-8.
4. Li M., Selvin PR. Amine-reactive forms of a luminescent diethylenetriaminepentaacetic acid chelate of terbium and europium: attachment to DNA and energy transfer measurements. *Bioconjug Chem*. 1977; **8**: 127-32.
5. Wang Y, Wang G, O'Kane DJ, Szalay AA. A study of protein-protein interactions in living cells using luminescence resonance energy transfer (LRET) from *Renilla* luciferase to Aequorea GFP. *Mol Gen Genet*. Accepted for publication.
6. Xu Y,Piston DW, Johnson CH. A bioluminescence resonance energy transfer (BRET) system: application to interacting circadian clock proteins. *Proc Natl Acad Sci U S A*. 1999; **96**: 151-6.
7. Angers S, Salahpour A, Joly E, Hilairet S, Chelsky D, Dennis M, Bouvier M. Detection of beta 2-adrenergic receptor dimerization in living cells using bioluminescence resonance energy transfer (BRET). *Proc Natl Acad Sci U S A*. 2000; **97**: 3684-9.

INDEX